# Offshore Geotechnical Engineering

Design practice in offshore geotechnical engineering has grown out of onshore practice, but the two application areas have tended to diverge over the last 30 years, driven partly by the scale of the foundation and anchoring elements used offshore and partly by fundamental differences in construction and installation techniques. As a consequence, offshore geotechnical engineering has grown as a speciality.

The book's structure follows a familiar pattern that mimics the flow of a typical offshore project. In the early chapters, it provides a brief overview of the marine environment, offshore site investigation techniques and interpretation of soil behaviour. It proceeds to cover geotechnical design of piled foundations, shallow foundations and anchoring systems. Three topics are then covered that require a more multi-disciplinary approach: the design of mobile drilling rigs, pipelines and geohazards.

*Offshore Geotechnical Engineering* serves as a framework for undergraduate and postgraduate courses, and will appeal to professional engineers specialising in the offshore industry.

It is assumed that the reader will have some prior knowledge of the basics of soil mechanics and foundation design. The book includes sufficient basic material to allow readers to build on this previous knowledge, but focuses on recent developments in analysis and design techniques in offshore geotechnical engineering.

**Mark Randolph** is the founding Director of the Centre for Offshore Foundation Systems at the University of Western Australia, he is a founding Director of the specialist consultancy, Advanced Geomechanics, and is a former Rankine lecturer.

**Susan Gourvenec** is a Professor at the Centre for Offshore Foundation Systems at the University of Western Australia and delivers under-graduate, post-graduate and industry courses on Offshore Geomechanics.

# Offshore Geotechnical Engineering

**Mark Randolph and Susan Gourvenec**

**With contributions from
David White and Mark Cassidy**

Centre for Offshore Foundation Systems
University of Western Australia

## CRC Press
Taylor & Francis Group
Boca Raton London New York

CRC Press is an imprint of the
Taylor & Francis Group, an **informa** business

A SPON PRESS BOOK

First published 2011 by Spon Press

2 Park Square, Milton Park, Abingdon, Oxon OX14 4RN
711 Third Avenue, New York, NY 10017, USA

*Spon Press is an imprint of the Taylor & Francis Group, an informa business*

First issued in paperback 2017

Typeset in Sabon by Glyph International

This publication presents material of a broad scope and applicability. Despite
stringent efforts by all concerned in the publishing process, some typographi-
cal or editorial errors may occur, and readers are encouraged to bring these
to our attention where they represent errors of substance. The publisher and
author disclaim any liability, in whole or in part, arising from information
contained in this publication. The reader is urged to consult with an
appropriate licensed professional prior to taking any action or making any
interpretation that is within the realm of a licensed professional practice.

*British Library Cataloguing in Publication Data*
A catalogue record for this book is available from the British Library

*Library of Congress Cataloging in Publication Data*
A catalog record has been requested for this book


ISBN 13: 978-0-415-47744-4 (hbk)
ISBN 13: 978-1-138-07472-9 (pbk)
ISBN 13: 978-1-315-27247-4 (ebk)

# Contents

# List of figures

# Preface

Design practice in offshore geotechnical engineering grew out of onshore practice, but the two application areas have tended to diverge over the last 30 years, driven partly by the scale of the foundation and anchoring elements used offshore, and partly by fundamental differences in construction (or installation) techniques. As a consequence, offshore geotechnical engineering has grown as a speciality, initially among professional engineers and researchers, but gradually spreading downwards through tertiary education in the form of specialist courses at postgraduate and undergraduate levels.

The origins of this book stem from a course on Offshore Geomechanics taught originally in 1998 as part of a Masters degree in Oil and Gas Engineering at the University of Western Australia (UWA). Since then, in addition to a Masters course offered every other year or so, there is now a final-year elective offered annually at undergraduate level within the Bachelor of Engineering at UWA. Most, but not all, students taking these courses will have taken previous courses covering the basics of soil mechanics and foundation design. The aim has therefore been to include sufficient basic material to allow readers to build on their previous knowledge, but to focus on issues relevant to modern offshore engineering developments, which have now reached water depths in excess of 2,000 m.

With one or two exceptions, notably *Marine Geotechnics* (Poulos 1988), this area of geotechnical engineering has received scant coverage, and it is hoped that this book will both serve as a framework for other undergraduate and postgraduate courses and appeal to professional engineers specialising in the offshore industry. As with any text book, the authors have had to make difficult decisions in respect of (a) prior knowledge of soil mechanics and geotechnical engineering assumed and (b) the extent to which subjects on the periphery (or beyond) the authors' expertise are addressed. Our approach has been to avoid mere repetition of material that is covered better elsewhere, both in terms of fundamentals, and with regard to specialist areas such as seismic design or the rapidly developing area of geohazard assessment. As a minimum, brief overviews are given, with suggestions for further reading, but without attempting to paraphrase material with which we have limited familiarity. A particular aim, however, has been to identify new research material and to focus on those areas where traditional design approaches – including some that form part of industry guidelines such as the American Petroleum Institute and International Standards Organisation series – appear deficient in the light of modern analysis.

The structure of the book follows a familiar pattern that mimics the flow of a typical offshore project. The early chapters start with a brief overview of the marine environment and the 'signature' geotechnical characteristics of the seabed sediments in the most active offshore hydrocarbon regions. The various means of investigating the seabed properties are then explored, starting with high-level geophysical approaches, and proceeding through the *in situ* geotechnical techniques in common use offshore to the different types of laboratory testing. A chapter on soil behaviour, paralleling the 'interpretive' report of an industry project, then attempts to identify how key soil characteristics that influence subsequent design can be quantified through the different site investigation techniques. The role of physical modelling is also discussed, addressing aspects of soil response or soil–structure interaction that are less amenable to analysis and where scaled models may provide special insight.

Three central chapters deal, respectively, with geotechnical design of piled foundations, shallow foundations and anchoring systems. The offshore industry has been at the forefront of new design approaches that have been developed through research initiatives funded by the offshore industry and subsequently been adopted within design guidelines. The lateral response models for piles developed in the 1970s, the approaches for axial capacity developed in the 1980s for clay, and in the last few years for sand, and the framework for assessing the impacts of cyclic loading on the performance of shallow foundations are obvious examples. The extent to which these design models capture reality is explored. Anchoring systems continue to evolve rapidly, in response to the relentless thrust by industry into deeper water. While this has forced a more scientific approach to the design of conventional drag anchors, it has also necessitated reliance on numerical and physical modelling as a basis for design methods with only limited validation through field-scale testing.

These three chapters where the geotechnical design can generally be separated from other aspects of the offshore development are followed by three chapters where a more multi-disciplinary approach is required. The first of these deals with the design of mobile drilling rigs, in particular so-called 'jack-up' rigs, which have always proved to be an area with perhaps the highest risk of unanticipated performance. New approaches to modelling the interaction of the 'spudcan' foundations with the soil, and integrating that response with structural and hydrodynamic modelling of the entire rig are outlined. The other two chapters in this group deal, respectively, with pipelines, and with geohazards, both of which have become of much greater importance in deep-water developments, which typically involve more hazardous seabed terrain and much longer pipeline tiebacks to shore. Geohazard assessment is largely the domain of engineering geologists and marine geomorphologists, while pipeline design is itself a specialist discipline. Each area, though, has geotechnical aspects, such as assessment of the stability of submarine slides and their impact on seabed infrastructure, and detailed interactions between pipelines and the seabed, and these are the main focus of these chapters.

The authors are indebted to a number of people who have contributed to this book, either directly in preparing course material that has been subsumed into the present text, or indirectly through research results that have been incorporated. Many of these are present or former members of the Centre for Offshore Foundation Systems (COFS), which was established and funded as a Special Research Centre by the Australian Research Council over the period 1997–2005. Particular acknowledgement goes to Professor David White who was primarily responsible for Chapters 5 and 9 on pile foundations and pipelines and to Professor Mark Cassidy who was the principal author of Chapter 8 on mobile drilling rigs. In addition to these chapter contributions, thanks are due to Dr Marc Senders for his contributions to the section on site investigation and Mr James Hengesh for his contribution to the section on geohazards. Particular thanks are also due to Mr Carl Erbrich of Advanced Geomechanics (AG), who was one of the main lecturers on the original Masters course in 1998, and the other lecturers on that first course including Dr Hackmet Joer (then COFS, now AG), Professor Martin Fahey (School of Civil and Resource Engineering, UWA), Dr Ian Finnie (AG) and Mr Paul Hefer (then AG).

Mark Randolph, Susan Gourvenec
Perth, February 2010

# Notation

| | |
|---|---|
| a | Foundation radius or semi-width |
| a | Soil attraction factor |
| a | Eccentricity of yield surface |
| A | Bearing area of foundation, cross-sectional area |
| A' | Effective area of foundation |
| $A_b$ | Effective bearing area of anchor line per unit length |
| $A_e$ | External cross-sectional area |
| $A_h$ | Hydrodynamic cross-sectional area (structural member) |
| $A_i$ | Internal cross-sectional area |
| $A_p$ | Frontal projected area of fluke anchor |
| $A_{plug}$ | Cross-sectional area of soil plug |
| $A_{r,eff}$ | Effective area ratio of pile |
| $A_s$ | Surface area |
| $A_s$ | Effective surface area of anchor line per unit length |
| $A_{se}$ | External surface area |
| $A_{si}$ | Internal surface area |
| $A_{tip}$ | Skirt tip bearing area |
| b | Effective width of link chain |
| B | Foundation breadth |
| B' | Effective foundation breadth |
| $c_v$ | Coefficient of consolidation (isotropic, or vertical) |
| $c_h$ | Coefficient of consolidation (horizontal) |
| CAU | Anisotropically consolidated undrained triaxial test |
| $C_c$ | Compression index |
| $C_d, C_m, C_s$ | Drag, inertia and shape coefficients |
| CIU | Isotropically consolidated undrained triaxial test |
| $C_r$ | Recompression index |
| $C_s$ | Swelling index |
| d | Drainage path length |
| d | Skirt tip depth |
| $d_c$ | Depth factor in classical undrained bearing capacity expression |
| $d_{cH}$ | Depth factor for pure horizontal load capacity |
| $d_{cM}$ | Depth factor for pure moment capacity |
| $d_{cV}$ | Depth factor for uniaxial vertical bearing capacity |
| $d_q, d_\gamma$ | Depth factors for drained bearing capacity |
| $d_{50}$ | Sieve size passing 50 per cent of sample by mass |
| D | Diameter (of foundation or pile) |

| | |
|---|---|
| $D_e$ or $D_0$ | External diameter |
| $D_{eq}$ | Equivalent diameter of pile |
| $D_h$ | Hydrodynamic cross-sectional diameter (structural member) |
| $D_R$ | Relative density |
| $e$ | Eccentricity of loading above mudline |
| $e$ | Void ratio |
| $e_c$ | Void ratio following consolidation |
| $e_{cr}$ | Critical void ratio |
| $e_{max}$ | Maximum void ratio |
| $e_{min}$ | Minimum void ratio |
| $e_0$ | *In situ* (or initial) void ratio |
| $E, E_u, E'$ | Young's modulus |
| $E^*$ | Modified Young's modulus accounting for anisotropic characteristics |
| $EI$ | Bending rigidity |
| $E_0$ | One-dimensional stiffness modulus |
| $f$ | Form factor for drag embedment anchor |
| $f_s$ | Cone penetrometer sleeve friction |
| $f_b$ | Buoyancy factor in pipe penetration analysis |
| $f_{lay}$ | Static stress concentration factor in pipe penetration analysis |
| $f_{dyn}$ | Dynamic embedment factor in pipe penetration analysis |
| $F$ | Factor of safety |
| $F$ | Bearing capacity modification factor for soil strength heterogeneity |
| $F$ | Unit soil friction acting on anchor line |
| FPS | Floating production system |
| FPSO | Floating production, storage and offloading (system) |
| $F_{wi}$ | Wind force (on projected area of structural member) |
| $F_1, F_2$ | Resultant of interslice forces in slope analysis |
| $g$ | Dimensionless constant for stiffness of sand |
| $G, G_u, G'$ | Shear modulus ($G_0$ for small strain shear modulus) |
| GBS | Gravity-based structure |
| $G_{cy}$ | Cyclic shear modulus |
| $h$ | Debris flow height |
| $h$ | Normalised horizontal capacity ($H/H_{ult}$) |
| $h$ | Soil layer thickness |
| $h$ | Wave height |
| $h$ | Distance above pile tip |
| $h_{cavity}$ | Limiting cavity depth of penetrating spudcan (also depth at onset of backflow) |
| $h_{max}$ | Maximum wave height in a storm |
| $h_o$ | Normalised horizontal dimension of yield surface |
| $H$ | Horizontal load |
| $H$ | Vertical elevation of debris flow source above deposit |
| $H_{max}$ | Maximum horizontal load |
| $H_s$ | Sand layer thickness in spudcan punch through analysis |
| $H_{ult}$ | Ultimate horizontal capacity |
| $H_1, H_2$ | Horizontal interslice forces in slope analysis |
| | |
| $i_c$ | Inclination factor for undrained bearing capacity |
| $I_p$ | Plasticity index |

| | |
|---|---|
| $i_q, i_\gamma$ | Inclination factors for drained bearing capacity |
| $I_r$ | Rigidity index, $G/s_u$ |
| $I_R$ | Relative density index |
| $I_\rho$ | Elastic influence factor for vertical displacement |
| IFR | Incremental filling ratio |
| $k$ | Undrained shear strength ratio ($s_u/\sigma'_{v0}$) |
| $k$ or $k_{su}$ | Rate of increase of undrained shear strength with depth |
| $k$ | Vertical penetration stiffness, $V/w$ |
| $k_v, k_h$ | Soil permeability (vertical, horizontal) |
| $k_v, k_h, k_m,$ $k_t, k_c$ | Dimensionless foundation–soil stiffness factors (vertical, horizontal, moment, torsion, coupled) |
| $k_{P-y}$ | Lateral pile–soil stiffness, $P/y$ |
| $k_{vp}$ | Vertical bearing modulus for pile penetration |
| $K$ | Dimensionless vertical penetration stiffness, $k/(V_{ult}/D)$ |
| $K, K_a, K_p$ | Earth pressure coefficient, $\sigma'_h/\sigma'_{v0}$ (active, passive) |
| $K_0$ | Earth pressure coefficient at rest, $\sigma'_{h0}/\sigma'_{v0}$ |
| $K, K_u, K'$ | Bulk modulus |
| $K_c$ | Undrained bearing capacity modification factor |
| $K_s$ | Punching shear coefficient in spudcan punch through analysis |
| $K_q, K_\gamma$ | Drained bearing capacity modification factors |
| L | Horizontal distance of (limit of) debris flow deposit from source |
| L | Foundation or pile length (plan length for surface foundation and embedded length for caisson or pile) |
| L | Height of centre of rotation of scoop mechanism above foundation level |
| L/H | Runout ratio of debris flow |
| $L_H$ | Height of line of action of H above load reference point |
| LRP | Load reference point |
| m | Normalised moment capacity ($M/M_{ult}$) |
| $m_E$ | Rate of increase of stiffness modulus with depth |
| $m_o$ | Normalised moment dimension of yield surface, $M_{ult}/V_{ult}$ |
| $m_v$ | Coefficient of volume compressibility |
| M | Moment |
| $M^*$ | Modified moment parameter |
| $M_{max}$ | Maximum moment |
| $M_r$ | Recompression modulus |
| $M_{ult}$ | Ultimate moment capacity |
| $M_p$ | Plastic moment capacity of pile |
| n | Gradient of ultimate lateral pile resistance with depth, $P_f/zD$ |
| $n_k$ | Gradient of lateral pile–soil stiffness with depth, $k_{P-y}/z$ |
| N | Number of cycles |
| N | Lateral pressure coefficient on pile, $P_f/D\sigma'_{v0}$ |
| $N_c$ | Undrained bearing capacity factor |
| $N_{cH}$ | Undrained horizontal bearing capacity factor |
| $N_{cM}$ | Undrained moment bearing capacity factor |
| $N_{cV}$ | Undrained vertical bearing capacity factor |
| $N_{eq}$ | Equivalent number of cycles |
| $N_f$ | Number of cycles to failure |
| $N_p$ | Lateral bearing capacity factor on skirted foundation or pile |
| $N_q$ | Drained bearing capacity factor for surcharge |

| | |
|---|---|
| $N_\gamma$ | Drained bearing capacity factor for self-weight |
| OCR | Overconsolidation ratio |
| $p, p'$ | Mean stress (total, effective) |
| $p_a$ | Atmospheric pressure (100 kPa) |
| $p'_c$ | Mean effective stress following consolidation |
| $p'_0$ | Effective overburden at foundation level, usually $\gamma' d$ |
| $p_0, p'_0$ | Initial mean stress (total, effective) |
| P | Passive lateral resistance at toe of slope |
| P | Lateral resistance on pile, per unit length |
| PI | Plasticity index |
| q | Deviator stress |
| $q_b$ | Unit pile base resistance |
| $q_c, q_{net}$ | Cone tip resistance (total, net) |
| $q_{peak}$ | Peak resistance at spudcan punch through |
| $q_0$ | Normalised torsional dimension of yield surface $T_{ult}/V_{ult}$ |
| Q | External load |
| Q | Normal force transmitted to anchor line from soil, per unit length |
| Q | Torsional load |
| $Q_{net}$ | Normalised cone tip resistance ($q_{net}/\sigma'_{v0}$) |
| $Q_b$ | Total pile base resistance |
| $Q_s$ | Total pile shaft resistance |
| r | Excess pore pressure ratio, usually $u_e/\sigma_v$ |
| r | Radial distance from axis |
| R | Radius of slip circle, pile radius |
| $R_a$ | Surface roughness |
| $\dot{s}, \ddot{s}$ | Structural velocity and acceleration |
| $s_c$ | Shape factor in classical undrained bearing capacity expression |
| $s_{cM}$ | Shape factor for pure moment capacity |
| $s_{cv}$ | Shape factor coefficient |
| $s_{cV}$ | Shape factor for uniaxial vertical bearing capacity |
| $s_q, s_\gamma$ | Shape factors in classical drained bearing capacity expression |
| $s_u$ | Undrained shear strength |
| $s_{u,p}$ | Peak undrained shear strength |
| $s_{u,r}$ | Fully remoulded undrained shear strength |
| $s_{u,ref}$ | Undrained shear strength at reference strain rate |
| $s_{u0}$ | Undrained shear strength at foundation level |
| $s_{uc}$ | Undrained shear strength following consolidation |
| $s_{uc}$ | Undrained shear strength from triaxial compression test |
| $s_{ue}$ | Undrained shear strength from triaxial extension test |
| $s_{um}$ | Undrained shear strength at mudline |
| $s_{u0}$ | Undrained shear strength at foundation level |
| $s_{uss}$ | Undrained shear strength from simple shear test |
| $\bar{s}_u$ | Average undrained shear strength |
| $\bar{s}_{u(t)}$ | Average undrained soil shear strength at time t after installation |
| $S_g$ | Specific gravity of pipeline |
| SS | Simple shear |
| $S_t$ | Sensitivity |
| t | Elapsed time |

| | |
|---|---|
| t | Pile or caisson wall thickness |
| T | Dimensionless time factor |
| T | Anchor line tension |
| $T_a$ | Anchor line tension at padeye |
| TLP | Tension-leg platform |
| $T_m$ | Anchor line tension at mudline |
| $T_n$ | Resisting force normal to anchor |
| $T_p$ | Resisting force perpendicular to anchor |
| TXC | Triaxial compression |
| TXE | Triaxial extension |
| $T_0$ | Horizontal component of tension in hanging pipe catenary |
| | |
| u | Horizontal displacement |
| $u, u_w$ | Pore water pressure |
| $u, \dot{u}$ | Horizontal water particle velocity and acceleration |
| $u_o$ | Hydrostatic pore water pressure |
| $u_g$ | Pore gas pressure |
| $u_e$ | Excess pore pressure |
| $u_{eo}$ | Initial excess pore pressure |
| U | Pore water force acting on slip plane |
| | |
| v | Penetration velocity |
| v | Normalised vertical capacity ($V/V_{ult}$, $V/V_o$) |
| v | Specific volume $(1 + e)$ |
| $v_o$ | *In situ* (or initial) specific volume |
| $v_{ref}$ | Reference wind velocity |
| V | Normalised penetration velocity, usually $vD/c_v$ |
| V | Vertical load |
| V | Slide pit volume |
| $V_o$ or $V_{max}$ | Current uniaxial yield load |
| $V_{ult}$ | Ultimate vertical bearing capacity |
| w | Vertical displacement, pipe invert embedment |
| w | Self-weight of anchor line |
| W | Weight |
| $W'$ | Submerged weight of foundation, or of pipeline (per unit length) |
| $W'_{plug}$ | Submerged weight of soil plug |
| $W'_{pile}$ | Submerged weight of pile |
| y | Lateral pile movement |
| z | Depth below soil surface or mudline |
| $z^*$ | Optimal padeye depth |
| $z_a$ | Depth to padeye |
| $\alpha$ | Seabed inclination |
| $\alpha$ | Adhesion factor or interface friction coefficient, $\tau_f/s_u$ |
| $\alpha$ | Anisotropy coefficient |
| $\alpha$ | Elastic influence factor for rotation |
| $\alpha$ | Assumed angle of spread for punching shear |
| $\alpha_i, \alpha_o$ | Internal, external (outer) wall interface friction coefficient |
| $\alpha_{ult,}$ | Ultimate slope angle |

| | |
|---|---|
| $\alpha_h, \alpha_m, \alpha_q$ | Association factors for plastic potential (in horizontal, moment and torsional planes) |
| $\beta$ | Slope angle, or subtended half angle of pipe–soil contact |
| $\beta$ | Angle of anchor flukes to horizontal |
| $\beta$ | Pile shaft resistance parameter ($\tau_{sf}/\sigma'_{v0}$) |
| $\beta_1, \beta_2$ | Curvature factor exponents in equation for yield surface |
| $\beta_3, \beta_4$ | Curvature factor exponents in equation for plastic potential |
| $\delta$ | Padeye depth adjustment |
| $\delta$ | Interface friction angle, pile–soil friction angle |
| $\varepsilon_1$ | Major principal strain |
| $\varepsilon_2$ | Intermediate principal strain |
| $\varepsilon_3$ | Minor principal strain |
| $\varepsilon_v$ | Vertical strain or volumetric strain |
| $\phi$ | Angle of internal friction |
| $\phi_{cr}$ | Critical state angle of internal friction |
| $\phi_p$ | Peak angle of internal friction |
| $\gamma$ | Unit weight of soil |
| $\gamma'$ | Effective unit weight of soil |
| $\gamma$ | Shear strain |
| $\gamma_a$ | Average shear strain |
| $\gamma_{cy}$ | Cyclic shear strain |
| $\gamma_m$ | Material factor on shear strength |
| $\gamma_w$ | Unit weight of water |
| $\gamma'$ | Shear strain rate |
| $\gamma'_{ref}$ | Reference shear strain rate |
| $\eta$ | Pile base enlargement ratio ($D_{base}/D$) |
| $\lambda$ | Pile–soil stiffness ratio ($E_p/G_L$) |
| $\mu$ | Viscosity coefficient |
| $\mu$ | Friction coefficient |
| $\nu^*$ | Modified Poisson's ratio accounting for anisotropic characteristics |
| $\nu, \nu_u, \nu'$ | Poisson's ratio |
| $\theta$ | Slip plane angle |
| $\theta$ | Rotation |
| $\theta$ | Orientation of anchor line element to the horizontal |
| $\theta_a$ | Anchor line angle at padeye |
| $\theta_m$ | Anchor line angle at mudline |
| $\theta_w, \theta'_w$ | Angle of resultant anchor force to the fluke (weightless, weighty anchor) |
| $\rho$ | Stiffness gradient ratio (average/at base, $G_{avg}/G_L$) |
| $\rho_a$ | Density of air |
| $\sigma'_{v0}$ | *In situ* vertical effective stress |
| $\sigma'_{vc}$ | Vertical consolidation stress |
| $\sigma_s$ | Normal stress on slip plane |
| $\sigma_v, \sigma'_v$ | Total, effective vertical stress |
| $\sigma_h, \sigma'_h$ | Total, effective horizontal stress |
| $\sigma, \sigma'$ | Total, effective stress |
| $\tau$ | Shear stress |
| $\tau_a$ | Average shear stress |
| $\tau_{cy}$ | Cyclic shear stress |
| $\tau_f$ | Shear stress at failure (in soil or on pile or foundation) |

| | |
|---|---|
| $\tau_{f,cy}$ | Cyclic shear stress at failure |
| $\tau_{mob}$ | Mobilised shear stress |
| $\tau_s$ | Shear stress on slip plane or pile shaft |
| $\tau_{ult}$ | Ultimate shear stress |
| $\tau_y$ | Yield stress |
| $\tau_{ya}$ | Apparent yield stress |
| $\xi$ | Base stiffness ratio (at base/below base), $G_L/G_{base}$ |
| $\psi$ | Angle of dilation |
| $\zeta$ | Cumulative plastic shear strain |
| $\zeta$ | Contact force enhancement ('wedging') factor in pipe–soil interaction |
| $\zeta_{95}$ | Plastic shear strain to achieve 95 per cent remoulding |

# 1 Introduction

## 1.1 Historical perspective of offshore development

The first offshore oil rig 'Superior' was installed in 1947, 18 miles from the coast of Louisiana in the United States, in just 6 m depth of water (Figure 1.1). Today, there are over 7,000 offshore platforms around the world located in water depths now starting to exceed 2,000 m. As late as the early 1970s, deep-water developments meant water depths of 50–100 m, with the majority of platforms still in water depths of less than 50 m. Nowadays, the terms 'deep water' and 'ultra-deep water' are generally taken to refer to around 500 m and 1,500 m, respectively.

For many years, offshore design was dominated by the initial Gulf of Mexico experiences where soft clays dictated driven pile foundations, although soil conditions in the North Sea, Middle-East and (ultimately) the Southern Hemisphere (Australia, Brazil and West Africa) have each proved radically different, leading to new foundation approaches even for fixed platforms.

The Norwegian sector, aided by strong government policy to retain high Norwegian input, pioneered the development of concrete gravity structures in the North Sea, where stronger clays and dense sands provided adequate bearing capacity for shallow foundations.

Specific problematic soils (e.g. calcareous soils) or environmental conditions (e.g. ice forces in Canadian waters) led to new foundation or platform concepts (e.g. grouted piles in calcareous soils, sand islands in the Beaufort Sea).

As hydrocarbon reserves close to continental landmasses have gradually been depleted, new fields have been developed at greater distances from land and, therefore, in deeper water. Development of fields in deep water has led to a variety of flexible structures (guyed towers) or floating facilities (tension-leg platforms and floating production units) tethered or moored by tension members and seabed anchors with perhaps the most significant being suction-installed deep caissons, such as those used for the Na Kika development (Figure 1.2, Newlin 2003a).

Greater distances separating the offshore development from land have necessitated either local storage, mainly for liquids, and periodic offloading onto tankers, or long export pipelines. There are also economic drivers to network adjacent fields, with pipelines bringing the product to a central facility for processing or export. These trends have led to much greater lengths of pipeline, both infield and for export, so that pipeline design has become an increasingly important economic factor in new developments.

*Figure 1.1* Superior – The first offshore installation: 1947, Louisiana coast (from *Leffler et al.* 2003)

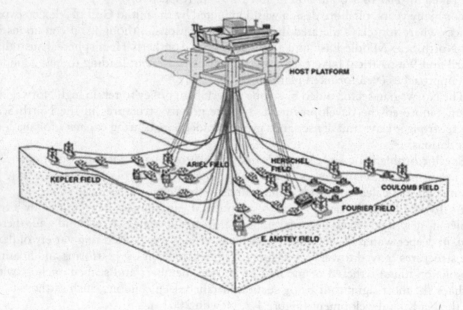

*Figure 1.2* Na Kika development, Gulf of Mexico, 2,000 m water depth (Newlin 2003a)

A further feature of developments in deep water is the increasing exposure to a range of geohazards, including dissociation of gas hydrates, migration of gas through the foundation soils, submarine slides emanating from the edge of the continental shelf and pipeline routing up steep and rugged terrain such as the Sigsbee escarpment in the Gulf of Mexico (Figure 1.3).

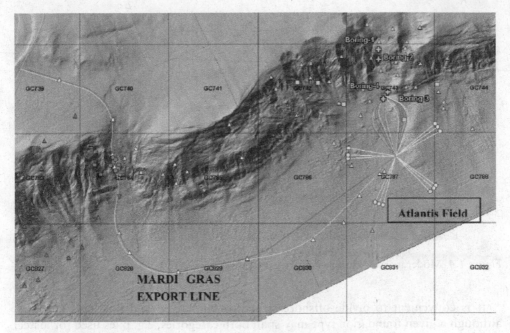

*Figure 1.3* Sigsbee escarpment in the Gulf of Mexico (Jeanjean *et al.* 2005)

## 1.2 Features of offshore engineering

There a number of distinguishing features of geotechnical engineering for offshore conditions:

1. Site investigations are extremely expensive, with mobilisation and hire costs of suitable vessels typically several million US dollars.
2. Soil conditions are often unusual, particularly in respect of carbonate soils and corals.
3. Applied loads are large, with a high component of environmental loading, and large moment loading relative to the weight of the structure.
4. Design modifications during construction are generally not possible or incur severe cost penalties.
5. Emphasis is focused more on capacity, or ultimate limit state, than on deformations although the foundation stiffness is important for the dynamic response of the structure.

The cost of installing foundations, in combination with the relatively high environmental loads to be withstood, leads to large foundation sizes, such as piles of 2–3 m in diameter, penetrating up to 200 m below the seabed.

A typical field development may extend over a wide area (many tens of square kilometres), potentially with several fixed structures, or anchoring locations, and infield flowlines linking wells. Export pipelines, to shore or feeding into a regional trunkline, may also be required. Site investigation requirements may therefore extend over a significant region, even if limited to only shallow depths through most of it.

64 m

*Figure 1.4* Moderate sized steel jacket structure during transport

It is convenient to divide offshore structures into fixed and floating structures, although a given foundation type may span both categories, e.g. piles used for a steel jacket structure, or as anchoring for tension-leg platforms or other types of floating systems. The evolving nature of foundation and anchoring systems to suit increasing water depths has necessitated considerable investment in research in order to validate new foundation and anchoring systems. In parallel, design guidelines or 'Recommended Practices' have been developed by the American Petroleum Institute (API) and, more recently, the International Standards Organisation (ISO). While these inevitably lag the advances in understanding achieved through research, industry has strived to keep pace by continuously updating the design codes through standing committees of specialists. Regulatory bodies such as the American Bureau of Shipping (ABS), Det Norske Veritas (DNV) and Lloyds, have also played an important role in validating new design approaches supported by research findings and helping to coordinate advances across the industry.

## 1.3  Foundations for fixed structures

The first offshore structures were a steel jacket, or template, design and soil conditions in the Gulf of Mexico comprising normally or lightly overconsolidated plastic clays dictated driven pile foundations. Initially, timber piles were used but were soon superseded by steel tubular piles for ease of construction. Simple platform designs incorporate the piles into the corners of the steel jacket (Figure 1.4), with the piles driven using either a steam hammer acting on a follower pile, or, in modern times, an underwater hydraulic hammer capable of fitting down the pile sleeves.

Typical soil conditions in the early development regions of the North Sea comprised heavily glaciated overconsolidated clays (with shear strengths generally 100–700 kPa), interbedded with dense sands. These conditions permitted the development of concrete gravity structures, sitting directly on the seabed with minimal skirts. More recent platforms, such as Gullfaks C (Figure 1.5) and Troll, are in deeper water (200–300 m) and with softer seabed conditions; this has necessitated much longer skirts, in

*Figure 1.5* Schematic of Gullfaks C concrete gravity structure

the range 20–30 m, and the use of suction (technically reduced pressure compared with the ambient hydrostatic pressure) to help penetrate the skirts into the seabed.

Steel-skirted foundations, variously referred to as plated or bucket foundations, or suction cans or caissons for individual elements, have proved to be an extremely versatile foundation approach, able to withstand compressive, tensile or lateral loading and to be installed using relatively lightweight support vessels. Suction caissons at the base of jacket structures have been used successfully as an alternative to gravity base structures in the North Sea.

Small structures can be based on simple tripod foundations (with a central tubular column supporting the deck) or a 'monopod' single pile. Hybrid structures (such as steel on a concrete base) are also becoming more common, as are less conventional approaches, for example, a mobile drilling (or 'jack-up') rig located more or less permanently on a steel or concrete mat foundation.

Jack-up units used as mobile drilling rigs (Figure 1.6) have three or four independent legs, each equipped with 'spudcan' foundations, which are shallow conical foundations with a central spigot on the underside. These may penetrate the seabed by several diameters as the hull of the unit is raised above the sea surface.

Piles, shallow foundations and spudcans are considered in detail in Chapters 5, 6 and 8, respectively.

## 1.4 Moorings for flexible and buoyant facilities

In deeper water, generally over 200 m, there is sufficient flexibility in conductor casing to allow relatively mobile or buoyant structures. At intermediate depths, up to perhaps 500 m, compliant towers (CTs) are a more attractive option than a

*Figure 1.6* Jack-up mobile drilling rig

conventional platform, for which the footprint would be very large. The CT operates with a large mass and buoyancy in the upper region giving a sluggish response to environmental loading. Typical 10–15-second-period waves pass through the structure before the structural frame can respond. Generally, CTs have a lower limit of around 300 m as in shallower waters the structure becomes too stiff to allow the concept to work.

Ronalds (2005) provides an overview of different types of floating systems, considering Floating Production Storage and Offloading vessels (FPSOs), semi-submersibles (semis), tension-leg platforms (TLPs – also mini-TLPs) and SPARs and the drivers behind choice of each system. All floating systems require moorings and ultimately, some form of anchor on the seabed.

Tension-leg platforms use vertical taut cables (or steel pipe) to apply tension between a seabed template and the buoyant facility. The seabed template is restrained either by driven steel piles or by a combination of dead load and suction caissons.

Other buoyant structures can be moored by catenary or (semi-) taut moorings. Catenary moorings are common for floating facilities, or to help stabilise flexible structures, in shallow to moderately deep water. The anchoring system can be a simple dead weight, i.e. gravity anchor, although higher capacity can be achieved by embedding anchors within the seabed.

For developments in deeper water, catenary moorings become less attractive than some form of taut or semi-taut cable system, making an angle of up to 45° at the seabed. This form of loading has led to the development of novel embedment anchors

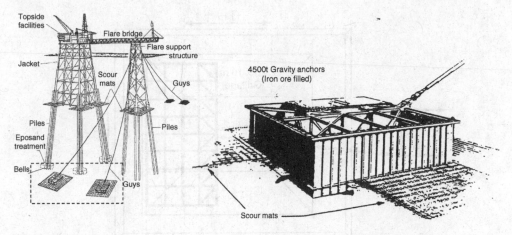

*Figure 1.7* Box anchors for North Rankin Flare Tower (Woodside Petroleum 1988)

that are designed to operate successfully with a high vertical component of load (e.g. embedment plate anchors such as the Vryhof Stevmanta), or suction caissons, which may also be ballasted by additional dead load.

### 1.4.1 Anchor types

The simplest type of anchor is just a dead weight (i.e. gravity anchor), most commonly achieved by use of a hollow box that is placed on the seabed and then filled with rock or (heavier) iron ore. Examples of this on the North-West Shelf of Australia include the anchor boxes for the guyed flare support tower at North Rankin (Figure 1.7), and the box anchors for the FPSO vessel for the Wanaea and Cossack fields.

An innovative type of gravity anchor is the use of a buried grillage, beneath a rock-fill or iron ore berm (Erbrich and Neubecker 1999). The grillage is placed towards the back of the berm, but such that the complete berm must be moved if the grillage starts to fail. This arrangement was used for the Apache Stag field development on the North-West Shelf (Figure 1.8).

Embedded anchors may be divided into five main types: driven (or drilled and grouted) piles; suction caissons; drag embedment anchors; plate anchors and dynamically embedded (torpedo) anchors. Anchor piles can give the highest absolute capacity, particularly when loaded in pure tension, as is usually the case for TLPs. Design of these different types of anchor is considered in Chapter 7, but the most common type currently used for permanent moorings in deep water are suction caissons (Figure 1.9).

## 1.5 Pipelines and geohazards

The final two chapters of this book deal with aspects of offshore geotechnical engineering that have become increasingly important over the last two decades, but generally fall somewhat outside the expertise of traditional geotechnical specialists.

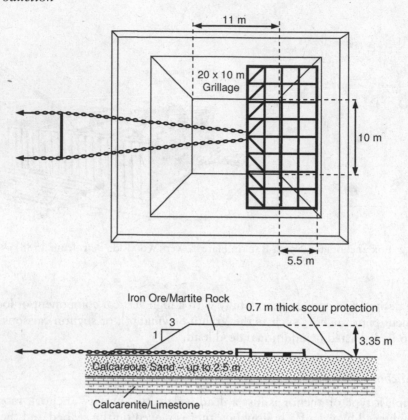

*Figure 1.8* Grillage and berm anchors for the Apache Stag field CALM buoy (redrawn from Erbrich and Neubecker 1999)

Pipeline design is a specialist discipline in the offshore world with the primary focus being to ensure successful transport of potentially multi-phase hydrocarbon product inside the pipe. As a structure, a pipeline must be designed to ensure that it has sufficient strength to withstand internal and external pressures, will not buckle during laying or operation, and is stable on the seabed. Geotechnical aspects of design are mostly concerned with the stability of the pipeline on the seabed, either buried within a trench or laid directly on the seabed, and what are the limiting interaction forces between the pipeline and seabed that govern stability.

Over the last decade, geotechnical input into pipeline design has become more sophisticated, driven by the increasing lengths of pipeline needed for remote deep-water developments. In deep water, pipelines are generally laid directly on the seabed and must be designed to withstand repeated cycles of thermal expansion and contraction, with consequential lateral buckling and other phenomena such as 'pipeline walking' (Bruton *et al.* 2006; Carr *et al.* 2006). These and other aspects of geotechnical pipeline design are dealt with in Chapter 9, with particular attention to estimating the embedment of pipelines in soft sediments, which influences subsequent response under transverse or longitudinal motion.

As evident from Figure 1.3, submarine slides from the escarpment at the edge of the continental shelf present a particular geohazard for pipelines or indeed, other

*Figure 1.9* Suction anchors assembled on load out vessel

infrastructure on the seabed. Closer to shore, mudslides induced by hurricanes can also cause damage to pipelines (Gilbert *et al.* 2007). Geohazard assessment relies on input from many different branches of geoscience, with geotechnical engineering playing a relatively minor role, but a brief overview is provided in Chapter 10, together with a discussion of methodologies to assess and mitigate the level of risk.

# 2  The offshore environment

## 2.1 Active geology, continental drift and plate tectonics

The earth is an active planet. Earth scientists began to understand the earth as a mobile, dynamic body only during the 1960s. This change represented a major scientific revolution, which, like the Copernican and Darwinian revolutions, should be named for its champion, German meteorologist Alfred Wegner. He was the first to collect much evidence in support of the theory of continental drift in 1915. As a meteorologist, he spent his winters on the ice in Greenland, and it was during these periods of isolation that he contemplated the Earth and wrote about continents that drift about like icebergs in the sea. He deduced this theory from the remarkable geometric fit of the east coast of South America and the west coast of Africa.

Wegener identified that parts of the Earth's crust slowly drift atop a liquid core, and hypothesised that there was a gigantic super-continent 200 million years ago, which he named Pangaea (meaning 'All-earth' in Greek). Pangaea started to break up into two smaller super-continents, called Laurasia and Gondwanaland, about 130 million years ago (MA). By 66 MA – when the dinosaurs were just dying out – the continents were separating into landmasses that look like the modern-day continents.

The continental plates that drift atop the soft mantle are made of rocks that vary in thickness between 80 and 400 km. The plates move both horizontally and vertically, at speeds of between 1 and 10 cm per year, and change in size as their margins are added to, crushed together or pushed back into the Earth's mantle. The theory of plate tectonics (meaning 'plate structure') explains the movement of the Earth's plates. Plate tectonics also explains the cause of earthquakes, volcanoes, oceanic trenches, mountain range formation and other geologic phenomena as most of the Earth's seismic activity occurs at the plate boundaries as they interact.

The top layer of the continental plates is called the crust – the 'continental crust' above sea level and the 'oceanic crust' beneath the oceans. Figure 2.1 shows a contour map of crust thickness across the world based on seismic refraction data collected by the US Geological Survey, which indicates that continental crust thickness typically ranges between 35 and 40 km while oceanic crust is thinner, typically less than 10 km thick.

Unlike continental crust, oceanic crust is actively created at various mid-oceanic ridges with new crust being pushed up into the seabed at 'spreading centres' and pushed down into the mantle at 'subduction zones'. Spreading can occur at a rate of up to 100 mm/year, but often less than 30 mm/year. Crust activity is illustrated in

*Figure 2.1* Continental and oceanic crust thickness (in km)

Source: US Geological Survey http://earthquake.usgs.gov/research/structure/crust/download.php This material is the property of the US Government but is in the public domain and is not subject to copyright protection.

Figure 2.2 through a digital model of the age of the oceanic lithosphere, compiled by the US National Geophysical Data Centre. The youngest crust indicates spreading centres and the oldest crust indicates subduction zones. Figure 2.3 shows a schematic of some of the dynamic processes associated with continental plates and the earth's continental and oceanic crust including sea floor spreading and subduction, mountain forming as plates collide and volcanoes.

## 2.2 Marine geology

### 2.2.1 Topographical features of ocean floors

Figure 2.4 illustrates the main topographical features common to all oceans: the continental margin, the continental rise and the abyssal plain of the deep ocean. Trenches and seamounts may also be present in the deep ocean. The continental margin comprises the continental shelf and the continental slope. The shelf is the submerged continuation of the adjacent land. The seaward extent of the continental shelf is the shelf break, or the continental ridge, leading into the continental slope. The toe of the continental slope is marked by the continental rise leading into the abyssal plain.

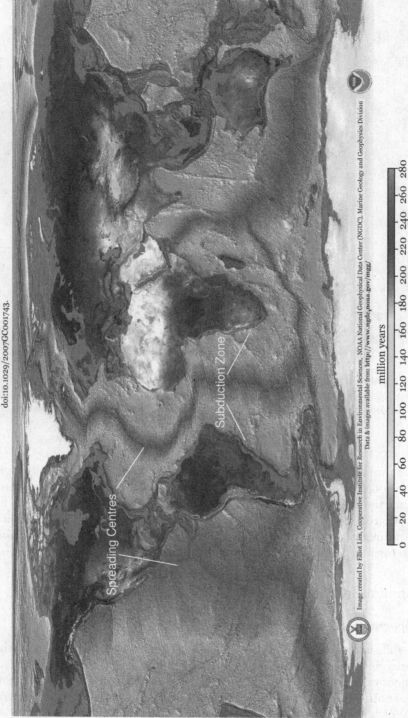

Age of Oceanic Lithosphere (m.y.)

**Data source:**

Muller, R.D., M. Sdrolias, C. Gaina, and W.R. Roest 2008. Age, spreading rates and spreading symmetry of the world's ocean crust,Geochem. Geophys. Geosyst., 9, Q04006, doi:10.1029/2007GC001743.

Image created by Elliot Lim, Cooperative Institute for Research in Environmental Sciences, NOAA National Geophysical Data Center (NGDC), Marine Geology and Geophysics Division
Data & images available from http://www.ngdc.noaa.gov/mgg/

million years

0  20  40  60  80  100  120  140  160  180  200  220  240  260  280

Spreading Centres

Subduction Zone

*Figure 2.2* Activity of the earth's crust (Muller *et al.* 2008)

Source: National Geophysical Data Centre http://www.ngdc.noaa.gov/mgg/image/crustalimages.html This material is the property of the US Government but is in the public domain and is not subject to copyright protection.

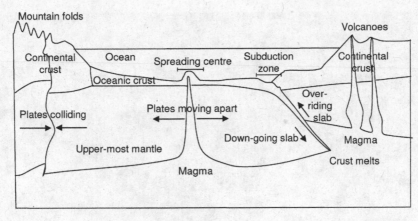

*Figure 2.3* Plate and crust tectonics

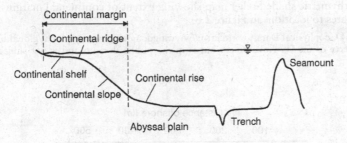

*Figure 2.4* Topographical features of ocean floors (after Poulos 1988)

The continental margins are extremely important as oil reservoirs and as such are of most interest to engineers concerned with harnessing offshore oil and gas resources. The continental margin covers approximately 20 per cent of the total ocean floor, an area of $74 \cdot 10^6$ km² – approximately ten times the size of Australia.

While the topography of all oceans can be described in general terms, regional variations naturally exist. Figure 2.5 shows a shaded relief map of land topography and ocean bathymetry compiled by the National Geophysical Data Centre that indicates, amongst other things, the extent of the continental margins. Figure 2.6 shows the topographical profile of the continental margin in a selection of offshore locations, illustrating the potential diversity of shelf and slope bathymetry.

The locations shown in Figure 2.6 are indicated on the world map in Figure 2.5, showing that significant variations in bathymetry can also be encountered locally, particularly in the North Pacific, the Gulf of Mexico and the North-West Shelf of Australia.

The continental shelf can be virtually non-existent or extend several hundred kilometres offshore with the shelf break occurring between depths of 10 and 500 m. Shelf breaks are deepest off glaciated areas and shallowest in areas with extensive coral growth. The average slope of the continental shelf is approximately 1:500 (0°07').

*Figure 2.5* Bathymetric shaded relief map showing extent of continental margins (number key relates to locations in Figure 2.6)

Source: National Geophysical Data Centre http://www.ngdc.noaa.gov/mgg/global/global.html This material is the property of the US Government but is in the public domain and is not subject to copyright protection.

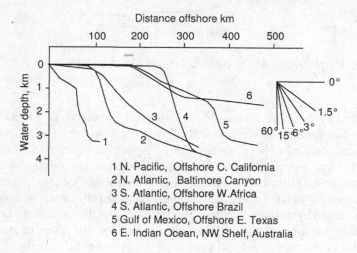

*Figure 2.6* Profiles of ocean floor topography at selected locations

Beyond the shelf break, the continental slope has a steeper gradient than the shelf, ranging from an average 1:40 (1.2°) in delta regions but up to 1:10 (6°) in faulted areas, and reaches water depths of 2,000–3,000 m. The continental rise, at the foot of the continental slope, has gradients ranging from 1:1,000–1:700. The abyssal plain adjacent to the continental rise is smooth with gradients between 1:1,000 and 1:10,000, and occur from depths of 2,500–6,000 m.

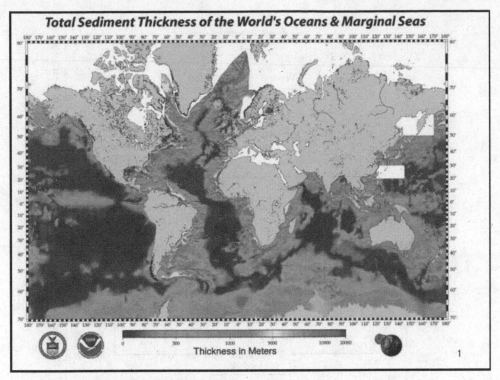

*Figure 2.7* Sediment thickness of the world's oceans and marginal seas (Divins 2009)

Source: National Geophysical Data Centre http://www.ngdc.noaa.gov/mgg/sedthick/sedthick.html This material is the property of the US Government but is in the public domain and is not subject to copyright protection.

## 2.3  Marine sediments

### 2.3.1  Distribution

Figure 2.7 shows a digital model compiled by the National Geophysical Data Centre showing the distribution of sediment in the world's oceans and marginal seas. Sediment is usually thickest near continents and thinnest on newly formed mid-oceanic ridges. Some areas of the ocean bottom are devoid of sediment, having been swept clean by strong bottom currents. Although the continental margins cover only 20 per cent of the seabed area, they contain nearly 75 per cent of the total marine sediments. In many areas, the continental rise is a depositional feature, formed mainly of sediment slurry, and reaching up to 1.6 km thickness in places. Canyons often cut across the rise and act as channels for the seaward transport of sediment. Abyssal plains are connected by canyons or other channels to landward sources of sediments, which are transported as dense slurries to the plains (Poulos 1988).

### 2.3.2  Origin and classification

Marine sediments are composed of detrital material from land or from the remains of marine organisms, which leads to the principal classification of sediment as either

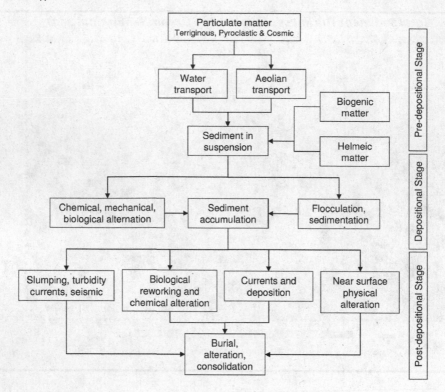

*Figure 2.8* Sedimentary process of marine deposits (after Silva 1974)

terrigenous (transported from land) or pelagic (sediments that settle through the water column). Pelagic sediments are deposited so slowly that nearshore and coastal areas are overwhelmed by terrigenous deposits.

Terrigenous material can be supplied by rivers, coastal erosion, aeolian or glacial activities. The sub-classifications used to describe terrigenous sediments are based on the size of the particles. Terrigenous sediments tend to be grains of silicate-based minerals such as quartz and feldspar, and are principally formed from the erosion of rocks – leading to the term 'lithogenous'.

Pelagic sediments are generally fine grained and are instead classified according to their composition. Organic, or biogenous, pelagic sediments derive from the insoluble remains of marine organisms, e.g. shells, skeletons, teeth and tests. Lithogenous pelagic sediments are formed when particles are transported by wind into the ocean, before settling through the water column.

Marine sediments can also form from biological and chemical reactions occurring in the water column or within sediments, and are referred to as hydrogenous sediments. Figure 2.8 summarises some of the processes involved in the formation of marine sediments.

## Lithogenous sediments

Rivers are the largest source of sediment in the ocean and contribute about 20 billion tonnes of sediment to the world's oceans each year. The settlement of particles is

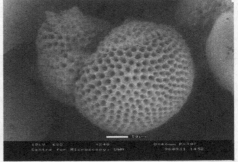

*Figure 2.9* Seabed foraminifera from a deepwater location on the NW shelf of Australia

controlled primarily by grain size such that generally finer particles are found with increasing distance from shore. Exceptions include material transported during submarine slides as debris flows and turbidity currents (heavy fluid flows). Turbidity currents, in particular, can transport sediment from the continental margins to the continental rise and onto the abyssal plain. Deposits formed by turbidity currents have a graded bedding, with larger particles overlain by progressively smaller particles (submarine slides, debris flows and turbidity currents are discussed in detail in Chapter 10).

Wind transports around 100 million tonnes of sediment into the world's oceans annually, and small particles can be carried for considerable distances – even around the world – before being deposited in the sea as pelagic sediments.

### Biogenous sediments

Biogenous sediments may either be siliceous or calcareous and are formed from planktonic organisms, which are the most abundant organism in the oceans (both now and through geological history). The two principal types of plankton are phytoplankton (which photosynthesise like plants) and zooplankton (which graze like animals). The insoluble shells of these creatures can be formed of calcite or silica. Calcite organisms are more common in shallow water and temperate tropical climates, whilst silica organisms are more common in the polar regions or at the equator and in very deep water. Calcium carbonate, the main constituent of calcareous deposits, is soluble at high pressure, such that calcareous sediments are not usually found in water depths greater than 4,000 m.

The most common organisms that form siliceous biogenous sediments are diatoms (a microscopic phytoplankton) and radiolarian (a microscopic zooplankton). Calcareous biogenous sediments are composed of large accumulations of the skeletal remains of plant and animal life such as coralline algae, coccoliths, foraminifera and echinoderms. The most common biogenic calcareous particles are shells or tests of foraminifera – micro-organisms generally less than 1 mm in diameter with tests consisting of a single chamber or many interconnected chambers. Figure 2.9 shows some seabed foraminifera from a water depth of approximately 1,000 m on the North-West Shelf of Australia. Coral may also be found within biogenous carbonate deposits, where the water was previously very shallow.

The common trend, of all these micro-organisms is that their tests, shells or skeletons are often perforated and angular, which, when coupled with the relative softness

(a)                                             (b)

*Figure 2.10* Micrographs of (a) calcareous sand and (b) silica sand

of calcium carbonate (of which they are made), leads to the fragility and high compressibility that are the hallmarks of calcareous sediments. Figure 2.10 compares micrographs of a calcareous sand, from Goodwyn on the North-West Shelf of Australia, with a quartz-grained silica sand, showing clearly the difference in particle shape. The calcareous sand has fragile, angular and hollow particles as opposed to the hard, rounded silica grains.

Carbonate soils are also prone to post-depositional alterations by biological and physiochemical processes that can result in a range of effects from light cementation between grains to the formation of lenses of highly cemented material that profoundly affect mechanical behaviour.

### 2.3.3  *In situ* stress state

The *in situ* stress state of seabed sediments may be normally consolidated, overconsolidated or incompletely consolidated (sometimes referred to as underconsolidated). Slowly deposited sediments that have been unaltered since deposition are normally consolidated. Note that although such deposits may be referred to as normally consolidated, aging effects result in a yield stress that generally exceeds the *in situ* effective stress; hence, from a geotechnical engineering standpoint, they are typically lightly overconsolidated with a yield stress ratio of 1.5 to 2. True overconsolidation may be a relic of glaciation or a consequence of more recent events, such as gradual marine erosion, the sudden removal of overburden due to a submarine slide or past wave loading increasing preconsolidation pressure. Incompletely consolidated materials occur when excess pore pressures have not had an opportunity to dissipate. Excess pore pressures may be generated as a result of rapid deposition, such as in deltaic regions or where run out from a submarine slide comes to rest downslope, due to dissociation of gas hydrates, exsolution of free gas or repeated wave loading.

When concerned with geotechnical engineering offshore, it is important to realise that the height of free water, or changes in the height of free water above the mudline, do not affect excess pore pressures in the seabed, since the water column above the mudline contributes equally to the total stress and pore pressure such that the effective stress within the seabed remains constant. Only changes in the pore water or pore gas regime within the seabed, relative to the total stress, affect the effective stress state and hence the sediment shear strength and the stability of the seabed.

Shelf and slope sediments occur in a range of *in situ* stress states depending on their historical and recent stress history. An understanding of *in situ* stress state and associated excess pore pressure conditions are critical to predicting engineering behaviour. Methods of identifying *in situ* stresses and pore pressures are discussed in Chapter 3, and the importance of excess pore pressures as triggering mechanisms for geohazards is discussed in Chapter 10.

### 2.3.4 *Geotechnical characteristics of some offshore regions*

Regions of continental shelf have characteristic soil conditions that arise from the depositional environment and the geomorphic and ocean processes that have shaped the bathymetry and have worked the sediments. These characteristic soil conditions in turn shape the forms of foundation system and field architecture that are technically and economically the most feasible.

In broad terms, finer-grained sediments predominate further from shore and in deeper water. This is because coarser terrigenous sediments cannot be transported this far, so pelagic sediments predominate. Certain geohazards are also more prevalent in deep water, including the steep scarps that are usually found at the margins of the continental shelf.

In extremely broad terms, some common seabed conditions in the major areas of oil and gas exploration can be characterised as:

- Gulf of Mexico – soft, normally consolidated, medium high plasticity clays (~30 < PI < 70), often with interbedded sand layers.
- Campos and Santos Basins offshore Brazil – sands and clays with high carbonate content.
- West Africa – soft, normally consolidated, very high plasticity clays (~70 < PI < 120, Puech *et al.* 2005), often deposited rapidly by river deltas.
- North Sea and other glaciated regions – stiff, overconsolidated clays and dense sands, with a recent drape of softer material.
- South-East Asia – desiccated crusts of stiff soil, which are remnants of low sea levels during the Pleistocene, with strengths one or two orders of magnitude greater than the underlying soil.
- Australia's North-West Shelf and the Timor Sea – carbonate sands, silts and clay-sized soils, often with variable cementation.

Each of these areas is vast, and different soil conditions are present across each region. However, the foregoing descriptions broadly characterise the type of deposit that dominates design decisions – such as field architecture and foundation solutions – of offshore oil and gas developments.

Relatively new and significant offshore frontiers include substantial deep-water discoveries on each coast of the South Atlantic – off West Africa (Gulf of Guinea and offshore Angola) and East Brazil (Campos and Santos Basins) and the North-West Australian continental slope and the Timor Sea where less familiar seabed sediments predominate (very high plasticity, highly compressible or carbonate, respectively). In addition to unfamiliar seabed sediments, an array of geohazards is associated with

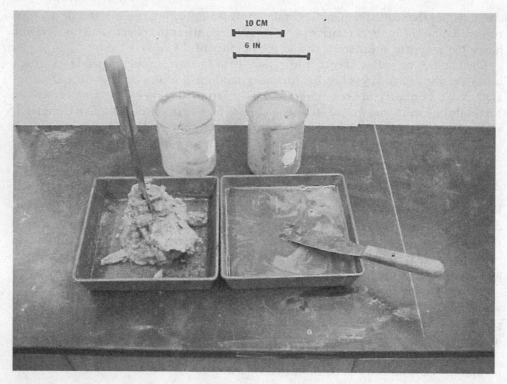

*Figure 2.11* Effect of saline pore water on soil strength: salt water (left) and freshwater (right) (Elton 2001)

deeper water. The Caspian Sea, in particular the super-giant oil fields of the Azeri Chirag Gunshali complex, 120 km off the coast of Azerbaijan, and the West Nile Delta offshore Egypt are emerging prospects in highly geohazardous terrain. The Arctic basin holds about 25 per cent of unrealised petroleum resources and potentially makes up the next great frontier, opening up not only a wealth of resources, but also a wealth of challenges associated with engineering in arctic conditions.

### 2.3.5 Comparison of marine and terrestrial sediments

Environmental and compositional differences between terrestrial and marine sediments affect their engineering behaviour. For example, the high-pressure, low-temperature environment of ocean sediments may affect microstructure and the behaviour of gases in pore fluid, and the high biogenic content of ocean sediments affects strength and compressibility and encourages post-depositional biological and chemical reactions. Properties of pore fluid differ in the ocean and on land. Ocean sediments are saturated with saline water, and terrestrial sediments are typically saturated with freshwater. The experiment illustrated in Figure 2.11 indicates a potential effect of salinity of pore fluid on the engineering behaviour of soil. The experiment involves mixing equal quantities of the same clay with equal quantities of water. One sample is mixed with salt water (shown on the left) and the other with freshwater (shown on the right). The clay mixed with salt water exhibits sufficient shear strength to hold its shape and support the palette knife while the clay mixed with salt water is a slurry.

## 2.4 Hydrodynamic regimes

Hydrodynamic forces, i.e. forces caused by the motion of the ocean, contribute to the environmental loading applied to an offshore structure, causing scour around an object installed or laid on the seabed (such as a foundation or pipeline), causing seabed mobility, and ultimately instability of the seabed in the free-field.

Hydrodynamic forces arise through ocean waves and currents that may impact on an offshore structure or that can transport sediment in suspension (often referred to as 'bedload'). Transport of seabed sediments leads to erosion and deposition, which may be manifested as, for example, as sand waves or ripples and scour channels (Figure 2.12). Hydrodynamic forces can lead to general surficial discontinuities across the seabed or in extreme cases can cause shear failure of the seabed. Uneven seabeds pose complications to siting foundations and other field infrastructure while, post-construction, hydrodynamic forces will contribute a significant proportion of environmental loading imposed on a foundation system or pipeline and may cause scour around the structure. The science of hydrodynamics is a specialist subject and will only be touched on briefly here. For further information, the reader is directed to specialist texts (e.g. Dean and Dalyrymple 1991, Duxbury *et al.* 2002). Reliable scientific websites are also a useful source of up-to-date specialist information; for example, the US National Oceanic and Atmospheric Administration (http://www.noaa.gov).

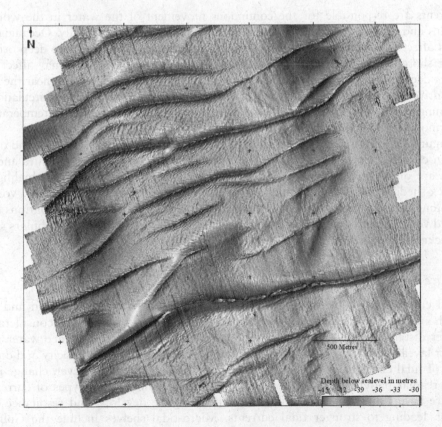

*Figure 2.12* Sand ripples

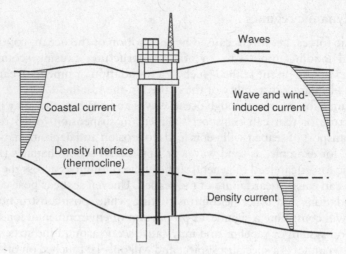

*Figure 2.13* Interaction of current systems (Gerwick 2007)

### 2.4.1 Currents

Currents are responsible for the continuous movement of the water in the world's oceans and flow in a complex pattern, both horizontally and vertically. Oceanic currents affect nearshore water and the open ocean, surface water and the deep ocean. Near shore, currents are driven by tides, local winds and waves and may affect the entire depth of the water column. Currents in the open ocean occur both near the surface driven by global winds, and in the deep ocean driven by thermohaline circulation – meaning circulation driven by density variations in the water due to temperature (thermo) and salinity (haline).

Figure 2.13 illustrates the location and extent of a selection of nearshore and open ocean currents. Currents can have a significant impact on offshore facilities and as such are of interest to offshore geotechnical engineers. Gerwick (2007) highlights vortex shedding on risers and piles and vibrations of wire lines and pipelines. Vortex shedding can result in scour in shallow water, and in cyclic dynamic oscillations (so-called vortex-induced vibration or VIV) of cables, mooring tethers and tubulars such as jacket legs and risers leading to fatigue.

### Surface currents

Tidal currents result from the gravitational attraction between the Moon and the Earth and to a lesser extent between the Sun and the Earth (due to the considerably longer distance). Tidal currents may be stratified, such that the surface water and lower layer of water are moving in opposing directions, hence the velocity and direction of tidal currents are constantly changing. Tidal currents, however, change in a very regular pattern and can be accurately forecast, unlike other types of currents. Where basins are semi-enclosed, a micro-tidal regime develops and resonance can occur, leading to stronger tidal currents. Micro-tidal shelves include the Gulf of California, the Arabian Gulf and the Gulf of Korea (Fookes *et al.* 2005).

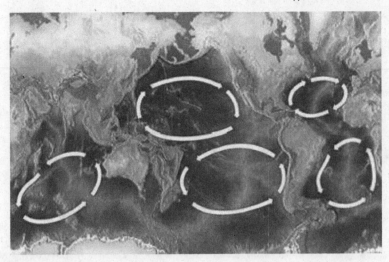

*Figure 2.14* Circulation of major ocean gyres

Source: National Oceans Service http://oceanservice.noaa.gov/ This material is the property of the US Government but is in the public domain and is not subject to copyright protection.

Coastal currents are driven by local winds and waves. Longshore winds, which blow along the shoreline, affect wave formations and, in turn, the coastal currents. The speed at which waves approach the shore depends on the sea floor, shoreline features and water depth. When the wave reaches a coastline, it releases a burst of energy, which generates a current that runs parallel to the shoreline, referred to as a longshore current.

While coastal currents are driven by local winds, surface currents in the open ocean are driven by a complex global wind system, commonly known as the Trade Winds and Westerlies. The rotation of the earth coupled with the Coriolis effect causes warm air from near the equator to rise and travel towards the poles in a clockwise direction in the Northern Hemisphere and counter-clockwise in the Southern Hemisphere. Some of the air cools around 30° North and South latitude, descends and travels back towards the equator, while some of the air travels on to the poles. These global wind systems drag the surface water of the oceans, causing near-surface circling currents, called 'gyres' that, due to the Coriolis effect, travel clockwise in the Northern Hemisphere and counter-clockwise in the Southern Hemisphere. There are five major ocean-wide gyres, and they are flanked by strong, narrow Western Boundary Currents and weaker, broader Eastern Boundary Currents. Figure 2.14 shows a map of the circulation of the major ocean gyres. The Gulf Stream is a Western Boundary Current, well known for maintaining the temperate Florida climate. The Gulf Stream Current is a fast-flowing current that travels between 2 and 5 km per hour.

Although winds drive surface currents, they do not necessarily travel in the same direction as each other, due to a phenomenon known as the Ekman spiral, a consequence of the Coriolis effect. When surface water is moved by the force of the wind, deeper layers of water are dragged along. Each layer of water is moved by friction from the shallower layer, and each deeper layer moves more slowly than the layer above it, until the movement ceases (at a depth of about 100 m). Like the surface water, the deeper water is deflected by the Coriolis effect – to the right in the

*Figure 2.15* Thermohaline circulation – The Global Conveyor Belt

Source: National Oceans Service http://oceanservice.noaa.gov/ This material is the property of the US Government but is in the public domain and is not subject to copyright protection.

Northern Hemisphere and to the left in the Southern Hemisphere. As a result, each successively deeper layer of water moves to the right or left, creating a spiral effect.

### Deep currents

Currents in the deeper ocean are driven by variations in density as a result of changes in temperature and salinity (colder, saltier water being denser than warm freshwater). The travel of density differential currents is referred to as thermohaline circulation. In the Polar regions, seawater is saltier than elsewhere as when water freezes the salt does not and the remaining water becomes saltier. The dense, cold, salty water sinks and surface water is pulled in to replace it, which in turn becomes cold and salty and dense enough to sink. This process initiates thermohaline circulation – the so-called 'global conveyor belt'. Joining the global conveyor belt in the northern polar region, cold, salty, dense water sinks and heads south along the Western Atlantic Basin. The current is recharged with more cold, salty water as it travels along the coast of Antarctica. The main current splits into two sections, one travelling northward into the Indian Ocean, while the other heads up into the western Pacific. The two branches of the current warm and rise as they travel northward, then loop back around southward and westward. As the water warms, it rises through the water column and the warm surface waters continue circulating around the globe. They eventually return to the North Atlantic where the cycle begins again. It is estimated that any given cubic metre of water takes about 1,000 years to complete the journey along the global conveyor belt. Figure 2.15 shows a schematic representation of the global conveyor belt of deep ocean currents.

### 2.4.2 Waves

Wave forces often constitute the dominant design criterion for fixed-bottom structures and cause motion of floating structures in all six degrees of freedom and as such their consideration in offshore design is essential. Specialist texts discuss the impact of waves on offshore construction activities and practical guidance on wave forecasting (e.g. Chakrabarti 2005, Gerwick 2007). The purpose of this section is simply to present a basic introduction to waves.

## Surface waves

Waves on the surface of oceans are primarily wind generated and propagate along the water–air interface. Waves form as the wind causes pressure and friction forces that perturb the equilibrium of the water surface and transfer energy from the wind into wave energy. Wind blowing over a calm sea forms small ripples. The undulations of the surface water provide a better surface for the wind to 'grip' and ripples grow into wavelets. When the wavelets become high enough to interact with the airflow, the wind becomes turbulent just above the surface of the water, and the wind transfers energy to the waves. As the sea becomes choppier, more energy is transferred to the waves, and they get bigger. The cycle continues and waves get bigger and steeper. The formation of ocean waves is affected by wind speed, wind duration and fetch – the distance over which the wind blows in a single direction, and these factors work together to determine the size of waves. A fully developed sea state occurs if, for a given constant wind speed, the depth, fetch and duration are sufficient that the wave travels at the same speed as the wind. In a fully developed state, the wave height and wavelength have reached their full potential: further energy cannot be transferred from the wind to the wave and waves will not continue to grow even if the depth, fetch or duration increases further. Waves that travel beyond the wind-affected zone (by distance or time) are referred to as 'swell'. Swells propagate across the ocean away from their area of generation and can propagate in directions that differ from the direction of the wind. Swells can travel for hundreds and sometimes thousands of kilometres – swells generated from storms off the Antarctic often reach the equator.

Ocean waves, as for any oscillation, require a restoring force to return equilibrium in order for them to propagate. Small ripples can be restored by surface tension, whereas waves are restored by gravity; hence, surface ocean waves are also known as 'gravity waves'.

Surface ocean waves can be characterised by their length, height and the water depth over which they propagate. All other parameters, such as wave-induced water velocities and accelerations can be determined theoretically from these quantities. The characteristic features of an idealised ocean wave are illustrated schematically in Figure 2.16. The highest point of a wave is termed the 'crest' and the lowest point the 'trough'. The wavelength L is the horizontal distance between two successive wave crests (or between two successive wave troughs) and the height of a wave H is the vertical distance between trough and crest. An ideal ocean waveform is characterised as sinusoidal, i.e. the crests and troughs have an identical shape and are separated by a fixed wavelength. Water depth, h, is measured from mean sea level to the mudline. The time for two successive crests or troughs to pass a particular point is called the 'wave period' T and the distance the wave travels per unit of time is called the 'celerity' of the wave C or alternatively the 'phase velocity' (or 'wave velocity').

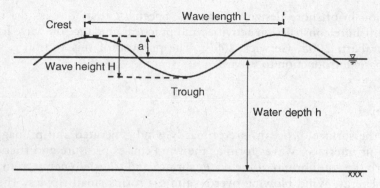

*Figure 2.16* Characteristics of an idealised ocean wave

From linear (Airy) wave theory, the surface elevation may be expressed in terms of time, t, and horizontal distance x as

$$\eta(x,t) = \frac{H}{2}\sin\left[2\pi\left(\frac{t}{T} - \frac{x}{L}\right)\right] \tag{2.1}$$

with the celerity C equivalent to L/T. The wavelength is often referred to in terms of the wave number k which is $2\pi/L$, while the angular wave frequency $\omega$ is equivalent to $2\pi/T$. For a given water depth h and angular frequency, the wave number may be determined implicitly from the 'dispersion' relationship

$$\omega^2 = gk\tanh kh \tag{2.2}$$

Although the wave form travels horizontally across the surface of the ocean (with celerity C), the water particles that comprise the wave do not translate in the direction of the wave – only their collective energy does. The water particles within the wave move in a nearly closed orbit. The diameter of the orbital paths of water diminishes with depth and, at some distance below the crest of the wave, the water motion may become essentially zero.

In shallow water, the water particles within the wave orbit an elliptical path (Figure 2.17b) and in very shallow water, $h \le L/20$, only the minor axis diminishes with depth, such that close to the seabed water motion is horizontal (Figure 2.17a). In deep water, the so-called wave base occurs at a depth below mean sea level of half the wave length, L/2 (Figure 2.17c), where the magnitude of the water motion is only a few per cent of that at the sea surface.

For estimating the hydrodynamic loading from waves, it is useful to evaluate the wave-induced water velocities at any depth z below the water surface. The horizontal and vertical components may be expressed as

$$v_x = \omega\frac{H}{2}\frac{\cosh[k(h-z)]}{\sinh(kh)}\sin(\omega t - kx)$$
$$v_z = \omega\frac{H}{2}\frac{\sinh[k(h-z)]}{\sinh(kh)}\cos(\omega t - kx) \tag{2.3}$$

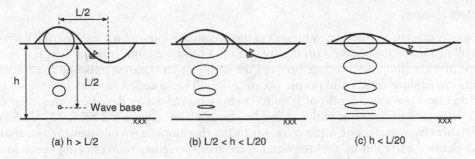

(a) h > L/2            (b) L/2 < h < L/20            (c) h < L/20

*Figure 2.17* Orbital paths of water trajectories beneath progressive waves

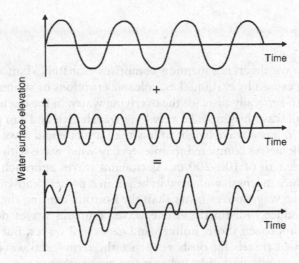

*Figure 2.18* Superposition of two sinusoidal wave forms (Dean and Darymple 1991)

Surficial seabed sediments may be affected by surface wave action in water depths up to 200 m in a severe storm. In normal conditions, seabeds in water depths of 50 m would be significantly affected by surface wave action, with potential liquefaction and scour of sandy deposits and strong hydrodynamic loading of exposed pipelines or other infrastructure.

In reality, there is a great amount of randomness in ocean wave forms and they rarely conform to the idealised periodically repeating sinusoid and do not necessarily propagate in the same direction. Ocean waves are often a combination of local wind waves from one or more directions and swell from another. Waves tend to create a chaotic sea state with component waves of varying direction as well as varying wavelength and period while swells tend to be more regular and can be more adequately described by a single wave form that repeats itself periodically. A less idealised sea state can be achieved by superposing various sinusoids. Figure 2.18 shows an example of a complex wave form resulting from the superposition of two sinusoids.

## Internal waves

Another class of gravity waves present in the oceans are internal waves. Internal waves oscillate within, rather than on the surface of, a fluid medium. Internal ocean waves typically act along the interface between the warm upper ocean waters and the colder, saltier, therefore denser and deeper ocean waters. The so-called thermocline typically occurs between water depths of 100–200 m but internal waves have been measured at water depths of 1,000 m with a wave height of 60 m (Gerwick 2007). The density variation in the layers of water adjacent to the thermocline is considerably less than the density variation at the air–water interface; therefore, the restoring force and hence the energy to initiate and propagate internal waves is less than for surface waves. Internal waves travel slower than surface waves, with typical periods of several minutes (although periods of several hours have been measured in the open ocean), compared with typical wind wave periods of 5–15 seconds and typical swell periods of 20–30 seconds.

## Tsunamis

A final class of wave deserving mention comprises tsunamis. Tsunamis are impact waves that can be caused by earthquakes, volcanic eruptions or submarine slides with sufficient impact to vertically displace the overlying water. In the open ocean, tsunami waves have a small wave height and a very long wavelength and can go unnoticed. A tsunami wave often has a wave height of less than a metre and a wavelength of tens or hundreds of kilometres compared to an everyday wind wave with a height of 2 m or so and a wavelength of 100–200 m. As tsunami waves approach shore, they are slowed down, reduce in length and grow in height in a process known as 'wave shoaling'. The top of the wave moves faster than the bottom, causing the sea to rise dramatically and produce a distinctly visible wave. Tsunami waves do not generally break when they approach shore, unlike wind generated waves, but act more like a fast flowing tide that travels far onshore (hence the term 'tidal wave', although the waves have nothing to do with tides). A tsunami may comprise several waves, which may reach shore over a period of hours due to the long wavelength, and the first wave is not always the most severe. About 80 per cent of tsunamis occur in the Pacific Ocean but have been recorded at most major subduction zones (see Figure 2.2). Detailed information on tsunamis is provided by the National Oceanographic and Atmospheric Administration (NOAA) Centre for Tsunami Research (http://nctr. pmel.noaa.gov), the US Geological Survey Centre for Tsunami and Earthquake Research (http://walrus.wr.usgs.gov/tsunami) and the International Centre for Geohazards (www.geohazards.no).

# 3  Offshore site investigation

## 3.1 Introduction

A reliable model of the seabed stratigraphy including quantification of engineering parameters for relevant layers is essential for engineering design of an offshore foundation. Often, preliminary design studies will be called for prior to any site investigation at the particular site, and estimation of the seabed characteristics must rely on regional knowledge. In any case, it is advisable to establish a site model that extends beyond the immediate location, as this allows for subsequent adjustments in the positions of facilities and indicates the spatial uniformity of seabed conditions across the region. Critical features such as faults, buried channels or other localised non-uniformities will need to be mapped.

Complete site characterisation involves three stages:

1. Desk study, including preliminary met-ocean data
2. Geophysical investigation
3. Geotechnical investigation.

Integration of these studies will lead to the engineering properties needed for the foundation design of the planned structures. Note that the met-ocean data is not part of the geotechnical characterisation of seabed conditions but will be needed during preliminary foundation studies to determine loads. Knowledge of wave and current conditions will also help in assessing potential seabed features such as sand waves, storm-induced mud slides or ice gouges.

### 3.1.1 Desk study

The desk study assembles existing data for preliminary assessment of site conditions and evaluating alternative conceptual designs. The data to be sought include:

- Approximate bathymetry
- Regional geology
- Potential geohazards
- Seabed obstacles or other features
- Regional met-ocean data.

In addition to providing a basis for evaluating preliminary designs, the desk study will formulate requirements for subsequent geophysical and geotechnical investigations.

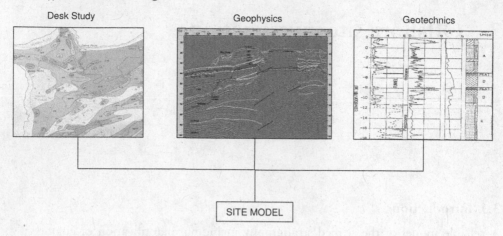

*Figure 3.1* Conceptual plan for site characterisation

It should identify what data required for the engineering design are missing, or insufficient, and also specific points of interest that may require targeted data acquisition. The study should include a risk analysis, both in terms of geohazards such as seismic and submarine slide risk or evidence of shallow gas seepage, and also in terms of cost and feasibility aspects of the preliminary designs and whether they are likely to provide robust and safe solutions.

In most cases, geophysical data acquired as part of the hydrocarbon field assessment will prove of sufficient resolution to identify key seabed features and major stratification changes for the desk study. The data will include approximate bathymetry (water depths and seabed gradients) and allow detailed planning of the next stages of investigation.

Regional met-ocean data will generally be available, particularly in mature regions of offshore development such as the Gulf of Mexico and North Sea. Important data include: currents in the upper water column and near the seabed; extreme wind and waves for design return periods; tidal information; temperature environment and seasonal weather patterns (hurricanes or cyclones, monsoons and sea ice). The data are important to determine the environmental loads on the planned structures and also for the timing of site investigations to match weather windows. More detailed information on currents, and wind and wave spectra, will be provided during the course of the project but the met-ocean data acquired during the desk study should be sufficient to allow specific design concepts to be judged as feasible or not.

Desk studies can be initiated at short notice and will typically occupy a period of one to several months, depending on the nature and size of the project. Costs will be an order of magnitude lower than subsequent stages of site investigation that involve acquisition of new data.

### 3.1.2 Geophysical investigation

Geophysical investigation comprising data from 3D, shallow seismic and side-scan sonar surveys will give an indication of the local seabed and soil conditions over a large area. Seabed features and seabed obstructions can be revealed and the main stratification boundaries and faulting within the soil column identified. The nature of the signals, and the extent they penetrate a given layer, may also reveal the presence of shallow gas or methane hydrates. A typical investigation will include (at least):

- Bathymetry for water depth and seabed gradients
- Seismic reflection data for sub-bottom stratigraphy
- Side-scan sonar for seabed topography such as pock marks and ice gouges.

Geophysical studies form the main framework for assessing geohazards associated with seismic activity of submarine slides. Historical features (on a geological timescale) such as faults, scarps, run outs from debris flows and non-conforming strata sequences allow a picture to be assembled of the frequency and magnitude of events that might form a geohazard risk.

Geophysical data generally requires 'ground truthing' to be useful in geotechnical design in a quantitative sense, but it is essential information in building a complete model of the seabed. Often, geophysical data will overlap other sites where site investigation has been carried out, allowing tracing of different soil strata across the region. This may enable quantitative assessment of specific soil properties to be estimated at the site in question, but, as a minimum, it will indicate the continuity of strata, the dip angle and direction of each layer and any internal structure from previous geological events. Final quantitative confirmation of soil layers and properties of each layer will require a separate geotechnical site investigation, after which the geophysical model may be used to interpolate between borehole or profiles of *in situ* testing.

The lead time for a geophysical survey may be up to three months and typical costs are of the order of a million US dollars. Further details of the range of equipment and illustration of typical results are given in a later section of this chapter.

### 3.1.3 Geotechnical investigation

Geotechnical investigations usually involve a combination of offshore work, followed by onshore laboratory testing of samples recovered from offshore and a subsequent interpretation of the test results to yield engineering design parameters. Work carried out offshore will generally include recovering samples from the seabed and may also include *in situ* testing such as penetration and vane shear testing.

Results of the geotechnical investigations will 'ground truth' the results of the geophysical investigation, providing much more detailed stratigraphy (at a limited number of locations) with a description of the various soil types throughout each of the layers and quantitative measurement of key quantities such as shear strength or penetration resistance. Subsequent treatment of borehole samples onshore will include geological logging and a full suite of basic and advanced soil tests from which design parameters may be deduced.

The raw borehole information from a geotechnical investigation should provide confirmation of the desk study and geophysical investigation by means of the vertical profiles of soil conditions at each borehole location. Integration with the geophysical data provides a means of extending the geotechnical study over a much larger area, obviously with a gradually decreasing level of quantitative accuracy as the distance from the borehole increases.

All offshore work will require support from vessels of a significant size, some of which are purpose-designed for site investigation work, and thus involves considerable cost.

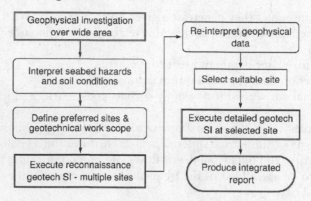

*Figure 3.2* Integration of desk study, geophysics and geotechnics

There is a wide variety of different working platforms (vessels of different size and type) and the choice will depend on needs, availability and costs. The investigation may be relatively primitive at preliminary stages, or where large areas have to be covered, perhaps with only disturbed material being recovered from relatively close to the seabed. More detailed investigations will involve drilling and recovery of minimally disturbed samples, plus *in situ* testing, and this work may be carried out entirely from the vessel or from a rig lowered to the seabed and operated via an umbilical cable.

Details regarding the geotechnical data required and the range of equipment available are described later in this chapter, together with how the results of the geotechnical investigation are reported and interpreted. Planning a geotechnical site investigation, particularly for the main (or sole) expedition requires significant lead time (a minimum of 3 months), both to ensure that the range of data required for the project can be acquired and also to source and reserve an appropriate site investigation vessel or drilling capability. Costs are high because of the length of time required to mobilise vessels to the site and then carry out the various drilling and testing operations, and they will in most cases amount to several million US dollars. The total elapsed time between starting to plan a geotechnical investigation, acquiring the data, carrying out onshore laboratory testing and interpreting the data in terms of engineering design parameters will often exceed a year.

For major projects, the geophysical and geotechnical are interwoven, and each aspect may require more than one offshore investigation. A typical example for a project where there is some uncertainty regarding the precise site for the offshore facility is shown in Figure 3.2. Although not specifically shown in the flowcharts, an appraisal of the geological conditions based on available information should be carried out at the start. At the end of the process, all available information should reappraised and a geological model for the site produced in addition to detailed engineering reports.

## 3.2 Geophysical investigation

### 3.2.1 Overview

The theory behind the various techniques of geophysical measurements and interpretation of data is extremely complex, forming a separate scientific discipline. The aim

*Table 3.1* Main types of geophysical investigation

| Purpose | Equipment |
|---|---|
| Bathymetry (water depth) mapping | Conventional echo sounding, swathe bathymetry |
| Sea floor mapping of seabed features | Side-scan sonar |
| Seismic profiling of 'sub-bottom' stratigraphy | Boomer, sparker, pinger, chirp profilers and high-resolution digital surveys (airgun) |

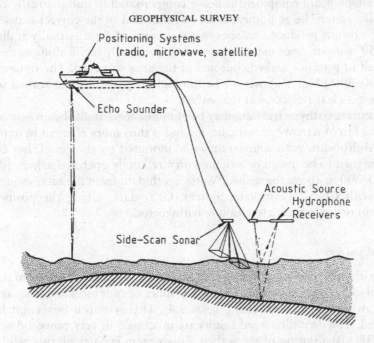

*Figure 3.3* Main types of geophysical survey: (1) bathymetry mapping by echo sounder, (2) sea floor mapping by side-scan sonar and (3) subseabed continuous seismic profiling by means of an acoustic source and hydrophone receivers (Sullivan 1980)

here is limited to a brief description of the different types of equipment that are available, and their capabilities. An overview is given of how geophysical investigations are planned, in terms of the layout of survey lines, and typical results for some of the equipment will be illustrated. The application of geophysical data to hazard assessment on a local and regional scale is also discussed briefly.

The main types of geophysical investigation are summarised in Table 3.1 and illustrated schematically in Figure 3.3. The three objectives focus on the water column, the detailed seabed features and the underlying sediment stratigraphy.

### 3.2.2 Bathymetric mapping

Bathymetric mapping is used to quantify the water depth and thus provide a visual 3D image of the seabed. This type of investigation gives important information about the seabed slope at the proposed location, but can also lead to detection of palaeo-slope failures or debris flows, geological features such as volcanoes, scarps from faulting and seabed obstructions. The most common types of bathymetry measurements are echo sounding and swathe bathymetry.

### Echo sounding

A conventional single-beam echo sounder is similar to echo sounders available on all ocean going vessels. However, in order to provide the accuracy necessary for survey work, this single-beam equipment is heave compensated to automatically correct for prevailing sea state. The equipment must be calibrated to the correct seawater sound velocity. The results produced are spot measurements of depth, usually at distances of about 25–50 m apart. Spot measurements are continually made along each track line and are used to produce seabed contours of the area surveyed. The frequency range (typically 30–300 kHz) is chosen to be too high to penetrate the seabed sediments, thus ensuring a clear reflection at the seabed.

More accurate bathymetric data may be obtained from multi-beam echo sounding, which uses a fan of narrow acoustic beams and is thus more efficient in terms of time and cost. Multi-beam echo sounders may be mounted on the vessel, but commonly they are designed to be towed or mounted on a remotely operated vehicle (ROV) at a depth of 50–300 m above the seabed. Water depth data from the echo sounder is then combined with Global Positioning System (GPS) data, giving the position of the instrument in order to evaluate absolute bathymetry.

### Swathe bathymetry

A swathe bathymetry system usually comprises a beam that sweeps from side to side as the vessel sails ahead, resulting in a large number of spot measurements as it sweeps that cover virtually the entire area (Figure 3.4). This system is heave, pitch and roll compensated. The digitally stored results are processed in very powerful software to produce a 3D visualisation of the seabed. This system is particularly useful for areas of rugged terrain and, if mounted on and ROV or towed platform, can simultaneously produce side-scan information. Typical width of coverage is about four times the water depth.

### 3.2.3 Sea floor mapping

Side-scan sonar is a method of underwater imaging using narrow beams of acoustic energy (sound) transmitted out to the side of the 'towfish' (or equivalent such as an ROV) and across the bottom. Sound is reflected back from the bottom and from objects to the towfish. Certain frequencies work better than others; high frequencies such as 500 kHz–1 MHz give excellent resolution but the acoustic energy only travels a short distance. Lower frequencies such as 50 kHz or 100 kHz give lower resolution but the distance that the energy travels is greatly improved.

The towfish generates one pulse of energy at a time and waits for the sound to be reflected back. The imaging range is determined by how long the towfish waits before transmitting the next pulse of acoustic energy. The image is thus built up one line of data at a time. Hard objects reflect more energy causing a dark signal on the image, soft objects that do not reflect energy as well show up as lighter signals. The absence

*Figure 3.4* Schematic of swathe bathymetry system

*Figure 3.5* Side-scan example: images of ship wrecks

of sound such as shadows behind objects show up as white areas on a sonar image (see example ship wreck images in Figure 3.5).

### 3.2.4 Subseabed continuous seismic profiling

Continuous seismic 'sub-bottom' profiling techniques use a towed sound source and receiver combination that produces a series of reflection records. After processing, a

*Table 3.2* Compromise between penetration and resolution in sub-bottom profilers

| Variable | Penetration | Resolution |
|---|---|---|
| Frequency increase | Decrease | Increase |
| Bandwidth increase | Decrease | Increase |
| Pulse duration increase | Increase | Decrease |
| Power increase | Increase | Decrease |

*Table 3.3* Various sub-bottom profiling tools

| Type | Penetration | Resolution | Comments |
|---|---|---|---|
| Boomer | ~ 100 m | 0.6–1 m | Low power |
| Sparker | ~ 100 m | 0.6–1 m | Difficult frequency curve |
| Pinger | 25 m | 0.3–0.6 m | High power |
| Chirp | 25 m | 0.3–0.6 m | Improved technology |
| Airgun | > 200 m | < 0.3 m | Multi-streamer |

seismic cross-section of the sediments may be displayed graphically through the area being surveyed.

The profiling system is made up of three components: a sound source, a 2D array of receivers (hydrophones or streamers for high-resolution surveys) and a recorder. The source and receivers are typically towed a few metres below the sea surface. Nowadays, recorders are digital, but most produce an 'analogue' output for immediate assessment. The ultimate goal is for a tool that provides maximum penetration (depth below seabed) and produces the highest resolution (definition of thin layers). These two goals are to some extent mutually exclusive and therefore a compromise is necessary. Modern very-high-resolution survey equipment has narrowed the gap between the two objectives of high penetration and high definition, but high-resolution equipment is still relatively expensive and would not necessarily be justifiable for a straightforward survey of a platform location. The main variables in sub-bottom seismic systems and how they affect the penetration and resolution of individual sediment layers are summarised in Table 3.2.

The degree of penetration is ultimately controlled by the composition of the sea floor and the underlying layers (strong layers near the surface will inhibit penetration). The sea floor both reflects and scatters the acoustic energy. Reflection loss determines the amount of energy that penetrates into the sediments and absorption into the sub-layers. Reflection loss encountered at sub-bottom horizons or interfaces will further influence the degree of penetration.

The energy emitted by the tool must therefore be high enough to achieve the required penetration, but the frequency curve of the energy must not limit resolution, thus making interpretation difficult. Ultimately, the type of tool must be selected to suit the specific application. The various seismic sources are referred to by names that reflect their power and frequency. Table 3.3 summarises the relative differences between the various tools available for sub-bottom profiling. The sources come in a variety of shapes and sizes, and the equipment is designed to be hydrodynamically efficient and such that the depth of tow can be controlled. A typical towing configuration can be seen in Figure 3.6.

High-resolution digital surveys (HRDS) utilise an airgun as a sound source. This is a much better source than the boomer/sparker/pinger family as it is smaller and

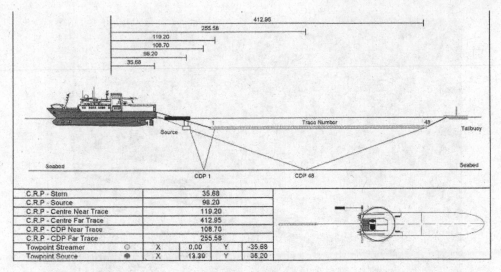

| | | | | |
|---|---|---|---|---|
| C.R.P - Stern | | 35.68 | | |
| C.R.P - Source | | 98.20 | | |
| C.R.P - Centre Near Trace | | 119.20 | | |
| C.R.P - Centre Far Trace | | 412.95 | | |
| C.R.P - CDP Near Trace | | 108.70 | | |
| C.R.P - CDP Far Trace | | 255.58 | | |
| Towpoint Streamer | ○ X | 0.00 | Y | -35.68 |
| Towpoint Source | ● X | 13.30 | Y | 35.20 |

*Figure* 3.6 Typical towing configuration for sub-bottom profiler (courtesy: Woodside Energy Ltd)

produces a 'cleaner' acoustic signal. Multi-channel receivers are aligned in streamers of various lengths, spacing and grouping combinations, designed specifically for the particular survey objectives including depth of penetration and resolution in the depths of interest. Readings of the reflection from each point on the seabed are recorded in each of the channels available and are added or 'stacked' in order to produce a single trace that provides, in the first instance, an increase in the signal to noise ratio for random noise.

High-resolution digital systems have an added advantage in that both the sound source and streamer are towed at a shallower depth than for other equipment (approximately 3 m versus the conventional 7 m). This means that reflections from the water surface are more easily managed. However, the downside is that this type of system is more affected by poor sea-states.

### 3.2.5 Autonomous underwater vehicles

The various systems described in the foregoing section are typically operated directly from a vessel, even though the transducers themselves may be mounted on towed devices or ROVs (the latter being connected to the vessel by an umbilical cable). However, the last decade has seen rapid growth in a new system for deep-water bathymetry, seabed mapping and sub-bottom profiling based on autonomous underwater vehicles (AUVs). In contrast to a traditional towed survey system that requires the use of a long umbilical cable to transmit information back to the vessel, AUVs store data from the various survey sensors internally for download upon vehicle recovery. In addition, for real-time data observation and quality control, the AUV can also transmit survey data back to the mother ship through the water column by use of an on-board acoustic modem (Figure 3.7). A similar modem arrangement is used to send operating instructions to the vehicle and to receive critical feedback on system performance.

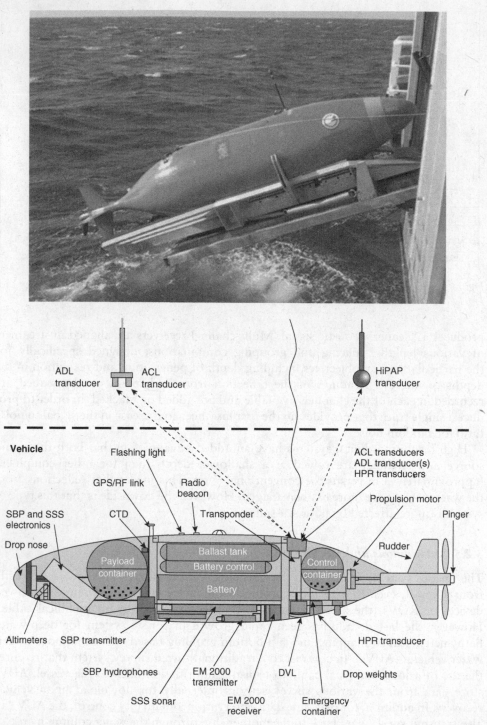

*Figure 3.7* Hugin 3000 AUV (photo courtesy Kongsberg Marine Ltd, diagram from Cauquil *et al.* 2003)

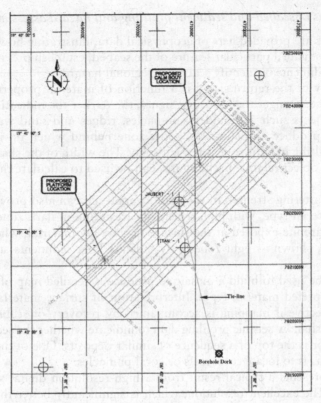

*Figure 3.8* Example of survey line pattern

### 3.2.6 *Survey line layout*

The plan of survey lines for a platform location is usually designed to capture the entire area of interest, i.e. infield flowlines, subsea facilities (if present), moorings and buoys and the possible locations of any platforms. Separate surveys would generally be conducted for routing of export pipelines, where the target will be narrow corridors of the ocean and seabed. Primary grid line spacings within the main field are typically 75–100 m with cross lines at 300–500 m (Figure 3.8). A vessel will sail over this grid and collect all necessary data. It is clear that a long tow with streamers makes such investigations relatively arduous and that AUV-based systems therefore have certain advantages over traditional equipment.

Where existing borehole information is available (even if only in the vicinity of the survey area), it is important that a survey tie-line is carried out from the borehole to the survey area. This enables geophysicists on board the vessel to calibrate survey data against the borehole data and then to trace layer and horizon continuity over the area (where such continuity exists). Having this information during the field operations makes it possible to rerun lines if necessary without significant additional expense. Grid spacing may be adjusted over areas of particular interest, for example, where known hazards exist. This allows better resolution and definition of areas of potential problems. An example of a survey grid can be seen in Figure 3.8.

### 3.2.7 Examples of seabed and sediment profiling and hazard assessment

A few examples are provided here of geophysical data, illustrating how the data may be used to investigate a particular feature of the seabed, establish preliminary stratigraphy of the sediments or identify seabed or regional hazards.

The intensity of the return signal is a function of material properties as well as topography. The stronger the returning signal, the darker the indication on the display. Large objects such as boulders, pinnacles, ridges and sand waves are good reflectors and produce an acoustic shadow zone behind which no reflections are generated, resulting in light zones on the display. The width of the shadow zone and position of the object relative to the source can be used to calculate the height of an object.

The back-scattering strength of the seafloor material can also provide an indication of the material type, with high reflectivity (resulting in dark zones) potentially indicating calcarenite exposed at the surface or gravelly material, whereas uniform low reflectivity, shown as light zones may indicate finer sediments such as muddy sands.

Images can be used to build a mosaic to produce a detailed map of sea floor features and interpreted material types. Interpretation of surface material type can be made less subjective if mapping is accompanied by recovery of seabed samples or cores, or using shallow seismic profiling data to indicate whether the sediment constitutes a veneer or is the top of a sequence of similar deposits. One of the most important applications is to locate coral reefs or coral pinnacles.

Figure 3.9 presents a typical result from a high-resolution digital survey using a mini air gun. The excellent resolution, which is significantly better than obtainable with older technology, is immediately apparent. The interpretation has benefited from extrapolation from an existing borehole, as one of the survey lines passed directly over the borehole location.

Bedforms such as sand mega-ripples and sand waves (e.g. Figure 2.12) are manifestations of prevailing currents and provide an excellent indication of conditions in a particular environment. Small pockmarks can also be observed and are often an indication of surface collapse due to escaping gas. Occasionally individual pockmarks can become major features, such as the one discussed by Cauquil *et al.* (2003) and attributed to lateral fluid migration along a particular subseabed interface, which was some 650 m in diameter and over 60 m deep (see Figure 3.10 and Figure 3.11, from data obtained using an AUV).

### Local hazard assessment

Local hazard assessments at specific sites have to be undertaken to ensure safety for temporary operations and also permanent infrastructure such as subsea facilities or fixed platforms.

Temporary operations include drilling exploration wells, temporary anchoring facilities and work-overs of existing facilities. It is important to know what the anchoring conditions are likely to be for drilling rigs moving into new areas. Anchoring can be affected by outcropping cemented layers, coral reefs, coral pinnacles, rugged topography and even unknown ship wrecks (in some parts of the world, even by debris simply dumped in the ocean). It is also important to know whether hazardous shallow gas will be encountered during drilling operations so that necessary precautions can be taken. Jack-up rigs positioned alongside existing facilities for

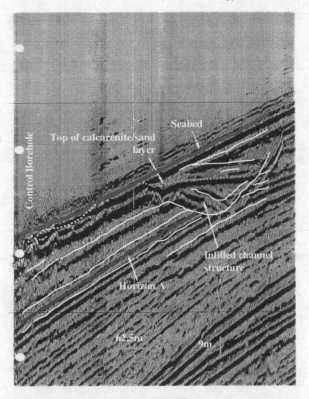

*Figure 3.9* Continuous seismic profiling – mini air gun record

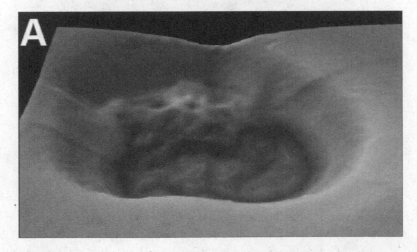

*Figure 3.10* Bathymetry of pockmark from combined side-scan sonar and multi-beam echo sounder (Cauquil *et al.* 2003)

the purpose of work-overs require information with respect to penetration of the spudcans, identifying any changes to the seabed since site investigation data was obtained previously.

Subsea facilities are relatively small in plan area and are generally sensitive to unevenness of the seabed. It is important that the survey coverage allows small outcrops,

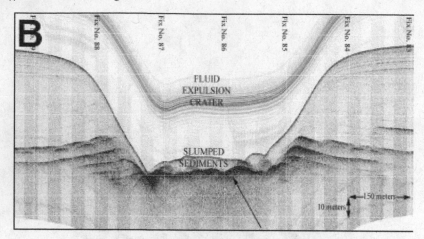

*Figure 3.11*  3D seismic profiling of pockmark (Cauquil *at al.* 2003)

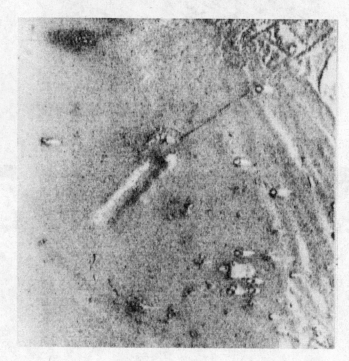

*Figure 3.12*  Typical scar of an anchor before it buries itself below the seabed

obstructions and seabed non-uniformities to be detectèd. Pipelines and flowlines spanning over scars or pockmarks can be a problem, and it is also preferable to avoid routing them over faults that could be active during the lifetime of the facility. Even scars from drag anchor embedment (Figure 3.12) may lead to problems, given the significant volume of soil disturbed by large drag anchors with fluke spans of more than 10 m. An anchor scar can require extra work on site during pipeline laying, affecting productivity.

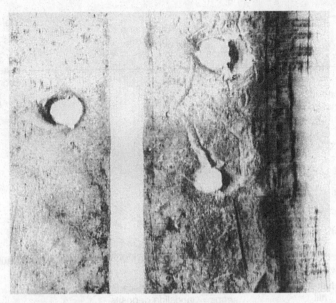

*Figure 3.13* Uneven seabed due to footprints left by jack-up rig

Fixed platforms comprising jackets (with mudmats) or gravity base options are generally subject to strict out-of-level tolerances, as the amount of adjustment possible when installing the topsides is limited. Local seabed unevenness can also result in excessive differential settlement over the life of the structure and, in the case of concrete gravity base structures, structural components of the base can be overstressed or may require sub-base grouting to prevent this from occurring. Unevenness can be created by offshore operations, such as associated with jack-up rig footprints (Figure 3.13), and indeed these can be a major hazard for repeat work-over operations to service unmanned platforms.

### Regional hazard assessment

Developments are moving into deeper waters, either closer to the continental shelf-edge or beyond. Areas of significant change in water depth, with high gradients, are particularly vulnerable to natural forces such as earthquakes and high ocean currents that can instigate slope instability or mudslides. Geohazard assessment is considered in detail in Chapter 10, but an example regional model for hazard assessment, based on geophysical investigations, is shown in Figure 3.14.

A geophysical survey undertaken for this purpose should gather good bathymetric data and high-resolution sub-bottom profiles as the basis for building a geological model of the area. This should be accompanied by the recovery of sufficient sample cores (drop or piston) along the survey lines, supplemented if necessary by a deep borehole. These samples are used to calibrate the geophysical data and can also be used to date sediments (e.g. by carbon 14 dating) to establish the frequency of identified slope failures and the extent to which debris from slope failures have been dispersed by ocean currents over geological time.

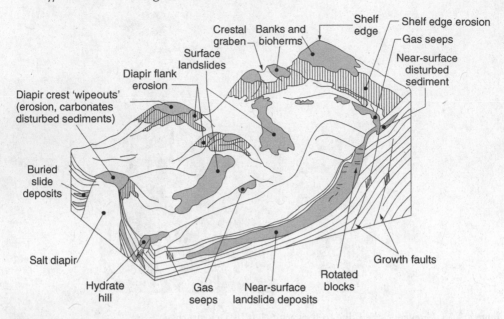

*Figure 3.14* Typical regional hazard features

## 3.3 Geotechnical investigation

### 3.3.1 *Choice of investigation system*

Typically, a geotechnical investigation is performed after the geophysical investigation has taken place and the results have been examined. Sometimes, however, only a geotechnical investigation is carried out in order to reduce costs or because of the small extent of the site to be investigated. The lack of geophysical data increases the risk from unforeseen problems and resolving differences across the site. The main goal of the geotechnical investigation is to characterise the soil properties by means of *in situ* tests and samples that are subsequently examined and tested. In most cases, the geotechnical investigation ground-truths geophysical interpretation and can thus enhance understanding of the seabed sediments on a regional basis. The geotechnical properties obtained from the field investigation and subsequent laboratory work provide the basis for the engineering parameters required for design of foundations and other offshore infrastructure.

A geotechnical site investigation is performed by specialised equipment from a purpose built platform. The investigation can be achieved by means of two alternative systems: (a) downhole; and (b) seabed (Figure 3.15). A downhole system refers to drilling of a borehole and testing or sampling of the soil by means of 'downhole tools' that fit through the drill string. This is comparable with the set-up used in the oil drilling industry, where a wide variety of downhole tools are used to characterise hydrocarbon reservoirs at depths of a few thousand metres down drill strings. However, the drilling in a geotechnical investigation needs to be performed in a much more sensitive manner compared with reservoir drilling, and the type of testing is quite different, with evaluation of soft sediments rather than reservoir rocks.

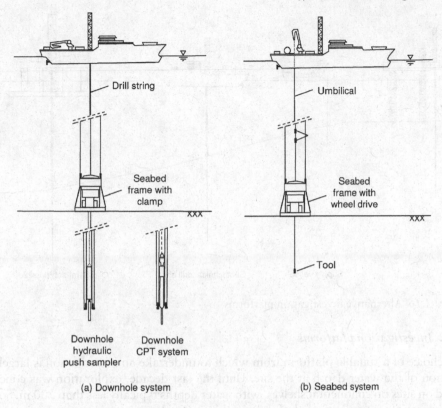

Figure 3.15 Downhole and seabed systems

The type of drilling will depend on whether the work is performed from a floating work area (i.e. a drill ship or other vessel) or from a stable platform such as a jack-up rig. Typical equipment used in the downhole drilling is described later in this chapter.

By contrast, a seabed system comprises equipment that rests on the seabed, and provides the means to carry out tests and recover samples from the underlying sediments. The maximum investigation depth is normally less than for a downhole system, and the presence of a hard layer may restrict this further unless the seabed system is able to drill through that layer and thus continue testing at greater depths. The flexibility and greater control of seabed systems bring many other advantages. Normally, the seabed equipment is simply lowered over the side of a vessel, and the testing and sampling operations are controlled via an umbilical cable. The operation is therefore much faster than a downhole system, as tools do not need to be repeatedly lowered and recovered through the water column.

Note that all equipment has to be able to withstand the high hydrostatic pressures at and below the seabed. Equipment used in shallow water (< 100 m) might therefore be quite different compared with equipment used in deep water (e.g. 2,500 m). The equipment also has to be capable of achieving the tests and sampling specified by the client, and thus needs to be matched to the sediment types since a tool may behave quite differently in soft clay than in dense sand.

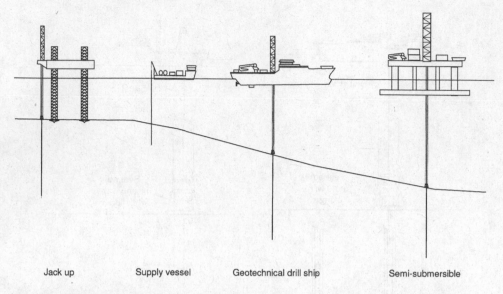

| Jack up | Supply vessel | Geotechnical drill ship | Semi-submersible |

*Figure 3.16* Alternative investigation platforms

### 3.3.2 Investigation platforms

The choice of a suitable platform from which to undertake an investigation is largely a function of the water depth at the site. Until the last decade, exploration was concentrated on sites on continental shelves, with water depths typically less than 200 m. Some of the possible investigation platforms in this water depth are shown in Figure 3.16. Since the investigation platform is generally the most expensive component, the economic advantages of various options must be carefully weighed up. The smaller the platform, the cheaper the option; however, smaller options will influence the ability to undertake an appropriate work scope. Options in these water depths include:

1. Diver operated from a support vessel or barge (water depth < 20 m)
2. Jack-up rig (water depths from 20 to 120 m)
3. Remote equipment from a support vessel (all water depths, depending on vessel draft)
4. Specialised drilling vessel (water depth > 20 m).

Anchoring capabilities of vessels are generally limited to depths of less than about 200 m, and therefore dynamically positioned (DP) vessels are more appropriate for deepwater locations. Mobilisation costs of specialised DP vessels are generally very high, as also are costs of semi-sub drilling rigs. Therefore, in remote areas, or regions such as Australia where there is insufficient activity for suitable specialised vessels to be stationed there permanently, greater reliance tends to be placed on remotely operated equipment, supported by construction vessels or smaller DP vessels.

### Low draft barges

Landing barge type vessels with a low draft and thus ability to work in shallow waters provide a reasonable working area for both equipment and diving support. They are particularly well suited to investigations for pipeline shore approaches. A small 'A'

*Figure 3.17* Low draft barge for shallow water investigations

frame facilitates lowering of diver-operated drilling rigs to the seabed (Figure 3.17). A multi-point mooring system is required to keep the vessel on station, usually with a minimum of three anchors. Ocean currents can be onerous in shallow reaches between islands and therefore, at times, more than three anchors are required. The size of anchors that can be deployed is, however, limited.

*Support vessels*

Support or supply vessels are suitable for deploying remotely operated equipment via an 'A' frame or deck crane (Figure 3.18). A multi-point mooring system usually has to be specially installed in order for the vessel to keep stationary. This also means that these vessels are generally limited to water depths of less than about 125 m without anchor handling assistance. However, winches available in Australia for this purpose are generally not large enough to carry sufficient wire to allow anchoring in greater water depths.

There are few support vessels that have true dynamic positioning capabilities. While there are a number of vessels that can be kept on position using a 'joy-stick', only low-risk equipment can sensibly be deployed from them, otherwise damage to equipment would prove excessively costly should the vessel suddenly move off location.

Typical costs for support vessels are in the low tens of thousands of US dollars per day.

*Dynamically positioned vessels*

Dynamically positioned (DP) vessels (Figure 3.19) are ideally suited to deep-water work, investigations that require relatively large areas to be investigated and pipeline investigations. Redundancies in the position-keeping equipment mean that there is less danger of damaging expensive equipment.

Construction DP vessels generally have a moon-pool, so that it is possible for drilling equipment to be used in addition to remotely operated equipment. Heave compensation

*Figure 3.18* Support (or supply) vessel

*Figure 3.19* Dynamically positioned site investigation vessel

*Figure 3.20* Specialised drilling vessel

equipment would have to be supplied for drilling from a DP vessel as such equipment is not standard for this type of vessel. Transportable drilling equipment that can be operated over the side or stern of a DP vessel is also available.

Typical day rates for DP vessels are in the high tens of thousands of US dollars, depending on the vessel capability. Vessels may not be readily available and mobilisation costs to more remote areas will be high. However, DP vessels are the most economic solution for deep-water work.

### Specialised drilling vessels

Specialised drilling vessels generally have self-anchoring capabilities or have a DP system and are therefore self-sufficient (Figure 3.20). It is therefore relatively cost-effective to undertake a number of boreholes. The motion compensated drilling system, which also has a high degree of drilling control, provides a system capable of producing the high-quality results required for geotechnical work, particularly for executing geotechnical investigations in deeper water. The moderate size of the vessel shown in Figure 3.20 ensures good response to the DP system and the purpose-designed deck layout and moon-pool location helps ensure efficient and high-quality operations. As the drilling system is equipped with a hard tie heave compensation system, (i.e. one that is linked to the seabed) minimal sample disturbance can be achieved even in soft clays.

Supply of freshwater is quite limited on specialised drilling vessels. This not only limits the total time the vessel can remain on location, but also limits the choice of drilling fluid since only fluids compatible with seawater can be used.

Specialised drilling vessels that have dynamic positioning capabilities currently operate in the North Sea, West Africa and Russia. Mobilisation costs of such vessels

*Figure 3.21* Jack-up drilling rig

to other areas are extremely high unless a number of different investigations can be carried out sequentially in the same area. Typical day rates for these vessels are in the high tens to low hundreds of thousands of US dollars.

### Jack-up rigs

Jack-up exploration rigs provide a stable platform (Figure 3.21), and therefore provide some of the advantages of site investigation from seabed rigs. However, they are subject to a number of limitations:

- Considerable effort is required to reposition the rig and therefore it is generally not cost-effective to drill more than one borehole at a site.
- The primary drilling rig is not suitable for geotechnical work and therefore a full spread of geotechnical drilling equipment is required to be mobilised each time.
- It is generally not possible to undertake an investigation whilst the rig carries out regular oilfield activities. Space limitations often require temporary removal of some oilfield equipment in order to accommodate geotechnical work.
- Jack-up rigs are limited in water depth to around 120 m, although this limit is gradually increasing as larger rigs are developed.
- Day rates are expensive, typically in the low hundreds of thousands of US dollars, and significant back-up support is required (e.g. helicopters, support vessels and high level of staffing).

Due to these limitations, jack-ups are rarely chosen as a platform for geotechnical site investigations apart from in shallow water near shore.

*Figure 3.22* Semi-submersible drilling rig

## Semi-submersible drilling rigs

Semi-submersible drilling rigs (Figure 3.22) can provide a reasonably stable platform in deep water, but their day rates are extremely high, often in excess of a quarter of a million US dollars per day. The vessels are anchored to the seabed, which provides an additional cost in terms of the time required to set and retrieve the anchors, with each operation taking up to a day to complete.

Because of the long anchor line lengths, it is possible to move the vessel on anchor (i.e. without moving the anchors themselves) over the footprint of a standard structure (jacket or gravity base) and therefore drill multiple boreholes. This is not the case if the mooring pattern of, say, a floating production storage and offloading (FPSO) vessel is to be investigated, as the anchoring radius of these is often greater than 1 km. However, the high day rate still limits the number of boreholes that can be drilled economically.

Other disadvantages in respect of the drilling system are similar to those listed for the jack-up rig. As such, this type of working platform is also rarely used for geotechnical investigations.

### 3.3.3 Drilling and coring systems

#### Manual and remote subsea systems

In shallow water, drilling and sampling operations can be carried out by divers, using equipment that is essentially similar to onshore systems. Drilling operations are generally limited to daylight hours, i.e. single 12-hour shifts, unless work is relatively near shore and two crews can be accommodated onshore. A significant shortcoming is that commercial divers are generally not drillers and therefore inexperience affects drilling quality. This problem may be partly overcome by the use of video surveillance and

(a)

(b)

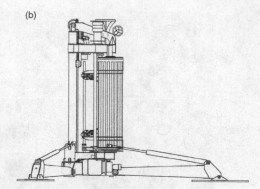

*Figure 3.23* Remote subsea drilling system (PROD) (a) launching PROD off stern of vessel and
(b) schematic of PROD after deployment of legs (courtesy of Benthic Geotech)

voice communications. However, the geotechnical supervisor is also responsible for
logging and packing core samples, and so cannot always oversee subsea activities
continuously. This problem is further exacerbated by limited on-bottom times, which
require a constant rotation of divers.

The depth of borehole achievable is limited by the power of the equipment, although
depths of about 40 m may be achieved if drilling conditions are favourable. Unstable
boreholes require casing and, whilst this is possible, it is not an easy operation. In
addition, drilling fluids are generally restricted to seawater.

A more sophisticated robotic subsea geotechnical investigation system has been
operational since about 2005. The PROD (portable remotely operated drill) system
is able to drill, take rock cores or soil samples and carry out *in situ* penetrometer
tests down to depths of some 100 m below the seabed, and in water depths of up to
2,000 m (Figure 3.23). The design of the tripod support allows it to operate on
relatively high seabed gradients. The original version incorporated a relatively small
diameter drill allowing recovery of 45-mm-diameter piston samples. A second
system was launched during 2009, which is capable of taking 75-mm-diameter
samples.

Rotary core drilling with PROD is achieved using modern thin Kerf core barrels
that recover cores of 44 mm diameter in lengths of up to 2.70 m. Aluminium liners are
used to capture the core during the sampling process and to protect the core samples
during extraction from the barrels and storage on the vessel. A core catcher is incorporated within the inner barrel (Figure 3.24).

### Jack-up rigs

Jack-up rigs have their own drilling equipment, designed for well drilling, and this
may be used for geotechnical investigation. A better approach, however, is to mobilise
conventional onshore (geotechnical) drilling equipment, either positioned on the drill
floor itself or cantilevered over the edge (Figure 3.25).

Since the jack-up provides a stable platform, normal onshore drilling techniques
can be applied by experienced drillers, allowing very-high-quality results to be
achieved. Coring, push sampling and *in situ* testing is also possible. Alternatively, a
cone penetrometer can be deployed through the drill string. Provided all the required
investigation tools can be deployed, this approach provides the ultimate approach for

*Figure 3.24* End views of PROD's Kerf core barrels (note core catcher on left)

*Figure 3.25* Conventional onshore drilling equipment operating over the side of a jack-up rig

offshore geotechnical investigation. However, as mentioned earlier, the cost of drilling more than one borehole at a location is high because of the time-consuming process of moving the rig. The size of jack-up rigs allows abundant storage of freshwater and other fluids and thus a wide choice of drilling fluids. Ample accommodation on most rigs allows a 24-hour operation.

*Figure 3.26* Seabed frame (SBF) being prepared

## Specialised drilling vessels and semi-submersibles

The majority of offshore site investigation is conducted from specialised drilling vessels or, less commonly, from semi-submersibles. The set-up comprises four critical elements:

- Seabed guidebase with clamp
- Motion compensation
- Open drag bit on bottom hole assembly (BHA)
- Umbilical or wireline tools.

These elements are generally standard on specialised geotechnical drilling vessels. On semi-subs and jack-up drilling rigs, most components are available but require slight modification to suit the geotechnical equipment. Transportable systems exist that can be temporarily fitted to other vessel types that have a moon-pool.

### SEABED GUIDEBASE AND HEAVE COMPENSATION

The seabed guidebase (Figure 3.26) is ballasted to provide approximately 10 tonnes of reaction. Peripheral skirts can be provided to limit settlement into softer seabeds. On a fully equipped drilling vessel, the guidebase is lowered on a separate winching system with its own heave compensation to ensure that the lowering lines remain in tension. On a semi-submersible, the rig's tension wires (all compensated) are used.

*Figure 3.27* Standard 5″ (127 mm) API drill string

A remotely operated hydraulic clamp is mounted on the guidebase and is used to immobilise the drill string when testing is carried out. This clamp effectively connects the drill string to the guidebase in order to provide a reaction force. In deep water, pressure losses over the hydraulic control line require that a subsea hydraulic power pack, mounted on the guidebase, is necessary to activate the clamp.

DRILLING BIT AND PIPE

The drill string is usually 5″ (127 mm) external diameter (Figure 3.27) and generally accords to the American Petroleum Institute (API) standard with flush internal joints to allow downhole tools to pass through the centre. Lengths are typically 9.2 m. A bottom hole assembly (BHA) forms the bottom of the drill string. The BHA comprises an open drag bit (Figure 3.28) and includes drill collars to provide the necessary bit weight allowing for reaction to the motion compensator. The BHA also contains latching sections into which downhole tools latch in order to take a sample or carry out an *in situ* test. Unlike an exploration drilling rig, it is necessary to be able to set the motion compensators to react sufficiently sensitively for the particular soil type and therefore minimise disturbance of the soil below the bit. Even with care, the upper few hundred millimetres of the soil to be sampled or tested may show signs of disturbance.

UMBILICAL CABLE

The downhole tool is deployed either on a simple wire-line, or (preferably, especially for *in situ* testing) one that comprises a hydraulic or electrical umbilical (Figure 3.29). With a simple wire-line system, any data acquired during a test must be stored within

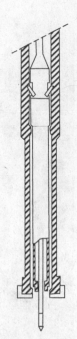

*Figure 3.28* Drill bit and schematic of lower part of borehole assembly

*Figure 3.29* Umbilical cable and insertion of tool into drill stem

the tool, and is subsequently downloaded once the tool is retrieved to the surface. The advantage of an umbilical system is that data can be viewed in real time (while the test is underway) and direct control of the test is possible.

On specialised drilling vessels, the top drive of the drilling unit usually incorporates a mud swivel, which is simply opened to allow the insertion and lowering of the tool. On semi-subs, it is usually necessary to break the drill string to allow the tool to be inserted (see Figure 3.29).

DOWNHOLE TOOL

The section of the downhole tool attached directly to the umbilical can be either hydraulically or electrically activated. This section of the tool contains an electronic unit that controls the actions of the equipment being used, captures the data and converts this into digital signals that are sent up the line to the control unit on the vessel. The bottom section of the tool is detachable and specific to the type of equipment being used (cone or ball penetrometer, vane shear or thin-walled push or piston sampler). This section usually contains a hydraulic ram or an electric motor. The ram or motor is used to thrust the equipment into the ground ahead of the drill bit. The vane shear also requires a motor to drive (twist) the vane blades. In the case of a wire-line tool, the drilling fluid is pressurised in order to force the equipment ahead of the drill bit. The detachable bottom half of the tool allows an array of equipment to be positioned on deck and therefore easy switching between tools as required.

Electrically driven tools form part of the new generation of equipment, but are not yet used widely. This type of equipment is more manageable and requires less space on deck in respect of power packs and bulky hydraulic umbilical reels. A further advantage is that anything that is thrust into the ground can be simply retracted by reversing the motor – the slim hydraulic rams required for downhole operations do not generally allow retraction.

ROCK CORING

It is possible to use this system, albeit in a modified fashion, to recover rock core. Recovery of weak rock, such as calcarenite, core (see Figure 3.30) requires a hard-tie system to ensure maximum recovery. In order to achieve a true hard-tie system, a separate drilling rig is used in a 'piggy back' fashion. A separate riser is run down to the guidebase and clamped. A platform is mounted on top of this riser and attached to the vessel's motion compensator to ensure that the riser remains in tension at all times (to prevent buckling). A smaller land-based drilling rig is mounted on the platform and drillers deploy a separate drill string through the riser. Essentially, this means that drilling is carried out from a motionless platform and therefore is 'hard-tied'. The operational constraints of this system are determined by the total movement allowed by the vessel's motion compensator.

Because the drill string has to pass through the riser, a smaller diameter pipe has to be used. The BHA at the end of the drill string is designed to accommodate both a wireline triple tube core barrel and the downhole tool described earlier. However, the downhole tool has to be modified slightly to allow it to work within the smaller diameter drill string. This usually means that the stroke available for a cone test (e.g. between zones of core recovery) is reduced by about one-half.

*Figure 3.30* Offshore classification of rock core

This system has been very successful on the North-West Shelf of Australia in recovering calcarenite, a notoriously difficult material type when it comes to obtaining high-quality core.

WHEEL DRIVE

Wheel drive equipment (Figure 3.31) was originally developed to undertake cone penetration tests (CPTs) from the seabed, as described later. It comprises a frame heavy enough to provide the required reaction to thrust the cone into the seabed to the target depth. A bank of hydraulic motors are mounted in series on top of the frame. Each motor drives a set of wheels, which turn against a continuous rod and therefore drive the rod into the seabed. The cone (or more typically piezocone) is attached to the end of the rod and the rod is strung together in sections before lowering the frame to the seabed with the rods kept under tension. Once lowered to the seabed, the motors are activated and the rod is driven into the seabed to target depth or refusal. The depth achievable is a function of the tip resistance on the cone and the friction acting along the length of the rod, and it is common to incorporate an expanded section of rod behind the cone to help reduce friction on the following rods. The rod is eventually retracted by reversing the direction of the wheel drive motors and the frame raised to the surface.

Test depths up to 65 m have been achieved on the North-West Shelf of Australia. The wheel drive is now also used for other tools such as the T-bar and ball penetrometers and vane shear (see later). The frame may also be modified to accommodate a remote rock coring unit that can be used in conjunction with the cone testing. A piston sampling device is sometimes mounted on the frame to recover samples in the upper 1.5 m below the seabed. Seabed frames incorporating wheel drive units or alternative

*Figure 3.31* Wheeldrive unit

robotic equipment are the most appropriate means of site investigation for deep-water developments, where the focus is generally on the upper 20–50 m of the seabed.

### 3.3.4 In situ *testing equipment*

#### Cone penetrometer

The cone penetration test (CPT) is currently the most performed *in situ* test during a geotechnical investigation, generally using a piezocone that incorporates pore pressure measurement. The piezocone is a cylindrical penetrometer with a 60° conical tip to which a load cell is attached (Figure 3.32). This measures the resistance to penetration, which is usually expressed as the force per unit area $q_c$. The industry standard piezocone has a cone area of 1,000 mm$^2$ (35.7 mm diameter) although 1,500-mm$^2$ cones (43.8 mm diameter) are being used increasingly in soft soils. A sleeve is mounted above the tip and a load cell attached to this sleeve measures the side friction, again expressed as force per unit area $f_s$. Piezocones also have a pressure transducer that measures the pore water pressure as the cone penetrates. The filter for the transducer is mounted either on the face of the cone (so-called $u_1$ position), on the shoulder (just above the tip – $u_2$) or above the sleeve ($u_3$). Cones are available that have filters at all three locations, the two lower locations or one at either one of the lower two locations.

Standard procedures for conducting cone penetration tests are provided by the International Reference Test Procedure (IRTP) published by the International Society

*Table 3.4* Standard dimensions and capacity of cone penetrometers

|  | *10 tonnes* | *15 tonnes* |
|---|---|---|
| Apex angle of cone | 60° | 60° |
| Diameter | 35.7 mm | 43.8 mm |
| Projected area of cone | 1,000 mm$^2$ | 1,500 mm$^2$ |
| Length of friction sleeve | 134 mm | 164 mm |
| Area of friction sleeve | 15,000 mm$^2$ | 22,500 mm$^2$ |
| Max. force on penetrometer | 100 kN | 150 kN |
| Max. cone resistance $q_c$ (if $f_s = 0$) | 100 MPa (100 kN) | 100 MPa (150 kN) |
| Max. sleeve friction $f_s$ (if $q_c = 0$) | 6.6 MPa (100 kN) | 6.6 MPa (150 kN) |
| Diameter of push rods | 36 mm | 36 mm |

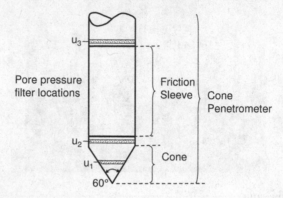

*Figure 3.32* Schematic of piezocone

for Soil Mechanics and Geotechnical Engineering (ISSMGE) (1999) and various national and international standards such as NORSOK Standard (2004), ASTM D5778-07 (2007) and ENISO 22476-1 (2007). The standard rate of penetration is 20 mm/s, or approximately half a diameter per second, although modern equipment is generally capable of varying the rate by at least one order of magnitude. A detailed description of the cone penetration testing, both in terms of equipment and interpretation, is provided by Lunne *et al.* (1997).

Dimensions and capacities of the two most common cone geometries are summarised in Table 3.4. The cone capacity is generally either 10 or 15 tonnes, although the maximum thrust may be limited by the weight of the seabed frame. Two different types of measuring systems for the tip and sleeve resistance are used, referred to as 'compression' and 'subtraction' cones. A compression cone measures the tip resistance and the sleeve friction separately, which allows the load cell for the sleeve friction to be tuned to the likely required maximum of about 1 MPa. A subtraction cone measures the total force on the penetrometer (sleeve + tip) and the tip resistance. The sleeve friction is then calculated by subtracting the tip resistance from the total force. From a system perspective, this is less attractive since the relatively small sleeve friction must be obtained by subtracting two much larger measurements.

The piezocone is primarily a profiling tool and the results can be used to determine the soil types of the various layers intersected (e.g. Robertson 1990). Since data are obtained continuously with depth, it is able to detect fine changes in stratigraphy,

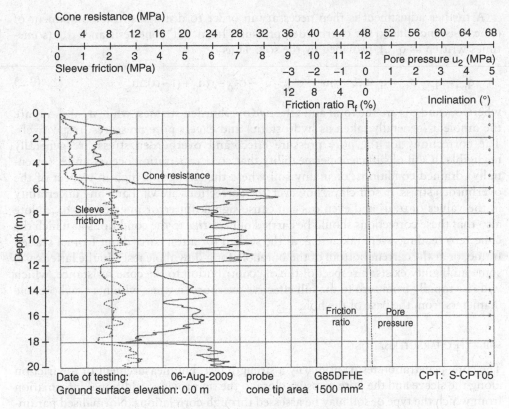

Cone resistance (MPa)

*Figure 3.33* Example data record from piezocone penetration test

as evident in Figure 3.33. It is therefore often used in the first phase of subsurface investigation, before recovery of high-quality samples, providing an essential tool to help ground-truth results of geophysical surveys.

CORRECTIONS TO MEASURED CONE RESISTANCE

The measured cone resistance $q_c$ requires correction because of pore pressure acting on the back face of the cone, either within the $u_2$ pore pressure filter, or acting on the seal at the base of the load cell protection sleeve. A similar correction is required for the sleeve friction measurement. The correction of measured cone resistance to total cone resistance $q_t$ is given by (Lunne *et al.* 1997):

$$q_t = q_c + (1 - \alpha)u_2 \tag{3.1}$$

where $\alpha$ is the 'area ratio' – ratio of the area at the shoulder of the cone (not acted on by pore pressure) to the total cross-sectional area of the cone shaft. Although the area ratio may be calculated from the physical dimensions of the cone, it is usually evaluated by calibration within a pressure chamber, plotting the ratio of the fluid pressure in the cell to the cone tip resistance, which is numerically equal to $\alpha$.

A further adjustment is then necessary in order to distinguish the component of cone resistance due to the overburden pressure, from 'net' cone resistance $q_{net}$ (sometimes written as $q_{cnet}$), provided by the soil. Thus

$$q_{net} = q_t - \sigma_{v0} = q_c + (1-\alpha)u_2 - \sigma_{v0} = q_c - \sigma'_{v0} - \alpha u_0 + (1-\alpha)\Delta u_2 \qquad (3.2)$$

where $\sigma_{v0}$ and $\sigma'_{v0}$ are the total and effective overburden stresses, while $u_0$ and $\Delta u_2$ are the ambient (generally taken as hydrostatic) and excess pore pressures, respectively. The corrections for the pore pressure effect and overburden stress are generally negligible in soil of sufficient permeability that cone penetration occurs under (essentially) drained conditions, or in any soil where the strength is high in respect of the overburden stress. In soft clays, however, the corrections are vital and any uncertainty in the values of $\alpha$, $u_2$ and even $\sigma_{v0}$ can cause significant error in assessing $q_{net}$. Note also that these corrections should be carried out *relative* to the conditions at which the cone zero readings were taken (e.g. at the seabed for a test carried out from a seabed frame, or at the current bottom of the borehole for a downhole test). In the latter case, some ambiguity exists in respect of the $\sigma'_{v0}$ contribution to the cone resistance, which will rise rapidly from zero to the full effective overburden stress within a few borehole diameters from the base of the hole.

SOIL TYPE CHARACTERISATION

The cone penetration test is primarily a strength test, but measurement of the friction along the sleeve and the pore pressure close to the tip provides additional information from which the type of soil may be assessed through correlations. Normalised parameters are used for such classification, principally:

- Normalised cone tip resistance, $Q = q_{net}/\sigma'_{v0}$ (note, $Q_t$ also used)
- Excess pore pressure ratio, $B_q = \Delta u_2/q_{net}$
- Friction ratio, $F_r = f_s/q_{net}$ (note $R_f$ also used).

An example set of classification charts is shown in Figure 3.34 (Robertson 1990). The general principles of such charts are clear, in that stronger soils (dense sands or cemented sediments) will show higher normalised cone resistance and low friction ratio compared with clays. Certain trends, for example, the effects of increasing overconsolidation ratio or increasing degree of drainage for silty soils, may be difficult to distinguish, since both characteristics lead to increasing $Q$, but decreasing $B_q$. This is partly due to the use of $q_{net}$ to normalised the excess pore pressure, and the trends may be separated by considering the excess pore pressure ratio, $\Delta u_2/\sigma'_{v0}$, instead. Modified charts illustrating this are shown in Figure 3.35 (Schneider *et al.* 2008a).

In general, soft- to medium-strength clays will give high positive pore pressures, well in excess of the effective overburden stress and with $B_q$ values typically 0.3–0.7, increasing with increasing sensitivity of the clay. Typical friction ratios will be 2–5 per cent, but decreasing with increasing sensitivity. Stronger materials, such as heavily consolidated clays, cemented material and dense sands may exhibit dilational characteristics, with low or negative excess pore pressures that tend towards zero as the consolidation coefficient increases. Particular signatures may be identified for

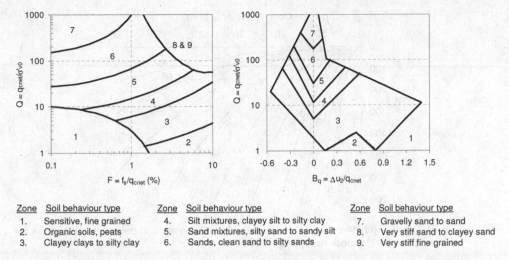

Zone | Soil behaviour type
1. | Sensitive, fine grained
2. | Organic soils, peats
3. | Clayey clays to silty clay

Zone | Soil behaviour type
4. | Silt mixtures, clayey silt to silty clay
5. | Sand mixtures, silty sand to sandy silt
6. | Sands, clean sand to silty sands

Zone | Soil behaviour type
7. | Gravelly sand to sand
8. | Very stiff sand to clayey sand
9. | Very stiff fine grained

*Figure 3.34* Soil classification charts (after Robertson 1990)

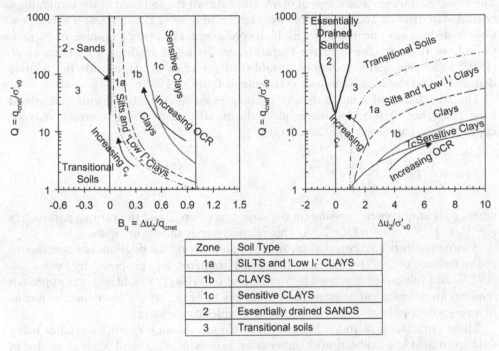

| Zone | Soil Type |
| --- | --- |
| 1a | SILTS and 'Low I$_r$' CLAYS |
| 1b | CLAYS |
| 1c | Sensitive CLAYS |
| 2 | Essentially drained SANDS |
| 3 | Transitional soils |

*Figure 3.35* Modified soil classification chart using excess pore pressure ratio (Schneider *et al.* 2008a)

specific conditions, for example, the presence of gas or gas hydrates, by matching profiles of normalised cone parameters against visual evidence from recovered samples (e.g. Sultan *et al.* 2007). In unfamiliar sediment types, it is important to calibrate charts such as those shown in Figure 3.34 and Figure 3.35 against evidence from borehole samples.

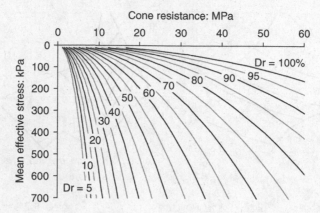

*Figure 3.36* Relationship between cone resistance, relative density and mean effective stress for silica sands (Jamiolkowski *et al.* 2003)

INTERPRETATION IN SANDS (FREE-DRAINING SEDIMENTS)

The cone resistance is a function of many attributes of the soil such as the mineralogy, critical state friction angle, ambient stress level and relative density, and it is not possible to deduce any one quantity with high confidence. However, a number of correlations have been developed in the literature, as discussed in detail by Lunne *et al.* (1997). The principal correlations, established for silica sands, are for the relative density $D_r$ and the state parameter $\psi$ (Been and Jefferies 1985).

The correlation of relative density $D_r$ (expressed as a fraction) with normalised cone resistance $Q = q_{net}/p_0'$ where $p_0'$ is the *in situ* mean effective stress, may be expressed as (Jamiolkowski *et al.* 2003)

$$D_r = \frac{1}{C_2} \lambda n \left( \frac{1}{C_0} \frac{q_c}{p_0'} \left( \frac{p_0'}{p_a} \right)^{1-C_1} \right) \tag{3.3}$$

where $p_a$ is atmospheric pressure (in the same units as $p_0'$), and the various parameters are: $C_0 = 25$; $C_1 = 0.46$; $C_2 = 2.96$. This relationship is shown in Figure 3.36.

A more sophisticated correlation, aimed at quantifying the dilational or contractive characteristics of sand by means of the state parameter, was proposed by Been *et al.* (1987), and subsequently refined by Shuttle and Jefferies (1998). Ideally, the approach requires an independent assessment of the gradient of the critical state line, by means of a series of triaxial tests on reconstituted samples of the sand.

These correlation methods have been developed through extensive studies using data from field and calibration chamber tests, mainly in silica sand. Care is needed in more compressible (but still free draining) sediments, such as carbonate sands, which tend to give much lower normalised cone resistance unless cemented.

A cone penetration test may be regarded as a form of model test, and design approaches are being developed increasingly based directly on the measured cone resistance. For shallow foundation design, the ultimate bearing pressure is typically taken in the range 0.1–0.2 times the cone resistance, with the lower limit corresponding to a limiting settlement of 5 per cent of the foundation dimension and relatively loose sand, and the upper limit corresponding to a limiting settlement of 10 per cent of the foundation dimension and relatively dense sand (Randolph *et al.* 2004).

*Table 3.5* Parameters for cone factor, $N_{kt}$

| Parameter | Teh and Houlsby (1991) | Yu et al. (2000) | Lu et al. (2003) |
|---|---|---|---|
| $C_1$ | $1.67 + I_r/1,500$ | 0.33 | 3.4 |
| $C_2$ | $1.67 + I_r/1,500$ | 2 | 1.6 |
| $C_3$ | 1.8 | 1.83 | 1.9 |
| $C_4$ | 2.2 | 2.37[1] | 1.3 |

[1]Friction ratio expressed in terms of the ratio of interface and soil critical state friction angles

A number of design methods based on cone resistance have been developed for driven piles in sand, and these are described in Chapter 5. Design guidelines published by the American Petroleum Institute (API) and International Standards Organisation (ISO) now express strong preference for cone-based design methods, rather than the traditional approach using shaft friction and end bearing parameters based on assessing the particle size (silt, sand or gravel) and relative density.

INTERPRETATION IN CLAYS (UNDRAINED CONE PENETRATION)

In fine-grained sediments, where cone penetration occurs under undrained conditions, interpretation of cone tests is mainly focused on assessing a shear strength profile from the net cone resistance, by means of a cone factor $N_{kt}$ according to

$$s_u = \frac{(q_t - \sigma_{v0})}{N_{kt}} = \frac{q_{net}}{N_{kt}} \tag{3.4}$$

The cone factor $N_{kt}$ may vary widely; Lunne *et al.* (1997) quote cone factors between 7 and 17, and even higher values are used for stiff, structured clays in the North Sea. Industry practice usually involves calibrating the cone resistance at each site against laboratory strength measurements, but it is useful to consider the theoretical basis for $N_{kt}$.

The theoretical value of $N_{kt}$ may be expressed in terms of the rigidity index $I_r = G/s_u$ a stress anisotropy factor $\Delta = (\sigma'_{v0} - \sigma'_{h0})/2s_u$ and the friction ratio on the cone face $\alpha_c = \tau_f/s_u$ according to:

$$N_{kt} \approx C_1 + C_2 l_n(I_r) - C_3\Delta + C_4\alpha_c \tag{3.5}$$

Values for the various constants in Equation 3.5 have been proposed by Teh and Houlsby (1991), based on a strain-path approach, and by Yu *et al.* (2000) and Lu *et al.* (2004), using different forms of large displacement finite element analysis. The values are summarised in Table 3.5 and the resulting values of cone factor are shown in Figure 3.37, adopting $\alpha_c = 0.3$ (and $\delta/\phi_{cs} = 0.7$ for the Yu *et al.* approach). There is good consistency between the two finite element approaches, once allowance is made for the different soil models. The trends indicate that, for lightly overconsolidated soils, where $\Delta$ will be greater than 0, cone factors would be expected to lie in the range 12 to about 15 or 16.

Empirical values of $N_{kt}$ range widely, as noted by Lunne *et al.* (1997). A recent worldwide study of penetration resistance factors for lightly overconsolidated clays (Low *et al.* 2010) recommended an average $N_{kt}$ value of 13.5, with a range of 11.5–15.5. The study showed a well-defined trend for $N_{kt}$ to increase with increasing small-strain rigidity index $I_r = G_0/s_u$. In some regions of the world, such as the Gulf of Mexico,

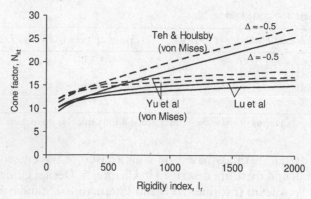

*Figure 3.37* Theoretical cone factors

a default $N_{kt}$ value of around 17 is generally used in design. It is not clear whether this is due to unusually high rigidity index (possible due to aging, for samples taken from depth), or is a historical figure reflecting strengths measured following slight disturbance of the *in situ* soil.

### PORE PRESSURE MEASUREMENT AND CONSOLIDATION CHARACTERISTICS

Pore pressure measurements during a cone test, generally through a filter positioned at the shoulder of the cone ($u_2$ position), provide additional information by which to assess the soil characteristics. They also play an important role in correcting the measured cone resistance due to unequal area effects (see Equation 3.1). The excess pore pressure $\Delta u_2$ may be normalised by the net cone resistance to give the pore pressure ratio $B_q$ or by the effective overburden stress $\sigma'_{v0}$ (see Figure 3.35) to help assess the soil type. Sudden reductions in $B_q$ or $\Delta u_2/\sigma'_{v0}$ allow detection of silt lenses (hence increased permeability), often accompanied by a corresponding spike in the cone resistance.

Quantitative measurement of the consolidation coefficient may be obtained by halting the cone and allowing the excess pore pressure to dissipate. Theoretical solutions (e.g. Teh and Houlsby 1991) may then be used to fit the against the measured dissipation response, and thereby estimate a consolidation coefficient. During pore pressure dissipation, soil close to the cone will be consolidating (reducing water content), but the major part of the surrounding soil will undergo swelling. The coefficient of consolidation deduced from the test therefore reflects both the (predominantly) horizontal flow paths, and also the swelling nature of the strain path (Levadoux and Baligh 1986). As a result, it is often referred to as $c_h$, to distinguish it from the $c_v$ value measured in a laboratory oedometer test.

A typical set of piezocone dissipation curves are shown in Figure 3.38 for lightly overconsolidated clay, along with a theoretical dissipation curve based on a rigidity index of 100. While the theoretical curve decays monotonically with time, the measured responses show an initial rise, which may be attributed either to incomplete saturation of the pore fluid filter, or to local dilation effects (Burns and Mayne 1998). In order to estimate an initial maximum excess pore pressure $\Delta u_{max}$ a root time plot has been used, extrapolating the early linear decay curve back to the origin (see inset). The theoretical $T_{50} = t_{50}c_h/d^2$ is close to unity, so that for $t_{50} \sim 4{,}000$ s, and $d = 35.7$ mm, the value of $c_h$ is about 10 m²/yr.

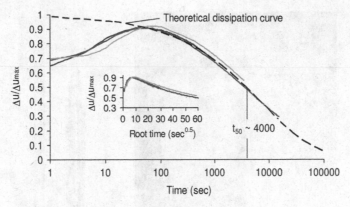

*Figure 3.38* Theoretical dissipation curves for piezocone

It should be noted that dissipation curves need to be extended to at least $t_{50}$ in order to allow a reasonable fit to be achieved, and hence confidence in the deduced $c_h$ value. This makes offshore dissipation tests in clay very expensive, with times of 1–3 hours potentially required. In siltier sediments, dissipation tests are much shorter duration, and also are of particular value, since they help to determine whether the cone penetration test itself was essentially undrained or whether the measured cone resistance may be affected by partial consolidation. Cone tests carried out at different penetration rates suggest that the normalised penetration rate $V = vd/c_v$ (where v is the penetration rate, typically 20 mm/s) should exceed 100 for the penetration to be essentially undrained (Randolph and Hope 2004). It is good practice during site investigations involving silty sediments to plot the normalised cone resistance as a function of V in order to identify layers where the cone resistance has been increased by partial consolidation during the test.

## T-bar and ball penetrometers

The T-bar penetrometer was first introduced at the University of Western Australia in order to improve the accuracy of strength profiling in centrifuge model tests (Stewart and Randolph 1991, 1994), and was first used offshore in the late 1990s (Randolph *et al.* 1998a). The probe consists of a short cylindrical bar attached at right angles to the penetrometer rods, just below a load cell (Figure 3.39). It has two major advantages over the cone. First, the load cell measures what is essentially a differential force (or net pressure) on the bar, so that minimal adjustment need be made for the overburden stress and ambient pore pressure. Second, the correlation between net pressure on the bar and the shear strength of the soil is underpinned by an exact plasticity solution, with a potential range of 'bar' factor of less than ± 10 per cent (due to different roughness of the bar surface), compared with cone factors that may vary from as low as 7 in sensitive clays, to over 17 – a range of ± 40 per cent. Both a downhole (20 mm diameter by 100 mm long) and a wheeldrive version (40 mm diameter by 250 mm long) have been developed, although the downhole version has not proved robust and has essentially been replaced by a piezoball penetrometer.

The piezoball penetrometer (Figure 3.40) is better suited to downhole testing because of the smaller protrusion relative to the shaft. Diameters vary between 60 and 80 mm, with the shaft just behind the ball sized so that the ball area is approximately

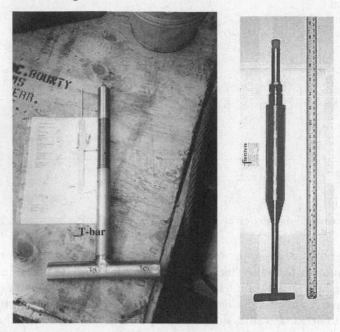

*Figure 3.39* T-bar penetrometer (a) wheeldrive and (b) downhole

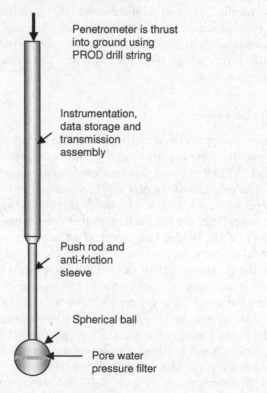

Penetrometer is thrust
into ground using
PROD drill string

Instrumentation,
data storage and
transmission
assembly

Push rod and
anti-friction
sleeve

Spherical ball

Pore water
pressure filter

*Figure 3.40* Piezoball penetrometer (Kelleher and Randolph 2005)

ten times that of the shaft. The filter for the pore pressure sensor can be placed either at the mid-depth (as indicated), or at discrete points elsewhere on the lower half of the ball. Again, the ball resistance factor is based on plasticity solutions, but for both the T-bar and ball corrections are required to allow for (a) strain rate effects and (b) gradual softening of the soil due to remoulding as it flows past the tool. Appropriate T-bar and ball factors will be discussed later.

Full-flow penetrometers such as the T-bar and piezoball can be used to measure the remoulded strength of the soil, by carrying out penetration and extraction cycles over a short depth interval (typically less than 0.5 m). A minimum of 10 cycles should be undertaken, which is usually sufficient to achieve close to fully remoulded conditions. This ability has meant that T-bar (and potentially ball) penetrometers have become the standard field testing tool to investigate the upper 1 or 2 m of the seabed for deep-water pipeline and riser design, both of which involve significant remoulding of the soil.

Standard practice for full-flow penetrometers is still being developed, but the following principles are generally agreed:

1. For field use, T-bar dimensions of 40 mm diameter by 250 mm long have become standard, giving a projected area of ten times that of the cone shaft.
2. Different T-bar dimensions may be used, but the cylinder should have a minimum length to diameter ratio of 4, with the area of the connecting shaft and load cell being no more than 15 per cent of the projected area of the T-bar.
3. A ball penetrometer should have a diameter in the range 50–120 mm (113 mm giving a projected area of ten times that of a cone) with a connecting shaft that occupies no more than 15 per cent of the projected area of the ball.
4. The penetration rate should be in the range 0.2–0.5 diameters per second, although specific tests may be undertaken at different rates in order to evaluate strain-rate dependency of the penetration resistance, or the onset of consolidation (House *et al.* 2001).
5. The extraction resistance should be recorded in addition to the penetration resistance throughout the profile, and at least one cyclic penetration and extraction test should be performed within each penetration test in order to provide confirmation of the load cell offset, in addition to assessing the penetration resistance for remoulded conditions.

INTERPRETATION OF FULL-FLOW PENETROMETER TESTS

As for the cone, some correction is required to the measured penetration resistance in order to obtain the required net resistance, but the correction is much smaller because of the large projected area of the penetrometer relative to the shaft. The correction is (Chung and Randolph 2004):

$$q_{\text{T-bar}} \text{ or } q_{\text{ball}} = q_m - [\sigma_{v0} - u_0(1-\alpha)]A_s/A_p \tag{3.6}$$

where $u_0$ is the ambient pore pressure, $\alpha$ is the area ratio (as for the cone in Equation 3.1) and $A_s/A_p$ is the ratio of the shaft and penetrometer cross-sectional areas. Since this ratio is typically around 0.1, it is often sufficient to ignore the correction, although it becomes more important as the soil strength reduces (e.g. in a cyclic test).

The undrained shear strength may be estimated from the corrected (net) penetration resistances $q_k$ by dividing by a bearing capacity factor $N_k$ (where the subscript

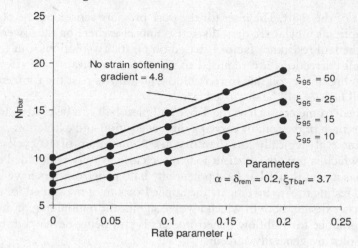

*Figure 3.41* Resistance factors for T-bar in rate dependent softening soil (Zhou and Randolph 2009a)

k refers to T-bar or ball). Originally, a factor of $N_{T-bar} = 10.5$ was recommended, based on the plasticity solution for a cylinder moving laterally through the soil (Randolph and Houlsby 1984, Martin and Randolph 2006). More refined analyses have taken into account the relatively high strain rates involved in field testing, of similar order of magnitude to the normalised penetration velocity of $v/D \sim 0.5$ s$^{-1}$, and also the gradual softening of the soil due to remoulding as it flows around the penetrometer. Typical results are shown in Figure 3.41 for a soil with sensitivity of 5, where $\mu$ is the proportional strength increase per log cycle of strain rate, and $\xi_{95}$ is the shear strain required to achieve 95 per cent remoulding (Zhou and Randolph 2009a). For $\xi_{95}$ in the range 15–25 (i.e. 1,500–2,500 per cent), and $\mu \sim 0.1$, the value of $N_{T-bar}$ would be in the range 11–12.5.

Low *et al.* (2010) quote T-bar factors from a number of onshore and offshore sites comprising lightly overconsolidated clays, with values mostly lying within the range 10.5–13, which is consistent with the theoretical range. The overall average T-bar factor was 11.9 (relative to the average strength measured in triaxial compression, triaxial extension and simple shear), but given the likelihood of some disturbance in the relatively soft samples recovered from offshore, a value closer to 11 may well be more representative.

Theoretical resistance factors for the ball penetrometer are around 20–25 per cent greater than for the cylindrical T-bar (Randolph *et al.* 2000, Zhou Randolph 2009a). However, field and laboratory data both indicate much closer resistances, as indicated in Figure 3.42 (Chung and Randolph 2004). Generally, the ball penetrometer resistance has been found to be between 0 and 10 per cent greater than the T-bar resistance.

Figure 3.42 shows that the extraction resistance for full-flow penetrometers is less than the penetration resistance, due to partial remoulding of the soil. In lightly overconsolidated clays of moderate sensitivity (3–6), the extraction resistance is typically about 60 per cent of the penetration resistance.

Cyclic penetration and extraction tests allow direct assessment of the remoulded shear strength. An example cyclic test for a ball penetrometer is shown in Figure 3.43(a). With each half cycle, the penetration or extraction resistance decreases in magnitude as

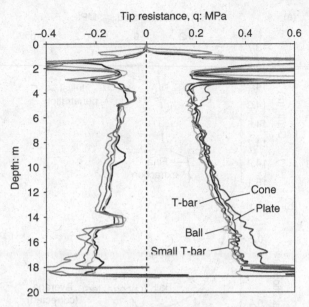

*Figure 3.42* Comparison of net penetration and extraction resistances for cone, T-bar, small T-bar (L/d = 4), ball and plate (Chung and Randolph 2004)

a zone of soil immediately around the penetrometer is remoulded. Generally, the response should remain symmetric around the zero resistance axis, but tests may sometimes show an asymmetric response with penetration resistance at each cycle exceeding the previous extraction resistance (or vice versa). This is primarily an indication of an incorrect load cell zero, although very slight asymmetry may arise from the mechanism itself.

The gradual degradation in resistance may be quantified by plotting the resistance normalised by the initial penetration resistance, with the initial cycle plotted as cycle number 0.25 (see Figure 3.43(b)). The cycle numbering reflects the fact that, treating penetration and extraction as one complete cycle, the average (partly softened) strength of soil during the first half cycle is equivalent to an elapsed quarter of a cycle (Randolph *et al.* 2007).

It is recommended that cyclic penetrometer tests should include a minimum of ten cycles, and that the cyclic resistance is taken as that during the tenth penetration stroke. Further slight softening may occur with additional cycles, but ten cycles appears a reasonable compromise between the overall duration of the test and obtaining a reasonable estimate of the resistance for fully remoulded conditions. It should also be noted that the final degradation factor will be greater than the inverse of the sensitivity, as the penetrometer factor $N_{k,rem}$ for remoulded conditions is greater than for intact conditions (Yafrate *et al.* 2009, Zhou and Randolph 2009b).

PORE PRESSURE MEASUREMENT

Pore pressure sensors may be built into both T-bar and ball penetrometers (Peuchen *et al.* 2005; Kelleher and Randolph 2005). The 'piezoball' penetrometer, in particular, shows promise at being able to ascertain soil type, and also to provide data on consolidation characteristics by means of pore pressure dissipation tests. Low *et al.* (2007)

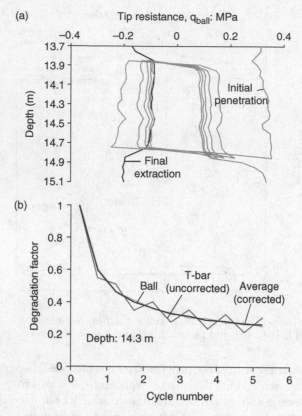

*Figure 3.43* Example cyclic T-bar and ball penetrometer tests (a) cyclic ball penetrometer test and (b) resistance degradation curves

present field data comparing dissipation rates for piezocone and piezoball penetrometers. For a given diameter, dissipation times for 50 per cent decay of excess pore pressure are about 2.5 times smaller for the piezoball compared to a piezocone. However, given that the piezoball will typically be 1.5–2 times the diameter of the cone, actual times will be similar or 50 per cent greater, since they are proportional to the square of the diameter.

*Vane shear*

Three different sizes of vane are used offshore, all with a height to width ratio of two and with the height ranging between 80 mm and 130 mm (Norsok 2004). The size of vane is chosen according to the strength of the soil to be investigated, with the largest size appropriate for soft sediments. The vane is thrust into the soil ahead of the drill bit by a minimum distance of 0.5 m, and then rotated at a speed of 0.1 or 0.2°/s. Additional tests may be undertaken by pushing the vane deeper, with a minimum separation of 0.5 m recommended. In onshore practice, a remoulded shear strength is achieved by rotating the vane rapidly for ten revolutions, and then reverting to the original test speed in order to obtain a remoulded strength. Offshore equipment, whether using a downhole tool or from a seabed frame, generally has an upper speed limit of less than 1°/s, so that it is not practical (economically) to rotate the vane more than 0.5–1 revolution. As such, reported values of remoulded strength will be higher than they should be, and the sensitivity too low. The test is relatively time-consuming (and thus expensive) with a typical test time of 20 minutes to obtain one measure of peak strength and one of (partially) remoulded strength (Peuchen and Mayne 2007).

Interpretation of vane tests to obtain peak and remoulded shear strengths follows the classical relationship between torque and undrained shear strength. For a vane of height h and diameter d, the torque is given by

$$T = \frac{\pi d^3}{6}\left(1 + 3\frac{h}{d}\right)s_u \tag{3.7}$$

For the usual aspect ratio h/d of 2 the contribution from the sides of the cylindrical shearing region is around 86 per cent of the total (and higher if allowance is made for non-uniform shear stress mobilisation across the ends of the cylinder – Chandler 1988). It should be noted that the vane test induces relatively high strain rates in the soil. For typical strain-rate dependency properties for soil, the maximum strain rate adjacent to the vane is approximately 30 times the rotation rate (in rad/s) – hence numerically equivalent to approximately 0.5 times the rotation rate in degrees per second (e.g. between 0.05 and 0.1 $s^{-1}$ for a rotation rate of 0.1–0.2°/s, so 3–4 orders of magnitude higher than in a typical laboratory simple shear test).

The strength measured in a vane test is sensitive to the precise testing procedure, in particular the delay between insertion and testing, and the rotation rate (Chandler 1988). Standardising the test helps in consistency of results in a given soil type but the various effects of disturbance, partial consolidation (leading to strength recovery) and enhanced strength due to the high strain rates compensate each other to some extent but may not prove consistent among soils of different plasticity and consolidation characteristics. Current practice, however, is not to apply any correction factor to offshore vane strengths (Quirós and Young 1988).

The vane is primarily used for strength measurement in soft offshore sediments, but in practice, it has few advantages compared with full-flow penetrometers, as indicated in Table 3.6. Although the cone is globally accepted as the primary offshore *in situ* testing tool, and is likely to remain so, its weakness lies partly in uncertainties in assessing the intact strength in low strength deep-water sediments and partly in its inability to measure the remoulded strength.

## Specialised tests

There are a number of more specialised tests that may be undertaken, and there is a general trend towards greater reliance on field testing rather than laboratory testing of recovered samples, particularly for soft, deep water, sediments. Examples of specialised *in situ* tests, which are summarised by Lunne (2001), include:

- Seismic piezocone tests, using a seismic source at the seabed and either a single geophone or (preferably) a pair of geophones built into the cone rods, to obtain profiles of small strain shear modulus $G_0$.
- Gas and pore water sampling (BAT probe, Rad and Lunne (1994)) and deep-water gas probe (Mokkelbost and Strandvik (1999)), with the ability to assess the

*Table 3.6* Comparison of *in situ* strength profiling tools in soft sediments

|  | Cone penetrometer | Full-flow (T-bar or ball) penetrometer | Vane shear |
|---|---|---|---|
| Undisturbed strength | Good | Excellent | Moderate |
| Remoulded strength, hence sensitivity | Unreliable | Excellent | Moderate |
| Undisturbed strength profiling rate | 20 mm/s + time for rod changes or wire line operations | 20 mm/s allowing for 2–10 m equivalent cyclic travel + time for rod changes etc | 1–3 minutes per measurement + penetration time + wait time |
| Time for remoulded strength measurement | N/A | 2–8 minutes | >1 hour |
| Influence zone of remoulded strength measurement | N/A | 0.2–0.5 m depth | ~0.1 m depth |
| Continuous profiling | Yes | Yes | No |
| Soil classification | Good | Potentially good | N/A |
| Consolidation characteristics | Good | Potentially good | N/A |

local permeability of the soil and also obtain 100-ml samples of filtered pore fluid together with any dissolved gas.

- Cone pressuremeter tests, with a cylindrical expansion pressuremeter situated just behind the cone, with the primary aim of measuring unload–reload shear modulus, though also able to provide an estimate of shear strength to corroborate the measured cone resistance.
- Electrical resistivity (in sand) and nuclear density probes, both of which provide an estimate of the soil density (Tjelta *et al*. 1985).
- Piezometer probe (or piezoprobe), to determine the ambient pore pressure (Whittle *et al*. 2001).
- Thermal probe, to measure both the *in situ* temperature, but also the thermal conductivity of the soil (Zelinski *et al*. 1986).

Devices such as the piezoprobe and thermal probe need to be designed in such a way as to minimise their 'inertia' from a diffusion perspective, so that they come to equilibrium in as short a time as possible. This is usually accomplished by locating the measurement transducer on a small diameter extension ahead of the main body of the probe.

More specific tests, which are generally aimed a particular foundation option in order to validate design parameters or a construction procedure, include:

- Plate load tests, generally conducted at the seabed
- Hydraulic fracture tests
- Open hole stability tests
- Grouted section tests.

Whilst equipment has already been developed for some of these tests, the cost of developing new equipment for a particular project can be very high.

*Table 3.7* Common types of sampler used in offshore investigations

| Type of sampler | Dimensions | Quality | Comments |
|---|---|---|---|
| Grab sampler | Typically ~0.1 m³ | Highly disturbed | Very shallow (0.5 m) |
| Box-core | Up to ~0.25 m² in plan by 0.5 m deep | Undisturbed soil in body of sample | Miniature vane or penetrometer testing |
| Drop and vibro-core | 85–150 mm dia. and up to 6 m long | Moderately disturbed (especially vibro-core) | Preliminary SI with geophysical survey |
| Gravity or jacked piston core | 110–170 mm dia. and 20–30 m deep | Undisturbed provided good piston control | Good triggering of piston crucial |
| Drilling and piston core | 45–85 mm dia. and lengths of 1–3 m | Undisturbed subject to sampling tube design | Sample quality reduced by drill pipe motion |

Another example of specialised field testing is the SMARTPIPE facility for measuring pipe–soil interaction performance (Looijens and Jacob 2008). This is a sophisticated seabed frame that can subject an instrumented segment of pipe to vertical, axial and lateral motions, under either force or displacement control on each axis. Penetrometer tests can also be undertaken by means of a separate actuator system.

### 3.3.5 Sampling equipment

Samples of the seabed are required for a number of reasons, ranging from developing a geological profile to obtaining high-quality samples for laboratory testing. The main types of sampler are summarised in Table 3.7, and range from highly disturbed, very shallow, grab samples to long piston cores that, in soft sediments, may recover 20–30 m of relatively undisturbed soil.

### Grab samples and box cores

Very shallow sediments may be recovered by grab samples or box cores, with particular application to pipeline routes where the upper 0.5 m of the seabed is critical. Grab samples provide disturbed material, allowing only simple mineralogical testing, although the material may also be useful for model testing. Box cores are largely undisturbed and the most effective manner of exploring the strength characteristics is by *in situ* miniature vane or (preferably) mini-penetrometer testing, as illustrated in Figure 3.44 (Low *et al.* 2008).

### Drop cores and vibro-cores

In its simplest form, a gravity drop corer comprises a steel tube, typically 6–8 m long and 100 mm external diameter with an 85 mm internal diameter core liner, a cutting shoe and core catcher at the tip and a mass of 500–1,000 kg at the upper end. The corer penetrates the seabed by free-fall from about 10 m above the seabed. In harder soil, or sand, the weight is replaced by an electrically driven vibrator (hence vibro-core) and the diameter may be increased.

*Figure 3.44* Box-core being landed and (right) set up for miniature T-bar penetrometer tests

Both types of corer result in relatively low-quality samples, with the main objective being to provide a stratification check in the upper few metres of the seabed, and to allow classification tests to be undertaken.

### Seabed piston corers

Sample quality can be much improved by incorporating a piston in the sampling tube. Ideally, the piston stays fixed in absolute terms while the sampling tube penetrates the soil. Suction developed between the piston and the soil aids progress of soil into the tube; it also helps to ensure full recovery of the soil when the corer is withdrawn, with the piston now fixed to the corer. Pistons may be incorporated into gravity samplers using a Kullenberg-style release (Kullenberg 1947). As the sampler is lowered, a tripping arm contacts the seabed and releases the (weighted) corer, which penetrates the sediments under free-fall. At the point where the sampler hits the seabed, the piston is arrested so that the sampling tube penetrates without further motion of the piston.

Different forms of remote seabed sampler have been described by Young *et al.* (2000) (Jumbo Piston Corer or JPC) and Borel *et al.* (2002) (STACOR sampler). The principle of the STACOR is shown in Figure 3.45. The piston is connected to the base plate, and so is arrested when the base plate contacts the seabed surface (or at a depth where the sediments have sufficient strength to carry the weight of the sampler). These devices have a steel barrel, with typically a 90–130-mm-diameter PVC liner. They penetrate the seabed under their own weight, and can retrieve samples of up to 20–30 m long. Recovery rates in excess of 90 per cent can be achieved, although there is always some uncertainty in respect of precisely where the piston is arrested, and hence the depth of the uppermost soil sampled. If the surface sediments are soft, the base plate of the STACOR may penetrate 0.5 m or so before it, and hence the piston, come to rest.

Disturbance assessed from radiographs appears to be limited to the edges of the core, and strength measurements show similar normalised parameters to those obtained from conventional high-quality sampling. For soft sediments, corers penetrating by self-weight have clear advantages in terms of the speed with which the samples can be obtained without the need for sophisticated drilling vessels, and in the continuous large-diameter samples obtained.

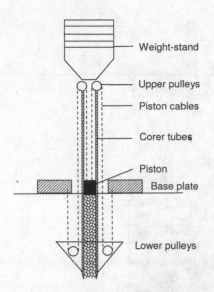

Weight-stand

Upper pulleys

Piston cables

Corer tubes

Piston

Base plate

Lower pulleys

*Figure 3.45* The STACOR gravity piston sampler. Reprinted from Borel, D., Puech, A., Dendani, H. and de Ruijter, M. (2002) 'High quality sampling for deep-water geotechnical engineering: the STACOR® Experience'. *Ultra Deep Engineering and Technology*, Brest, France, by kind permission of PennWell Conferences and Exhibitions.

Recently, a statically penetrated seabed piston sampler has been developed (Lunne *et al.* 2008). This provides greater control over the sampler penetration, at the cost of a more time-consuming operation. A useful feature of the design is that the liner is made up of pre-cut sections of 1 m length, with O-rings fitted between sections to maintain a seal.

### Borehole samples

Push or piston sampling can be carried out down drilled boreholes using either thick-walled or thin-walled sample tubes with or without a piston. In a conventional borehole, the sampler will be operated via a wireline and the quality of the heave-control system has a strong influence on the quality of the recovered samples. The sample tubes are generally about 75 mm internal diameter and a thin wall is defined as about 2 mm. The highest-quality samples are obtained using piston samplers with a thin cutting shoe and an external taper that is less than about 5° (Siddique *et al.* 2000). Any significant thickening of the sampler wall should be set back by three or four diameters. Down-hole piston samplers may be up to 3 m long, generally with liners that are in 1 m segments (Figure 3.46).

### 3.3.6 *Summary comments and reporting*

As noted at the start of the chapter, it is important to integrate the results of a geotechnical site investigation into the geological framework obtained from the geophysical investigation. For major projects, at least two separate geotechnical investigations will usually be undertaken, with the first aimed at ground-truthing the geological model

*Figure 3.46* Removing the sampling tube from the piston corer

and providing preliminary data on soil properties, and a second more detailed site investigation, once the development options for the project have been refined in terms of location of facilities and types of foundation and anchoring systems. The preliminary site investigation will be restricted to a few strategically sited boreholes, from which a geological log may be derived from the samples, and adjacent cone penetrometer tests to provide further stratigraphic details together with a quantitative measure of the soil strength.

The field phase of the geotechnical investigation will be summarised in a factual report, which is used as a basis for planning and execution of the laboratory testing programme. Samples representative of the various layers need to be selected and a test schedule drawn up to produce design information relevant to all facility concepts being considered. The schedule also needs to consider the various stress regimes to be imposed within the seabed, and hence the relevant stress ranges for the laboratory tests.

Eventually, results from the field and laboratory phases of the geotechnical investigation will be synthesised into an interpretative report, which provides the geotechnical design basis for the planned development. Ideally, the interpretative report should start by integrating the geophysical and the geotechnical data at a regional and local scale.

As soon as the geotechnical field investigation is finished a preliminary field report should be prepared covering the following aspects

- Results of all *in situ* tests
- Results from any laboratory tests conducted offshore
- Borehole logs
- Descriptions and pictures of any cores exposed offshore (such as rock cores).

The samples should be offloaded and transported to the laboratory of the client's choice or into a store where they can be preserved under appropriate humidity and temperature conditions.

A typical borehole log (see Figure 3.47) should include the following:

- Details of the drilling and sampling operation, including the proportion of core recovered in the sample
- A visual legend to distinguish the main strata
- A brief description of each layer
- Thickness and (relative) depth for each layer and changes from one to another
- Results of any simple tests undertaken to determine either chemical composition (e.g. the carbonate content) or strength (hand vane, annotations of cone resistance etc).

Photographs of any recovered (cemented) core must be taken and kept as a record of how the material appeared immediately upon extraction from the borehole (see Figure 3.48). This is an important part of the process as the appearance of materials changes rapidly once exposed to the atmosphere.

Note that, for offshore site investigations in uncemented sediments, where laboratory testing of the samples is planned, a photographic record would rarely be taken unless a spare borehole was incorporated in the site investigation. Geological logging would be limited initially to inspection of the ends of each sample, prior to waxing, with more detailed logging undertaken at the time samples were extruded for testing.

## 3.4 Soil classification and fabric studies

### 3.4.1 Introduction

Laboratory studies cover both qualitative assessment of the mineralogy, geologic origins, fabric and classification properties of the soil and also quantitative element tests to determine soil and rock properties. Once the soil and rock samples have been brought to the laboratory and the offshore results have been processed, tests need to be conducted to develop appropriate data for analysis. Some of these tests focus on the sample's composition and structure and some measure mechanical properties of soil, such as consolidation and shear strength. The purpose of laboratory soil element tests and *in situ* tests is to determine the mechanical properties of soils (i.e. the relationships between stress, strain and time) so that these properties can be used for design or analysis. In addition, bulk samples of soil may be used to conduct physical model tests, which focus on the performance of a structure as a whole, rather than on the basic properties of the soil at an element level.

A comprehensive description of laboratory testing for soil is given by Head (2006), and a brief description of the most common tests conducted for offshore projects is provided in the following pages.

### 3.4.2 Soil composition and classification

Classification testing is an important feature of every site investigation programme in order to establish geologic origin, mineralogy, grain size, plasticity index and so forth for the various stratigraphic layers. Standard tests that should be undertaken in every

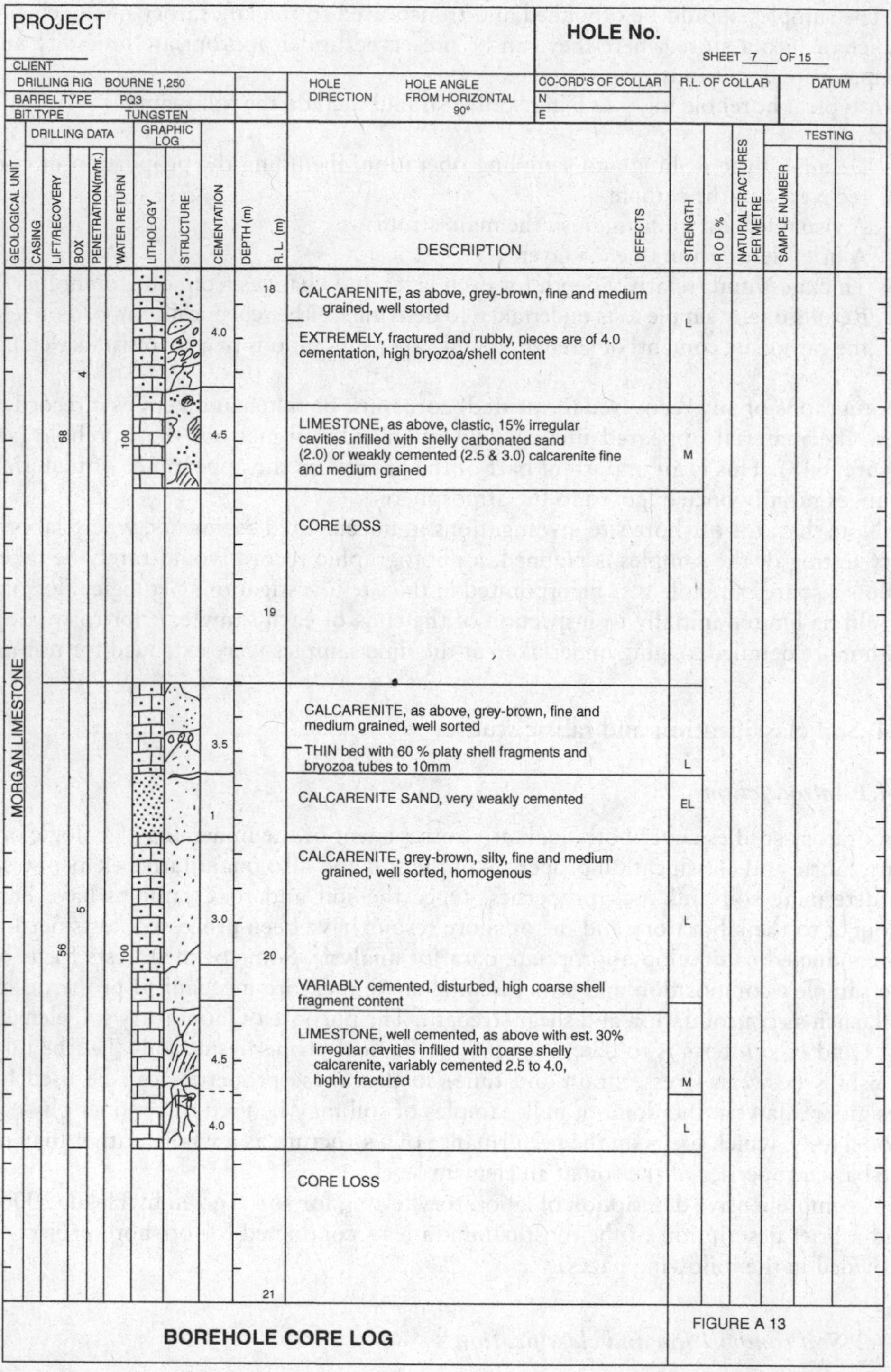

| PROJECT | | | | | | | | HOLE No. | | | |
|---|---|---|---|---|---|---|---|---|---|---|---|
| CLIENT | | | | | | | | SHEET 7 OF 15 | | | |
| DRILLING RIG   BOURNE 1,250 | | | | HOLE DIRECTION | HOLE ANGLE FROM HORIZONTAL 90° | | CO-ORD'S OF COLLAR  N | R.L. OF COLLAR | DATUM | | |
| BARREL TYPE   PQ3 | | | | | | | | | | | |
| BIT TYPE   TUNGSTEN | | | | | | | E | | | | |

Figure shows a detailed borehole core log. Columns: GEOLOGICAL UNIT, CASING, LIFT/RECOVERY, BOX, PENETRATION (m/hr.), WATER RETURN, GRAPHIC LOG (LITHOLOGY, STRUCTURE), CEMENTATION, DEPTH (m), R.L. (m), DESCRIPTION, DEFECTS, STRENGTH, R.O.D %, NATURAL FRACTURES PER METRE, SAMPLE NUMBER, TESTING.

Geological unit: MORGAN LIMESTONE

Descriptions from top:

18 — CALCARENITE, as above, grey-brown, fine and medium grained, well storted  (4.0) — L

EXTREMELY, fractured and rubbly, pieces are of 4.0 cementation, high bryozoa/shell content

LIMESTONE, as above, clastic, 15% irregular cavities infilled with shelly carbonated sand (2.0) or weakly cemented (2.5 & 3.0) calcarenite fine and medium grained  (4.5) — M

CORE LOSS

19

CALCARENITE, as above, grey-brown, fine and medium grained, well sorted  (3.5) — L

THIN bed with 60 % platy shell fragments and bryozoa tubes to 10mm

CALCARENITE SAND, very weakly cemented  (1) — EL

CALCARENITE, grey-brown, silty, fine and medium grained, well sorted, homogenous  (3.0) — L

20

VARIABLY cemented, disturbed, high coarse shell fragment content

LIMESTONE, well cemented, as above with est. 30% irregular cavities infilled with coarse shelly calcarenite, variably cemented 2.5 to 4.0, highly fractured  (4.5) — M  (4.0) — L

CORE LOSS

21

Drilling data values: Penetration 4, Lift/recovery 68, Water return 100; Penetration 5, Lift/recovery 56, Water return 100.

**BOREHOLE CORE LOG**          FIGURE A 13

*Figure 3.47*  Typical borehole log

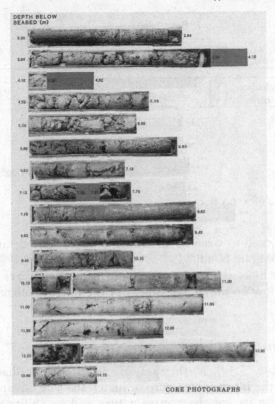

*Figure 3.48* Photograph of cores, showing depth ranges

case include moisture content, dry density, particle specific gravity, particle size distribution, Atterberg limits and carbonate content. More specialised classification tests include X-ray diffraction for assessing particle mineralogy and electron microscopy to enable detailed geological classification. These tests are most commonly undertaken where carbonate sediments are encountered.

The high cost of offshore samples demands minimising any wastage. While the ideal method of logging a profile of samples would be to extrude them and split the cores in half, logging is more often achieved through a combination of x-raying the cores while they are still within the sample tubes, with detailed examination of material at the ends of each tube and between sections retained for testing. Different types of composition and classification tests include:

- X-ray techniques are used first in order to 'visualise' the content of the tubes and evaluate the degree of disturbance.
- X-ray diffraction permits determination of the type and quantity of the various components (mineral) of a soil sample.
- Palaeontology provides data concerning the fossils, burrows, algae and various life forms, which existed in the area.
- Geochronology (e.g. Carbon 14 dating) is used to determine the age of the sample, which is particularly useful in evaluating sedimentation rates and gaps in the chronology, for example due to erosion or mass transport events.

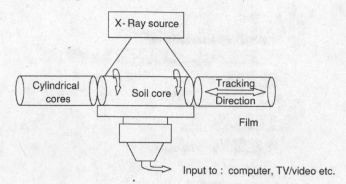

*Figure 3.49* Schematic of X-ray test

- Fabric study is used to visualise at the microscopic scale details of the particles and any grain to grain bonding.
- Index tests such as the carbonate content, particle sizes and plastic and liquid limits allow an initial assessment of the likely mechanical behaviour of the soil.

### 3.4.3 X-ray examination

Radiography (Figure 3.49) is the use of penetrating radiation to produce shadow images of the internal structure of materials. The advantage of radiography (x-raying) lies in its non-destructive nature and its ability to reveal features that sometimes cannot be seen in any other way. It must be kept in mind that there are limits to the application of radiography and that radiography should be used in combination with other techniques of the geological and soils engineering laboratories.

X-raying offshore samples prior to extruding them is now a standard procedure, allowing:

- Identification of size and location of any inclusions, such as shells, in the sample
- Determination of the extent of any sample disturbance
- Determination of sample layering or variation in density along the sample.

The x-rays provide a basis for selection of the best parts of the core sample to be used in the element tests.

The example in Figure 3.50a shows radiographs of an unopened core sample and various sections of the same core. The best detail is obtained from the thinnest slice of soil. However, the result obtained from the unopened core is nonetheless sufficient to make a satisfactory interpretation. The last two figures are the radiograph and corresponding photograph of a slice of the sample (both 10 mm thick). As expected, the radiograph shows more detail than the photograph. A second example (Figure 3.50b) shows an interface between sand and clay. Clear stratification within the upper layer of sand is evident and there is a very distinct interface between the sand and clay. Faint burrows may be observed within the clay layer.

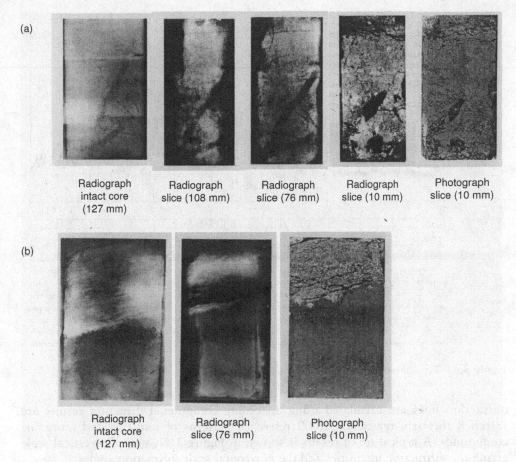

(a)

Radiograph intact core (127 mm)   Radiograph slice (108 mm)   Radiograph slice (76 mm)   Radiograph slice (10 mm)   Photograph slice (10 mm)

(b)

Radiograph intact core (127 mm)   Radiograph slice (76 mm)   Photograph slice (10 mm)

*Figure 3.50* Example radiographs (a) internal sample disturbance and (b) sand over clay sediment interface

### 3.4.4 X-ray diffraction: Mineral composition

X-ray diffraction (XRD) is a quantitative analysis of the crystal structures present in the geological sample under study. When illuminated by an X-ray source, crystalline material will generate X-ray diffraction peaks. The peak positions are described by the crystal unit cell parameters, and the peak intensities are given by the placement of the atoms in the unit cell. The peak widths are a result of two parameters, finite crystallite sizes and micro-stress within the crystallites. As such, the parameters that define a different crystal structure can be simply accessed from an x-ray diffraction pattern. Each mineral type is defined by a characteristic crystal structure, which will give a unique x-ray diffraction pattern, allowing rapid identification of minerals present within rock or soil samples.

Samples are first pulverised with a McCrone micronising mill to homogenise the samples and reduce particle sizes. Randomly orientated specimens are then prepared for quantitative XRD analysis by pressing the pulverised samples onto aluminium specimen holders making disks of depth 2 mm and diameter 27 mm. The samples are then placed into the XRD device for analysis, and positions and intensities of the

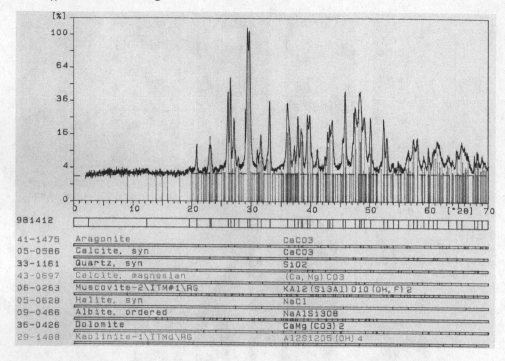

*Figure 3.51* Typical result from XRD analysis

diffraction lines are calculated using an online computer. Diffraction results are searched through, typically, 32,000 reference patterns of minerals and inorganic compounds. A typical set of results is shown in Figure 3.51, with the vertical scale giving the diffraction intensities and the horizontal scale diffraction angles.

### 3.4.5 Palaeontology and geochronology

Figure 3.52 shows a photograph of a section of a soil sample of 10 mm thickness. In this case, the layering and burrows can be seen clearly from the photograph of the sample. Tracking the traces of fossils and previous life forms can be a time consuming process, although x-ray techniques can reduce the processing time for palaeontology studies.

The information from palaeontology studies, which reflect the deposition environment of the sediments, can be supplemented by geochronology studies to evaluate the rates of sediment accumulation and any discontinuities in the profile. The most common approach used to determine the age of a soil specimen is Carbon-14 dating, which is most reliable in the range 10,000–40,000 years old. An alternative technique, called 'varve chronology', may prove superior for younger sediments.

### 3.4.6 Fabric study

Details of the soil fabric can be observed through a scanning electronic microscope (SEM) on an undisturbed sample or on a thin section of an undisturbed sample. Undisturbed samples can be observed directly under an environmental scanning

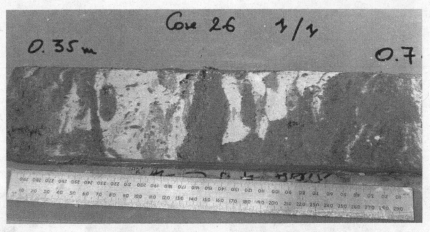

*Figure 3.52* Split soil sample revealing fossils and previous life forms

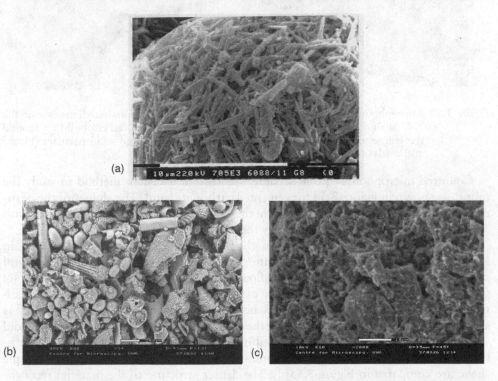

*Figure 3.53* ESEM images of calcium carbonate sediments from the NW Shelf of Australia (a) aragonite crystals, (b) sand from near the Goodwyn gas field and (c) muddy carbonate silt from near the Gorgon gas field

electronic microscope (ESEM). A thin section specimen is prepared by gluing a thin section of an undisturbed soil sample onto a glass plate. Examples of ESEM images of carbonate sediments from the North-West Shelf of Australia are shown in Figure 3.53, ranging from aragonite crystals of calcium carbonate to examples of carbonate sand and muddy silt.

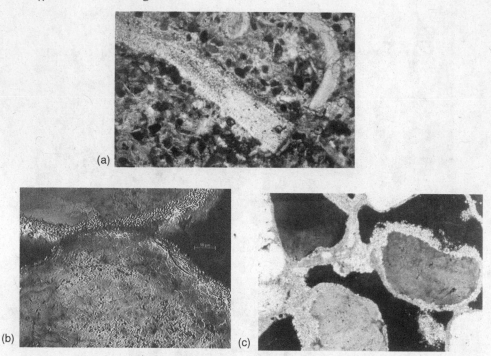

*Figure 3.54* Microphotos from thin sections of cemented calcium carbonate sediments from the NW Shelf of Australia (a) thin section of calcarenite, (b) cement build up around the fringes of two particles and (c) cement fringes around solid particles (black) and voids (grey)

Coloured microphotos of thin sections are also an excellent method to study the fabric, as shown in Figure 3.54 for cemented carbonate sand. The shapes of the particles can be clearly identified. The voids in the sample in Figure 3.54(c) were filled with a blue liquid, in order to clearly identify them.

The influence of the soil fabric on the stress–strain response is illustrated in Figure 3.55. ESEM images of carbonate silt reconstituted from remoulded natural material using (a) water or (b) synthetic flocculent may be compared with an image (c) from undisturbed natural material after extrusion from a sample tube. The reconstituted material resedimented in the presence of a flocculent, the effect of which was subsequently removed by gently heating the consolidated sample, has a more open void structure than the material resedimented in water, reflecting the fabric of the undisturbed material. The stress–strain responses of the three materials, measured in simple shear, are compared in Figure 3.55(d). The denser structure of the material reconstituted with water leads to a dilatant response, with generation of negative excess pore pressure when sheared under undrained conditions. This results in a steadily increasing shear stress, in comparison with more plastic responses from the undisturbed material and that reconstituted with a synthetic flocculent (Mao and Fahey 1999).

### 3.4.7 Particle size classification methods

The particle size distribution (PSD) gives information regarding the size, grading and uniformity of the particles in the soil sample. Several systems have been developed to

*Table 3.8* ASTM soil classification based on particle size.

| Sieve Size (passing) | Particle size (mm) | Soil Classification |
|---|---|---|
| 3 in | 19.0–75.0 | Coarse gravel |
| ¾ in | 4.75–19.0 | Fine gravel |
| #4 | 2.00–4.75 | Coarse sand |
| #10 | 0.425–2.00 | Medium sand |
| #40 | 0.075–0.425 | Fine sand |
| #200 | < 0.075 | Fines (silt + clay) |

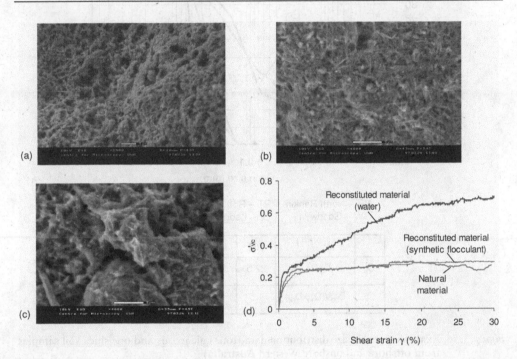

*Figure 3.55* Effect of soil fabric on stress strain response of carbonate silt (a) material reconstituted with water, (b) material reconstituted with synthetic flocculent, (c) natural material and (d) simple shear stress–strain responses

classify soil particles based on their size, and the ASTM system is shown in Table 3.8. Traditional approaches for obtaining the PSD are:

1. Sieving: measures the grain-size distribution of a soil by passing it through a series of sieves. This is achieved by shaking the soil (using a mechanical shaker) through a stack of wire screens with openings of known sizes. The definition of particle diameter is the smallest side dimension of the square holes in the sieve through which the particle will pass. The test is suitable for d > 0.074 mm.
2. Hydrometer: based on Stoke's equation for the velocity of a freely falling sphere; the definition of particle diameter for a hydrometer test is the diameter of a sphere of the same density that falls at the same velocity as the particle in question. The test is used on particles with d < 0.074 mm.

*Table 3.9* ASTM soil classification based on particle size

| Effective size | $D_{10}$ | |
|---|---|---|
| Uniformity coefficient | $C_u = D_{60}/D_{10}$ | $C_u < 3$ for uniformly grading soil $C_u > 5$ for well-graded soil |
| Coefficient of grading | $C_c = D_{30}2/(D_{60} \times D_{10})$ | For most well-graded soils, $0.5 < C_c < 2$ |

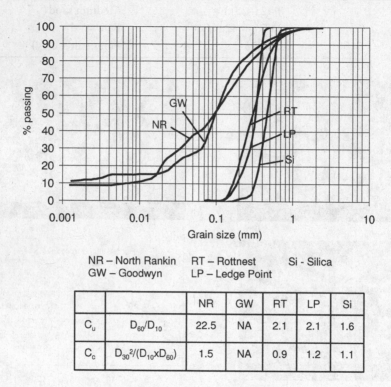

|   |   | NR | GW | RT | LP | Si |
|---|---|---|---|---|---|---|
| $C_u$ | $D_{60}/D_{10}$ | 22.5 | NA | 2.1 | 2.1 | 1.6 |
| $C_c$ | $D_{30}^2/(D_{10} \times D_{60})$ | 1.5 | NA | 0.9 | 1.2 | 1.1 |

NR – North Rankin   RT – Rottnest   Si - Silica
GW – Goodwyn   LP – Ledge Point

*Figure 3.56* Example particle size distribution data (four calcareous and one silica soil samples from offshore and onshore Western Australia)

The test procedure depends on the soil in question. If nearly all the grains are large enough not to pass through square openings of 0.074 mm (No. 200 screen), the sieve analysis is preferable. For soils that are nearly all finer than a No. 200 screen, the hydrometer test is recommended. For silts, silty clays, etc., which have a measurable portion of grains both coarser and finer than a No. 200 sieve, a combined approach is followed, first sieving out the larger particles and then carrying out a hydrometer test.

Over the last decade, the hydrometer test is increasingly being replaced by optical methods based either on laser diffraction, such as the Malvern Mastersizer 2000, or single-particle optical sizing, such as the Accusizer 780 system. These techniques tend to give slightly different results, since there is no absolute definition of particle size for irregularly shaped particles. Their advantages are that smaller volume samples are sufficient, with faster measurement and more consistent results (White 2003).

Real soils almost always contain a variety of particle sizes mixed together, as illustrated in Figure 3.56 for four carbonate soils and a silica sand (Si).

A quantitative analysis of grading curves may be carried out using certain geometric values known as grading characteristics. Often, the median particle size $D_{50}$ is quoted, with values in Figure 3.56 ranging from 0.1 mm to just under 0.5 mm. Three other points on the grading curve are also useful to describe the relative uniformity and the permeability of the soil; these are the characteristic sizes $D_{10}$, $D_{30}$ and $D_{60}$ (the maximum sizes of the smallest 10, 30 and 60 per cent of the sample). Permeability is most closely linked to $D_{10}$, with the Darcy permeability estimated using Hazen's formula:

$$k \approx C_k D_{10}^2 \, \text{m/s} \tag{3.8}$$

with $C_k$ typically 0.01–0.015 for $D_{10}$ expressed in mm.

Two common grading characteristics are the uniformity coefficient $C_u$ and the coefficient of grading $C_c$ (see Table 3.9 and Figure 3.56), both of which become unity for a single-sized soil. (Note, these symbols should not be confused with $c_u$, formerly used for shear strength instead of $s_u$, and $C_c$, the compression index during one-dimensional compression.)

### 3.4.8 Liquid and plastic limits

In offshore conditions, soils may be assumed fully saturated (with the exception of parts of the seabed with active gas seepage), although it is not uncommon for dissolved gas to come out of solution when samples are recovered from deep water. The behaviour of soil is closely related to the void ratio or specific volume, and so it is useful to identify values of these quantities that give similar behaviour for different soils. For sands and other soils of high permeability, these limits are the maximum and minimum void ratios (or minimum and maximum densities). For clays and soils of low permeability, the key values are expressed in terms of water contents.

Atterberg defined the boundaries of four states (liquid, plastic, semi-solid and solid) in terms of 'limits' as follows:

1. Liquid limit: the boundary between the liquid and the plastic states
2. Plastic limit: the boundary between the plastic and the semi-solid states
3. Shrinkage limit: the boundary between the semi-solid and solid states.

The shrinkage limit has most relevance for partially saturated soils, and standard classification tests focus on Atterberg's liquid and plastic limits, which may be interpreted as representing particular values of shear strength.

Methods to determine the liquid and plastic limit are covered in various standards (see Head 2006). The traditional Casagrande cup for determining the liquid limit is increasingly being replaced by the fall cone test (Hansbo 1957; Budhu 1985), where the liquid limit is the water content at which a cone of a given angle and mass (generally 30°, 80 g or 60°, 60 g), released for a position with its point just touches the clay surface, penetrates a particular distance (10–20 mm depending on the standard).

The fall cone test provides a measure of strength and is commonly used to measure the sensitivity of offshore soils, in addition to the liquid limit. The shear strength $s_u$

for a cone of weight Q and penetration depth, h, may be expressed as (Hansbo 1957),

$$s_u = K\frac{Q}{h^2}$$ (3.9)

where K is a constant that ranges from around 1.33 for a 30° cone down to 0.3 for a 60° cone. Koumoto and Houlsby (2001) provide an excellent analysis and discussion of the fall cone test, and verify that the strength at the liquid limit is in the region of 2 kPa. They also point out that the fall cone test involves extremely high strain rates, in the range 1–10 s$^{-1}$.

The plastic limit is defined as the water content at which the soil begins to crumble when rolled (by hand) into threads of a specific size (about 3 mm in diameter). By the nature of the test, it is quite operator dependent, resulting in significant variability for a given soil when tested in different laboratories. It has been suggested that it would be preferable to replace the traditional plastic limit test by one that was more objective, and linked to a certain shear strength. A (static) conical indentation test is an obvious candidate (Stone and Phan 1995).

The liquid and plastic limit tests are both forms of strength tests (increasingly so with the move towards cone-based methods of assessment) with a strength ratio of ~100 between them (Wroth and Wood 1978). The difference in water content between the two limits, which notionally span the range within which the soil behaves plastically, is referred to as the plasticity index $I_p$ defined as

$$I_p = w_L - w_P$$ (3.10)

Critical state soil mechanics provides a framework that links the shear strength, and the dilatational or contractive response of soils when sheared, to the water content and effective stress level. Interpreting the plasticity index as the change in water content for a given ratio of shear strength (~100) allows the compressibility to be expressed in terms of the plasticity index (Wroth and Wood 1978).

Soil may be classified according to where its plasticity index and liquid limit falls within the plasticity chart shown in Figure 3.57, where what is referred to as the A-line divides soil broadly into clays (C) and silts (M), of varying degrees of plasticity. With some exceptions, the friction angle of clay soils shows a decreasing trend with increasing plasticity index. A notable exception is Mexico City clay, which has extremely high plasticity but contains highly frictional diatoms. Also many offshore West African soils show similarly high plasticity indices (~100) and high friction angles (35–40°) when subjected to triaxial or simple shear testing. However, the latter soil shows much lower residual friction angles (typically 10–20°).

Just as relative density relates the density of a coarse-grained soil to the minimum and maximum extremes, so the liquidity index, LI provides a linear interpolation between (and outside) the liquid and plastic limits, taking values of 1 and 0, respectively, and the two limits. Near surface soils, or material involved in a submarine debris flow, may well have a liquidity index exceeding unity, while glaciated clays from the North Sea will typically have negative liquidity indices.

The mechanical response of soils is affected by factors related to the mineralogy, deposition environment and age in addition to its water content, plasticity

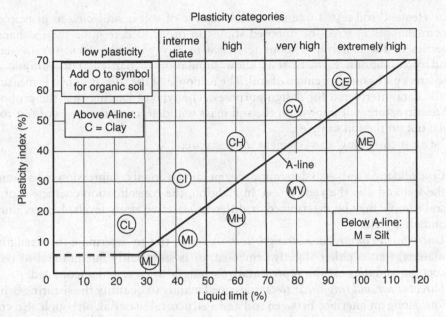

*Figure 3.57* Plasticity chart for classification of soils

and level of confining effective stress. From a design perspective, the response may be assessed through different laboratory tests as described in the following sections. More detailed treatment of the soil chemical and physical properties at the particle level is provided in texts such as Mitchell (1993) and Santamarina *et al.* (2001).

## 3.5 Laboratory element tests

Soil samples retrieved offshore may in the first instance be subjected to simple laboratory tests, carried out offshore, to assess moisture content, unit weight and shear strength. For major projects where the majority of laboratory testing is carried out onshore, these tests would mostly be limited to material at the end of sample tubes. For other applications, such as site assessment for temporary foundations of a mobile drilling (or jack-up) unit, samples would be extruded offshore and estimates made of the undrained shear strength by means of simple, relatively crude, tests such as torvane, pocket penetrometer and miniature (manual or motorised) vane shear tests, and also fall cone tests. Results from the first two tests are very subjective to the technician carrying out the tests, whereas the latter two are more objective but still tend to give somewhat variably, and thus approximate, estimates of the shear strength. Estimates of jack-up foundation (spudcan) penetration may rely entirely on the results of this type of testing, often without even the benefit of an *in situ* penetrometer test, but this is not a recommended practice. In this section, the more common soil element soil tests carried out (primarily) in an onshore laboratory following the field site investigation will be discussed. These more sophisticated tests, together with results from *in situ* testing, form the basis of estimating the engineering properties of the soil for use in design.

An 'element' soil test is one in which a sample of soil is subjected in principle to uniform changes in stress or imposed strains in order to determine the mechanical properties of the material without reference to any particular structure or set of boundary conditions. Element tests allow simulation of particular monotonic and cyclic stress paths on a specimen of soil. The response of the soil specimen is measured and the data interpreted for design purposes. The type of loading that the proposed offshore structure will impose on the soil mass will determine the types of test to be carried out on the soil samples.

The most commonly used element soil tests are:

- *Consolidation tests* – to determine the one-dimensional compression behaviour of the soil and also the yield stress. In addition, the consolidation coefficient of the soil sample may be determined at different effective stress levels, before, during and after yield.
- *Unconfined compressive strength test* – a quick test to determine the strength of material that is either slightly cemented, or is sufficiently fine-grained to retain suction, and thus exhibit representative strength when tested unconfined.
- *Direct shear and ring shear tests* – designed mainly to quantify the shearing behaviour along an interface between soil and a structural material, although also commonly used to obtain friction and dilation angles for shearing along a soil–soil interface. A more sophisticated form of the test allows the normal stress to be varied in accordance with any dilation or compression of the soil during shearing.
- *Triaxial test* – the most widely used laboratory test for soil, allowing application of confining effective stresses that reflect *in situ* conditions and then shearing the sample following a specified total stress path, with or without allowing drainage from the sample.
- *Simple shear test* – a form of hybrid between a direct shear test and a triaxial test, attempting to invoke uniform shear strain within a low profile sample under confining stresses. This test is commonly referred to as a 'direct simple shear test', but the 'direct' has been omitted here in order to avoid confusion with the direct shear test.

The foregoing tests are described in greater detail in the following sections, with particular reference to application in offshore design. More detailed descriptions of these and other laboratory tests may be found in Head (2006).

### 3.5.1 Consolidation tests

There are two basic types of consolidometer in current use: the standard oedometer (Figure 3.58) used by many laboratories, the loads generally being imposed by a lever device; and the 'Rowe cell' (Figure 3.59) in which the vertical load is imposed by a pressurised fluid acting through a membrane above the sample. Both consolidometers are used to investigate the stress–strain behaviour of low permeability soils (silts or clays) in one-dimensional compression (or swelling). Consolidation is permitted by two-way drainage through porous discs sandwiching the sample.

In a standard oedometer test, the sample is usually 75 mm in diameter and 20 mm in height, while the Rowe cell is considerably larger being typically 250 mm in diameter but can be up to 1,000 mm in diameter. In both cases, the sample is set in a steel confining ring preventing lateral strain. The larger sample size is beneficial in capturing

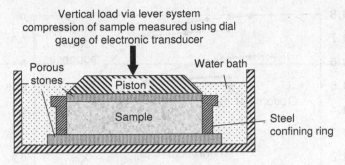

*Figure 3.58* Standard oedometer apparatus

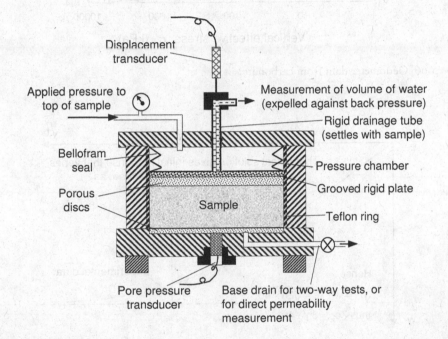

*Figure 3.59* Rowe cell oedometer

macro-features in the sample. Additional advantages of the apparatus include control and measurement of pore pressures, application of back-pressure (similar to the triaxial apparatus) and various combinations of drainage control for determining the permeability of soils by flow tests.

In a basic consolidation test, the time-settlement response is monitored for a series of incremental loads, or vertical stresses, from which the one-dimensional stress–strain response of the soil may be determined (see Figure 3.60). This is usually interpreted in terms of compression and swelling (or reloading) indices, $C_c$ and $C_s$, denoting the change in void ratio with each log-cycle increase in vertical effective stress. The soil stiffness may also be expressed in terms of a one-dimensional modulus $E_{1-D} = \Delta\sigma'_v/\Delta\varepsilon_v$ or compressibility $m_v = 1/E_{1-D}$. The data in Figure 3.60 have been fitted with the theoretical model of Pestana and Whittle (1995).

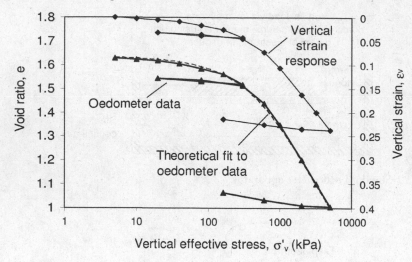

*Figure 3.60* Oedometer data from carbonate silt

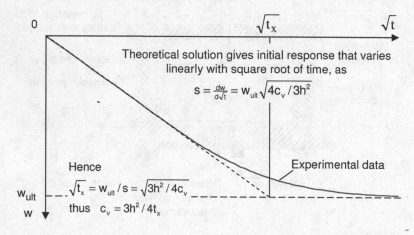

*Figure 3.61* Deduction of consolidation coefficient from root time compression response

A further important parameter is the apparent preconsolidation stress, or yield stress $\sigma'_{vy}$ which is often associated with a local minimum stiffness of the soil and the transition point to being 'normally consolidated'. The overconsolidation (or yield stress) ratio is then obtained as OCR = $\sigma'_{vy}/\sigma'_{v0}$ where $\sigma'_{v0}$ is the *in situ* vertical effective stress for the soil sample in question.

The rate at which consolidation occurs is defined by the coefficient of consolidation $c_v$ expressed most conveniently for clay soils in units of $m^2$/year (or $mm^2$/s or $m^2$/s), with typical values that can range from 1 to 10 $m^2$/year for a soft clay at shallow depth to values in excess of 10,000 $m^2$/year for silty soil. The theoretical one-dimensional consolidation solution results in a linear variation of settlement with the square root of time, with a gradient s as given in Figure 3.61. The consolidation

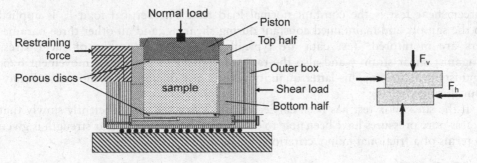

*Figure 3.62* Direct shear apparatus

coefficient may be deduced from the time intercept $t_x$, at which the extended linear fit crosses the horizontal line for ultimate settlement $w_{ult}$, for that load increment, with

$$c_v = \frac{3h^2}{4t_x} \qquad\qquad (3.11)$$

where h is the maximum drainage path, usually half the sample height, assuming double drainage.

A variation on the conventional oedometer test is a constant rate of strain test, where drainage is only allowed at the upper surface. The sample is strained at a constant rate, with measurement of the excess pore pressure at the bottom face, with the strain rate chosen to give an excess pore pressure around 10 per cent of the current vertical stress. This form of test has the advantage that it allows a continuous measure of consolidation coefficient as a function of the average (current) vertical effective stress within the sample. The test may be carried out in modified versions of a standard oedometer or Rowe cell, or preferably in purpose designed apparatus that allows a back-pressure to be applied to the sample.

### 3.5.2 *Direct shear test*

The direct shear test, also referred to as a shear-box test, is one of the simplest and most commonly used apparatus to investigate the shear strength and shear stress–strain behaviour of a soil sample. A low aspect ratio cylindrical or rectangular soil sample (typically 60 mm · 60 mm in plan and 20 mm thick) is placed inside a split box (Figure 3.62). A vertical stress is applied by dead-weights on a hanger and shearing is achieved by a motor driven ram acting on the bottom half of the box. The lateral force $F_h$ required to hold the top half of the box stationary is measured by a load cell or proving ring.

The sliding motion forces a shear plane through the sample. Drainage of the pore water into and out of the narrow shear zone cannot easily be prevented during shearing. If an undrained soil strength is required then the test may be carried out rapidly. However, if drainage is to be guaranteed, shearing must take place sufficiently slowly to allow dissipation of pore water, which for clays may mean a test takes several days.

The vertical and horizontal loads, $F_v$ and $F_h$, along with the vertical and horizontal $\delta_v$ and $\delta_h$ displacements are monitored during shearing. The most common form of direct shear test is the constant normal load test: The vertical load $F_v$ is applied to the sample and maintained constant during shearing, and all other three parameters are monitored. Test data are typically presented as graphs of shear stress $\tau$ against shear strain $\gamma$ and also the ratio of vertical to horizontal movement $\delta_v/\delta_h$ against shear strain. This latter quantity reflects any dilatant or compressive behaviour during shearing.

If the shear box test has been carried out on a dry soil, or sufficiently slowly that excess pore pressures have been able to dissipate, the mobilised shear strength is given in terms of a frictional failure criterion:

$$\tau = \sigma'_v \tan\phi' \tag{3.12}$$

where $\sigma'$ is the applied normal stress ($F_v$ divided by the sample cross-sectional area A) and $\tau$ is the measured $F_h/A$. Peak and residual values of $\phi'$ may therefore be determined.

If a test on a low permeability soil has been carried out rapidly enough to prevent any pore pressure dissipation, the test may be considered 'undrained' and the soil strength may be determined according to:

$$s_u = \tau_{max} = F_h/A \tag{3.13}$$

Other tests that can be performed in the direct shear apparatus include:

(a) Constant volume: in this case, the height of the sample is maintained constant ($\delta_v = 0$) during shearing.
(b) Constant normal stiffness: the vertical load ($F_v$) is adjusted in order to maintain the normal stiffness ($K_n = \Delta\sigma_v/\Delta h$) constant during shearing.

The latter form of test has proved very useful for assessing the monotonic and cyclic shear response of soil for the design of axially loaded piles, particularly drilled and grouted piles where a steel tubular is grouted into a hole drilled in the soil (Johnston *et al.* 1987). Dilation or compression within the shearing zone can have a significant effect on the confining stress, and hence on the frictional strength. The stiffness $K_n$ should be chosen to represent the cavity expansion stiffness around the pile, so $K_n = 4G/d_{pile}$ where G is the shear modulus of the soil and $d_{pile}$ the pile diameter.

Interface tests can also be carried out in a direct shear apparatus, by forming the sample from two different materials (one in each half), taking care that the shear plane is located along the interface between the two materials.

One of the major drawbacks with the direct shear apparatus is that the state of stress and strain within the sample is non-uniform, particularly near the ends of the box and at relatively large strains. As such, while it provides data on the peak and residual friction angles mobilised, it does not provide useful information on the stiffness of the soil.

Some of the limitations of non-uniformity may be overcome in a form of the test whereby an annular soil sample is sheared by twisting about an axis through the centre of the annulus. The so-called 'ring shear' apparatus has proved of particular value in determining residual friction angles for clays, for application to re-activated landslides and, in the offshore world, for assessing the shaft capacity of driven piles (Lupini *et al.* 1981).

Instruments for
local strain
measurement

General view of equipment

*Figure 3.63* Triaxial test apparatus and sample with local strain measuring devices

### 3.5.3 Unconfined compressive strength test

Unconfined compression tests are typically used for the measurement of the remoulded strength of clays and to test soft rocks and very stiff or hard soils. During an unconfined compression test, a cylindrical soil sample is subjected to a steadily increasing axial load until failure occurs. The axial load and the corresponding variation in sample length $\Delta h$ are the only numerical data that are recorded. A stress–strain curve can be obtained from the data, relating axial stress given by $\sigma_a = F_v/A$, to the strain $\varepsilon_a = \Delta h/h_{sample}$.

The unconfined compressive strength, UCS, is the compression strength at failure of a specimen subjected to a compression test without lateral confinement. Note that the shear strength at failure will be half of the UCS, since the lateral confining stress is zero.

### 3.5.4 Triaxial test

The triaxial test enables a variety of stress or strain controlled tests to be carried out and is suitable for all types of soil; as a result, it is the most widely used shear strength test (Figure 3.63).

A cylindrical soil sample (traditionally 38 mm in diameter and 76 mm long, but in the offshore industry more commonly 76 mm diameter and 150 mm long, or 100 mm diameter and 200 mm long) is placed on a pedestal and enclosed between rigid end caps inside a thin rubber membrane. Rubber 'O-rings' are fitted over the membrane to provide a seal (Figure 3.64). The cell is closed and the specimen is subjected to a stress by filling the cell with water and raising the pressure to a prescribed value. A back pressure is generally used in order to ensure full saturation of the sample, so that the effective confining stress is the difference between the cell pressure and back pressure. The cell pressure and axial load (via a loading ram) can be varied to apply different loading paths.

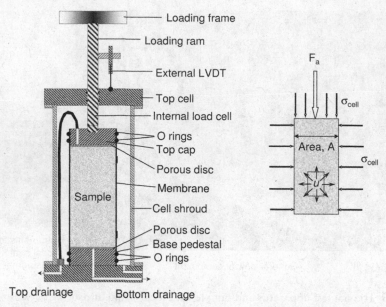

*Figure 3.64* Schematic of triaxial test apparatus and stress conditions

The first phase of the test consists of bringing the sample to an equilibrium effective stress state, either under isotropic stresses (with no additional load applied via the axial loading ram) or under anisotropic stresses. In the latter case, compressive or tensile load must be applied by the loading ram during the consolidation phase, in order to arrive at $K_0$ (= $\sigma'_h/\sigma'_v$) conditions that are less than or greater than unity.

The most common test is the conventional triaxial compression test (TC) in which the cell pressure is held constant while the loading ram is displaced downwards to increase the axial load. The test continues until the axial strain has reached a specified limit (typically 30 per cent). Alternatively, the axial ram can be withdrawn, applying a tensile component of load to the sample, while the cell pressure is increased. This is known as a conventional triaxial extension test (TE). During shearing, drainage taps may be left open or closed to allow (drained test) or prevent (undrained test) volume change of the sample.

As described earlier, monotonic triaxial compression and tension tests are essentially displacement controlled. For offshore applications, cyclic tests are also required and these tend to be stress controlled, applying cycles of axial stress between pre-defined limits until the sample accumulates the specified maximum value of strain (typically 15 or 20 per cent).

Throughout monotonic and cyclic tests, the variation in sample height $\Delta h$, axial load $F_a$, cell pressure $\sigma_c$ and pore pressure u are monitored during loading. These are converted to axial and horizontal effective stresses, excess pore pressure and axial (or shear) strain by dividing by appropriate quantities such as the nominal cross-sectional area (allowing for lateral straining) and the initial height of the sample. Triaxial test data is conventionally expressed on graphs of deviator stress q or stress

ratio q/p′ against axial or shear strain $\varepsilon_a$, $\gamma$; volumetric strain $\varepsilon_{vol}$ or change in pore pressure $\Delta u$ against axial or shear strain $\varepsilon_a$, $\gamma$, and void ratio, or specific volume e, v against shear strain $\gamma$ or the logarithm of p, or p′.
where

| | |
|---|---|
| Deviator stress: | $q = \sigma_a - \sigma_c = F_a/A$ |
| Average or mean stress: | $p = (\sigma_a + 2\sigma_c)/3$ |
| Total axial stress: | $\sigma_a = \sigma_c + F_a/A$ |
| Axial strain: | $\varepsilon_a = \Delta h/H_o$ ($H_o$ is initial height of sample) |
| Pore pressure variation: | $\Delta u = u - u_o$ ($u_o$ is initial pore pressure) |
| Volumetric strain: | $\varepsilon_{vol} = \varepsilon_a + 2\varepsilon_r$ |
| Shear strain: | $\gamma = 0.5(3\varepsilon_a - \varepsilon_{vol})$ |
| Void ratio: | $e = \omega_f G_s$ (water content × specific gravity of solids) |

The end point of a triaxial tests represents the critical state where the soil continues to deform at constant stress ratio q/p′ and constant specific volume (v − 1 + e). A typical presentation of results from a monotonic triaxial compression test on a sample that was initially overconsolidated to a ratio of about four is shown in Figure 3.65.

If the drainage taps are closed during a triaxial test (i.e. an undrained test) the sample deforms under constant volume with no change in void ratio. The undrained shear strength of the sample is given by half the maximum deviatoric stress, $s_u = q_{max}/2$, equal to the radius of the maximum diameter Mohr circle for stress. The maximum deviatoric stress and undrained shear strength will be a function of the void ratio of the sample at the start of the shear phase of the test – which is a function of the preconsolidation pressure and stress history. In addition to the undrained shear strength, the maximum and critical state friction angles, $\phi_{max}$ and $\phi_{cs}$ may also be evaluated from an undrained test.

The shear strength, and thus maximum deviator stress, is a function of the void ratio of the sample; therefore in drained tests the shear strength is defined by a maximum stress ratio criterion, to give a frictional strength $\phi_{max}$. A critical state friction angle $\phi_{cs}$ may also be measured.

Triaxial tests also allow measurement of the modulus of the sample, either directly from the external measurements of stress and deformation, or more precisely through internal measurements. An example of instrumentation to measure strains within the triaxial cell is shown in Figure 3.63. Such instrumentation, which avoids errors associated with local deformation near the ends of the sample, allows evaluation of the small strain modulus $G_0$ (equal to one-third of the small strain Young's modulus $E_0$ for undrained conditions). An alternative approach for the small strain modulus is to measure the shear wave velocity through the sample. This can be accomplished by using so-called bender elements (piezoelectric wafers that can transmit and receive shear waves) situated in the top and bottom end caps of the triaxial equipment.

### 3.5.5 Simple shear test

The simple shear apparatus is a derivative of the direct shear apparatus, with the aim of providing more uniform shear strain within the sample by allowing the sides to rotate. The original versions were square in plan, but modern equipment is now mostly based on cylindrical samples. Two variants are in common use. One of these

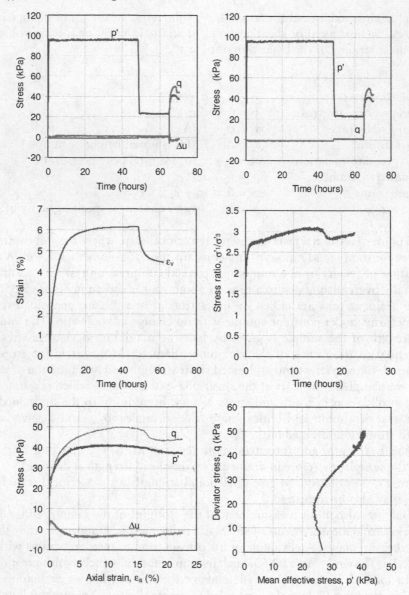

*Figure 3.65* Typical presentation of results from triaxial test

follows a similar approach to a triaxial apparatus, with the sample contained by a normal rubber membrane and subjected to cell pressure and with similar connections for pore pressure etc., but with the base of the sample mounted on a platform that can slide horizontally (Figure 3.66). The alternative approach, adopted in the Norwegian Geonor simple shear apparatus, is to contain the sample in a spiral wire-reinforced membrane to provide lateral stress and prevent sample distortion, but without applying an external cell pressure (see central schematic in Figure 3.67).

Simple shear tests are generally performed on cylindrical soil samples, typically 50 mm or 75 mm diameter and 20–30 mm tall. The base pedestal on which the sample

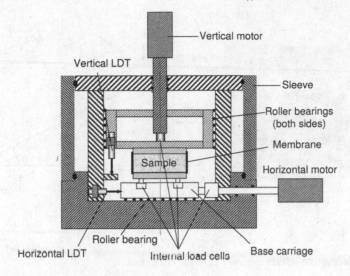

*Figure 3.66* Simple shear apparatus

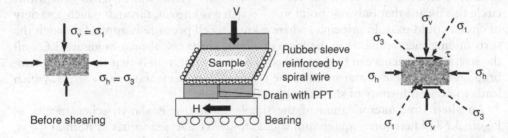

*Figure 3.67* Schematic of simple shear test apparatus and stress conditions

is mounted rests on a carriage that can move laterally on a linear bearing. For the apparatus shown in Figure 3.66, the sample is consolidated in an identical way to a triaxial sample, with cell pressure and back pressure applied and either isotropic or anisotropic stress conditions. For the Geonor apparatus, the sample is consolidated one-dimensionally (because of the wire-reinforced membrane) under an applied vertical stress.

Vertical load $F_v$ is applied via a loading ram. A lateral loading ram connected to the base carriage allows horizontal force to be applied to the sample $F_h$. Both the vertical $\delta_v$ and horizontal $\delta_h$ displacements are measured during loading of the sample. The main parameters that can be deduced are:

| | |
|---|---|
| Shear stress: | $\tau = F_h/A$ |
| Vertical stress: | $\sigma_v = F_v/A$ |
| Pore pressure variation: | $\Delta u = u - u_o$ ($u_o$ = initial pore pressure) |
| Axial strain: | $\varepsilon_a = \delta_v/H_o$ ($H_o$ = initial height of sample) |
| Shear strain: | $\gamma = \delta_h/H_o$ |

A drawback of the Geonor-style of simple shear test (also true of the direct shear test) is that only the shear stresses and normal stresses on the horizontal planes are

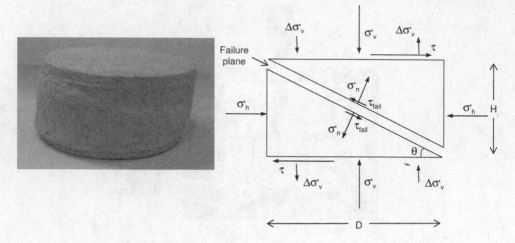

*Figure 3.68* Alternative interpretation of simple shear test

measured and the stresses on the vertical planes are unknown. In terms of a Mohr circle this means that only one point in $\tau - \sigma$ space is known, through which a variety of circles could pass. In principle, where a known cell pressure is applied through the surrounding membrane, as in the form of simple shear test shown in Figure 3.66, all the stress components can be determined. The only limitation is that the lateral membrane is not capable of transmitting the required complementary shear stress, which leads to non-uniformity of stresses within the sample.

An alternative interpretation of the simple shear test is shown schematically in Figure 3.68, where force equilibrium is used to derive average values of normal stress, $\sigma'_n$, and shear stress, $\tau_{fail}$, on a putative failure surface inclined at an angle, $\theta$, to the horizontal (Joer *et al.* 2010). Typical results of a simple shear test are shown in Figure 3.69. The soil properties and consolidation history are identical to those for the example triaxial compression test shown in Figure 3.65, which gave a measured shear strength of 25 kPa. This compares with a maximum shear stress ($\tau_{fail}$ or $\tau_{xy}$) of 18 kPa from the simple shear test. The ratio of shear strengths is typical for lightly overconsolidated clays, with $s_{uss}/s_{uc}$ of 0.7–0.8. A consistent value of friction angle, with $\phi \sim 30°$, is obtained from the q/p' ratio in Figure 3.65, and the t/s' and $\tau_{fail}/\sigma'_n$ ratios in Figure 3.69.

### 3.5.6 *Cyclic testing*

Particular attention must be paid to the effects of cyclic loading in the design of offshore structures, and the bulk of most laboratory testing programs is focused on the cyclic stress–strain response of the soil. Since a number of samples in close depth proximity are required to establish a consistent set of test data, simple shear testing tends to be favoured over triaxial testing in order to develop the cyclic fatigue diagram.

Cyclic shearing tests may be performed at different levels of mean and cyclic shear stress. Pure 2-way cyclic loading corresponds to zero mean shear stress, while pure 1-way cyclic loading corresponds to equal mean shear stress and cyclic (amplitude) shear stress, so that the shear stress ranges between 0 and twice the cyclic amplitude. Eventually, data such as the gradual development of shear strain, or of excess pore

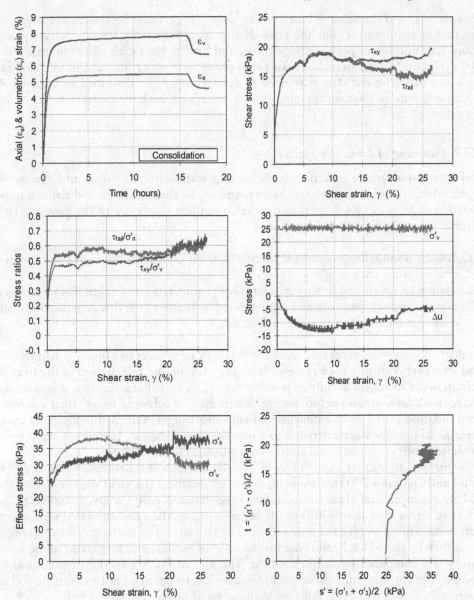

*Figure 3.69* Typical presentation of results from simple shear test

pressure, are plotted in contour diagrams, indicating the number of loading cycles at a given level of cyclic shear stress (expressed either as $\tau_{cyc}/\sigma'_{vc}$, or as $\tau_{cyc}/s_u$) for a certain degree of damage to occur.

Equivalent amounts of damage may be incurred from a large number of cycles at low shear stress levels, or a small number of cycles at high shear stress levels. Miner's Rule is often used to assess what equivalent number of cycles at a high shear stress level (such as the maximum subjected during a design storm) would give the same degree of damage as a full sequence of cyclic loading at different levels. The cyclic stress levels, and the relative bias in terms of $\tau_{cyc}/\tau_{mean}$, must be chosen according to the design requirement, but the tests should span failure in 10–20 cycles (typical equivalent number of peak loading cycles), and further tests with failure in, say, 100–200 cycles and failure in no less than 1,000 cycles. For major projects, more complete contours of cyclic response may need to be developed, as described in Chapter 4, which will require a significant number of tests varying both the cyclic stress level and the relative bias.

### 3.5.7 *Planning of laboratory tests*

Soil response depends critically on the current state (effective stress and density or water content) and previous stress history, and so the aim is that the soil element to be tested replicates as closely as possible the conditions that exist in the ground. Two distinct types of test may be identified:

1. Testing nominally undisturbed specimens from core samples, especially for fine-grained soils, such as silts and clays.
2. Testing reconstituted or reconsolidated soil, for instance soil taken from a dredged sample or disturbed core and reconstituted into a test cell; this is particularly relevant for coarse-grained soils, such as sands.

Undisturbed samples recovered from the field will be subject to sampling disturbance and stress relief during their removal from the ground and installation in the triaxial cell. In order to regain as fully as possible the *in situ* soil characteristics, it is essential to reconsolidate samples before testing, which may be achieved using either isotropic consolidation (CIU) or one-dimensional consolidation (CAU). Although many commercial laboratories are restricted to isotropic consolidation, laboratory testing for offshore projects almost always adopts anisotropic consolidation.

A further alternative is to follow a so-called SHANSEP procedure (Ladd 1991, Ladd and DeGroot 2003). This involves first establishing the yield stress ratio for the soil in question, and then consolidating (one-dimensionally) a given sample to an effective stress level that is at least 20 per cent higher than the estimated yield stress before unloading to give the required yield stress ratio. The measured shear strength must then be scaled back, allowing for the ratio of vertical effective stress in the laboratory test compared to that in the field. The aim of the SHANSEP procedure is to compensate for sampling disturbance, but care has to be taken not to take the soil to a stress level that will cause collapse of the internal fabric (Burland 1990).

The type of tests to be carried out in the laboratory depend on the type of loading the soil mass is going to be subjected to, either during construction or after completion of the structure. Two examples are given here: one for a shallow foundation (gravity base structure – GBS) and one for a pile foundation.

Example 1: Soil conditions under a GBS

For a complete investigation for a gravity structure, various types of strength tests are required, because the modes of shearing in different areas under the platform are different. This is illustrated in Figure 3.70. It can be seen that, in some areas, the simple shear test (SS) is the best test to replicate the shearing mode, whereas in other areas either triaxial compression (TC) or triaxial extension (TE) tests are appropriate.

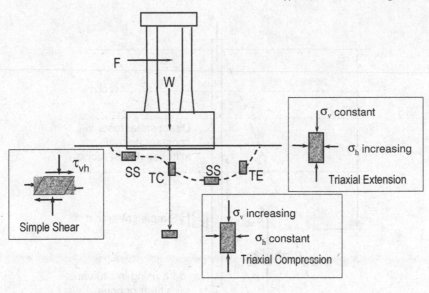

*Figure 3.70* Stress conditions under a gravity base structure

For most soils, the simple shear strength $s_{uss}$ corresponds closely with the average of all three strengths (triaxial compression $s_{uc}$ simple shear $s_{uss}$ and triaxial extension $s_{ue}$). The triaxial extension strength is usually the lowest strength, with typical strength ratios (relative to that in triaxial compression) of 0.5–0.8. For this platform investigation, it is decided to perform several triaxial tests, simple shear tests and oedometer tests. The strength testing will include tests to replicate the *in situ* effective stress conditions, but also some tests at a stress state that represents the state after placing the GBS (e.g. *in situ* + 100 kPa). Both monotonic and cyclic tests will be undertaken, with sufficient of the latter to establish relevant cyclic contour diagrams.

The amount and rate of settlement of the GBS after installation can be assessed by use of data from oedometer tests carried out on undisturbed samples recovered from different depths. For a given (average) applied pressure, elastic theory is first used to estimate the distribution of vertical stress changes at different depths down the centreline of the structure. The corresponding strains at each level are then estimated from the oedometer results and then integrated to provide an overall settlement. The yield stress profile for the soil is important, as the yield stress marks the transition from high stiffness (low strains for a given stress change) to low stiffness.

### Example 2: Axial capacity of a pile

The shearing mode in the case of a pile foundation is shown in Figure 3.71. The axial capacity in compression is obtained by integrating the limiting shaft friction $f_s$ and end-bearing $q_b$ over the shaft and base area, as detailed in Chapter 5.

The shaft friction $f_s$ between the pile and soil is a function of the local (normal) effective stress and the interface friction angle between pile and soil. The latter may be measured in direct shear tests, preferably carried out under conditions of constant normal stiffness (CNS) as discussed previously, or simple shear tests. The choice of test depends to some extent on the pile construction method. CNS testing is generally

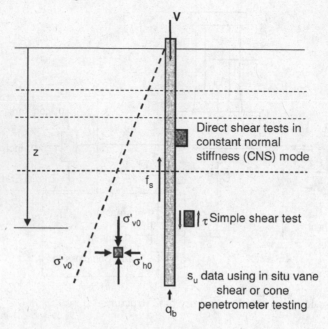

*Figure 3.71* Shear modes along an axially loaded pile

preferred for drilled and grouted (i.e. cast *in situ*) piles, while a combination of simple shear (to evaluate stress paths and undrained strength) and ring shear tests (to establish peak and residual interface friction angles at the pile–soil interface) are appropriate for driven piles.

The end bearing $q_b$ also depends on the pile construction method. Limiting values may be estimated either indirectly from strength measurements (e.g. estimates of $s_u$ from *in situ* or laboratory testing) or more directly from cone penetration tests. The cone resistance, $q_c$ or $q_{net}$, will need to be factored down to allow for the pile displacement required to mobilise base resistance, as detailed in Chapter 5 for different types of pile.

## 3.6 Physical model tests

### 3.6.1 Introduction

The purpose of laboratory element tests and *in situ* tests is to determine the basic mechanical properties of soil (i.e. the relationships between stress, strain and time) so that these properties can be used for design or analysis. Model tests, on the other hand, focus on the overall performance of the structure rather than on the basic properties of the soil. In many cases, the determination of the various engineering characteristics of a site does not enable the design problem to be solved with an appropriate degree of accuracy or certainty. In such situations, it may be appropriate to perform physical model tests, particularly in cases such as:

* Where existing soil models are insufficiently developed to provide good predictions
* Where the boundary conditions make the analysis prone to error

- Where the analysis requires the simultaneous solution of a number of problems (e.g. where pore pressure generation and consolidation and soil deformation mechanisms must be considered simultaneously for analysis of seismic or cyclic loading).

The prediction of pile capacity is an example where existing soil models do not provide the required answer. Pile foundations are widely used to support offshore structures. The capacity of the pile (bearing and shaft) can be determined using model tests. Rod shear and calibration chamber tests are commonly used to determine the capacity of grouted piles. Centrifuge model testing is another tool available to the geotechnical engineer, allowing results of model tests conducted at elevated g-levels to be extrapolated to a prototype situation.

It is important to understand that although model tests can be used to understand the interaction between soil and structure under given loading conditions, they are limited by the extent to which the exact soil conditions can be reproduced in the laboratory without some simplification, and also may suffer from certain scale effects. Nevertheless, the data obtained from model tests may be extremely useful in assessing aspects such as the effect of cyclic loading and provide data that may be used to verify a particular design procedure. If a design procedure cannot match data obtained under well-controlled laboratory conditions, its reliability for prototype design must be questioned.

### 3.6.2 Rod shear test

A rod shear test is essentially a model grouted pile test, with the grouted pile contained in a cylindrical sample (Figure 3.72). The apparatus is similar to a triaxial apparatus with the exception that the cell pressure acts on the lateral boundary only. Undisturbed samples can be used directly in this test. The lateral pressure is applied to the sample, via a tap system mounted on the side of the chamber. The chamber is filled with water and the pressure applied to the sample. During the test, the flow of water into or out of the cell is prevented by closing the tap, and the lateral pressure is monitored to assess any tendency for dilation or compaction of the sample as the rod is sheared. Axial load (and hence average shaft friction) and axial displacement are measured during monotonic and cyclic loading tests, with the aim to develop design guidelines for the prototype grouted pile. Test data are often interpreted in conjunction with data from CNS tests.

### 3.6.3 Calibration chamber test

Theories and approaches for interpreting *in situ* tests such as the cone penetration test in terms of soil parameters need to be verified by experimental data. The interpretation methods are also often completely or partly based on correlations with experimental data. For cohesive material, the basic soil characteristics (for instance, strength and deformation parameters) can be established from laboratory tests on undisturbed samples. For sandy soils, the problems of sample disturbance normally prevents this approach from being used. Calibration chamber testing has therefore been developed as the most efficient means of verifying interpretation theories and establishing engineering correlations for sands (Jamiolkowski *et al.* 2003). Calibration chamber

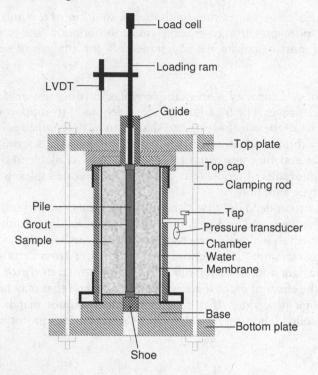

*Figure 3.72* Rod shear apparatus

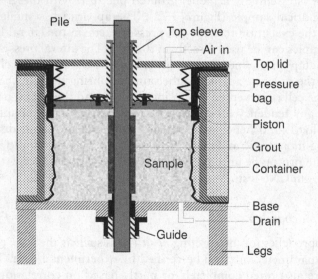

*Figure 3.73* Calibration chamber test for grouted piles

testing can also be used for model tests such as model piles, suction caissons and so forth (Figure 3.73).

The apparatus shown, which is 400 mm in diameter and 800 mm tall, permits application of independent vertical and horizontal pressures to the soil sample. Single or group pile tests may be carried out using an appropriately designed lid and piston.

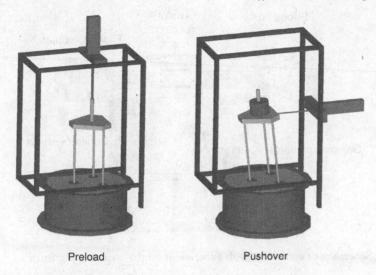

Preload                    Pushover

*Figure 3.74* Model jack-up rig for laboratory floor testing (Vlahos 2004)

As for the rod shear apparatus, the lateral pressure is applied to the sample, via a tap system mounted on the side of the chamber, which is closed during the test. Axial load, axial displacement and lateral pressure at the sample boundary are measured during loading.

### 3.6.4 Laboratory floor (1 g) model test

An example of a 1 g test (i.e. carried out on the laboratory floor) is shown in Figure 3.74. The loading is applied to a jack-up simulated as a frame resting on three independent spudcan foundations, in order to investigate the overall response of the jack-up rig and the transfer of vertical, horizontal and moment loading to each spudcan (Vlahos 2004). The soil bed for these experiments was clay, preconsolidated to an appropriate strength to obtain the correct relative stiffness of the jack-up structure and the soil.

### 3.6.5 Centrifuge modelling

Soils have stress-dependent properties: their strength and stiffness and hence the deformation and failure mechanisms are affected by the level of effective stress. Therefore, ideally, model tests should be conducted at the same effective stress levels as in the prototype situation, preserving scaling of the strength ratio $s_u/\sigma'_{v0}$ (or equivalent for free-draining soils) and stiffness ratio between structure and soil.

Geotechnical centrifuges have been developed to test models at scales up to two orders of magnitude smaller than the prototype, but at an elevated acceleration level. A typical centrifuge is shown in Figure 3.75. Soil models are placed on a swinging platform at the end of the centrifuge arm and then accelerated so that they are subjected to an inertial radial acceleration field of N times earth's gravity g acting normal to the surface of the platform. If an acceleration of N times earth's gravity g is applied

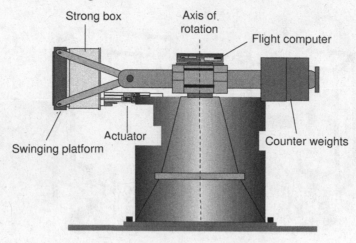

Figure 3.75 Schematic of centrifuge with experiment on the swinging platform

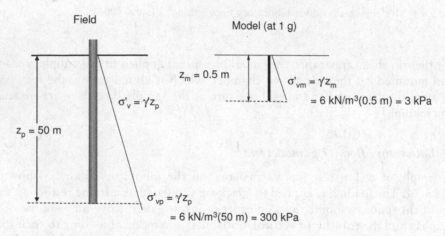

*Figure 3.76* Principle of 1:1 scaling of stress between prototype and centrifuge model

to a material of density $\rho$, then the vertical stress $\sigma_v$ at depth $h_m$ in the model is given by:

$$\sigma_{vm} = \rho(Ng)h_m = \rho g(Nh_m) = \rho g h_p = \sigma_{vp} \qquad (3.14)$$

where $h_p$ is the corresponding prototype depth. Hence the acceleration level, N, is chosen equal to the inverse of the linear dimension scaling ratio $h_m/h_p$ (Figure 3.76).

The size and capacity of geotechnical centrifuges is often quoted in g-tonnes, being the product of the maximum acceleration level and the maximum package mass at that level. For example, the centrifuge at the University of Western Australia (shown schematically in Figure 3.75), which can accelerate a package mass of 200 kg to 200 g, is 40 g-tonnes. At the opposite extreme are centrifuges such as at the University of California, Davis or the Laboratoire Centrale des Ponts et Chaussées in Nantes, with capacities in excess of 400 g-tonnes.

*Table 3.10* Scaling factors for centrifuge modelling

| Quantity | Scale factor |
| --- | --- |
| Acceleration | N |
| Stress and strain | 1 |
| Linear dimension | 1/N |
| Velocity | 1 |
| Area | $1/N^2$ |
| Mass | $1/N^3$ |
| Force | $1/N^2$ |
| Energy | $1/N^3$ |
| Time (consolidation) | $1/N^2$ |

*Figure 3.77* Example centrifuge models: (a) suction caisson (upper) and (b) drag anchor (lower)

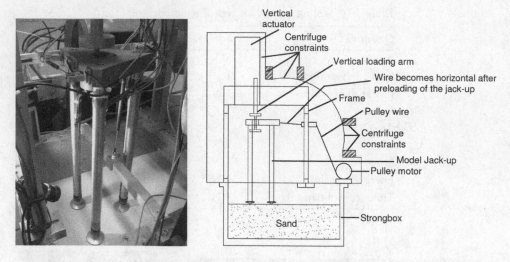

*Figure 3.78* Example centrifuge modelling of jack-up rig (Bienen et al. 2009)

General scaling factors for different quantities may be derived from those for acceleration (N), stress and strain (1) and linear dimensions (1/N) as summarised in Table 3.10 and more extensively by Garnier *et al.* (2007). An example of 'equivalent' centrifuge model and prototype conditions are indicated below for a test carried out at 200 g.

| Centrifuge model | Prototype |
|---|---|
| 0.5 m soil layer | z = 100 m |
| 10-mm-diameter pile | D = 2 m |
| $A_m = \pi D^2/4 = 7.85 \times 10^{-5}$ m$^2$ | $A_p = 3.14$ m$^2$ |
| Consolidation: 1 hour | t = 40,000 hours = 4.57 years |

Two examples of models used in centrifuge tests, and their prototypes are shown in Figure 3.77: a suction caisson (Tran 2005) and a drag anchor (O'Neill 2000). These scale models were subjected to varying g-levels between 120 and 200 g during testing. A more sophisticated model, of a complete jack-up rig, is shown in Figure 3.78 (Bienen *et al.* 2009). In contrast to the 1 g model tests referred to in Figure 3.74, these tests were conducted on a sand seabed, hence it was essential to test at an elevated acceleration in order to obtain similitude of the structure–soil stiffness ratio.

# 4  Soil response

## 4.1  Compression and shear

### 4.1.1  *Overview of classical theories*

Soil response can be broadly sub-divided into compressive and shear behaviour, unified through the framework of critical state soil mechanics (Roscoe *et al.* 1958, Schofield and Wroth 1968). In this section, the principles of classical theories of compressive and shear behaviour are summarised followed by an overview of the critical state framework.

#### *Compression*

When a soil is subjected to an increase in compressive stress, due to, for example, a foundation load, the resulting soil compression consists principally of three parts:

1. Immediate elastic compression
2. Primary compression or 'consolidation'
3. Secondary compression or 'creep'.

The priority of the geotechnical engineer is to determine how much compression will occur and at what rate the compression will occur.

In a sandy deposit, because of the high permeability, all compression (except creep) is usually assumed to take place immediately. In clay soils, the calculation is usually divided, using elastic theory to predict the immediate undrained compression and consolidation theory to predict the time-dependent primary compression. The term compression is used more generally to describe changes in volume due to changes in effective stress without reference to the time scale over which they occur. The time related process of soil deformation due to the dissipation of non-equilibrium pore water pressure is described as consolidation. When a load is applied to a low permeability material, the response is undrained, i.e. there will be no immediate volume change. Initially, the load will be carried wholly by the pore fluid, not by the soil skeleton. With time – how much time depends on the permeability of the material – drainage occurs and pore fluid is expelled from the soil skeleton allowing it to compress. Under drained conditions the external load is carried wholly by the soil skeleton. Consolidation is the reduction in void ratio e as a result of the time-dependent dissipation of excess pore water pressure as drainage takes place within the soil skeleton due to a change in

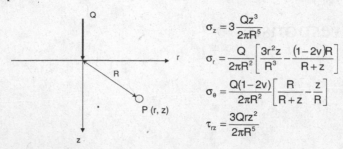

$$\sigma_z = 3\frac{Qz^3}{2\pi R^5}$$

$$\sigma_r = \frac{Q}{2\pi R^2}\left[\frac{3r^2z}{R^3} - \frac{(1-2v)R}{R+z}\right]$$

$$\sigma_\theta = \frac{Q(1-2v)}{2\pi R^2}\left[\frac{R}{R+z} - \frac{z}{R}\right]$$

$$\tau_{rz} = \frac{3Qrz^2}{2\pi R^5}$$

*Figure 4.1* Stress changes in an elastic half space due to a vertical point load at the surface (Boussinesq 1885)

external load. Consolidation is the process that defines the transition from undrained to drained conditions.

ELASTIC THEORY

Elastic theory is based on a set of governing equations for the determination of elastic stress changes within a half space due to a surface point load (Figure 4.1) (Boussinesq 1885). The solution for a single point load can be integrated for a collection of point loads acting over any given area of the surface of a half space to determine the stresses that result from the application of any pattern of applied load. Strains within an elastic half space are ascertained from the solution for elastic stress change in conjunction with the elastic properties of the soil.

The elastic response of an isotropic material is fully described by Young's modulus E and Poisson's ratio $v$. Young's modulus is given by the ratio of change in deviatoric stress (in a standard triaxial test) to change in vertical strain,

$$E = \frac{\Delta q}{\Delta \varepsilon_1} \tag{4.1}$$

Young's modulus of an isotropic material can be determined from a triaxial test from the slope of a graph of deviator stress against axial strain.

Poisson's ratio is given by the ratio of an increment of lateral strain to vertical strain,

$$v = \frac{-\Delta \varepsilon_3}{\Delta \varepsilon_1} \tag{4.2}$$

For small strains under triaxial conditions (with $\varepsilon_2 = \varepsilon_3$), volumetric strain can be expressed as $\Delta \varepsilon_v = \Delta \varepsilon_1 + 2\Delta \varepsilon_3$, leading to re-expression of Poisson's ratio as

$$v = 0.5\left(1 - \frac{-\Delta \varepsilon_v}{\Delta \varepsilon_1}\right) \tag{4.3}$$

Young's modulus is denoted by $E_u$ to refer to undrained conditions and $E'$ for drained conditions. Similarly, Poisson's ratio is denoted by $v_u$ for undrained conditions and $v'$

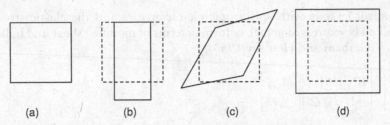

(a)        (b)        (c)        (d)

*Figure* 4.2 Elastic deformations (a) undeformed (b) Young's modulus describing change in
length and Poisson's ratio describing change in width (c) shear modulus describing
change in shape at constant volume (d) bulk modulus describing change in volume
at constant shape (Muir Wood 1990)

for drained conditions. For undrained conditions, volumetric strain $\varepsilon_v = 0$ such that
the undrained value of Poisson's ratio $v_u = 0.5$. Typical values of drained Poisson's
ratio fall in the range $0.1 < v' < 0.3$.

Shear modulus G is an alternative elastic parameter that couples the Young's modulus and Poisson's ratio,

$$G = \frac{E}{2(1+v)} = \frac{E_u}{2(1+v_u)} = \frac{E'}{2(1+v')} \tag{4.4}$$

Shear modulus does not distinguish between undrained and drained conditions
since water cannot carry shear. The slope of a graph of deviator stress against shear
strain from a triaxial test has a gradient of 2G, applicable to isotropic or anisotropic
conditions.

The elastic stress–strain relationship can be written in terms of shear modulus G
and bulk modulus K (rather than Young's modulus E and Poisson's ratio $v$) in order
to separate the shear and volumetric effects. Bulk modulus K is a measure of the volumetric strain due to a change in mean stress. The undrained bulk modulus $K_u$ is nominally infinite for fully saturated soils, if the finite compressibility of water is ignored.
The drained bulk modulus $K'$ is defined as

$$K' = \frac{\Delta p'}{\Delta \varepsilon_{vol}} \tag{4.5}$$

and can be determined for isotropic or anisotropic conditions from the gradient of a
graph of mean effective stress against volumetric strain from a triaxial test.

Elastic deformations described by Young's modulus, Poisson's ratio, shear modulus
and bulk modulus are illustrated schematically in Figure 4.2.

The stress–strain relationship for an isotropic elastic material can then be written
in matrix notation in terms of G and K,

$$\begin{bmatrix} \Delta p' \\ \Delta q \end{bmatrix} = \begin{bmatrix} K' & 0 \\ 0 & 2G \end{bmatrix} \begin{bmatrix} \Delta \varepsilon_{vol} \\ \Delta \gamma \end{bmatrix} \tag{4.6}$$

The off-diagonal zeros show that for isotropic conditions shear and volumetric
deformations may be considered independently. Shear and volumetric effects are

interdependent in soils with anisotropic characteristics and the elastic stress–strain behaviour can be expressed in matrix form in terms of modified shear and bulk moduli $G^*$ and $K^*$ (Graham and Houlsby 1983),

$$\begin{bmatrix} \Delta p' \\ \Delta q \end{bmatrix} = \begin{bmatrix} K^* & J \\ -J & G^*/2 \end{bmatrix} \begin{bmatrix} \Delta\varepsilon_{vol} \\ \Delta\gamma \end{bmatrix} \tag{4.7}$$

where

$$K^* = \frac{E^* (1 - v^* + 4\alpha v^* + 2\alpha^2)}{9(1 + v^*)(1 - 2v^*)} \tag{4.8}$$

$$G^* = \frac{E^* (2 - 2v^* - 4\alpha v^* + \alpha^2)}{6(1 + v^*)(1 - 2v^*)} \tag{4.9}$$

$$J = \frac{E^* (1 - v^* + \alpha v^* - \alpha^2)}{3(1 + v^*)(1 - 2v^*)} \tag{4.10}$$

$E^*$ and $v^*$ represent modified values of Young's modulus and Poisson's ratio that account for anisotropic characteristics. The constant $\alpha$ is a measure of the degree of anisotropy; $\alpha = 1$ indicates isotropy, $\alpha > 1$ indicates the soil is stiffer horizontally than vertically and $\alpha < 1$ indicates the soil is stiffer vertically than horizontally. Cross-anisotropic conditions, in which both horizontal directions exhibit the same characteristics but differ from the vertical direction, require five independent parameters to describe elastic behaviour ($E_v$, $E_h$, $G_{vh}$, $v_{vh}$, $v_{hh}$). Fully anisotropic conditions require 21 independent parameters.

Elastic solutions have been derived for a variety of load conditions, boundary value problems and layered, heterogeneous and anisotropic deposits. A comprehensive selection of elastic solutions is provided by Poulos and Davis (1974).

CONSOLIDATION THEORY

Traditional methods of predicting time-dependent settlements rely on one-dimensional consolidation theory (Terzaghi 1923), based on the assumption of one-dimensional flow, one-dimensional strain and the rate of expulsion of water at the drainage boundaries being equal to the progress of displacement. One-dimensional conditions, which are commonly used to determine consolidation parameters for design calculations, are achieved in an oedometer test, in which a confining ring prevents lateral strain during vertical loading. Consolidation data can be presented in terms of the stress–strain relationship and as time histories of excess pore pressure, effective stress or settlement. Time is typically expressed on a logarithmic scale by a dimensionless time factor,

$$T = \frac{c_v t}{d^2} \tag{4.11}$$

where
  $c_v$ = representative coefficient of consolidation
  t = time since change in total stress
  d = representative drainage path length, or other significant dimension

The coefficient of consolidation of a soil varies with stress level such that a value representative of the field conditions should be selected. Likewise, engineering judgement is required to select a drainage path length that is appropriate to the conditions under consideration for design.

The one-dimensional coefficient of consolidation may be expressed as a function of the coefficient of vertical permeability $k_v$, the one-dimensional modulus $E'_0$ and the unit weight of water $\gamma_w$ as

$$c_v = \frac{k_v E'_0}{\gamma_w} \tag{4.12}$$

The one-dimensional modulus is defined as the increment of vertical stress to vertical strain as

$$E'_0 = \frac{\Delta \sigma_v}{\Delta \varepsilon_1} = \frac{E(1-v')}{(1+v')(1-2v')} \tag{4.13}$$

and can be determined from oedometer test data.

The one-dimensional modulus is sometimes expressed as its reciprocal, referred to as the coefficient of volume compressibility,

$$m_v = \frac{1}{E'_0} \tag{4.14}$$

Compression index is an alternative stiffness parameter that defines change in volume in terms of void ratio e as a function of an increment of log stress,

$$C_c = \frac{\Delta e}{\Delta \log \sigma_v} \tag{4.15}$$

A similar parameter, the swelling index $C_s$, is used to define the stiffness of an over-consolidated soil during unloading.

One-dimensional conditions are rarely encountered in field situations, where three-dimensional flow and strains generally govern the consolidation process. The theory of three-dimensional consolidation (Biot 1935, 1956) accommodates three-dimensional flow, three-dimensional strains and provides coupling between changes in effective stress and reduction in excess pore pressure via volume continuity and the material effective stress–strain response (unlike one-dimensional theory in which the total stress is constant, so the change in effective stress is equal and opposite in sign to the change in pore pressure).

Development of effective stresses (as the strain producing component) rather than the dissipation of excess pore pressure are of greatest interest during three-dimensional consolidation. A significant feature of three-dimensional consolidation is the

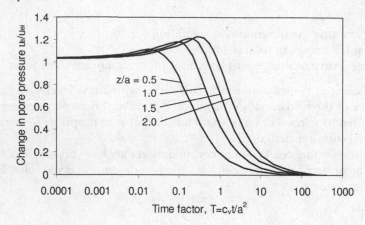

*Figure 4.3* Mandel-Cryer effect; excess pore pressure response with depth beneath a strip load (Schiffman *et al.* 1969)

Mandel-Cryer stress-transfer effect (Mandel 1950, Cryer 1963). As soil near to the free surface drains more rapidly than deeper soil, the strain associated with drainage of the nearer-surface material 'squeezes' the deeper undrained material leading to an increase in total stress over the initial increase in total stress due to the externally applied load. Therefore, in the early stages of consolidation, parts of the soil experience an increase in excess pore pressure over the initial value. As the drainage front advances, the deeper soil drains and total stress reduces. The early rate of development of effective stress is small, gradually increasing with time. At the point when the rate of development of effective stress becomes greater than the rate of increase of total stress, the excess pore pressure, having achieved a maximum, will start to dissipate. Figure 4.3 shows time histories of excess pore pressure dissipation in an element of soil along the centre-line beneath a uniform strip load resting on an elastic half space (Schiffman *et al.* 1969), with the half-width of the load, taken as the representative drainage path length for calculation of the time factor T. It can be seen that the Mandel-Cryer effect advances in magnitude and time with increasing depth.

Maximum shear stress varies during three-dimensional consolidation, contrary to one-dimensional consolidation in which the total stress components and, therefore, maximum shear stress are constant, such that local plastic zones and eventual instability can potentially occur during consolidation. In addition, contrary to one-dimensional consolidation theory, the rate of consolidation is affected by the value of Poisson's ratio (Figure 4.4). Generally, for three-dimensional consolidation problems the coefficient of consolidation is still based on the one-dimensional modulus, as defined in Equations (4.12) and (4.13).

Analytical solutions have been derived for three-dimensional consolidation for various loading regimes and soil conditions (e.g. McNamee and Gibson 1960, Gibson *et al.* 1970, Booker 1974, Chiarella and Booker 1975, Booker and Small 1986). Three-dimensional consolidation with respect to predicting foundation settlements, using established analytical solutions and more recent numerical solutions, is discussed in more detail in Chapter 6.

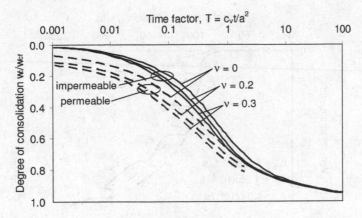

*Figure 4.4* Effect of Poisson's ratio on three-dimensional consolidation response beneath a rigid circular raft of radius, a (Chiarella and Booker 1975, permeable raft and Booker and Small 1986, impermeable raft)

CREEP

Deformation of a soil mass is generally considered to be associated with a change in effective stress, although many soils continue to deform at constant effective stress. Change in void ratio at constant effective stress is referred to as creep. Creep settlement is due to viscous resistance between soil particles although detailed mechanisms of creep are poorly understood. Creep is usually taken to refer to volumetric strains, but creep shear strains may also be important. Experimental evidence suggests creep strain is proportional to the logarithm of time, i.e. the same magnitude of creep strain occurs in sequential cycles of log time (Bishop and Lovenbury 1969). Creep is often a reasonably stable process with the rate of increase of strain in real time decreasing although creep may be unsustainable in sensitive or structured soils or soils that are close to failure, where creep rupture may occur. Normally consolidated, particularly highly plastic, clays, organic soils and calcareous soils are generally the most susceptible to creep. Creep effects may be largely ignored in many soils, for example, most sands, silts and heavily overconsolidated clays. Creep can affect the results of laboratory tests carried out at a constant rate of strain giving a false impression of soil stiffness. The slower a test is carried out, the more opportunity there is for creep strains to develop, resulting in a softer response. If the rate of strain is changed during a test, the soil will respond according to the new strain rate (Figure 4.5).

COMPRESSIBILITY OF SILICA AND CALCAREOUS SAND

Figure 4.6 shows the stress–strain relationship of a silica sand (Fontainbleau sand) and a calcareous sand (from the Iroise Sea, the Bay of Biscay). The initial void ratio in both cases is the same ($e_o = 0.93$), quite a loose state for the silica sand but a dense state for the calcareous sand. The compressibility of the silica sand might, therefore, be expected to be greater than that of the calcareous sand, but clearly, this is not the case. The higher compressibility of calcareous soils is due to the low hardness of calcium carbonate, the main constituent of carbonate soils, compared to quartz, leading to particle crushing at relatively low stress levels. The high compressibility of the calcareous sands

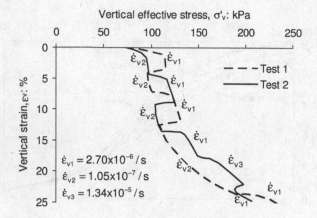

*Figure 4.5* Effect of strain rate on constant rate of strain oedometer test (Leroueil *et al.* 1985)

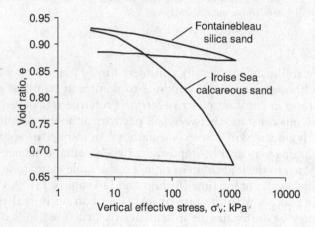

*Figure 4.6* Comparison of compressibility of silica and calcareous sands

at the North Rankin A site on the North-West Shelf of Australia was a key factor contributing to the very low shaft friction observed for driven piles, with many of the piles for this platform free-falling under their own weight through most of the top 100 m of the seabed during installation.

## Shear behaviour

The shear response of a soil depends on its void ratio and the normal effective and shear stresses acting in the soil mass. The exact definition of the normal and shear stress parameters varies depending on the conditions under consideration, but the underlying requirement to quantify the key independent state variables (e, $\sigma'$ or $p'$, $\tau$ or q) holds for all conditions.

### FRICTIONAL STRENGTH AND EFFECTIVE STRESS

The shear strength of all uncemented soils (sands or clays) is purely a frictional strength. The basic parameter is the coefficient of internal friction $\mu$ or the effective

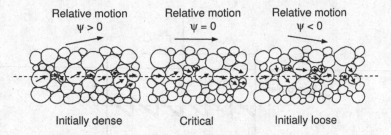

Relative motion
$\psi > 0$

Relative motion
$\psi = 0$

Relative motion
$\psi < 0$

Initially dense          Critical          Initially loose

*Figure 4.7* Dilation and contraction during shearing (Bolton 1991)

angle of internal friction $\phi$. For a frictional material, the shear strength $\tau$ on a potential failure plane is a function of the normal effective stress $\sigma'$ on the failure plane,

$$\mu = \tan\phi = \frac{\tau}{\sigma'} \tag{4.16}$$

Alternatively expressed as the effective stress failure criterion,

$$\tau_f = \sigma' \tan\phi \tag{4.17}$$

For a saturated soil, i.e. a two-phase material comprising soil particles and water, the normal stress that is effective in producing the frictional strength is the stress between the particles. The principle of effective stress states that the effective stress $\sigma'$ is the total stress $\sigma$ less the pore water pressure u (Terzaghi 1943).

$$\sigma' = \sigma - u \tag{4.18}$$

The behaviour of a partially saturated soil, i.e. three-phase soils, with a gaseous component, is governed by two stress parameters $(\sigma - u_g)$ and $(u_g - u_w)$, where $u_w$ is the pore water pressure and $u_g$ is the pore gas pressure. A single effective stress parameter for partially saturated soil remains elusive.

DILATION, CONTRACTION AND THE CRITICAL STATE

Volumetric changes take place during shearing of a soil, either dilation (an increase in volume) or contraction (a decrease in volume). The change in volume of a soil as it shears arises because soil is essentially a particulate material and in order for a portion of the soil mass to move relative to another part of the soil mass, the particles must take up a suitable arrangement of packing. The volume of the soil mass at which continuous shearing can take place without further volumetric strain is defined by a critical void ratio $e_{cr}$. If particles are initially more densely packed than their critical void ratio then some loosening of the matrix, i.e. dilation, has to take place before steady state shear can occur. Conversely, if particles are initially more loosely packed than their critical void ratio then some densification of the matrix, i.e. contraction, has to take place before steady state shear can occur (Figure 4.7).

The saw-tooth model provides a simple analogue for dilation (Figure 4.8). The frictional resistance along the sliding surface of the saw teeth represents the internal

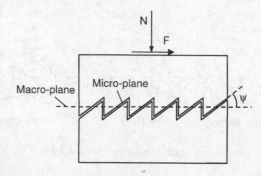

*Figure 4.8* Saw tooth model of dilation (Bolton 1991)

critical friction angle $\phi_{cr}$. When the two halves of the block are sheared with a vertical force N the observed shearing resistance F is greater than if the interface was horizontal, as the top saw-teeth must ride up over the underlying arrangement. The additional resistance depends on the angle of the saw-tooth planes with the horizontal, representing the angle of dilation $\psi$ (as indicated in Figure 4.7). The maximum frictional resistance of a soil is thus mobilised when the angle of dilation is greatest. According to the model, the difference between the apparent peak friction angle $\phi_p$ and the fundamental (critical state) friction angle $\phi_{cr}$ is given by the dilation angle $\psi$.

$$\phi_p = \phi_{cr} + \psi \tag{4.19}$$

Experimental evidence suggests Equation (4.19) slightly overestimates the apparent friction angle, and a modified relationship is often used in practice (Bolton 1986).

$$\phi_p = \phi_{cr} + 0.8\psi \tag{4.20}$$

Figure 4.9 shows typical results from two drained direct shear (shearbox) tests on a sand at two relative densities, sheared under the same normal effective stress. In the test on the initially dense sand, the shear stress $\tau$ gradually increases with increasing shear strain $\gamma$ before reaching a peak value and dropping off to a steady state that is maintained with continued shearing. Some contraction (positive $\varepsilon_v$) occurs at the start of shear after which the soil dilates (negative $\varepsilon_v$) until reaching a critical state marked by no further volume change with continued shearing. The initially loose sample exhibits no peak strength but undergoes gradual contraction during shearing until reaching a critical state marked by no further volume change with continued shearing. The initially loose sample eventually reaches the same critical shear stress and critical void ratio as the initially dense sample.

All soils when sheared will eventually reach a critical void ratio $e_{cr}$. This condition, at which unlimited shear strain can be applied without further changes in void ratio (i.e. volume) e, shear stress $\tau$ and normal effective stress $\sigma'$, is known as the critical state.

The tangent to the $\varepsilon_v$–$\gamma$ plot $d\varepsilon_v/d\gamma$ represents the current rate of dilation, which governs the shear strength at any stage of the test, expressed as

$$\psi = -\arctan\frac{d\varepsilon_v}{d\gamma} \tag{4.21}$$

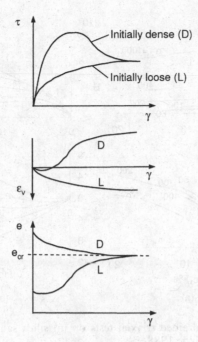

*Figure 4.9* Idealisation of dilation and contraction in a direct shear test

The tendency for a soil to dilate or contract depends not only on its relative density but also on the confining stress. The relative density $D_r$, which is a measure of the closeness of packing of the soil's grains, is expressed in terms of maximum ($e_{max}$) and minimum ($e_{min}$) void ratios, as

$$D_R = \frac{(e_{max} - e)}{(e_{max} - e_{min})} \tag{4.22}$$

Relative density alone is insufficient to determine whether a sample will dilate or contract when sheared at constant normal stress $\sigma'$ since effectively, for an initially dense sample, if the normal effective stress is sufficiently high, the tendency for dilation to occur will be suppressed and shearing will be accompanied by some grain breakage to allow the failure plane to develop. Conversely, for an initially loose sample, if the normal effective stress is sufficiently low, dilation can occur. The relative density index $I_R$ (Bolton 1986) accounts for confining stress $p'$ and, therefore, can indicate whether a soil will dilate or contract under shearing. It is expressed as

$$I_R = 5D_R - 1 \quad \text{for } p' \leq 150 \text{ kPa}$$
$$I_R = D_R \left( 5.4 - \ln \frac{p'}{p_a} \right) - 1 \quad \text{for } p' > 150 \text{ kPa} \tag{4.23}$$

where $p_a$ is atmospheric pressure = 100 kPa.

An alternative, though largely equivalent, approach for assessing the tendency of sand to dilate or contract is the state parameter $\Psi$ (Been and Jefferies 1985), which

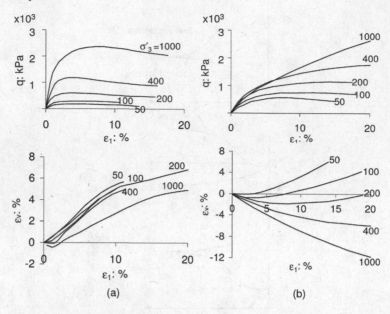

*Figure 4.10* Comparison of drained triaxial tests on (a) silica sand and (b) calcareous sand (Gólightly and Hyde 1988)

measures the distance (in terms of void ratio or specific volume) that the soil state lies from the critical state line (CSL).

COMPARISON OF RESPONSE OF SILICA AND CARBONATE SAND IN SHEAR

Figure 4.10 show the results of drained triaxial compression tests on a silica sand and a calcareous sand. The silica sand is well-documented Leighton Buzzard sand, a solid quartz grained sand with negligible calcium carbonate content; the calcareous sand is from Dog's Bay on the west coast of Ireland, a very angular material with an *in situ* void ratio $e_0 \sim 2$. The results are presented as deviator stress q and volumetric strain $\varepsilon_v$ against axial strain $\varepsilon_1$. The silica sand tends to dilate at all stress levels, although the stress level affects the rate of dilation, while the calcareous sand dilates at low stresses and contracts at high stresses. The difference in behaviour is due to the more angular nature and greater compressibility of the calcareous sand compared to the silica sand, as previously shown in Figure 4.6.

UNDRAINED SHEARING

Volume changes in a soil mass occur instantaneously if drainage takes place sufficiently quickly and changes in external load will be carried immediately by the soil skeleton. Undrained conditions occur if the rate of shearing is sufficiently fast that the time for any significant drainage is longer than the time for load application. Undrained conditions typically relate to fine-grained soils where the time for drainage and, therefore, consolidation may be very long (see Section 4.1.1). Undrained conditions can also apply to relatively coarse-grained soils for very rapid rates

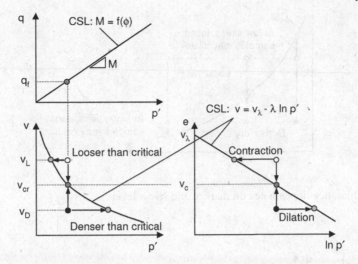

*Figure 4.11* The critical state concept

of loading, such as during an earthquake or wave loading. If a soil is sheared undrained, i.e. without volume change, the maximum shear stress to cause failure depends only on the initial void ratio of the soil and is denoted by the undrained shear strength $s_u$

$$\tau_f = s_u \tag{4.24}$$

Equation (4.24) defines the undrained failure criterion valid for soils sheared to failure without change in volume. Undrained shear strength $s_u$ is not a material constant but depends on the void ratio of the soil, which is related to its water content. The shear strength will vary significantly through a soil deposit, in normally or lightly overconsolidated clay deposits typically increasing with depth at a rate of 1–2 kPa/m. When a dense soil is sheared undrained, negative excess pore pressures are generated, increasing the mean effective stress in order to maintain the zero volume change condition. Conversely, positive excess pore pressures are generated, with a reduction in mean effective stress, when a loose soil is sheared undrained.

### 4.1.2 The critical state framework

Critical state soil mechanics is a theoretical framework through which the shear and volumetric response during yielding of saturated, isotropic, plastic materials can be described (Roscoe *et al.* 1958, Schofield and Wroth 1968). The basis of the critical state family of soil models is that, when sheared, a soil will tend to migrate towards ultimate states in $(q, p', v)$ space that lie along a single line: the critical state line (CSL). The projection of the CSL in $(q, p')$ space is a straight line with gradient linked to the critical state friction angle. The projection of the CSL in $(v, p')$ space is curved, but becomes close to a straight line with gradient $\lambda$ (the same gradient as the normal consolidation line – NCL) if plotted in $(v, \ln p')$ space (Figure 4.11) . If a sample lies on

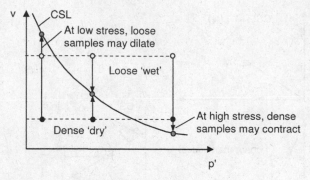

*Figure 4.12* Dilatancy dependency on density and stress level

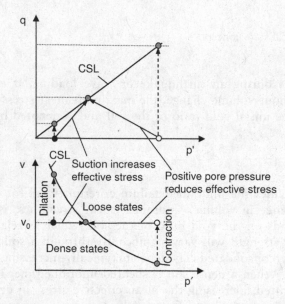

*Figure 4.13* Stress and state paths during drained and undrained shearing

the 'dense' side of the CSL in the v, p′ plane, dilation will occur during shearing; if a sample lies on the 'loose' side of the CSL in the v, p′ plane, then contraction will occur during shearing (Figure 4.12). The terms dry side and wet side of the CSL are used interchangeably with dense and loose since an initially dense sample will take in water during dilation, i.e. starting off on the 'dry' side of critical, while an initially loose sample will expel water during contraction, i.e. starting off on the 'wet' side of critical. Undrained shearing, i.e. with zero volume change, can also be represented in the critical state framework. The effective confining stress p′, given by the total direct stress p minus the pore pressure u, changes during shearing as positive or negative excess pore pressures are generated to maintain the zero volume change condition (Figure 4.13).

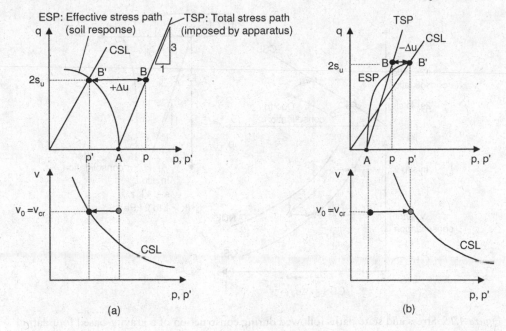

*Figure 4.14* Stress and state paths during an isotropically consolidated undrained triaxial compression test (a) wet of critical and (b) dry of critical

Critical state soil mechanics is conventionally described in terms of the first two stress invariants: mean normal stress p′ and deviatoric stress q, expressed in terms of the principal stresses, $\sigma_1$, $\sigma_2$ and $\sigma_3$ as,

$$p' = \frac{1}{3}(\sigma_1' + \sigma_2' + \sigma_3')$$

$$= \frac{1}{3}(\sigma_1' + 2\sigma_3') \quad \text{for triaxial conditions with } \sigma_2' + \sigma_3' \tag{4.25}$$

$$q = \frac{1}{\sqrt{2}}\sqrt{(\sigma_1' - \sigma_2')^2 + (\sigma_1' - \sigma_3')^2 + (\sigma_2' - \sigma_3')^2}$$

$$= \sigma_1' - \sigma_3' \quad \text{for triaxial conditions with } \sigma_2' = \sigma_3' \tag{4.26}$$

In the following examples, stress and state paths followed during an undrained triaxial compression test and two hypothetical offshore construction processes are illustrated within the critical state framework represented in (q, p′, v) space.

*Undrained triaxial compression test*

The most common type of triaxial test involves isotropic consolidation (q = 0) to some initial effective stress state p′₀ (point A in Figure 4.14) followed by undrained compression in which the cell pressure is kept constant and the axial stress is increased to failure (point B). The total stress path (TSP) is predetermined (A–B) and has a gradient

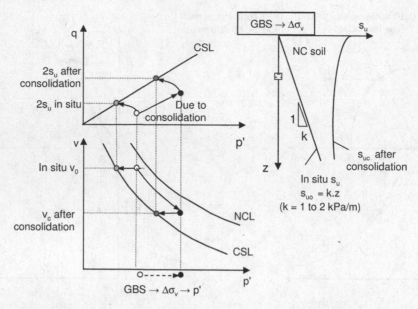

*Figure 4.15* Stress and state paths followed during construction of a gravity-based foundation

of 3:1 in (q, p) space. The soil response, or more precisely the pore pressure response, determines the offset of the effective stress path (ESP) from the TSP. If the sample tends to contraction, but is prevented because the test is undrained, positive excess pore pressures $u_e$ are generated and the ESP will lie to the left of vertical and of the TSP (A–B′) (see Figure 4.14a). Conversely, if the sample tends to dilation, suction will be generated and the ESP will lie to the right of vertical (i.e. increasing p′) and even to the right of the TSP (A–b′) (see Figure 4.14b). The effective stress path is generally curved, as shown in Figure 4.14, although is sometimes idealised as a straight line with $\Delta u_e$ proportional to $\Delta q$. Failure occurs when the ESP intersects the CSL. The undrained shear strength, corresponding to initial void ratio and consolidation stress, is given by the deviator stress at failure. In (v, p′) space, the mean effective stress reduces with positive excess pore pressure at constant volume until intersection with the CSL.

## Consolidated undrained strength following installation of a gravity-based platform

If a large gravity-based structure is placed on the seabed, the total stress acting on the soil increases, eventually leading to an increase in effective stresses and shear strength of the soil and hence an increase in stability of the structure with time after construction. This is illustrated in a very simplistic way in Figure 4.15 for an initially normally consolidated deposit with a linearly increasing undrained shear strength profile and zero shear strength at the mudline. It may take a very long time for the strength increase to occur, especially in very-fine-grained materials (i.e. with a low $c_v$) or beneath large structures (i.e. long drainage paths). Consolidated undrained strengths are relied on where peak loads are applied sufficiently rapidly for the soil response to be undrained, for example, wave loading, but the soil has consolidated and strengthened, for example, under the weight of a platform, before the peak loads are applied.

*In situ* conditions and the stress and state paths followed due to the applied foundation load are illustrated within the critical state framework in Figure 4.15. The initial state of an element of soil at depth z lies between the (isotropic) normal consolidation line (NCL) and the CSL in (v, p′) space with *in situ* mean effective stress $p'_0$, deviator stress $q_0$ and void ratio $e_0$. The *in situ* stress state lies in (q, p′) space at the *in situ* deviator and mean effective stresses $q_0$ and $p'_0$ (determined by the effective overburden stress, $\sigma'_{v0}$ and horizontal effective stress, $\sigma'_{h0} = K_0\sigma'_{v0}$). At failure, the mean effective stress corresponds to the value on the CSL in (v, p′) space at the *in situ* void ratio $e_0$. The *in situ* undrained shear strength $s_{u0}$ then corresponds to half the deviator stress on the CSL in (q, p′) space at that value of mean effective stress. Application of the foundation load leads to an increase (over time) in mean effective stress to $p'_c$, and reduction in void ratio to $v_c$. The undrained shear strength following consolidation $s_{uc}$ is obtained from the higher mean effective stress on the CSL in (v, p′) space at the new void ratio, $v_c$, and the corresponding deviator stress on the CSL in (q, p′) space at that mean effective stress. The shear strength at any stage depends on the change in mean effective stress, as determined by the position of the soil state in (v, p′) space relative to the CSL, and then the gradient of the CSL in the q, p′ plane.

### Staged construction to avoid undrained failure during installation of a gravity-based platform

Placing a structure on the seabed leads to an increase in the mean effective stress p′ after consolidation has occurred and an increase in soil strength, as seen in the previous example. However, load application also increases the deviator stress (shear stress) immediately, which may lead to failure if the shear stresses reach the undrained shear strength $s_u$ at sufficient locations for a failure mechanism to develop. To prevent an immediate undrained failure during installation, the load can be applied in stages with time allowed for consolidation between increments of load. The case illustrated in Figure 4.16 involves placing the total load in two stages. *In situ* conditions are represented by point 1, on the NCL line at ($p'_0$, $q_0$, $e_0$). (1–2) is the effective stress path resulting from application of the first increment. The load after application of the first increment is less than that required to cause failure (i.e. does not reach the CSL). Consolidation is then allowed (2–3). The second load increment (3–4) takes the stress state close to failure but subsequent consolidation (4–5) results in a further increase in shear strength, leaving a reasonable margin of safety under the eventual total load. Undrained shearing at this stage would cause failure at point 6.

## 4.2 Cyclic loading

### 4.2.1 Soil response under cyclic loading

The significant feature of offshore geotechnical engineering is that offshore structures are subjected to severe wave and storm loads. Therefore, cyclic behaviour of soils supporting offshore structures is of the utmost importance to offshore geotechnical engineers. To design for repeated loading, it is necessary to account for the significantly different behaviour of soils loaded under cyclic or repeated stresses. Cyclic loading generates excess pore pressures, reducing effective stresses in the seabed and causing average and cyclic shear strains to develop with continued cycling, ultimately leading

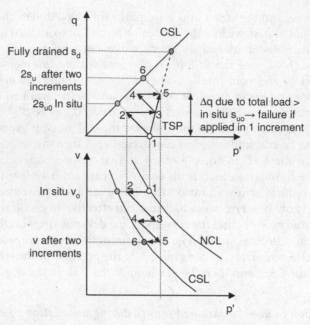

*Figure 4.16* Stress and state paths followed during staged loading

to a loss of shear strength or stiffness of the seabed sediments. No all-embracing constitutive model has been developed to date that captures all the key characteristics of soil response under cyclic loading and for practical purposes it is necessary to make use of simple approaches for estimating the cyclic behaviour of soils, relying on laboratory testing in order to determine the necessary soil parameters.

The response of soils under undrained cyclic loading is considerably different from that during monotonic loading, although the behaviour of sands and clays under repeated loading is similar in many respects, and it is, therefore, possible to approach the problem of cyclic response of soils in a reasonably unified manner. Nonetheless, different features are of interest when considering sand and clays. Assessment of the response of a sand to cyclic loading will typically involve consideration of the likelihood of liquefaction, the magnitude of the excess pore water pressure generated by the cyclic loading, cyclic strains and resulting displacements and the permanent (residual) strains in the soil. Assessment of the response of clays to cyclic loading will typically involve consideration of the possible loss of undrained shear strength, generation of excess pore water pressures and their subsequent dissipation, cyclic stiffness characteristics and the accumulation of permanent strains. The response of any soil to cyclic loading, whether in the laboratory or the field, depends on the mode, amplitude and frequency of the cyclic loading.

### 4.2.2 Modes of cyclic loading

The amplitude and frequency of waves, winds and storms are irregular although cyclic loading tests with constant stress amplitude and frequency are mostly used in investigating cyclic behaviour of soils, as depicted in Figure 4.17. Cyclic stress $\tau_{cy}$ and average

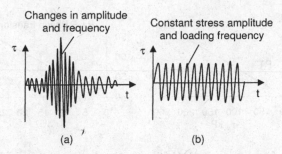

*Figure 4.17* Amplitude and frequency of cyclic loading of (a) a typical storm sequence and (b) a laboratory cyclic loading test

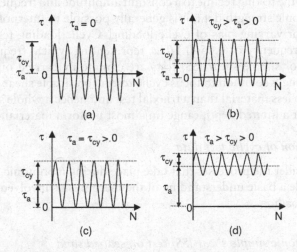

*Figure 4.18* Modes of cyclic loading (a) two-way, $\tau_a = 0$ (symmetric) (b) two-way, $\tau_a > 0$ (unsymmetric) (c) one-way, $\tau_a = \tau_{cy}$ (d) one-way, $\tau_a > \tau_{cy}$

(or mean) stress $\tau_a$ are defined, respectively, as the amplitude of cyclic stress and the average of the applied stress around which cyclic loading is applied. Four generic patterns of cyclic loading can be defined depending on the level of average shear stress $\tau_a$:

1. 'Two-way' cyclic loading is used to denote cycling in such a way that zero stress is crossed, i.e. cycling from negative to positive values of stress (Figure 4.18a and b).
2. 'One-way' cyclic loading is denoted as cycling in a range in which no zero stress is crossed (Figure 4.18c and d).
3. 'Symmetric' cyclic loading is a particular case of two-way loading with zero mean stress and is also called zero mean stress cyclic loading (Figure 4.18a).
4. 'Unsymmetric' cyclic loading denotes cycling around non-zero mean stress and is also called non-zero mean stress cyclic loading (Figure 4.18b, c and d).

### 4.2.3 Cyclic loading tests

Cyclic tests should be performed with cyclic to average shear stress ratios $\tau_{cy}/\tau_a$ and frequency that reflect the conditions appropriate for the design situation although it is

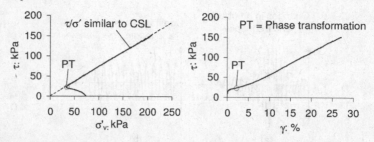

*Figure 4.19* Results of a monotonic CAU SS test on seabed sand, $\sigma'_{hc}$ = 75 kPa, $\sigma'_{hc}$ = 30 kPa (Mao and Fahey 2003)

typical to simplify the testing regime to a constant amplitude and frequency. Combining cyclic and monotonic strength data, it is generally possible to interpolate appropriate cyclic strengths for varying bias of cyclic loading. Cyclic loading tests are typically carried out at a frequency of 0.05–0.1 Hz, representing typical frequencies of wave loading. Triaxial or simple shear (SS) tests are commonly carried out although it is becoming increasingly common to focus cyclic testing on simple shear tests, since each test requires much less material than a triaxial test and hence a whole suite of tests can be performed over a limited depth range on almost uniform material.

### 4.2.4 Interpretation of cyclic test data

It is useful to consider the processes that take place during monotonic and cyclic loading tests to provide a basic understanding of the mechanisms involved in the response of soils to cyclic loading.

#### Undrained monotonic simple shear (SS) test on seabed sand

Figure 4.19 shows results from an anisotropically consolidated undrained (CAU) simple shear test on a saturated calcareous seabed sand, consolidated under vertical and horizontal consolidation stresses $\sigma'_{vc}$ = 75 kPa and $\sigma'_{hc}$ = 30 kPa, respectively. In the shearing phase, the total vertical stress $\sigma_v$ was maintained constant throughout the test and shear stress was applied to the sample until the test was terminated (in this case, failure was not reached and the test was stopped at a shear stress of approximately 150 kPa). During shear, initially the sample contracts and excess pore pressure $u_e$ builds up leading to a reduction in effective vertical stress $\sigma'_v$ with continued shearing. After reaching a maximum, pore pressure starts to decrease indicating dilative behaviour and effective stress starts to increase. The transition point that separates the soil response into dilation and contraction is called the phase transformation (PT). Following phase transformation, the stress path proceeds at an approximately constant stress ratio $\tau/\sigma'_v$. This stress path is sometimes labelled as the critical state line (CSL) as the stress ratio is similar to that of the CSL, but this is not actually the CSL – just its projection in $(\tau, \sigma')$ space. If the test had continued until excess pore pressure and shear strain had stabilised i.e. a plateau had been achieved in $(\sigma'_v, \tau)$ and $(\gamma, \tau)$ space, then the critical state would have been reached. The critical state is defined by deformation taking place at constant stress and constant volume and, therefore, cannot correspond to a state where continued dilation is taking place. A relatively stiff response is observed during initial shearing (shear modulus G = d$\tau$/d$\gamma$) followed

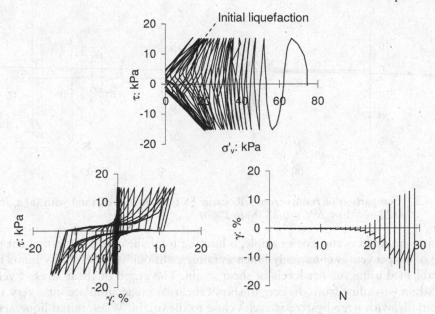

*Figure 4.20* Results of a CAU cyclic SS test on seabed sand, $\sigma'_{hc} = 75$ kPa, $\sigma'_{hc} = 30$ kPa, $\tau_{cy} = 15$ kPa (Mao 2000)

by a much softer response before approaching a constant gradient beyond the phase transformation.

### Undrained cyclic simple shear (SS) test on seabed sand

Figure 4.20 shows results from a CAU cyclic simple shear test carried out on the same saturated sand and consolidated under the same vertical and horizontal consolidation stresses as the monotonic test described earlier. Two-way symmetric cyclic loading was applied with a cyclic shear stress $\tau_{cy} = 15$ kPa, i.e. a much lower level than that would cause failure under monotonic loading. As in the monotonic test, the total vertical stress was held constant while changes in pore pressure $u_e$ and effective vertical stress $\sigma'_v$ were observed as a result of the cyclic loading. Reading the $\sigma'_v$ versus $\tau$ graph from the initial vertical effective stress state ($\sigma'_v = 75$ kPa), the first portion of the cyclic test looks similar to the first stage of the monotonic test as positive excess pore pressures are generated as the sample contracts, leading to a reduction in vertical effective stress. Shearing continues until a shear stress $\tau = 15$ kPa is reached, the shear stress is then reduced to $-15$ kPa and cycling at this amplitude continues. Excess pore pressure continues to build up with each cycle of loading unlike in the monotonic test where excess pore pressure decreased after initially increasing. Eventually, the excess pore pressure at the mid-point of each cycle reaches the total applied vertical stress $\sigma_v$, such that the effective vertical stress $\sigma'_v$ becomes zero. The moment when the effective stress first falls to zero is called 'initial liquefaction'. Note, however, that after this point has been reached, the sample tends to dilate (with reduction in excess pore pressure) as it is sheared, leading to the butterfly shaped stress paths, and the S-shaped cycles in ($\gamma$, $\tau$) space.

Failure under cyclic loading is not necessarily taken to correspond with the onset of liquefaction but is usually defined at a predetermined level of shear strain, before

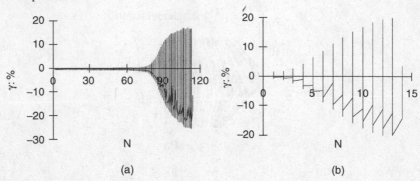

*Figure 4.21* Comparison of results from CIU cyclic SS tests on seabed sand with (a) $\tau_{cy}/\sigma'_{vc} = 0.23$ and (b) $\tau_{cy}/\sigma'_{vc} = 0.33$ (Mao 2000)

liquefaction would occur. For example, a limiting total shear strain (permanent plus cyclic) of 15 per cent is commonly taken as failure, although serviceability limits may be established using smaller levels of shear strain. The graph of shear stress $\tau$ versus shear strain $\gamma$ (reading from the origin) shows shear strain accumulates at a very slow rate initially, with steep hysteresis cycles close to the origin. When initial liquefaction occurs, at about the 20th cycle (seen in the graph of $\sigma'_v$ versus $\tau$), shear strain increases and stiffness reduces rapidly.

### Effect of cyclic shear stress $\tau_{cy}$

Figure 4.21 shows results from isotropically consolidated undrained (CIU) cyclic simple shear tests on the same seabed sand as described above, carried out at two different cyclic shear stress ratios $\tau_{cy}/\sigma'_v = 0.23$ and 0.33. The overall responses of the two tests are similar and as described for the foregoing CAU SS test on seabed sand – excess pore pressure is generated rapidly initially and then increases at a slower rate, eventually reaching the value of total applied vertical stress and the soil liquefies. Shear strain increases initially at a very slow rate until initial liquefaction occurs after which the soil loses shear strength and shear strain increases rapidly. The significant difference between the two tests is the number of cycles to reach failure $N_f \sim 100$ for $\tau_{cy}/\sigma'_v = 0.23$ and $\sim 7$ for $\tau_{cy}/\sigma'_v = 0.33$.

### 4.2.5 *Cyclic resistance curves*

Cyclic resistance curves are used to define the cyclic shear stress $\tau_{cy}$ to reach a given value of shear strain $\gamma$ after a certain number of cycles N. Shear stress is typically normalised by the consolidation stress $\tau_{cy}/\sigma'_{vc}$ or by the monotonic shear strength, e.g. $\tau_{cy}/s_{uss}$. Results of cyclic loading tests (whether triaxial or simple shear tests) can be used to construct cyclic resistance curves of strain and pore pressure contours, which can then be used in design.

### *Strain contour diagrams*

Figure 4.22 shows a strain contour diagram constructed with results from a monotonic and four undrained symmetric cyclic simple shear tests carried out with

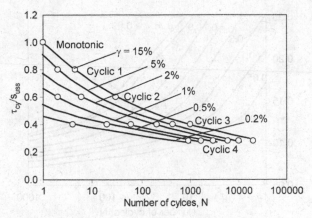

*Figure 4.22* Strain contour diagram, $\tau_a = 0$

components of cyclic shear stress $\tau_{cy}$ equal to 80, 60, 40 and 28 per cent of the monotonic undrained shear strength under simple shear $s_{uss}$. For each cyclic test, the number of cycles to reach 0.2, 0.5, 1, 2, 5 and 15 per cent shear strain $\gamma$ is determined from the test data from the shear strain development curve in $(N, \gamma)$. These points are then plotted in $(N, \tau_{cy}/s_{uss})$ space and the points corresponding to each shear strain level are connected to give contours of equal shear strain. The number of cycles that would cause a given shear strain can then be identified for any value of $\tau_{cy}/s_{uss}$. For example, from Figure 4.22, for this material subjected to a cyclic shear stress of $0.5s_{uss}$, failure defined by a shear strain $\gamma = 15$ per cent would be reached at around 200 cycles.

It is intuitive to expect a reduction in strength with cycling. However, the cyclic strength $\tau_{cy}/s_{uss}$ for small numbers of cycles can be greater than the monotonic strength due to the strain rate dependence of shear strength of most soils. Strength may be lower at the slow rate of shearing used in a monotonic test (typically 1–5 per cent per hour) than at the faster rate of loading used in a cyclic test (with cyclic periods of 10–20 seconds). Hence, failure during cyclic loading would be reached in a time that is some three orders of magnitude lower than during a monotonic test. For clay soils, shear strength has been found to increase by about 10 per cent with each order of magnitude increase in strain rate, hence the shear strength during a cyclic test might be 30 per cent higher than measured in a slow monotonic test.

## Pore pressure contour diagrams

The magnitude of excess pore water pressures generated under cyclic loading can be expressed as a function of the number of cycles of loading, in the same way as described previously for strains, in order to assess the potential for a soil to liquefy under operational conditions. Pore pressure contour diagrams are constructed in the same way as strain contour diagrams. The number of cycles $N$ required to generate a certain pore pressure ratio, typically normalised by the vertical consolidation stress as $u_e/\sigma'_{vc}$, is determined from a variety of cyclic tests and these data points can be joined up to give a series of contour lines for different pore pressure ratios (Figure 4.23).

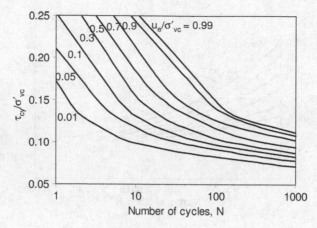

*Figure 4.23* Pore pressure contour diagram, $\tau_a = 0$

*Table 4.1* Hypothetical storm (Andersen *et al.* 1992)

| Number | $h/h_{max}$ (%) | $\tau/\sigma'_{vc}$ |
|---|---|---|
| 1 | 100 | 0.200 |
| 2 | 95 | 0.190 |
| 4 | 86 | 0.172 |
| 15 | 70 | 0.140 |
| 30 | 61 | 0.122 |
| 50 | 49 | 0.098 |
| 400 | 40 | 0.080 |
| 700 | 33 | 0.066 |

### 4.2.6 *Shear strain or excess pore pressure accumulation procedure*

Wave height, and therefore, stress intensity varies during a storm, starting off small and increasing to a maximum before decreasing. As a result, cyclic loading applied in a storm is not uniform, unlike in laboratory tests, but increases as the storm reaches its peak and reduces as the storm passes. Rather than model the complete loading sequence, offshore design often makes use of an 'equivalent' number of cycles of the peak load to represent the cumulative damage that occurs under the full sequence. Either the shear strain or excess pore pressure contour diagram may form the basis for establishing the equivalent number of cycles.

Table 4.1 shows a hypothetical storm made up of packets of waves of different heights expressed as a percentage of the height of the maximum wave $h/h_{max}$ and the cyclic shear stress ratio $\tau_{cy}/\sigma'_{vc}$ imposed by each height of wave (Andersen *et al.* 1992). Figure 4.24 illustrates the procedure for excess pore pressure accumulation and the definition of an equivalent number of cycles based on the design storm given in Table 4.1.

Starting with the smallest (most frequent) loading level, the excess pore pressure ratio that would develop under that cyclic shear stress is estimated by plotting the data point at the corresponding values of $\tau_{cyc}/\sigma'_{vc}$ and the number of cycles. The notional contour for that magnitude of excess pore pressure ratio is then traced back (parallel to the closest actual contour) to reach the next higher cyclic shear stress level

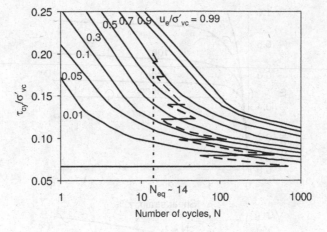

*Figure 4.24* Pore pressure accumulation procedure

in the storm sequence. That point represents an equivalent number of cycles at this cyclic shear stress level to give the same degree of damage as the larger number of cycles at the lower cyclic shear stress level. The process is now repeated, but with the end point from each stage plotted by adding the number of cycles at the new cyclic shear stress level to the (deduced) equivalent number of cycles obtained so far (from the previous loading levels). The process finishes with the peak design load level, and the final point represents the equivalent number of cycles for that design load level, and that particular storm sequence. Typically, the equivalent number of cycles will be in the range 10–20 cycles. The example shown in Figure 4.24 indicates an equivalent number of cycles $N_{eq} \sim 14$, i.e. it would take approximately 14 cycles of the maximum wave (given in Table 4.1) to achieve the same excess pore pressure ratio as would be generated by the entire storm of varying waves.

### 4.2.7 Construction of a cyclic stress–strain curve

The stress ratios $\tau_{cy}/s_{uss,c,e}$ (or $\tau_{cy}/\sigma'_{vc}$) required to achieve various shear strains $\gamma$ in N cycles can be determined from a strain contour diagram (such as shown in Figure 4.22) and used to construct a cyclic stress–strain curve (Figure 4.25). The cyclic shear modulus $G_{cy}$ can be determined from the tangent of a graph of cyclic shear strain to cyclic shear stress $d\tau_{cy}/d\gamma_{cy}$.

### 4.2.8 Unsymmetric cyclic loading

The strain contour and pore pressure contour diagrams described in the preceding section are constructed based on symmetrical cyclic loading tests, i.e. two-way cyclic loading about zero average stress $\tau_a$. In reality, the stress conditions in the soil beneath structures subjected to combinations of static and cyclic loads are more complex and the effect of the level of average shear stress on the cyclic load response should be considered. The load path and resulting strain and pore pressure time histories of a soil element subjected to a combination of non-symmetric average and cyclic shear

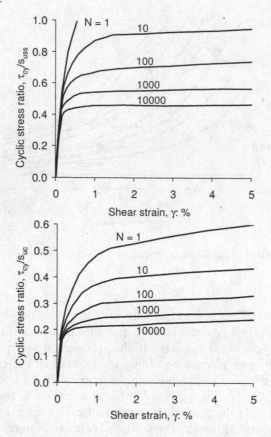

*Figure 4.25* Cyclic stress–strain curves as a function of number of cycles in (a) simple shear and (b) triaxial compression tests on normally consolidated Drammen clay (Andersen 2004)

stresses are illustrated in Figure 4.26. When average shear stress is increased by an increment $\Delta\tau_a$ to $\tau_a$, the soil will experience an increment in average shear strain $\Delta\gamma_a$ and average pore pressure $\Delta u_a$. The cyclic shear stress $\tau_{cy}$ will cause cyclic and average shear strains $\gamma_{cy}$ and $\gamma_a$ and cyclic and average pore pressures $u_{cy}$ and $u_a$ that increase with number of cycles of loading.

Data from four cyclic simple shear tests on a calcareous muddy silt are shown in Figure 4.27. The soil samples were consolidated under $\sigma'_{vc} = 150$ kPa and $\sigma'_{hc} = 60$ kPa and shear tests were carried out under different average stress conditions: (1) $\tau_a = 0$, (2) $\tau_a < \tau_{cy}$, (3) $\tau_a = \tau_{cy}$, and (4) $\tau_a > \tau_{cy}$ (essentially the stress conditions illustrated in Figure 4.18). The common behaviour of the four tests is that excess pore pressure is generated rapidly at the beginning of each test and then increases slowly while significant differences are observed in the stress–strain and shear strain development response. For the symmetrical test, Test 1, shear strain develops symmetrically (i.e. average shear strain $\gamma_a$ remains close to zero); problems of stability would, therefore, arise only due to the degradation of cyclic shear strength. As the component of average shear stress $\tau_a$ increases, the component of average (or permanent) shear strain $\gamma_a$ increases and the component of cyclic shear strain $\gamma_{cy}$ decreases. In Test 2, in which a

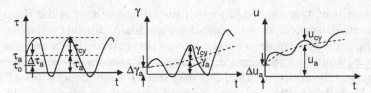

Figure 4.26 Development of shear strain and pore pressure during unsymmetric cyclic loading (Andersen 2009)

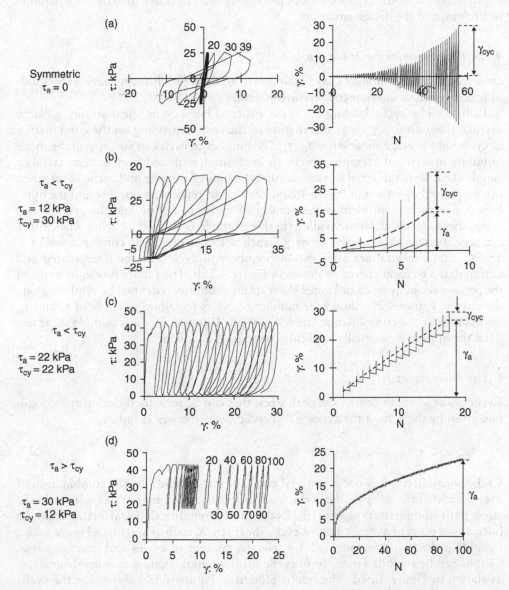

Figure 4.27 Effect of average shear stress $\tau_{av}$ on cyclic behaviour (a) Test 1: $\tau_a = 0$, (b) Test 2: $\tau_a < \tau_{cy}$, (c) Test 3: $\tau_a = \tau_{cy}$ and (d) Test 4: $\tau_a > \tau_{cy}$ (Mao and Fahey 2003)

small component of $\tau_a$ is applied, although the component of cyclic shear strain $\gamma_{cy}$ decreases, some permanent strain accumulates gradually during cycling. In Test 3, in which cycling takes place under a higher average shear strain, the component of cyclic shear strain $\gamma_{cy}$ continues to decrease, but a greater component of permanent shear strain is observed. Problems with stability may start to be influenced more by accumulated strain than degradation of shear strength under these conditions. In Test 4, in which $\tau_a > \tau_{cy}$, the total shear strain $\gamma$ consists almost entirely of average shear strain $\gamma_a$ with very little cyclic shear strain $\gamma_{cy}$ induced. The variation in response to different modes of cyclic loading, as shown in Figure 4.27, indicates the importance of selecting appropriate cyclic stress conditions in laboratory tests in order to determine parameters relevant to the design situation.

### 4.2.9  *Failure under cyclic loading*

Data from NGI's Drammen Clay database are used here to illustrate various methods of presentation of unsymmetric cyclic test data.

Failure under cyclic loading can occur either as large cyclic shear strains $\gamma_{cy}$, large average shear strains $\gamma_a$ or a combination of the two, depending on the combination of cyclic and average shear stress $\tau_{cy}, \tau_a$. The number of cycles to failure and the mode of failure in terms of average or cyclic shear strain determined from either triaxial or simple shear tests subjected to various combinations of average and cyclic shear stress can be plotted as shown in Figure 4.28a. Each point represents one test and the numbers beside each point identify the number of cycles to failure and the average and cyclic shear strains at failure. Failure in this example is defined as being when either cyclic or average shear strain, $\gamma_{cy}$ or $\gamma_a$, reaches 15 per cent. The combinations of $\tau_a$ and $\tau_{cy}$ that cause failure after various numbers of cycles can be interpolated and extrapolated to form curves as shown in Figure 4.28b. The failure mode, in terms of the percentage of average and cyclic shear strain at failure, is defined by symbols along the curves. Figure 4.28c shows the number of cycles to failure in terms of $\tau_a$ and $\tau_{cy}$ normalised by the consolidation stress $\sigma'_{vc}$. While Figure 4.28 shows simple shear test data, the approach is equally applicable to triaxial test data.

### *Cyclic shear strength*

Cyclic shear strength defines the peak stress that can be mobilised during cyclic loading, given by the sum of the average and cyclic shear stresses at failure,

$$\tau_{f,cy} = (\tau_a + \tau_{cy})_f \tag{4.27}$$

Cyclic shear strength is not a material constant but depends on the combination of average and shear stresses, the cyclic loading history (the number of cycles) and the stress path (simple shear or triaxial). Cyclic shear strength can be calculated from the normalised plots of average against cyclic shear stress, such as shown in Figure 4.28c, by adding the two components for a selected number of cycles and average stress. Graphs can be replotted in terms of cyclic shear strength against average shear stress as shown in Figure 4.28d. The results plotted in Figure 4.28d show that the cyclic shear strength and failure mode depend on the level of average shear stress $\tau_a$ and the number of cycles N. Cyclic shear strength and failure mode also depend on the type

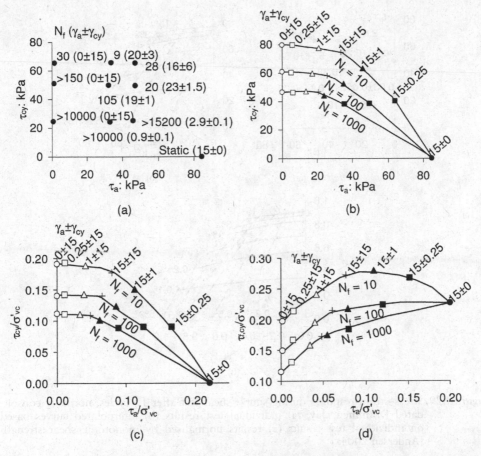

*Figure 4.28* Presentation of unsymmetric cyclic simple shear test data normally consolidated Drammen clay, number of cycles to failure defined by $\gamma_a$ or $\gamma_{cy} = 15$ per cent (a) individual test results; (b) interpolated curves based on individual test results; (c) results normalised by consolidation stress; (d) cyclic shear strength (Andersen 2004, 2009)

of test, for example, simple shear or triaxial. For triaxial tests, there is a difference between cyclic shear strengths in compression and extension. Compression failure occurs with average shear strain $\gamma_a > 0$ at failure and extension failure with $\gamma_a < 0$. Diagrams of this type, for different forms of laboratory test, have been presented by Andersen (2004, 2009).

### 4.2.10 Deformations under cyclic loading

Deformations due to cyclic loading are due to increased permanent shear strains due to the cyclic loading and dissipation of cyclically induced pore pressures. Determination of the stress–strain and stress–pore pressure relationships under cyclic loading is required to calculate cyclically induced displacements.

### Stress–strain relationship

It has been shown that both average strain $\gamma_a$ and cyclic strain $\gamma_{cy}$ depend on the combination of average and cyclic shear stress, $\tau_a$ and $\tau_{cy}$. Figure 4.29a shows average and cyclic shear strains after 10 cycles of loading in a simple shear test plotted as a

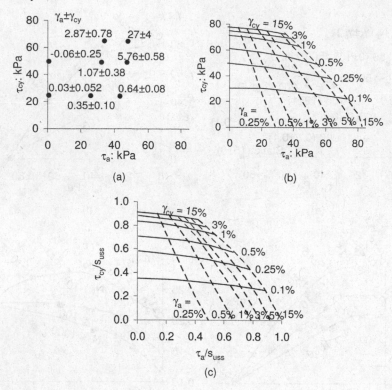

*Figure 4.29* Stress–strain response during simple shear test after 10 cycles, normally consolidated Drammen clay, (a) individual test results (b) interpolated curves based
on individual test results (c) results normalised by monotonic shear strength
(Andersen 2004)

function of the applied average and cyclic shear stress. Each point represents one test
and the numbers written beside each point give the measured average and cyclic shear
strains, $\gamma_a \pm \gamma_{cy}$, after 10 cycles. Interpolation and extrapolation of the data lead to
curves showing average and cyclic shear strains (Figure 4.29b). The solid lines represent the cyclic shear strains $\gamma_{cy}$ and the broken curves represent the average shear
strains $\gamma_a$. The stress–strain relationship as shown in Figure 4.29 can be transformed
to a general form by normalising the shear stresses by the monotonic simple shear
strength $s_{uss}$ after N cycles of loading, as shown in Figure 4.29c. Alternatively, the
shear stresses can be normalised by the vertical effective consolidation stress $\sigma'_{vc}$.

Cyclic shear strains $\gamma_{cy}$ for a given constant average shear stress $\tau_a$ can be plotted as a
function of the number of cycles as strain contour diagrams, as shown in Figure 4.22 for
symmetric cyclic loading. Similarly, cyclic stress–strain curves, as shown in Figure 4.25,
can be constructed for a given constant average shear stress $\tau_a$ from the strain contour
diagrams. Displacements under cyclic loading can be predicted by numerical analyses
using the stress–strain relationship identified in a cyclic stress–strain curve as the constitutive soil model. Analytical solutions using a best-estimate constant soil modulus may
be relied on for preliminary assessment if accurate predictions are not required.

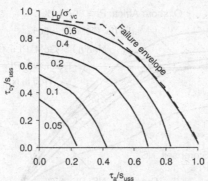

*Figure 4.30* Normalised permanent pore pressure ratio during simple shear tests on normally consolidated Drammen Clay for N = 10 cycles of loading (Andersen 2004)

### Stress–pore pressure relationship

Permanent pore pressures generated by undrained cyclic loading depend on the combination of average and cyclic shear stresses, the number of cycles and the type of test in a similar manner to the stress–strain response. It is, therefore, useful to plot the stress–pore pressure response in the same type of diagram as for shear strains (Figure 4.29c). Figure 4.30 shows permanent excess pore pressures $u_p$, normalised by monotonic shear strength, plotted against average and cyclic shear stresses after 10 cycles of loading in simple shear tests. The pore pressure diagrams are constructed based on the individual test results in the same way as described for the shear strain diagrams. Pore pressure ratio $u_p/\sigma'_{vc}$ for a given constant average shear stress $\tau_a$ can be plotted as a function of the number of cycles as pore pressure contour diagrams, as shown in Figure 4.23 for symmetric cyclic loading.

Pore pressure measurements on clay samples in cyclic tests are difficult to perform with a high degree of accuracy due to non-uniform stress conditions within the sample and the time required to reach pore-pressure equilibrium within the sample. The compliance of the pore-pressure measurement system must also be considered. Nonetheless, reasonably accurate values of permanent pore pressures may be obtained provided care is taken to obtain high-quality measurements.

### 4.2.11 Cyclic load test database

A very large programme of laboratory tests is required to develop a complete picture of the cyclic response of a given soil, and few offshore projects justify sufficient tests to achieve this given that the testing programme must also address variations due to depth (different strata) and also a region that may span an area of several square kilometres. Over the last three decades, an enormous quantity of data has been compiled by the offshore industry, and it is now common to minimise new test programmes by performing a limited number of tests and comparing these results to standard databases to provide a basis for generic adjustment of the failure contours. The Norwegian Geotechnical Institute (NGI) have compiled the most extensive database available on the simple shear and triaxial cyclic performance of clays, underpinned by the Drammen Clay database (Andersen *et al.* 1980, Andersen *et al.* 2009) and subsequently

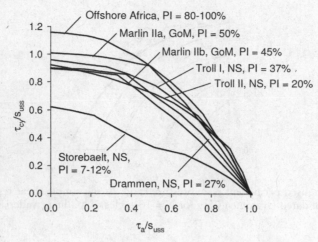

*Figure 4.31* Cyclic simple shear test data for various marine clays for $N_f = 10$, OCR $= 1$
(Andersen 2004)

augmented with various other marine clays (Andersen 2004). A summary of cyclic
simple shear test results on several marine clay soils is shown in Figure 4.31 giving
combinations of average shear stress $\tau_a$ and cyclic shear stress $\tau_{cy}$, normalised by the
monotonic shear strength in simple shear $s_{uss}$ required to cause failure ($\gamma = 15$ per cent)
in 10 cycles. The high plasticity clays tend to show greater strain rate dependency, and
hence show greater cyclic shear strength, while the reverse is true for low plasticity
clays. Under simple shear conditions, there is remarkable consistency in the cyclic
shear strength envelopes of all clays, with the only exception being the extreme example of Storebaelt, with a plasticity index of 7–12 per cent.

# 5    Piled foundations

## 5.1  Introduction

### 5.1.1  Structure of Chapter 5

This chapter covers piled foundations. First, piled foundations are introduced and the construction methods for different offshore pile types are explained. After considering the soil behaviour during installation, design methods for predicting axial capacity are presented. This analysis is then extended to the full axial load-settlement response. Finally, the analysis of a pile under lateral loading is described.

The design methods for piled foundations balance the conflicting need for a design to (a) account for the complex processes occurring during installation and loading of a pile and (b) be carried out using the restricted information available from a geotechnical site investigation.

Most methods for assessing the axial or lateral capacity of a pile have empirical elements, although recent research has improved the understanding of pile behaviour. In this book, recent data of the stress changes induced by pile installation and loading are used to explain and assess the empirical methods used to estimate pile strength and stiffness. These stress changes are too complex to be modelled explicitly in routine design. However, empirical methods should not be relied on without an appreciation of the underlying governing mechanics.

### 5.1.2  Offshore applications of piled foundations

Deep piled foundations are favoured over shallow foundations in situations where soft soil is at the surface and where high horizontal loads are present (which would cause a surface foundation to slide). At a particular site, the types of piled foundation that can be constructed depend on the geotechnical conditions. Offshore piles vary in diameter from around 30″ (0.76 m) for wellhead conductors to over 4 m for large monopile foundations, with typical diameter to wall thickness ratios of 25–100.

Driven steel piles are the traditional method for supporting offshore steel platforms. Small jacket structures typically have one pile at each corner of the platform, aligned with the main structural members of the jacket. Intermediate structures may include 'skirt' piles – situated along the longer sides of a rectangular jacket structure – while major platforms may involve a cluster of piles at each corner. For example, on the North-West Shelf near Australia, the North Rankin A platform was designed with

---

Primary author of this chapter was David White.

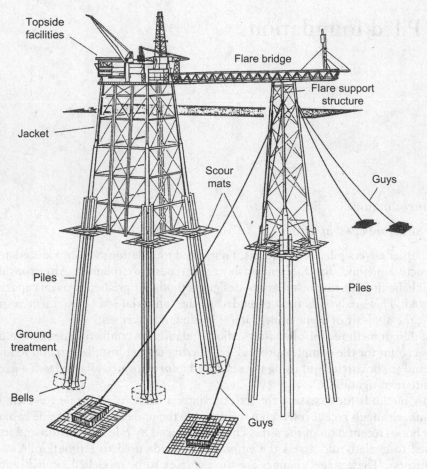

Topside
facilities

Flare bridge

Flare support
structure

Jacket

Scour
mats

Guys

Piles

Piles

Ground
treatment

Bells

Guys

*Figure 5.1* North Rankin A foundation arrangement (Woodside Petroleum 1988)

eight piles at each of the four corners (Figure 5.1), while the nearby Goodwyn A platform has five piles per corner.

Significant cost savings can be made if the number of piles supporting a jacket can be reduced. The lower fabrication and installation costs (plus the associated reduction in programme time) make careful optimisation of the design worthwhile.

However, any over-estimation of the pile capacity can prove expensive, due to the difficulty in providing additional foundation capacity. During construction of Woodside's North Rankin A platform in 1982, it was found that the shaft capacity of the driven piles was considerably lower than estimated using conventional design methods. Previous worldwide design experience was primarily in siliceous sands and proved unconservative for the calcareous sands found offshore Australia.

The NRA platform was initially uncertified for long return period storms, and Woodside de-manned the platform when cyclones approached. After remedial works costing AU$340 million (at 1988 prices), the platform was fully certified in February 1988 (Jewell and Khorshid 1988).

To avoid this low capacity in calcareous sands, the alternative of drilled and grouted piles has been adopted. A hole is drilled and a steel tubular pile is grouted into the

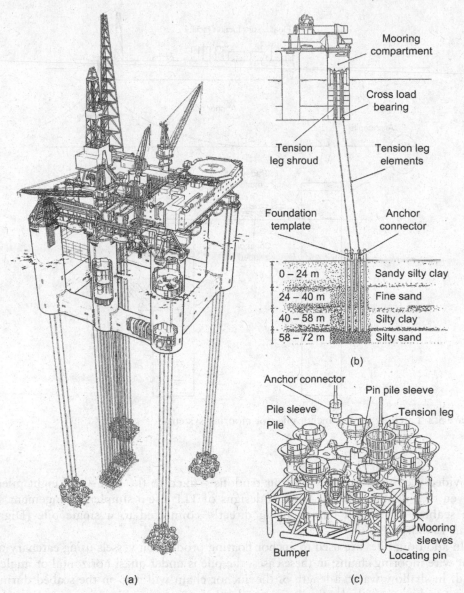

*Figure 5.2* Hutton tension-leg platform foundation arrangement (Tetlow *et al.* 1983)

hole. This is a time-consuming (and thus expensive) operation. The alternative, of driving a pile but then pressure grouting along the shaft surface (a 'grouted driven' pile), is attractive economically, but there are problems of quality control since the grout coverage is not easy to monitor.

Piles can also be used as anchors for floating structures, such as for tethering tension-leg platforms. In these cases, the pile is under vertical upwards load. The first tension-leg platform was installed at the Hutton field in the North Sea in 1982, and was supported by driven piles (Tetlow and Leece 1982) (Figure 5.2). A template was placed at the seabed beneath each corner of the TLP hull. The templates

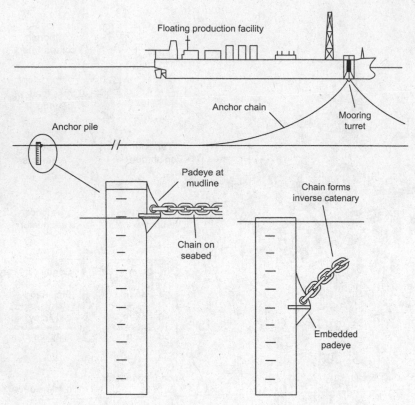

*Figure 5.3* Laterally loaded anchor piles for mooring systems

provided a connection between four tendons – fixed to the hull – and eight piles, driven to a depth of 58 m. Modern designs of TLP use a simpler arrangement at the seabed, with each tendon being directly connected to a single pile (Digre *et al.* 1999).

In addition, piles are used to anchor floating production vessels using catenary or taut wire mooring chains; in these cases, the pile is under quasi-horizontal or angled load. In shallow water, a length of the anchor chain will rest on the seabed during calm conditions, with the small mooring loads being resisted by friction against the seabed. Under storm loading, the chain applies a one-way horizontal cyclic load to the anchor pile. The chain may be connected to the anchor pile at the soil surface, or it may be connected via an embedded pad-eye, which provides a more efficient arrangement. In deep water, a taut line mooring is common, with the mooring line angled at 35° to the horizontal or steeper. Illustrations of typical anchor pile moorings are shown in Figure 5.3.

A further offshore application of piles is to support wind turbines. For this application, the design load is dominated by overturning moment, so piles are short and stubby, with a large diameter to provide sufficient lateral stiffness. Steel tubular 'monopiles' of up to 4 m diameter have been installed in the North Sea to support wind turbines.

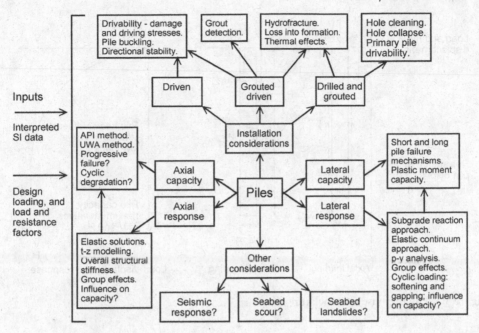

*Figure 5.4* Some design considerations for piled foundations

### 5.1.3 Basis of design for piled foundations

The design of a piled foundation must consider all aspects of the installation and performance of the system. Figure 5.4 summarises the main aspects that might need to be considered for a piled foundation and the primary analysis tools for this purpose.

For any given application, the geotechnical design process would involve (a) assessment of the site characteristics and the design conditions, then (b) addressing each aspect of the basis for design, grouping them appropriately. For example:

- Installation (drivability, hole stability and grouting)
- Axial capacity and performance under axial cyclic loading
- Lateral capacity and performance under lateral cyclic loading
- Group effects (leading to an overall foundation stiffness)
- Other considerations (seismic response, local seabed stability and scour).

The design of a piled foundation should ensure that the piles can be reliably installed or constructed to the target penetration, and that the resulting foundation has sufficient stiffness and strength to resist the design loads. The foundation should be optimised for cost by minimising the number and length of the piles (hence less material requirement and a shorter installation time)

In some respects, pile design is similar to shallow foundation design. As for shallow foundations, pile foundation strength is usually linked to undrained strength in clay and friction angle or cone resistance in sand, although there is an increasing trend for pile design to be based on cone resistance for all soils. In addition, pile foundation

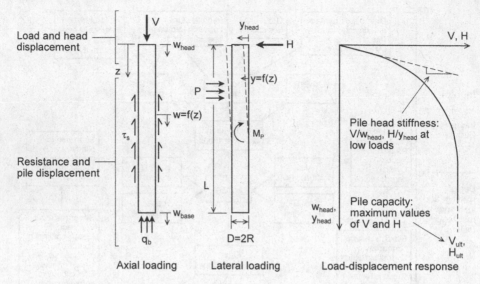

*Figure 5.5* Pile response nomenclature

stiffness is calculated from an operative soil stiffness using elastic solutions. However, piled foundations differ from shallow foundations in that:

1. The analyses that link soil strength and foundation capacity are less rigorous and more empirical for piles. This is partly because the failure mechanism of a pile – especially at the base – cannot be captured by analytical solutions. In addition, the soil properties and stress states that are inputs to these analyses are often modified due to the installation process.
2. Analyses for the strength and stiffness of piles, accounting for non-linear response and layering of the soil, can rarely be applied in closed-form, and often require numerical implementation due to the variation in soil properties through layered strata.

A simplification of piles compared to shallow foundations is that the interaction between combined loads is less significant: the application of a horizontal load does not significantly affect the vertical capacity and vice versa. This is because horizontal load is resisted over the upper few diameters of the pile, whereas vertical load is resisted on the lower part of the pile, where the stress and soil strength is generally higher.

### 5.1.4 Notation for loads, resistance and geometry

The notation used to describe components of pile load and resistance, and the geometry of piles, is summarised in Figure 5.5.

- *Applied loads.* As for shallow foundations, the loads acting at the pile head are divided into vertical V and horizontal components. If there is fixity at the pile head, or the horizontal load is applied to the pile other than at the soil surface, a moment load M may also be present. The maximum loads (ultimate capacity, failure) are denoted by the subscript 'ult'.

- *Displacements and rotations.* Vertical displacement is denoted by w and varies along the pile length due to compression of the pile. Horizontal displacement is denoted y and varies along the pile length due to bending of the pile. The rotation at the pile head is denoted $\theta$.
- *Axial resistance.* The axial capacity arises from shaft resistance and base resistance. The unit shaft resistance (a shear stress) is denoted $\tau_s$, and the unit base resistance (a stress) is denoted $q_b$.
- *Lateral resistance.* The lateral capacity arises from the normal and shear stresses acting horizontally on the shaft of the pile. These stresses vary around the circumference of the pile in a complex manner. In design, they are lumped into the force per unit length of pile which can then be treated as a distributed load varying along the length of the pile. The ultimate value is denoted $P_{ult}$. Note that P is not a stress (hence the upper case notation).
- The *bending strength* or plastic moment capacity of the pile is denoted $M_P$. Lateral pile failure can arise if either (a) the ultimate lateral soil resistance is exceeded ('geotechnical failure'), with the pile failing as a (notionally) rigid body or (b) the pile fails in bending ('structural failure').

## 5.2 Pile types

### 5.2.1 Design considerations between pile types

Two types of offshore pile are described: conventional driven steel piles and grouted piles (which are generally drilled and grouted, although a hybrid form of grouted driven pile has also been proposed). Driven piles are far more prevalent worldwide, so are given greater attention in this chapter. Grouted piles are favoured in cemented sediments and rock, so are used in some circumstances offshore Australia and in the Middle-East. Design methods for these types of pile are described briefly in this chapter.

### 5.2.2 Driven steel piles

Open-ended driven steel pipe piles are by far the most common offshore platform foundation. They are a type of *displacement* pile: the pile is installed by displacing rather than removing the soil. They are installed nowadays using hydraulic hammers that can operate underwater, although previously steam and diesel hammers were common, which were restricted to operating above water. The piles are usually fabricated in lengths of up to 100 m and spliced during installation if necessary.

To support a jacket structure, the piles are driven through sleeves attached to the structure. In shallow water, the piles can be driven using above-water hammers mounted on pile extensions (so-called 'followers'). In deep water, extensions are impractical, and underwater hammers are used. Modern underwater driving hammers are designed to follow the pile down through guiding sleeves on the jacket structure. After being driven to the required depth, the piles are welded or grouted to the sleeves at the base of the jacket legs. For anchor piles, support during driving is provided by a temporary frame at the seabed.

Piles are conventionally driven open-ended, with soil flowing into the pile and forming a 'plug'. A thickened wall is often adopted near the pile toe – termed a 'shoe' – to reinforce the tip and to reduce the driving resistance along the shaft in hard soils. If a stiffer base response is needed, a welded steel plate or (for larger piles) a conical tip is used to create a closed-end pile, with increased driving resistance. This approach has been used in the compressible calcareous sediments off the coast of Brazil, in order to provide high end-bearing capacity at moderate displacements, with the additional benefit of higher shaft resistance (De Mello *et al.* 1989, De Mello and Galgoul 1992).

Design considerations specific to this type of pile relate to drivability (especially through cemented sediments or dense sands) – see Section 5.4 – and to the consolidation (or 'set-up') period required to achieve full capacity. Drivability issues include refusal and tip damage. Refusal is when the required penetration cannot be reached due to the resistance exceeding the hammer capacity. If the pile tip is damaged before or during driving then the pile tip may buckle and collapse during driving (Barbour and Erbrich 1995). The resulting loss of pile shape may reduce the capacity, or it may lead to premature refusal. A case in which the latter occurred, requiring significant remedial works, is described by Alm *et al.* (2004).

Limitations of driven piles include the presence of a cemented caprock layer that might impede driving and damage the pile tip. In compressible and cemented soils, such as calcareous sands, driven piles can give very low shaft resistance.

### 5.2.3 Drilled and grouted piles

Drilled and grouted piles usually comprise a steel tubular pile inserted into an oversize drilled hole, which is filled with grout. These piles are similar to onshore bored piles, except that a steel tubular pile is used in place of a reinforcement cage.

This type of pile is used as an alternative to driven piles in rock (where driving is not possible), or in calcareous sediments (where driven piles suffer low shaft capacity). Drilled and grouted piles are costly to install, because of the long construction period and in some cases due to the need for primary driven piles through any soft overlying sediments.

The installation sequence of a drilled and grouted pile is as follows (Figure 5.6):

1. A 'primary' pile, comprising a standard steel pile, is driven through any shallow soft sediments, which might not stand open during drilling.
2. A rotary drilling rig is used to excavate beyond the primary pile to the required pile tip depth.
3. A steel tubular 'insert' pile is dropped into the excavated hole, and the annulus between the pile and hole is filled with grout, supplied via a cementing string at the bottom of the bore.

Drilled and grouted piles are expensive, due to the multi-stage construction process, and the relatively long construction period. Delays can occur if construction problems are encountered. The detailed design of drilled and grouted piles must consider:

1. Hole stability: if the bore is likely to collapse, support can be provided by a drilling mud. However, this can reduce the friction coefficient of the bore, even if the hole is flushed with sea water prior to grouting.

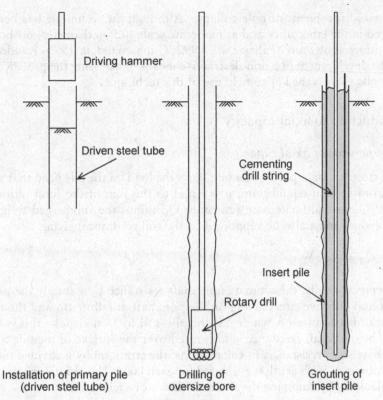

Installation of primary pile
(driven steel tube)

Drilling of
oversize bore

Grouting of
insert pile

*Figure 5.6* Installation stages of a drilled and grouted pile

2. The need (or not) for a primary pile (and the drivability of this pile).
3. The grouting operation.
4. Head of grout (or grout pressure): grouting at high pressure or using an expansive grout can lead to higher horizontal stress and hence greater shaft resistance; a high grout pressure will help to ensure the annulus around the insert pile is completely filled and may increase the shaft resistance, but too high a grout pressure will cause hydraulic fracturing of the formation.
5. Base cleaning: if soft drill cuttings fall to the bottom of the bore, the base response will be compliant.

Grouted driven piles represent a hybrid of a driven piles and a grouted pile case that has been proposed for calcareous soils (Rickman and Barthelemy 1988). This technique involves injection of grout along the pile–soil interface of a driven pile. The aim is to combine the simple installation process of a driven pile with the high shaft capacity of a grouted pile. This approach is similar to the onshore technology of post-grouting, but is yet to be applied to a major offshore project.

The grout is injected under pressure down grouting conduits, through holes drilled through the pile shaft prior to installation. The injection is completed in stages at a series of positions along the shaft. This technique has potentially significant cost advantages over the drilled and grouted pile, owing to the reduced construction time, and also avoids problems of hole collapse. Although the technique has been widely investigated in the laboratory and at moderate scale (0.9 m diameter) onshore, it has not been adopted offshore (Fahey *et al.* 1992, Gunasena *et al.* 1995, Randolph *et al.* 1996). The development of a non-destructive means of assessing the presence of grout along the pile shaft is the key to field use of this technique.

## 5.3 Introduction to axial capacity

### 5.3.1 Components of axial capacity

The axial strength, or capacity, of a pile $V_{ult}$ is the load on the pile head that will cause failure. From vertical equilibrium, it is equal to the sum of the total ultimate shaft resistance[1] $Q_{sf}$ and ultimate base resistance $Q_{bf}$ minus the submerged weight of the pile $W'_{pile}$ (which must also be supported by the soil resistance) giving

$$V_{ult} = Q_{sf} + Q_{bf} - W'_{pile} \tag{5.1}$$

In uncemented soils, the ultimate unit shaft resistance $\tau_{sf}$ is simply the product of the horizontal effective stress acting on the pile shaft at failure $\sigma'_{hf}$ and the mobilised coefficient of friction (tan $\delta$, where $\delta$ is the pile–soil friction angle) – this is Coulomb friction. The unit shaft resistance is integrated over the surface of the pile to find the total ultimate shaft resistance. In calculations, the stratigraphy is divided into layers. The variation in $\tau_{sf}$ with depth is evaluated for each layer. The ultimate shaft resistance is then calculated by summing the shaft resistance of each layer:

$$Q_{sf} = \pi D \int_0^L \tau_{sf} \, dz = \pi D \int_0^L \sigma'_{hf} \tan \delta dz \tag{5.2}$$

In clay soils, it is common to assess the shaft resistance using a correlation with the *in situ* undrained strength. There is no direct theoretical link between $\tau_{sf}$ and $s_u$; the ratio of these parameters is influenced by the roughness (hence friction angle) of the pile–soil interface and changes in stress and soil strength caused by the loading, re-moulding and consolidation processes that accompany pile installation. Since these effects can not easily be quantified, simple correlations have been developed with the general form

$$\tau_{sf} = \alpha s_u \tag{5.3}$$

It must be recognised that the friction factor varies depending on the underlying mechanisms of softening and consolidation, in order that an appropriate value of this empirical parameter is adopted in design. It also varies depending on the particular measure of undrained strength that has been used to define $s_u$.

In cemented soils the ultimate shaft resistance can be assessed based on some proportion of the intact strength (accounting for the disturbance during pile construction), through a correlation with cone penetration resistance, or through constant

---

[1]Also termed 'shaft friction' or 'skin friction'.

normal stiffness (CNS) direct shear tests that replicate the tendency of soil to dilate or contract as shearing occurs at the pile–soil interface (see Chapter 3).

The ultimate base resistance $Q_{bf}$ is the maximum stress that can be mobilised on the pile base $q_{bf}$ multiplied by the base area $A_b$ giving

$$Q_{bf} = \frac{\pi D^2}{4} q_{bf} \qquad (5.4)$$

The ultimate base resistance may only be mobilised after high pile settlement – perhaps in excess of one diameter for drilled and grouted piles in compressible carbonate sand. This may be an impractical settlement, prior to which the structure would suffer catastrophic failure anyway, so the 'ultimate' base resistance is often defined by the mobilised resistance at an allowable settlement – such as 10 per cent of the pile diameter (D/10).

On an open-ended pile, the base resistance is often assessed from two components acting on the pile wall and on the soil plug. At allowable settlements, the resistance on the pile wall is usually higher than on the soil plug. This is because the soil within the pile compresses as the axial load is mobilised, causing the soil plug response to be more compliant.

Estimation of the parameters $q_{bf}$ and $\tau_{sf}$ (or $\sigma'_{hf}$) in Equations 5.2–5.4 is challenging. Pile installation and loading is a process that causes complex stress changes in the soil around the pile from the *in situ* conditions to the conditions at failure. A design method is more robust if it has some basis in the underlying mechanics of the process (rather than being wholly empirical).

It is necessary to estimate the net result of installation and loading effects based only on knowledge of the *in situ* conditions prior to pile installation as identified during the site investigation – such as the *in situ* stresses and *in situ* strengths ($s_u$, $\phi$, $\delta$) or the results of an *in situ* test such as the cone penetration test (CPT) ($q_c$, $f_s$ etc.). Additional correlations involving simple lab tests such as plasticity index $I_p$ may also be used, and advanced laboratory tests can provide additional information on the cyclic behaviour.

The shaft resistance is affected by the rate of loading and cycles of loading. Therefore, the expected loading cycles due to environmental conditions (wind, waves) as well as the maximum single applied load are inputs to the design.

### 5.3.2 *Plugging of tubular piles*

A further consideration for tubular piles is the possibility of plugging. Alternative failure mechanisms are possible:

- *Unplugged penetration.* The soil column in the pile remains stationary ('coring').
- *Plugged penetration.* The soil column moves downwards with the pile.
- *Partial plugging.* The soil column moves downwards, but slower than the pile.

The degree of plugging is defined by the incremental filling ratio (IFR), which is illustrated in Figure 5.7.

$$\text{IFR} = \frac{\delta h_p}{\delta L} \qquad (5.5)$$

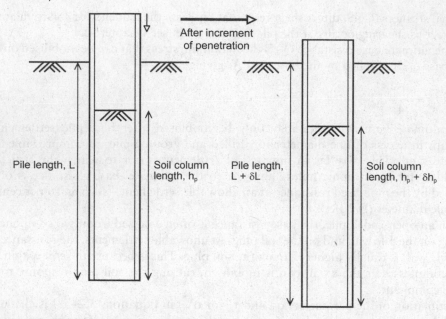

*Figure 5.7* Terms used to define incremental filling ratio

An IFR of zero corresponds to plugged penetration; an IFR of unity corresponds to unplugged (coring) penetration.

When calculating the capacity of an open-ended tubular pile, two calculations should be conducted: one considering unplugged failure and one considering plugged failure. The different components of resistance that must be overcome for each failure mechanism are shown in Figure 5.8. The mechanism with the lowest resistance will govern the pile capacity.

Plugging rarely occurs during driving. Piles usually drive in an unplugged manner, with the soil level in the pile remaining roughly at ground level. This is because the inertia of the soil column creates an additional component of resistance during driving, so that the unplugged penetration resistance is usually lower. However, under static loading piles usually fail in a plugged manner.

When plugging does occur during driving, it is usually when the pile tip passes from strong material into weak. At this point, the internal shaft resistance will be high (since the soil column comprises strong material) but the base resistance will be low (due to the underlying soft material). Plugging occurs when $V_{ult-p} < V_{ult-u}$. From Figure 5.8, (ignoring inertial effects) plugging would occur if

$$Q_{sf-i} > Q_{bf-p} - W_p \tag{5.6}$$

## 5.4 Pile drivability and dynamic monitoring

The detailed design of a driven pile must include a drivability study to assess the hammer energy required to install the pile to the required depth, and to provide an indication of whether refusal may be encountered. A drivability assessment leads to

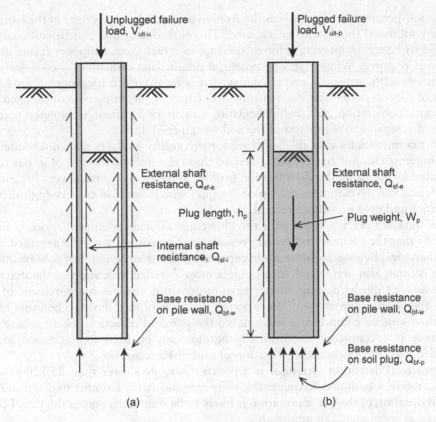

*Figure 5.8* Open-ended pile failure mechanisms (a) unplugged (IFR = 1), (b) plugged (IFR = 0)

the selection of a hammer, and a prediction of the anticipated blowcounts that will be required to install the pile. Hammer energies may exceed 3,000 kJ (e.g. 300 tonne force by 1 m stroke).

The drivability study must also ensure that the pile wall is not overstressed under the combined static load of the pile and hammer weights (allowing for any bending of the pile) and dynamic load from the pile driving. The driving process also causes fatigue that can represent a significant contribution to the fatigue capacity of a pile. The drivability assessment indicates the likely number of blows and the resulting cycles of stress that will be endured by the pile during installation.

A drivability analysis involves two stages. First, the soil resistance to driving (SRD) is assessed for short increments of pile penetration until the target embedment. Second, the dynamic driving process is simulated by a numerical method, using the adopted profiles of SRD. The same methods used to assess the static capacity of the pile (see Sections 5.5 and 5.6) are also used to assess the soil resistance to driving (SRD) in the drivability assessment. These approaches are sometimes modified to recognise the differing conditions during driving compared to subsequent static loading, although the enhanced resistance during driving due to rate effects is countered by the lack of set-up.

The dynamic driving process is simulated by software that discretises the pile as a system of springs, with the soil resistance being incorporated as spring-sliders and viscous dashpots. The most widely used software is GRLWEAP (PDI 2005).

The self-penetration of the pile under its own weight (plus the weight of the hammer and any follower) should also be assessed. This embedment may not be sufficient for the pile to be self-supporting prior to driving, in which case a support frame at the seabed is required. When assessing potential pile driving difficulties such as refusal, insufficient self-penetration and tip damage, it is necessary to use upper bound estimates of the soil strength. It is also prudent to consider the worst case of plugged and unplugged penetration, although penetration usually occurs in an unplugged manner during driving due to the inertia of the soil within the pile.

Most commercially available software for drivability analyses does not model the soil plug explicitly for open-ended piles, so that the effect of the soil plug has to be estimated by adjusting the shaft friction profile. In sands and gravels, very high internal friction can develop near the tip of the pile, which can lead to very high driving resistance and even premature refusal.

It is standard practice to monitor the blowcounts during pile installation, in order to verify that the encountered driving resistance is consistent with the assumed SRD and, therefore, by extension the static capacity adopted in design. Stress wave monitoring is often also performed, to provide a more detailed indication of the dynamic soil resistance during driving. Stress wave monitoring involves measurement of the strain and acceleration at one or more points above the seabed during a hammer blow. The blow sends a compression wave down the pile and reflections occur where soil resistance is encountered. Numerical programs can be used to back-analyse the recorded waves to assess the distribution of mobilised resistance.

A practical definition of refusal is a penetration rate slower than 250 blows per quarter metre. Continued driving at this slow rate may cause hammer damage and the overall duration of the pile installation is likely to be impractical given the vessel costs if refusal is encountered or approached.

Pauses during pile driving are inevitable if lengths of pile must be welded together and delays can also occur due to adverse weather or equipment problems. The possibility of set-up increasing the SRD of a partly installed pile during these periods should be considered.

## 5.5  Capacity of driven piles in clay

### 5.5.1  Stages in the installation and loading of a driven pile in clay

The shaft resistance of a driven pile in clay depends on the *in situ* conditions and the complex changes that take place during installation, excess pore pressure dissipation and subsequent loading. The life cycle of a driven pile in clay can be divided into a number of stages in order to understand the resulting shaft resistance (see Figure 5.9). Each stage is examined separately. It is useful to note the influence of the *in situ* over-consolidation ratio (OCR) of the clay when considering these stages, since this has an influence on the final capacity. Soft clay, with a low OCR, tends to contract when sheared, whereas stiff clay, with a high OCR, tends to dilate.

The following notation is introduced in Figure 5.9 to differentiate between the different stages:

- Subscript 0: *In situ* conditions
- Subscript i: Immediately post-installation

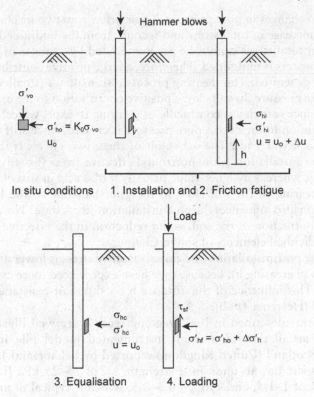

*Figure 5.9* Stages in the installation and loading of a driven pile in clay

- Subscript c: Post-equalisation – i.e. when any excess pore pressure around the pile has dissipated (the letter c is from consolidation, but the term equalisation is used to emphasise that different zones of soil will be either consolidating or swelling during pore pressure dissipation)
- Subscript f: Failure, as introduced previously.

The *in situ* hydrostatic pore pressure is denoted $u_0$, and excess pore pressures (relative to hydrostatic) are denoted $\Delta u$. The distance h describes the position of a point above the pile tip.

### 5.5.2 Stress changes during installation of a driven pile in clay

When a pile is driven into clay, the soil close to the pile undergoes high shear strains as the pile tip passes. There is a high degree of remoulding and a residual shear plane may form adjacent to the shaft. This process is usually considered to be undrained (no volume change). In high-permeability silty clays, some drainage may occur, particularly if driving is interrupted.

The total stress increases ($\sigma_{hi} > \sigma_{h0}$) as the soil is pushed away to accommodate the pile. The magnitude of this increase depends on the undrained strength of the soil, and is typically 4–6 $s_u$.

There are also changes in pore pressure. These arise from two mechanisms: first, in response to the increase in total stress and second, from the undrained shearing. The increase in mean total stress cannot be accommodated by changes in mean effective stress since the process is undrained. Therefore, a large positive contribution to excess pore pressure is generated. The shearing process also makes a (smaller) contribution to the excess pore pressure. In soft clay, a positive contribution to the excess pore pressure is created, since soft clay is contractile (it is trying to expel water). In stiff clay, a negative contribution to the excess pore pressure is created, since stiff clay is dilatant (it is trying to suck water in). The net result of these two effects is that during and immediately after installation a low horizontal effective stress (usually $< \sigma'_{h0}$) acts on a pile in soft clay, whereas a higher value (usually $> \sigma'_{h0}$) acts in stiff clay.

The cyclic shearing of the soil close to the pile shaft due to the hammer blows during installation also influences the post-installation stress state. The cycles of shearing encourage contraction of the soil, and a reduction in the effective stress level, as discussed for individual elements of soil in Chapter 4.

As a result, the post-installation horizontal effective stress is lower at points further from the pile tip (increasing h), because they have experienced more cycles of loading during driving. This influence of the distance h on the shaft resistance is known as 'friction fatigue' (Heerema 1980).

The mechanisms described in the foregoing section are well illustrated by field data from the installation of a small instrumented model pile in soft clay at Bothkennar in Scotland (United Kingdom) reported by Lehane and Jardine (1994). The clay at this site has an undrained strength, $s_u$, of 15–25 kPa (increasing with depth), an OCR of 1–1.5, and $s_u/\sigma'_{v0} \sim 0.3$–0.5, which is typical of an offshore soft clay site.

Figure 5.10a shows the *in situ* horizontal stress at the site and a profile of the pressuremeter limit pressure with depth. Also shown are the total stresses recorded on the base of the pile and at two elevations above the pile tip (h/D = 4, 7). During installation, $\sigma_{hi}$ exceeds $\sigma_{h0}$, although only by about 100 kPa. These values decrease with increasing h, due to friction fatigue. High positive excess pore pressures are evident (Figure 5.10b), but also decreasing with h. The resulting post-installation values of $\sigma'_{hi}$ are lower than $\sigma'_{h0}$.

### 5.5.3 Stress changes after installation of a driven pile in clay – equalisation

After installation, there exists a field of excess pore pressure around the pile. In soft clay, these pressures are entirely positive. In very stiff clay, close to the pile shaft where the dilatant shearing was greatest during installation, negative pore pressure may exist. However, in the far field, where less shearing occurs, the pressures are positive, to match the increase in total stress.

After installation, there is a process of equalisation, during which the excess pore pressures dissipate. This process is known as 'set-up' and it results in changes in the effective stress acting on the pile and, therefore, changes in the available shaft resistance. In silty clays, these changes can be significant in the first minutes after the end of driving (or after a pause in driving). In less permeable clays, the equalisation process can take months.

As positive pore pressures dissipate, water flows away radially (in the direction of decreasing excess pore pressure) and the soil close to the pile undergoes consolidation

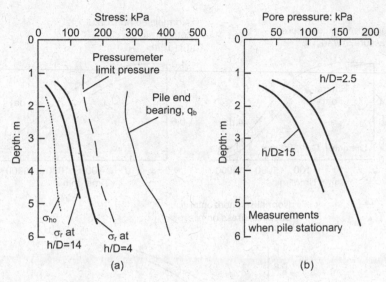

*Figure 5.10* Pile installation in soft clay (after Lehane and Jardine 1994a)

with a decrease in moisture content. At the same time, the total stress on the pile decreases (since the soil is contracting away from the pile).

In soft clay (with high initial $\Delta u$), the net effect is a significant increase in $\sigma'_h$. In stiff clay, the changes in $\Delta u$ and $\sigma_h$ are often comparable, leading to a relatively small increase in $\sigma'_h$ during equalisation. A reduction in $\sigma'_h$ is sometimes found after driving in soils that are strongly dilatant such as heavily overconsolidated clays and silts.

It is important to recognise that the process of pore pressure equalisation involves changes in total stress as well as pore pressure. In a one-dimensional consolidation test, the applied total stress is held constant, so during the pore pressure dissipation process any reduction in $\Delta u$ is balanced by an equal increase in effective stress, and the sample contracts. In contrast, during pile equalisation, the pile size is constant. It does not expand to maintain the same total stress on the surrounding soil when that soil contracts due to the outward flow of water. Therefore, for a typical soil response the total stress on a pile tends to fall during equalisation, as well as the excess pore pressure. Changes in shaft resistance, though, depend only on the effective horizontal stress.

The same instrumented pile test at Bothkennar provides a useful illustration of the equalisation process. The changes in pore pressure and total and effective horizontal (radial) stress after the end of installation are shown in a normalised manner in Figure 5.11. After a period of four days, the excess pore pressure dissipation is effectively complete (Figure 5.11a). Over the same period, the total stress falls (Figure 5.11b), but the horizontal effective stress increases by a factor of 3 (Figure 5.11c). At the end of equalisation, the effective stress on the pile shaft is approximately 2–3 times the intact undrained strength.

### 5.5.4 Duration of equalisation and 'set-up'

The time for equalisation to occur depends on the volume of soil from which dissipation must occur. The dissipation time, therefore, increases with the (square of the)

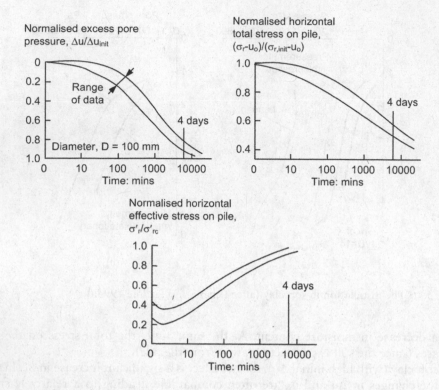

*Figure 5.11* Pile equalisation in soft clay (after Lehane and Jardine 1994a)

diameter of the pile, D. It also decreases with the coefficient of horizontal consolidation, $c_h$ (with flow assumed to be only in the radial direction, since L ≫ D).

For tubular piles of wall thickness t and diameter D, which generally penetrate into clay in an unplugged manner, it is more useful to consider a solid pile of equivalent area, which will generate the equivalent initial field of excess pore water pressure. The cross-sectional area of a tubular pile area is approximately $\pi Dt$, so the equivalent solid pile has a reduced diameter of $D_{eq} = 2\sqrt{Dt}$.

The relevant non-dimensional time factor for pore pressure equalisation around piles is $T_{eq} = c_h t/D^2_{eq}$ (where t is the time since the onset of dissipation). Simple models exist for the dissipation of excess pore pressure during equalisation. Linear–elastic perfectly plastic cavity expansion theory can be used to generate the initial excess pore pressure field, and consolidation theory can then model the dissipation. The resulting dissipation curves, which are shown in Figure 5.12, can be used to quantify the ratio of current excess pore pressure to post-installation excess pore pressure, $\Delta u/\Delta u_{,max}$, at dimensionless time $T_{eq}$. These two figures (Figure 5.12) are for a typical operative rigidity ratio $G/s_u = 100$. Higher stiffness would lead to a larger initial region of excess pore pressure, and hence slower normalised dissipation.

For practical design, it is difficult to predict the changes in effective stress during equalisation since the change in total stress is not well understood. This is a key obstacle to assessment of pile shaft capacity in clay based on an explicit calculation of the post-installation and then post-equalisation stresses on the pile. However, the time for equalisation to occur is a more straightforward quantity to assess, and is often a very

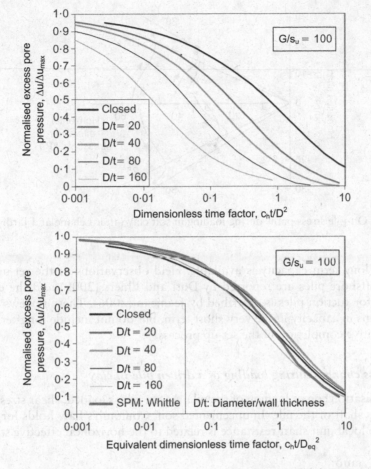

*Figure 5.12* Equalisation time for closed and open-ended piles (Randolph 2003a)

important consideration in design. In soft clays, the value of $T_{eq}$ when the piles will first be loaded should be evaluated to ensure sufficient dissipation of excess pore pressure has occurred. Any delay between pile installation in soft clay and application of the working load will increase the available shaft resistance.

From Figure 5.12, the dimensionless period for 90 per cent dissipation (i.e. $\Delta u/\Delta u_{,max} = 0.1$), denoted $T_{90}$, for all pile types is:

$$T_{90} = c_h t_{90}/D_{eq}^2 \approx 10 \qquad (5.7)$$

These curves can be used to estimate the relative dissipation times of different types of pile. Consider two piles of the same diameter: one closed-ended and one open-ended ($D/t = 40$). Since the curves of $T_{eq}$ closely overlie each other, it can be shown that $D/D_{eq} = 3.2$, so $t_{90,closed}/t_{90,open} \sim 10$. The equalisation period for a coring open-ended pile ($D/t = 40$) is ten times less than for a closed-ended pile of the same diameter in the same soil. This analysis is also applicable to suction caissons, to assess whether significant 'set-up' is likely during the suction installation process, and the likely period

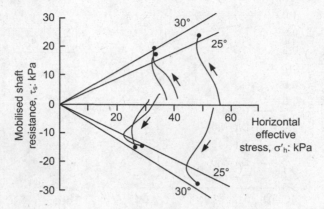

*Figure 5.13* On-pile stress paths during loading in soft clay (after Lehane and Jardine 1994a)

before the long-term capacity is available. Field observations of the set-up of large diameter offshore piles are reported by Dutt and Ehlers (2009) and the equivalent behaviour for suction piles is described by Jeanjean (2006). These reviews show significant gains in capacity in the very short term, but do not include long-term data to show definitive completion of the set-up process.

### 5.5.5 Stress changes during loading of a driven pile in clay

After equalisation comes loading of the pile. As the pile is loaded, shear stress is mobilised on the shaft of the pile. In uncemented soil, Coulomb's Law holds for the interface, so the local unit shaft resistance is related to the horizontal effective stress by:

$$\tau_{sf} = \sigma'_{hf} \tan\delta \tag{5.8}$$

By the end of consolidation, the radial stress will be higher than the vertical and circumferential stresses, so a reduction is expected on loading. Field data suggest this reduction is typically 20 per cent, so $\sigma'_{hf} \approx 0.8\sigma'_{hc}$. Further illustrative data from the model test reported by Lehane and Jardine (1994) are shown in Figure 5.13.

### 5.5.6 Interface friction angles in clay

In clay, two different values of the interface friction angle are usually considered: the peak value, $\delta_{peak}$, and the residual value reached after continued shearing, $\delta_{res}$. In monotonic shearing, $\delta_{peak}$ is usually mobilised within a few millimetres of interface displacement (typically 0.5–1 per cent of the pile diameter), followed by a reduction to $\delta_{res}$ over 10–50 mm. Tests show that $\tan\delta_{res}$ can be less than half of $\tan\delta_{peak}$, so the shaft resistance response in these types of clay can show a very brittle post-peak response.

It would be overly conservative to assume that the residual shaft resistance acts along the entire length of a pile, so it is common to quantify the degree of progressive failure that occurs due to the elastic compression of a pile. When the top part of the pile has sheared beyond peak resistance, the lower part will have not yet reached the peak value.

Therefore, the ultimate capacity of the pile is not governed by the peak interface friction angle, since $\delta_{peak}$ is not mobilised simultaneously along the pile length. The analysis of progressive failure is described in Section 5.8.9.

Interface friction angles for new sites should be determined from laboratory tests. This angle is affected by (i) clay mineralogy (e.g. rock flour, $\delta_{res} \sim 33°$, montmorillonite, $\delta_{res} \sim 7°$), (ii) normal effective stress level ($\delta$ reduces as stress level increases), (iii) previous loading (if the clay has been previously sheared, the peak strength may be reduced or not evident), (iv) the rate of loading and (v) the interface properties (rough interfaces push the failure plane into the surrounding soil, whereas smooth or painted interfaces have reduced friction).

### 5.5.7 API method for piles in clay

This section describes the calculation method for axial pile capacity in clay that is set out in the latest (21st) Edition of the API Recommended Practice (RP2A) for fixed structures.

ESTIMATION OF ULTIMATE SHAFT RESISTANCE

The sequence of stress changes described in the preceding section leading to $\sigma'_{hf}$ at failure is too complex to model explicitly in routine design. Instead, it is easier to correlate $\tau_{sf}$ with either $s_u$ or $\sigma'_{v0}$. These two approaches are known as the 'alpha' (total stress) and 'beta' (effective stress) methods, respectively:

$$\tau_{sf} = \alpha s_u \tag{5.9}$$

or

$$\tau_{sf} = \beta \sigma'_{v0} = K \sigma'_{v0} \tan \delta \tag{5.10}$$

defining K so that $\sigma'_{hf} = K \sigma'_{v0}$

In high-OCR stiff clays, $\alpha$ is typically low (0.4–0.6) and $\beta$ is high (0.8–1.2). In low OCR soft clays, $\alpha$ is typically high (0.8–1) and $\beta$ is between 0.2 and 0.3. Neither correlation is intrinsically better as a design method. The effective stress $\beta$-approach is couched in effective stresses, which is appealing. However, the $\beta$ parameter links two stresses that do not act at the same point concurrently ($\sigma'_{v0}$ acts in the undisturbed soil prior to installation and $\tau_{sf}$ acts on the pile at failure). Therefore, there is not necessarily a better direct link between $\tau_{sf}$ and $\sigma'_{v0}$ than $\tau_{sf}$ and $s_u$.

Since the stress changes described previously will be influenced by both $s_u$ and $\sigma'_{v0}$, it is useful to introduce the strength ratio, $s_u/\sigma'_{v0}$, which links $\alpha$ to $\beta$:

$$\beta = \alpha \left( \frac{s_u}{\sigma'_{v0}} \right) \tag{5.11}$$

The API method, which is widely used for offshore pile design, is an 'alpha' method, in which $\alpha$ is correlated with strength ratio $s_u/\sigma'_{v0}$. This approach was calibrated using a database of pile tests (Randolph and Murphy 1985). The correlations are:

$$\text{For } s_u < \sigma'_{v0}: \quad \alpha = \frac{1}{2} \left( \frac{s_u}{\sigma'_{v0}} \right)^{-1/2} \Rightarrow \tau_{sf} = \frac{1}{2} \sqrt{s_u \sigma'_{v0}} \tag{5.12}$$

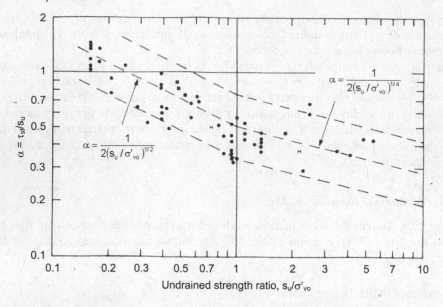

*Figure 5.14* API (2000) α-correlations for ultimate unit shaft resistance in clay (after Randolph and Murphy 1985)

$$\text{For} \quad s_u > \sigma'_{v0}: \quad \alpha = \frac{1}{2}\left(\frac{s_u}{\sigma'_{v0}}\right)^{-1/4} \Rightarrow \tau_{sf} = \frac{1}{2}s_u^{0.75}\sigma'^{0.25}_{v0} \tag{5.13}$$

It is assumed that equal shaft resistance acts inside and outside open-ended piles.

The resulting design chart, with the original load test database superimposed, is shown in Figure 5.14. This chart captures in a simple way the changes in horizontal stress acting on a pile. In soft clay, (low $s_u/\sigma'_{v0}$) the high excess pore pressure created during installation is converted to high horizontal stress during equalisation (i.e. high $\sigma'_{hc}$). Therefore, $\tau_{sf}$ (which is $\sigma'_{hc}\tan\delta$) is a high fraction of $s_u$, implying a high α on the left side of Figure 5.14.

In stiff clay, the excess pore pressures during installation are reduced due to dilation, so the resulting values of $\sigma'_{hc}$ are a lower fraction of $s_u$ implying a low α on the right hand side of Figure 5.14.

More recent data from sophisticated instrumented pile tests has augmented the original API database. The results generally support the original correlations, although low α-values have been encountered in sensitive low plasticity clays – reflecting the greater drop in effective stress that occurs during the remoulding associated with installation (Karlsrud *et al.* 1993).

Surprisingly low α-values are also recorded in soft clays if installation is very slow, or is interrupted. In these cases, partial equalisation occurs in parallel with installation. Between installation stages, the pore pressure dissipation and effective stress generation begins. On additional penetration, the soil is sheared again. This shearing interrupts the build-up of effective stress by regenerating excess pore pressure. By retarding the development of effective stress in this manner, the final α-value is lower. This behaviour is seen in some quick-draining silty clays, where installation time is comparable to equalisation period.

ESTIMATION OF ULTIMATE BASE RESISTANCE

Base resistance contributes a small part of the overall capacity of a pile in clay, so sophisticated analyses are less worthwhile than for shaft resistance.

The estimation of ultimate base resistance is more straightforward than shaft resistance. The conventional bearing capacity equation reduces to a single term (since $L \gg D$):

$$q_{bf} = N_c s_u \tag{5.14}$$

Skempton's (1951) empirical deep bearing capacity factor, $N_c = 9$, is commonly used, which is close to the theoretical values that can be derived from cavity expansion solutions (e.g. Vesic 1977). As the load is applied at the pile base, the underlying soil will consolidate and gain strength. It is, therefore, conservative to use the *in situ* value of $s_u$.

## 5.5.8 Limitations of API design method in clay

The API design method is relatively simple to apply, but does not consider the influence of friction fatigue, and the influence of OCR is considered in a rudimentary way using the $\alpha$–$s_u/\sigma'_{v0}$ correlations. The resulting predictions are surprisingly reliable for such a simple approach and the method remains very widely used. The mechanisms described in Sections 5.5.1–5.5.6 are not wholly captured by this method, so there are some circumstances in which particular caution should be exercised when applying the method:

- Long piles: The database used to calibrate the method generally comprised short piles. Friction fatigue and progressive failure become more important for longer piles.
- Sensitive low-plasticity soft clays: The drop in strength and loss of effective stress when these clays are sheared can lead to very low $\alpha$-values.
- Interrupted installation in permeable silty clays: Re-shearing of soft clays part-way through equalisation leads to reduced $\alpha$-values.
- Low friction angle clays: The correlations do not explicitly include the interface friction angle. Clays with a mineralogy that forms residual shear surfaces may show lower shaft resistance.

In order to capture some of these effects better, more complex calculation procedures have been proposed. Some of these aim to overcome the first limitation described earlier by including friction fatigue via a dependency of $\tau_{sf}$ on pile length (Kolk and Van de Velde 1996) or the position h/D (Lehane and Jardine 1994b, Jardine *et al.* 2005). It is also possible to include the sensitivity of the soil, $S_t$, as a factor that contributes a reduction to the $\alpha$ or $\beta$ parameter to capture the second limitation. Jardine *et al.* (2005) provide a correlation for K in Equation 5.15 that includes some of these considerations:

$$K = [1.7 + 0.011R - 0.6\,Log(S_t)]\,R^{0.42}\left(\frac{h}{D_{eq}}\right) - 0.2 \quad \text{for} \quad h/D_{eq} \geq 4 \tag{5.15}$$

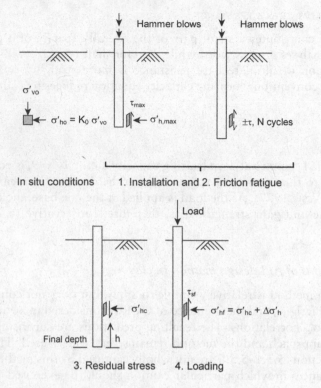

*Figure 5.15* Stages in the installation and loading of a driven pile in sand (White 2005)

where R is the yield stress ratio (or apparent overconsolidated ratio from an oedom-eter test). There are also correlations of a similar form that can be used to assess shaft resistance in clay directly from cone penetration resistance (Lehane *et al.* 2000).

## 5.6 Capacity of driven piles in sand

### 5.6.1 Stages in the installation and loading of a driven pile in sand

As for clay, the shaft resistance of a driven pile in sand depends on the *in situ* condi-tions and the complex changes that take place during installation and subsequent loading. The installation and loading of a driven pile in sand can be divided into a number of stages in order to understand the resulting shaft resistance (Figure 5.15). Typically, sand is sufficiently free draining that no equalisation stage exists, unlike in clay. Any excess pore pressure generated by the driving process dissipates prior to the end of installation. This simplifies the analysis of piles in sand, and removes the need to wait for a period of set-up following installation before the pile can be loaded.

As for clay, since shaft resistance is governed by Coulomb's Law, $\tau_{sf}$ can be estimated from the horizontal effective stress at failure, $\sigma'_{hf}$, and the pile–soil interface friction angle, $\delta$. From an initial value of $K_0\sigma'_{vo}$, the horizontal effective stress changes during installation, friction fatigue, and finally loading, to the value at failure, $\sigma'_{hf}$. These stages are described in the following sections. The framework of behaviour at each stage is then used to introduce two design approaches: the API and UWA methods for sand.

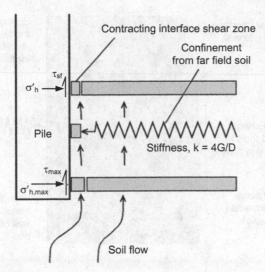

*Figure 5.16* Schematic diagram of friction fatigue mechanism (White and Bolton 2004)

### 5.6.2 *Stress changes during installation of a driven pile in sand*

During installation, as the pile tip advances towards a given element of sand, the stress level rises significantly to push the sand radially away from the pile tip. The stress level as the sand passes the pile tip will be comparable to the cone resistance $q_c$ which is typically 5–100 MPa in siliceous sands (ignoring dynamic stresses due to driving), or two orders of magnitude above the *in situ* stress. In uncemented carbonate sands, $q_c$ can be as low as 1–5 MPa, or can rise beyond the capacity of the cone if the carbonate sands have cemented into calcarenite.

As the sand passes around the pile tip and reaches the pile shaft, the stresses drop. The vertical penetration load is now no longer acting on the sand, which has been displaced laterally for the pile to penetrate. This reduced stress level behind the tip can be compared with measured values of cone sleeve friction $f_s$ although measurements of $f_s$ can be unreliable, depending on the condition of the sleeve and the dimensional tolerance of the cone assembly. For most sands, $f_s$ is 0.5–2.5 per cent of $q_c$. Therefore, the stress level has dropped by two orders of magnitude. The value of $f_s$ in silica sands is typically 100–1,000 kPa, but in carbonate sands is an order of magnitude lower, reflecting the low $q_c$ values.

As for clay soil, the process of friction fatigue leads to a reduction in unit shaft resistance with increasing distance behind the pile tip. As installation continues, with many hammer blows, the sand adjacent to the pile is sheared back and forth. This cyclic shearing leads to contraction of the sand since the permeability is high enough for volume change to occur. The contraction adjacent to the pile allows relaxation to occur in the surrounding cylinder of soil, causing a reduction in the horizontal stress acting on the pile shaft: a process known as friction fatigue, which is shown schematically in Figure 5.16. As h increases, $\sigma'_h$ decreases, and hence the available shaft resistance decreases.

Two examples of this behaviour are shown: one from a laboratory simulation and one from an instrumented pile test. An interface shear box test can be viewed as a short element of pile shaft. A feedback system mimicking a constant spring stiffness

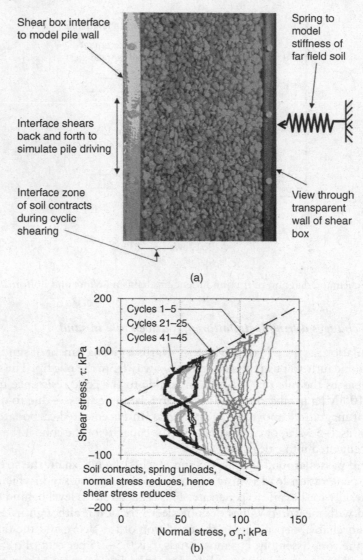

*Figure 5.17* Friction fatigue modelled in a transparent shear box (data from DeJong *et al.* 2003)

is used to control the normal stress, replicating the confinement provided by the far field soil (Boulon and Foray 1986, DeJong *et al.* 2003). In cyclic tests, as the interface is displaced back and forth, the sand sample contracts. This causes unloading of the normal stress, resulting in a reduction of the shear stress on the interface. The confinement provided by the far field soil is equivalent to a spring stiffness of $k = 4G/D$. In the example test shown in Figure 5.17, a spring stiffness of 250 kPa/mm was adopted, which corresponds to a typical soil stiffness of $G \sim 100$ MPa on a 1.5-m-diameter pile. The shear box was fitted with a transparent sidewall, which allowed the thin shearing zone to be identified. This zone, which was < 10 particles thick, contracted by ~250 μm (approximately one-third of a particle diameter). This slight contraction is sufficient to cause a halving of the confining stress and, therefore, the available shearing

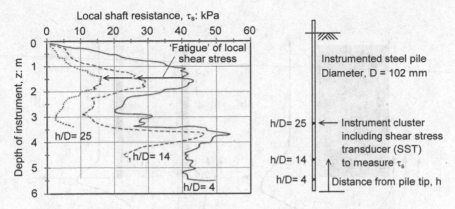

*Figure 5.18* Friction fatigue during installation of an instrumented pile (Lehane *et al.* 1993)

resistance. This simple example highlights the sensitivity of the normal stress on a pile to changes in volume in the adjacent soil (which can be caused by cyclic loading of the pile during installation, or under in-service loads). This sensitivity is due to the stiff constraint provided by the soil in the far field.

The instrumented pile test data shown in Figure 5.18 also illustrates the friction fatigue effect. This data is from model tests that were performed in Labenne, France using the same instrumented pile described in Section 5.5.2 (Lehane *et al.* 1993). The shear stress acting on the pile surface is measured by three sensors located at different elevations above the pile tip. As each instrument reaches a given soil horizon, the shear stress (and thus horizontal stress) has reduced compared to when the previous instrument was at that point. Each profile of local shaft resistance shows the same trend as the profile of cone resistance $q_c$ showing that $q_c$ can correlate well with shaft resistance.

### 5.6.3 Behaviour of open-ended piles during driving

As was discussed earlier for clay, plugging of open-ended piles rarely occurs during driving due to the inertia of the soil plug. This inertia component increases the resistance to plugged penetration, and the pile penetrates in a partially plugged or unplugged ('coring') manner instead. Full plugging is rare during installation.

The degree of plugging during installation affects the subsequent shaft resistance, because it affects the volume of soil that is displaced by the pile and, therefore, the magnitude of the stresses set up in the surrounding soil.

A closed-ended pile penetrates like a CPT, opening a cylindrical cavity from nothing, using a pressure of $q_c$ to generate the required radial soil displacement (Figure 5.19a). A thin-walled open-ended pile creates a smaller amount of radial displacement, so generates a lower pressure (Figure 5.19b). A pile penetrating in a partially plugged manner lies between these extremes (Figure 5.19c).

To quantify this effect, the degree of plugging is quantified based on the effective area ratio $A_{r,eff}$ which is the ratio of the displaced volume of soil (i.e. the gross pile volume minus any soil entering the plug) to the gross pile volume:

$$A_{r,eff} = 1 - IFR \frac{D_i^2}{D_o^2}$$

(5.16)

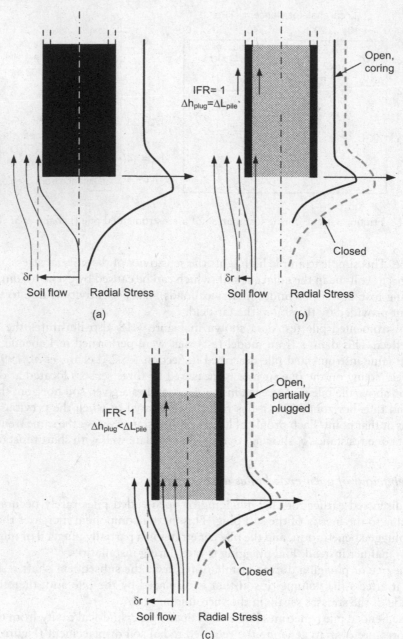

*Figure 5.19* Plugging: schematic streamlines of soil flow and radial stress (a) closed-ended pile ($A_r = 1$), (b) coring (unplugged) open-ended pile (IFR = 1, $A_r \sim 0.1$), (c) partially plugged open-ended pile (IFR < 1, $0.1 < A_r < 1$) (White *et al.* 2005)

where $D_0$ is the pile outer diameter, $D_i$ is the pile inner diameter and IFR is the incremental filling ratio, $\Delta h_p / \Delta L$ (Figure 5.7 and Figure 5.19).

Typical offshore piles have a wall thickness ratio of $D/t \sim 40$, for which $A_{r,eff} \sim 0.1$ during unplugged penetration. Large-diameter thin-walled caissons, which are increasingly being adopted in new offshore frontiers, displace a minimal volume of soil compared to their gross area ($A_{r,eff} \sim 0.01$), leading to only a small stress increase in the surrounding soil. Conversely, an open-ended pile penetrating in a plugged manner causes the same soil displacement as a closed-ended pile ($A_{r,eff} = 1$).

The area ratio, $A_{r,eff}$, can be incorporated in assessments of pile capacity to account for the influence of the pile end condition, as described in Section 5.6.9.

### 5.6.4 Stress changes during loading of a driven pile in sand

As for clays, a small change in $\sigma'_h$ is observed during loading of a pile in sand. In sands, an increase is measured, which can be attributed to (i) Poisson's ratio strains in the pile, which cause outward expansion against the surrounding soil under compression loading and (ii) dilation of the soil close to the pile–soil interface. The influences of Poisson's ratio strains were explored by De Nicola and Randolph (1993) and Lehane *et al.* (1993) review dilational effects. The confining stiffness imposed by the surrounding soil is 4G/D. The dilation of the shearing zone close to the pile will be affected more by the particle size than the pile diameter, so the resulting stress increase will be lower for larger piles. The dilation effect is usually insignificant for the sizes of pile used offshore, but it must be taken into account when interpreting small-scale pile tests to support design.

### 5.6.5 Stress changes during plugging of piles in sand

During static loading to failure, the internal shaft resistance is usually higher than the base resistance that can be mobilised on the base of the soil column so plugging occurs (see Equation 5.6).

The variation in internal shaft resistance along the plug of a pile in sand is different to the external shaft resistance. A simple analysis based on the vertical equilibrium of horizontal slices of the soil column shows that internal shaft resistance is very high in soil columns longer than a few diameters. This effect is known as 'arching'. The free body diagram shown in Figure 5.20 illustrates the equilibrium of a soil element within the plug of a tubular pile. From vertical equilibrium of the horizontal slice:

$$\frac{d\sigma'_v}{dz} = \gamma' + \frac{4}{D}\beta\sigma'_v \qquad (5.17)$$

where the parameter $\beta$ links $\tau_{sf}$ and the *local* vertical effective stress in the soil plug, $\sigma'_v$, not the *in situ* value, $\sigma'_{v0}$, at that depth.

The Mohr's circle of stress close to the pile wall represents a lower bound on $\beta$ (and, therefore, offers a conservative estimate of $\tau_{sf}$ when the plug fails) (Figure 5.20b). During loading, the soil column is pushed upwards into the pile, so the soil is at failure with $\sigma'_v > \sigma'_h$, i.e. active conditions. However, $\sigma'_v/\sigma'_h$ is not quite as low as the active earth pressure coefficient, $K_a$, since there is shear on the pile wall so the vertical and horizontal stresses are not principal.

From the Mohr's circle in Figure 5.20b it can be shown that:

$$\beta = \frac{\tau_{sf}}{\sigma'_v} = \frac{\sin\phi\sin(\Delta-\delta)}{1+\sin\phi\cos(\Delta-\delta)} \qquad (5.18)$$

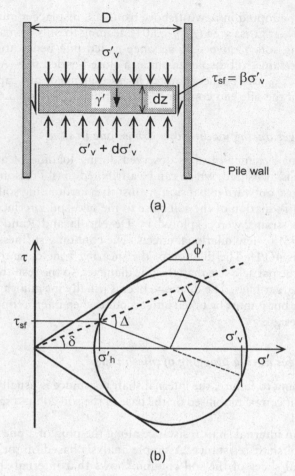

*Figure 5.20* Analysis of soil column in tubular pile (a) equilibrium of horizontal slide of soil column (b) Mohr's circle of stress close to pile wall

where

$$\sin\Delta = \frac{\sin\delta}{\sin\phi} \tag{5.19}$$

Using conservative minimum values of $\delta = 23°$ for sand-on-rough steel friction and a typical $\phi_{cv}$ value of $33°$ gives $\beta = 0.14$.

Integration of Equation 5.17 leads to an expression for the vertical stress on the base of a soil column of height $h_p$ (i.e. maximum internal shaft resistance plus plug weight):

$$q_{bf-plug} = \gamma' h_p \left( \frac{e^\lambda - 1}{\lambda} \right) \tag{5.20}$$

where

$$\lambda = 4\beta \frac{h_p}{D} \tag{5.21}$$

The stress at the base of the soil plug, $q_{bf-plug}$, is very sensitive to $h_p/D$ and $\beta$ due to the exponential term. As the plug lengthens, a sharp increase in $q_{bf-plug}$ occurs. For example, in a 1-m-diameter pile, using the conservative value of $\beta = 0.14$, mobilised over a plug length of only 10 m (10D), Equation 5.21 gives $\lambda = 5.6$. Therefore, in saturated sand of unit weight $\gamma' = 10$ kN/m$^3$, sufficient internal shaft resistance is available to mobilise $q_{bf-plug} = 4.8$ MPa.

Plugging will occur if the internal shaft resistance (and self-weight) of the soil column is sufficient to overcome the base resistance under the soil plug. Equation 5.20 gives $q_{bf-plug}$, which is the resultant vertical stress acting at the pile base from the internal shaft resistance and self-weight. If this number exceeds the ultimate base resistance on the plug area – which will approach the cone resistance, $q_c$ for large displacements – then plugging will occur under static loading. The foregoing example shows that relatively short plugs can overcome typical values of penetration resistance. Hence, even short soil columns comprising sand can form plugs. However, it is important to recognise that to mobilise the resulting high stresses may require significant settlement of the pile in order to accommodate the resulting compression in the soil beneath the pile base and within the soil plug (Lehane and Randolph 2002).

### 5.6.6 Interface friction angles in sand

The pile–soil interface friction angle depends primarily on the roughness of the interface, $R_a$, relative to the size of the sand particles, and is also influenced by the particle mineralogy and shape. If the sand particles are large relative to the interface, the interface is termed 'smooth' and the shearing resistance is governed by sliding of the particles over the interface surface. If the sand particles are sufficiently small ($d_{50} < \sim 10R_a$) that they engage with the asperities on the interface then the interface is 'rough' and the frictional resistance is identical to the soil–soil friction response. There is a transition range between the 'smooth' and 'rough' extremes (Uesugi and Kishida 1986a,b, Subba Rao *et al.* 1998).

For a typical steel roughness of $R_a = 10$ μm, a fully 'rough' condition applies for fine sand, and typical interface friction angles of 28–32° are found. For coarse sand, the reduced relative roughness means that angles of 20–24° are more typical. In practice, site-specific interface shear tests should be conducted to identify soils that lie outside typical trends.

### 5.6.7 Design method 1: API (American Petroleum Institute) method

The latest (21st) edition of the API Recommended Practice (RP2A) for fixed structures[2] includes two forms of calculation methods for the assessment of pile capacity in sands. There is a 'Main Text' method, which has evolved slowly over the past thirty years. This method uses a 'β-approach' for shaft resistance. The Commentary of RP2A also presents a family of methods that are based on cone penetration data. The Main Text method is set out in this section and the CPT-based methods are described in Section 5.6.8.

---

[2]API (2000) plus modifications that were included in the supplement issued in April 2008 – API (2008).

*Table 5.1* Coefficients for API Main Text method for piles in sand

| Soil relative density[1] | Soil type | Shaft friction factor, $\beta$ (–) | Shaft friction limit, $\tau_{s,lim}$ (kPa) | End bearing factor, $N_q$ | End bearing limit, $q_{b,lim}$ (MPa) |
|---|---|---|---|---|---|
| Medium dense | Sand-silt | 0.29 | 67 | 12 | 3 |
| Medium dense | Sand | | | | |
| Dense | Sand-silt | 0.37 | 81 | 20 | 5 |
| Dense | Sand | | | | |
| Very dense | Sand-silt | 0.46 | 96 | 40 | 10 |
| Very dense | Sand | 0.56 | 115 | 50 | 12 |

[1]The relative density descriptions are defined as: very loose: 0–15 per cent; loose: 15–35 per cent; medium dense: 35–65 per cent; dense: 65–85 per cent; very dense: 85–100 per cent.

ESTIMATION OF ULTIMATE SHAFT RESISTANCE

The API (2000) Main Text method for predicting ultimate unit shaft resistance in sand is a 'β-approach'. The basic equation for ultimate unit shaft resistance is, therefore:

$$\tau_{sf} = \beta\sigma'_{v0} \leq \tau_{s,lim} \qquad (5.22)$$

where values of β and of the limiting shaft resistance, $\tau_{s,lim}$, vary with soil type and density according to Table 5.1. The quoted values of β are for open-ended piles and may be increased by 25 per cent for closed-ended piles. This adjustment aims to reflect the additional stress created by the greater volume of soil displaced. The limiting values are typically reached at a depth of 20–30 m, so apply over much of the length of typical offshore piles.

The API Main Text method specifies that internal shaft resistance should be assumed to equal external shaft resistance, when assessing whether the pile will plug during static loading. This is highly conservative, given the behaviour described in Section 5.6.5.

The API Main Text method was calibrated using a database of load tests on mainly uninstrumented piles. This database suggested that the mean unit shaft resistance did not increase proportionally with depth, as would be implied by Equation (5.22) if limits were not applied. Without measurements of shaft resistance distribution, the assumption of a limiting unit shaft resistance was used as a reasonable approach to match the measured total capacity.

The specification of limiting values $\tau_{s,lim}$ is not consistent with newer field observations. Values of $\tau_{sf} \gg \tau_{s,lim}$ close to the base of piles are widely reported. Instead, field testing during the 1990s showed that friction fatigue reduces the shaft resistance on the upper part of the pile during installation, as the tip of the pile penetrates further, leading to the observed trend of the mean unit shaft resistance not increasing in proportion to pile length.

Although the concept of a limiting shaft resistance is misleading (Kulhawy (1984), Randolph (1993), Fellenius and Altaee (1995)), the use of limiting values of unit shaft resistance helps to improve the accuracy of the API method, by giving a trend of mean unit shaft resistance that increases slowly with, but not in proportion to, pile length.

Due to this limitation, the API Main Text method for assessing pile shaft resistance is not as reliable as more modern CPT-based methods and can be inaccurate when extrapolated beyond the database of load tests used for its derivation (Schneider *et al.* 2008b).

ESTIMATION OF ULTIMATE BASE RESISTANCE

To evaluate base resistance, the API (2000) method uses bearing capacity theory to estimate $q_{bf}$ from $\sigma'_{v0}$ with limiting values being applied. The bearing capacity equation is as for shallow foundations, with the $N_\gamma$ term ignored since D << L:

$$q_{bf} = N_q \sigma'_{v0} \leq q_{b,lim} \qquad (5.23)$$

In shallow foundation design, $N_q$ is calculated from $\phi$ using a plasticity solution. For piles, the API method specifies $N_q$ depending on the soil conditions and applies limiting values of $q_{bf} = q_{b,lim}$ (see Table 5.1). Field data indicates that $N_q$ decreases with increasing stress level, so the use of limiting values prevents unconservatism.

SUMMARY: API DESIGN METHOD IN SAND

The API (2000) design method is very straightforward to apply, but does not capture the governing mechanisms. Since the method was calibrated against a database of (mainly uninstrumented) pile load tests, there is a reasonable fit to these older data. However, the method has proven unreliable in some cases when extrapolated beyond the original database to long large-diameter piles, as are often used offshore (Schneider *et al.* 2008b). On the other hand, the simplicity of the method allows preliminary assessments of pile capacity to be made using only very rudimentary stratigraphic information.

### 5.6.8 CPT-based methods for pile capacity in sand

Due to the complicated stress history described in Sections 5.6.1–5.6.5, there is no easy link between the *in situ* horizontal stress, $K_0\sigma'_{v0}$, and the horizontal stress acting at failure $\sigma'_{hf}$. Pile design in sand is increasingly based on CPT data. Methods for estimating the axial capacity of piles on the basis of cone resistance $q_c$ have been in circulation for many years (Bustamante and Gianeselli 1982, Jardine and Chow 1996, De Cock *et al.* 2003). This approach is increasingly attractive for piles driven into sand, since traditional approaches such as the API Main Text method are known to have a weak theoretical basis and give poorer correlation with load test data than CPT-based methods. Correlation of end-bearing capacity with the cone resistance has a direct logic, although care is needed in allowing for differences between solid (or closed-ended) piles and the open-ended pipe piles commonly used offshore. Estimation of shaft friction from cone tip resistance requires an adjustment for friction fatigue to be introduced.

Since 2007, the Commentary of the API RP 2A code has included four CPT-based methods for the assessment of pile capacity in sand based on the cone resistance $q_c$. These methods were all developed in the past ten years, and they have many similarities in the formulations adopted. In this book, we have chosen to focus on the UWA-05 method for three reasons. First, because, of the four CPT-based methods, UWA-05 captures most clearly the governing mechanisms outlined in Sections 5.6.1–5.6.6; second, because it shows marginally better accuracy than the other methods when compared with the latest database of load test results; and finally, because the method was developed by our colleagues Prof. Barry Lehane and his (then) research students, Dr James Schneider and Dr Xiangtao Xu at the University of Western Australia (Lehane *et al.* 2005a,b, Schneider *et al.* 2008b, Xu *et al.* 2008).

Section 5.6.10 discusses the reliability of predictions made using these four different methods.

### 5.6.9 Design method 2: UWA-05 method

The UWA-05 method for calculating offshore pile capacity in sand was developed in 2005 (Lehane *et al.* 2005a,b).

ESTIMATION OF ULTIMATE SHAFT RESISTANCE

The design expressions for shaft resistance have a form that can be linked to the stress history around a pile in sand, and the separate components can be related to different mechanisms acting during:

(i) The stress drop around the pile tip (initially for a closed ended pile), from $q_c$ to the horizontal on-pile stress:

$$\sigma'_{h,tip,closed} = aq_c \tag{5.24}$$

(ii) The effect of an open end, leading to a reduced volume of soil displacement and a lower stress level:

$$\frac{\sigma'_{h,tip,open}}{\sigma'_{h,tip,closed}} = A^b_{r,eff} \tag{5.25}$$

The soil displacement is characterised using the effective area ratio, which was introduced in Section 5.6.3, and a value of $b \sim 0.3$ can be derived from cavity expansion analysis (White *et al.* 2005).

(iii) Friction fatigue along the pile shaft:

$$\frac{\sigma'_h}{\sigma'_{h,tip}} = \left( \max\left[ \frac{h}{D}, 2 \right] \right)^c \tag{5.26}$$

where h is the distance upwards from the pile tip.

(iv) A small increase in the horizontal stress on the pile during loading, principally due to constrained dilation at the interface (although $\Delta\sigma'_{hd}$ is generally negligible for offshore piles, so is not included in the 'offshore' version of UWA-05):

$$\sigma'_{hf} = \sigma'_h + \Delta\sigma'_{hd} \tag{5.27}$$

*Table 5.2* Input parameters for UWA-05 expression for shaft resistance

| Index | Idealised origin | Value |
|-------|------------------|-------|
| a | Stress drop from $q_c$ around pile tip | 0.03 |
| b | End condition: degree of displacement | 0.3 |
| c | Friction fatigue | –0.5 |

(v) Coulomb friction at the pile–soil interface, mobilising the steady state (constant volume) interface friction angle:

$$\tau_{hf} = \sigma'_{hf} \tan \delta_{cv} \tag{5.28}$$

These individual expressions lead to the full equation in the UWA-05 offshore method, which features three indices, a–c, that were calibrated based on a database of load test results:

$$\tau_{sf} = a q_c A_{r,eff}^b \left( \max \left[ \frac{h}{D}, 2 \right] \right)^c \tan \delta_{cv} \tag{5.29}$$

A lower limit of h/D = 2 is specified for equation 5.29, hence the square bracket term. The shaft resistance is assumed to be constant over the 2 diameters closest to the tip. For offshore piles, installation usually takes place in an unplugged mode, so the conservative assumption that IFR = 1 is made, hence:

$$A_{r,eff} = 1 - \frac{D_i^2}{D_o^2} \tag{5.30}$$

Comparison of Equation 5.29 with a database of > 75 load tests showed that the parameters shown in Table 5.2 give the most accurate design method for compression loading.

For tension loading, the shaft resistance is suggested as 75 per cent of the value in compression, which is broadly consistent with the expression proposed by De Nicola and Randolph (1993). The interface friction angle $\delta_{cv}$ should be found from interface shear box tests using a representative roughness and the constant volume value used.

*Estimation of ultimate base resistance*

In the UWA-05 method, the ultimate unit base resistance, $q_{bf}$, is defined at a settlement of D/10, and is calculated on the gross area of a tubular pile (including the plug) as (Xu *et al.* 2008):

$$q_{bf} = (0.15 + 0.45 A_{r,eff}) q_{c,avg} \tag{5.31}$$

where $q_{c,avg}$ is assessed using the 'Dutch method' of vertical averaging (Schmertmann 1978, Xu *et al.* 2008), which takes account of variations in $q_c$ over a range $8D_{eff}$ above the pile tip and $4D_{eff}$ below, where $D_{eff} = D \sqrt{A_{r,eff}}$.

This expression gives values of base resistance that are significantly below the 'continuum' solution of $q_b = q_c$, which would be expected during steady penetration. These values of $q_b/q_c < 1$ can be attributed to partial mobilisation, since ultimate capacity is defined according to a settlement criterion of D/10. Piles only mobilise a fraction of their ultimate (or 'plunging') base capacity at a settlement of D/10. For closed-ended driven piles, this fraction is typically 60–70 per cent. The response of open-ended piles is more compliant due to compression of the soil in the plug and below the pile tip, and lower residual stresses developed during installation. Therefore, lower values of $q_b/q_c$ are appropriate for design. For a typical offshore piles with $A_{r,eff} = 0.1$, $q_{bf}/q_c \sim 0.2$. For thicker-walled piles or piles that plug during installation (even partially) a higher proportion of the base resistance arises from the pile wall. For a closed-ended pile, $A_{r,eff} = 1$, so $q_{bf} = 0.6q_c$ at a settlement of D/10.

With additional penetration, $q_b$ is expected to increase steadily and ultimately reach the cone resistance $q_c$ so, leaving aside settlement considerations, design methods with $q_b/q_c < 1$ are conservative (Randolph 2003a, White and Bolton 2005).

Piles are often installed at only a shallow embedment into a strong bearing stratum. A significant embedment – several diameters – is required to mobilise the 'full' strength of that stratum. Since a pile has a greater diameter than a cone penetrometer, a deeper embedment into a hard layer is required to mobilise the 'full' strength of that layer. Prior to sufficient penetration, $q_{bf}$ will be less than the local $q_c$ since the previous layer will still be 'felt' by the pile tip.

A similar mechanism occurs when a soft layer is approached: it is 'felt' earlier by the pile than by the cone, so low values of $q_c$ below the pile tip should be included in any weighting of the design $q_c$.

At sites with significant vertical heterogeneity, it is necessary to convert the profile of $q_c$ into a profile of $q_{c,avg}$ for use in an assessment of base capacity (e.g. Equation 5.31).

This mechanism of low $q_{bf}/q_c$ is known as partial embedment. Meyerhof (1983) recommended that a penetration of 10 diameters into a harder layer was required to mobilise the full capacity. A simple expression to derive the relevant averaged cone resistance $q_{c,avg}$ at shallow embedment from a softer sand layer (with penetration resistance $q_{c,soft}$) into a harder layer (with $q_{c,hard}$), is, therefore:

$$q_{c,avg} = q_{c,soft} + (q_{c,hard} - q_{c,soft}) \frac{z_b}{10D} \quad \text{for} \quad \frac{z_b}{D} < 10 \tag{5.32}$$

where $z_b/D$ is the penetration of the pile base into the hard layer.

Xu and Lehane (2008) described a more detailed approach for averaging cone resistance profiles to account for nearby weak layers, either above or below the pile tip. They showed that the depth of this transition zone is greater for a greater differential in strength between the layers. Meyerhof's 10D transition zone is conservative for soft-over-stiff conditions, but a transition zone as shallow as only 4D may be sufficient if the penetration resistance of the two layers differs by a factor of 2 or less.

### 5.6.10  Comparison of CPT-based methods for pile capacity in sand

UWA-05 is just one of the four CPT-based methods that have recently been incorporated in a revised commentary to the API RP 2A guidelines for the design of fixed steel structures. The expressions for shaft resistance $\tau_{sf}$ in three of these methods – Fugro-05, ICP-05 and UWA-05 – follow similar forms, and may be expressed as

$$\tau_s = a\bar{q}_c \left( \frac{\sigma'_{v0}}{p_a} \right)^p A_r^b \left[ \max\left( \frac{L-z}{D}, v \right) \right]^{-c} (\tan\delta_{cv})^d \left[ \min\left( \frac{L-z}{D} \frac{1}{v}, 1 \right) \right]^e \tag{5.33}$$

*Table 5.3* Parameter values for shaft friction estimation in Equation 5.33

| Method | | *a* | *b* | *c* | *d* | *e* | *p* | *v* |
|--------|--|-----|-----|-----|-----|-----|-----|-----|
| | | | | *Parameters* | | | | |
| Fugro-05 | compression | 0.043 | 0.45 | 0.90 | 0 | 1 | 0.05 | |
| | tension | 0.025 | 0.42 | 0.85 | 0 | 0 | 0.15 | |
| Simplified ICP-05 | compression | 0.023 | 0.2 | 0.4 | 1 | 0 | 0.1 | |
| | tension | 0.016 | 0.2 | 0.4 | 1 | 0 | 0.1 | |
| Offshore UWA-05 | compression | 0.030 | 0.3 | 0.5 | 1 | 0 | 0 | 2 |
| | tension | 0.022 | 0.3 | 0.5 | 1 | 0 | 0 | 2 |

where $\sigma'_{v0}$ is the vertical effective stress, normalised by atmospheric pressure $p_a$ (100 kPa), $\delta_{cv}$ is the steady state interface friction angle between pile and soil, (L–z) is the distance, h, from the tip of the pile, and the parameters a to e, u and v are given in Table 5.3 for the three methods that are based directly on $q_c$ (Jardine *et al.* 2005, Schneider *et al.* 2008b, Kolk *et al.* 2005). The final method – NGI-05 – uses $q_c$ indirectly, via a conversion firstly to effective relative density, then to unit shaft resistance. The different methods also include varying approaches for assessing base resistance from $q_c$.

A comparison of these three methods with the traditional approach for estimating the capacity of driven piles in sand (i.e. the API Main Text method) showed significant improvements in accuracy, with much lower coefficients of variation of the ratio of predicted to measured capacity, for the current database of pile load tests in the public domain (Schneider *et al.* 2008b).

Various database studies have been conducted by the research groups involved in the development of the methods listed in Table 5.3. Database exercises are a necessary stage in the calibration of axial pile capacity design methods, due to the empiricism of the recommended equations. The methods listed in Table 5.3 all give broadly similar performance when compared to the field load test measurements, which is to be expected since all were calibrated against similar databases. There is a trend for the more recently developed methods to give better agreement with the current database, partly because of having been calibrated against the more recent versions. Typically, the database studies reveal a mean ratio of predicted to measured capacity of $1 \pm 0.1$ with a standard deviation of ~0.2–0.3.

However, these statistics should not be used as the basis for assessing factors of safety and reliability for these methods when used to design offshore piles. It is a common misconception that these statistics can be used to estimate the reliability arising from a particular design method combined with a given factor of safety. Offshore piles are generally larger than virtually all piles within the existing databases and are subjected to considerably higher loads. Therefore, the use of these methods involves extrapolation beyond the range of pile geometry and loading against which each method was calibrated.

The methods all adopt broadly the same form of expression for shaft and base resistance, with some differences in the mechanisms accounted for. However, the empirical fitting parameters differ, as shown in Table 5.3. As a result, the methods give significantly different capacity predictions for field scale piles – varying by up to 50 per cent, with a scatter that depends on the mean value and shape of the cone resistance profile and the pile geometry (Lehane *et al.* 2005b). A designer can then be faced with a design load that exceeds the capacity calculated according to one of the new design methods, but which is acceptable according to another, even after reducing this capacity by a factor of safety based on the database statistics.

This conflict presents difficulties for designers, highlighting the need for caution when applying these new design methods. The API Commentary notes that 'more experience is required with all these new methods before any single one can be recommended for routine design'. A conservative approach – perhaps overly so – would be to adopt the lowest of the capacities predicted by each method.

Statistical analysis can be used to estimate the increased uncertainty associated with extrapolation away from the database conditions (Zhang *et al.* 2004, Schneider 2007). This type of analysis suggests that when extrapolating to large offshore piles, the factors of safety currently in use may be inadequate. However, since experience shows that current methods exhibit acceptable reliability, it is likely that additional factors such as time effects (the apparent long-term increase in pile capacity with time that is often observed) and structural redundancy are already contributing – perhaps unintentionally – to provide a reserve of capacity.

There remains no consensus on the 'correct' values of the empirical parameters in Table 5.3, and it is unlikely that one will ever be reached. Instead, it should be acknowledged that the installed shaft capacity of an offshore pile is influenced by aspects of the soil response that cannot be determined solely from CPT data and by the driving procedure adopted. These effects underlie our view that axial pile capacity predictions in sand will always have a significant uncertainty, which is larger than found in most other areas of geotechnical engineering.

### 5.6.11 Shaft resistance in carbonate sands

Following the upgrading works required for the piled foundations at the North Rankin gas platform offshore Australia, it has become more widely recognised that very low shaft resistance is mobilised on driven piles in carbonate sands (Jewell and Khorshid 1988). Conventional design methods as described for sand earlier can lead to significant unconservatism.

Collated data of mean $\tau_{sf}$ from load tests at the North Rankin A site and in other carbonate sands worldwide is shown in Figure 5.21. A mean $\tau_{sf}$ of approximately 10 kPa is evident, even for pile lengths greater than 100 m. This is almost two orders of magnitude lower than would be expected in silica sands.

Values of $q_c$ in carbonate sands are typically one order of magnitude lower than in silica sands. However, the difference in $q_c$ alone does not account for the low $\tau_{sf}$. The mechanisms described in Sections 5.5.1–5.5.5 clarify the underlying behaviour. Since $q_c$ is approximately one order of magnitude lower for carbonate sand compared to silica sand, and $f_s/q_c$ is comparable, the additional order of magnitude loss of stress is linked to friction fatigue.

This is consistent with the contractile behaviour of the carbonate sand when sheared. Cycles of loading at the pile–soil interface lead to greater contraction of carbonate sand than silica sand, and hence greater loss of $\sigma'_h$. One way of capturing this behaviour is to adopt a higher value of the friction fatigue index 'c' (in Equation 5.29), coupled with a minimum value below which $\tau_{sf}$ cannot decay (Schneider *et al.* 2007).

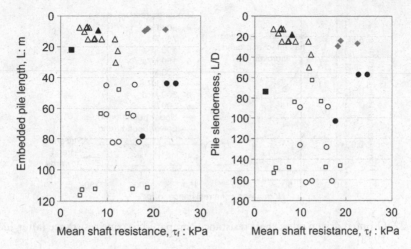

*Figure 5.21* Mean unit shaft resistance on piles in carbonate sands (Schneider *et al.* 2007)

## 5.7 Axial capacity of grouted piles

### 5.7.1 Estimation of ultimate shaft resistance

In carbonate soils and where socketing into rock is required, grouted piles are sometimes used. There is limited code guidance for these conditions. Design methods commonly rely on local experience and should be tailored to the particular construction technique that is used (see Section 5.2.3).

In uncemented sediments, the shaft resistance on a grouted pile depends on the grouting pressure, which governs the lateral stress that acts against the formation, and, therefore, the friction that can be mobilised. In cemented sediments and rocks, it is the strength of the pile–soil bond that governs the capacity. In both cases, it is known that dilation during the initial loading phase creates a significant increase in the peak shaft resistance. Unlike the interface around a driven pile, the grouted pile–soil interface has not been previously sheared, so has a more dilatant tendency.

A conservative approach, which ignores the dilation effect, is to adopt a horizontal stress equal to the grouting pressure – which can often be assumed to be the hydrostatic grout pressure, if the pile will be grouted quickly enough that the full bore is filled prior to any hardening. If the grout contracts when hardening, the final horizontal stress will be lower than the grouting pressure whereas an expansive grout helps to increase the horizontal effective stress as it sets.

The approaches linked to grouting pressure neglect the beneficial effect of dilation, which causes a significant, but strain-softening, additional component of shaft resistance. In the cemented calcareous sediments found offshore Australia, Abbs (1992) recommended adopting a peak shaft resistance of 2 per cent of the cone resistance $q_c$. This provides a robust lower bound to available field measurements but can be over-conservative in lightly cemented deposits with low cone resistance. An alternative approach was suggested by Joer and Randolph (1994), with the peak shaft resistance expressed as

$$\frac{\tau_{sf}}{q_c} = 0.02 + 0.2e^{-0.04q_c/p_a} \qquad (5.34)$$

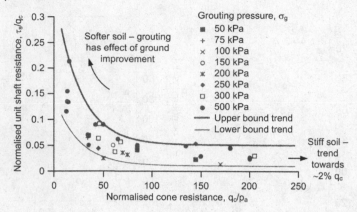

*Figure 5.22* Grouted driven piles: shaft resistance compared to cone resistance (after Joer *et al.* 1998)

This relationship, which was derived from model tests, gave good agreement with field data from drilled and grouted piles in similar soils. Grouted driven piles showed slightly higher capacity on average. The increase in $\tau_{sf}/q_c$ with reducing $q_c$ can be linked to compression and consolidation of the softer soil during the grouting process. There is significant scatter within the data that provided this trend (Figure 5.22). This is because $q_c$ alone does not capture all of the controlling parameters, which include the dilatancy of the grout–soil interface and the confining stiffness of the surrounding soil. In design, it is usual for grout–soil shear box tests to be conducted under CNS conditions (to replicate the constraint of the surrounding formation) in order to assess the peak shaft resistance. Due to the brittleness of the shaft response of a grouted pile, it is also necessary to perform load transfer analyses to assess the degree of progressive failure along the pile. The ultimate shaft capacity of a grouted pile is generally well below the sum of the peak shaft resistance along the length.

### 5.7.2 *Estimation of ultimate base resistance*

The base resistance of a grouted pile is often neglected, to allow for softening during the drilling process or debris falling in the hole, which would cause the base response to be too compliant for significant resistance to be mobilised at acceptable settlements.

Alternatively, if modest settlements can be accommodated, the unit base resistance can be estimated from the cone resistance $q_c$, and it is typical to assume that $q_{bf}/q_c = 0.15$–$0.2$ in design, with the lower limit being for denser soil (i.e. higher $q_c$).

## 5.8 Axial response

### 5.8.1 *Design relevance*

Foundation design is usually divided into ultimate limit state (ULS) and serviceability limit state (SLS) assessments, where the ULS relates to foundation strength, and the SLS to the settlements under working loads (i.e. stiffness). In onshore design, where the foundation must typically support a brittle concrete or masonry structure with windows and service connections, it is usually the SLS that governs the design.

In offshore design, settlement is of more limited concern. Offshore structures can tolerate a considerable uniform settlement during installation due to the large air gap above sea level. Limited differential settlement is also permissible since steel jacket structures are ductile.

However, it is necessary to consider the load-settlement response of an offshore piled foundation for the following aspects of design:

- *Progressive failure: reduced 'ideal' capacity.* Any compression of the pile causes the relative movement between the pile and the soil to vary along the pile length. If the shaft resistance response is strain-softening (i.e. drops from a peak to a residual value) then progressive failure can reduce the 'ideal' pile capacity.
- *Cyclic loading: reduced 'ideal' capacity.* Strain-softening of the shaft resistance response is exacerbated by cyclic loading. Significant movement of the pile head during cyclic loading can lead to a sharp reduction in the available shaft resistance, reducing the strength (and stiffness) of the foundation.
- *Dynamic response.* The stiffness of the foundation influences the natural frequency of the structure, and, therefore, the susceptibility to resonance and dynamic amplification during cyclic loading.
- *Relative stiffness of foundation components.* Some jacket structures include large mudmats, which may be relied on to take significant in-service loads, as well as providing temporary stability prior to the installation of piles. The relative stiffness of the piles and the mudmat response will affect the load distribution between the two foundation components.

During retrofit strengthening of structures, the added components must be sufficiently stiff to attract load from the original foundations. An example of this scenario is the external strut strengthening systems installed on the first generation Bass Strait platforms.

One of the external strut schemes installed on a Bass Strait platform is shown in Figure 5.23. These struts provide additional strength to supplement the original piled foundations. They are stiff enough to provide this additional strength before the existing foundations reach maximum capacity. The design would be ineffective if the struts were too compliant.

Because of these considerations, it is usually necessary to perform some form of load-displacement modelling of a piled foundation during the design process, in both the axial and lateral directions. This analysis will indicate not only the stiffness of the pile but also the true ultimate capacity, which will be affected by progressive failure and cyclic loading if the shaft resistance is brittle.

### 5.8.2 Load transfer analysis

Pile load-settlement behaviour is usually assessed using load transfer analysis. This technique simulates the pile–soil interaction using a series of 'springs' distributed down the length of the pile and at the base. Instead of explicitly modelling the elements of soil from the pile away to the far field, the soil response is integrated into a simple relationship between pile movement and resistance.

At any location, a pair of springs represents the axial and lateral responses. The simplest form of such springs would be elastic, or elastic-perfectly plastic (i.e. with a limiting

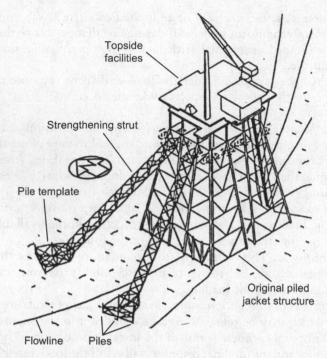

Topside
facilities

Strengthening strut

Pile template

Original piled
jacket structure

Flowline     Piles

*Figure 5.23* Bass Strait platform strut strengthening scheme (Wiltsie *et al.* 1988)

maximum value of local shear stress, or lateral pressure). However, it is customary to make use of non-linear 'springs', with a gradual reduction of secant stiffness as the load level increases, and a potential for strain-softening beyond the peak load transfer.

One of the advantages of the load transfer approach is that the complete jacket and pile foundations may be analysed in a standard structural analysis package, replacing the soil with the non-linear load transfer springs. A limitation of the approach is the loss of direct modelling of interaction between adjacent piles (through the soil continuum) and a tendency towards empiricism in the choice of load transfer parameters, divorced from the actual continuum behaviour of the soil.

This form of modelling is shown schematically in Figure 5.24 using simple non-linear load transfer springs (without strain softening). The relationships between the initial spring stiffness in the axial and lateral directions and the shear modulus, G, of the soil continuum are indicated. These values are derived from elastic solutions in Sections 5.8.4 and 5.8.5.

The relationship between axial pile settlement, w, and mobilised shear stress, $\tau_s$ is known as the 't–z' curve. The notation for pile settlement varies between sources. In this book, w is used. However, the generic term 't–z' is used regardless of the actual notation. The relationship between lateral pile movement, y, and soil resistance per unit length, P, is known as the 'p–y' curve.

There are various techniques used to derive the input parameter for t–z and p–y analyses. The ultimate axial resistance can be derived using the techniques given in Sections 5.4–5.7. The initial stiffness of the response can be assessed from measurements of soil stiffness, combined with elastic solutions for the associated deformation

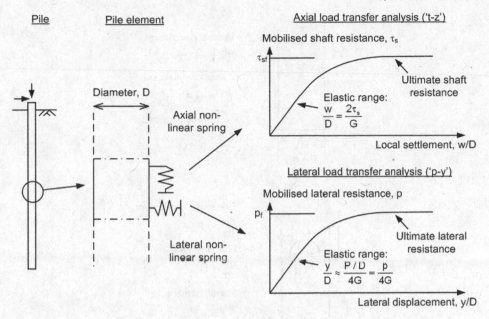

*Figure 5.24* Load transfer analysis of a single pile element

pattern (e.g. Section 5.8.4). The non-linear and cyclic aspects can be derived from advanced soil tests – particularly tests that mimic a short section of pile–soil interface, such as CNS tests – or they can be back-calculated from load test results.

### 5.8.3 Numerical solution methods

If non-linear 't–z' or 'p–y' curves are used, the load-settlement response cannot be calculated in closed form, and software must be used to calculate the deformation incrementally using either a finite difference code or a dynamic relaxation method.

This modelling of the pile response is normally achieved using a computer code that implements the beam-column equations for axial or lateral loading. The governing differential equations are standard from structural engineering, and a variety of programs are available in the industry.

The governing equation for the axial response of a short element of pile is shown in Figure 5.25a. This equation is derived from vertical equilibrium combined with elasticity for the compression of the pile. The governing equation for the lateral response of a short element of pile is shown Figure 5.25b. This equation is derived from horizontal and moment equilibrium combined with elastic bending of the pile.

The axial stiffness of the pile $(EA)_p$ must be calculated for an equivalent solid circular pile, hence keeping the perimeter (upon which shaft resistance acts) and the axial compressibility, $F/\varepsilon_z$, constant. For a steel tubular pile with wall thickness, t:

$$(EA)_p = E_p\left(\frac{\pi D^2}{4}\right) = E_{steel}A_{steel} \approx E_{steel}(\pi Dt) \tag{5.35}$$

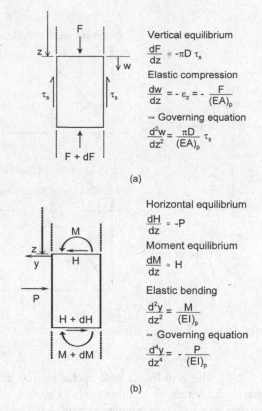

Vertical equilibrium

$$\frac{dF}{dz} \approx -\pi D\, \tau_s$$

Elastic compression

$$\frac{dw}{dz} = -\varepsilon_z = -\frac{F}{(EA)_p}$$

⇒ Governing equation

$$\frac{d^2w}{dz^2} = \frac{\pi D}{(EA)_p}\, \tau_s$$

(a)

Horizontal equilibrium

$$\frac{dH}{dz} \approx -P$$

Moment equilibrium

$$\frac{dM}{dz} \approx H$$

Elastic bending

$$\frac{d^2y}{dz^2} = \frac{M}{(EI)_p}$$

⇒ Governing equation

$$\frac{d^4y}{dz^4} = -\frac{P}{(EI)_p}$$

(b)

*Figure 5.25* Analysis of a short pile element (a) axial response (b) lateral response

Popular load transfer analysis programmes that have been developed at UWA are RATZ (Randolph 2003b) for axial 't–z' analysis and PYGMY (Stewart 2000) for lateral 'p–y' analysis.

### 5.8.4 Elastic solutions for rigid pile response

The deformation mechanism around a pile comprises shearing around the shaft and compression below the base. The initial slope of the $\tau_s$–w response of a pile can be found from an elastic solution for the deformation of a horizontal circular slice comprising a short element of pile and the surrounding soil. For the case of a rigid pile with a linear t–z response, a simple expression linking soil stiffness to pile head stiffness can be derived as follows.

PILE SHAFT RESPONSE

It can be assumed that the shaft resistance on an element of pile is supported by shear stress on concentric rings (Figure 5.26). From equilibrium, this shear stress decreases in proportion to the distance from the pile shaft:

At radius r:

$$\tau = \tau_s \frac{R}{r} \tag{5.36}$$

The deformation can be approximated as shear strain, $\gamma$, on vertical planes:

$$\gamma \approx \frac{dw}{dr} \tag{5.37}$$

In order to link equilibrium and compatibility (deformation), elasticity is used:

$$\frac{\tau}{\gamma} = G \tag{5.38}$$

A further assumption is needed in order to integrate the foregoing equations. Since shear stresses and strains do not decay to zero until infinity, a limit to the zone of deformation must be chosen, in order to give a finite stiffness, $\tau_s/w$. A 'magical radius' $r_m$ is introduced, beyond which it is assumed that strains are zero (Randolph and Wroth 1978). The resulting integration (combining Equations 5.36–38) is:

$$w = \int_{r=R}^{r=r_m} \frac{\tau_s R}{Gr}\, dr = \frac{\tau_s R}{G}\ln\left(\frac{r_m}{R}\right) = \frac{\tau_s R}{G}\zeta \tag{5.39}$$

The log of the magical radius divided by the pile radius defines a dimensionless zone of influence, $\zeta$. Finite element analyses indicate that the limiting radius is comparable to the pile length. The dimensionless zone of influence is typically in the range $\zeta = 3$–$5$ (Baguelin and Frank 1979) and a value of $\zeta = 4$ is often assumed, which makes the initial slope of the t–z response $\tau_s/w = G/4R = G/2D$.

   More precise relationships for $\zeta$, checked against numerical analysis are:

$$\zeta = \ln\left\{[0.5 + (5\rho(1-v)-0.5)\xi]\frac{L}{D}\right\} \quad \text{or} \quad \zeta = \ln\left\{5\rho(1-v)\frac{L}{D}\right\} \quad \text{for} \quad \xi = 1 \tag{5.40}$$

where $\xi$ is the end-bearing modulus ratio (see later, Figure 5.26).

   The above equations show that:

- Mobilised shear stress decays with $1/r$; only the soil very close to the shaft carries significant load.
- Settlement decays with $\ln(1/r)$; significant settlement extends some distance from the pile shaft, so the settlement trough will be large.

For a rigid pile, equal settlement w will occur along the entire length of the pile shaft so the stiffness of a spring representing the shaft resistance of the entire pile is:

$$\frac{Q_s}{w} = \frac{2\pi L G_{avg}}{\zeta} \tag{5.41}$$

Since this is an elastic analysis, the mean soil stiffness along the pile length, $G_{avg}$, can be used, allowing non-homogenous soil to be considered. The notation used for non-homogenous soil, including a higher base stiffness (if the pile is tipped into a bearing stratum), is shown in Figure 5.26.

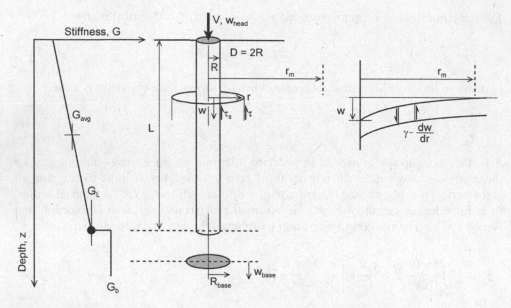

*Figure 5.26* Notation for analysis of deformation around a rigid pile

PILE BASE RESPONSE

The initial stiffness of the response of the pile base can be calculated by considering the pile base as a rigid punch acting on an elastic half-space. The classical solution (as used previously for shallow circular foundations) is:

$$w_{base} = \frac{Q_b}{R_{base} G_{base}} \frac{(1-v)}{4}$$
(5.42)

Different parameters for base radius and stiffness are introduced to allow adjustments for enlarged (under-reamed) bases and pile that are tipped into stiffer layers.

PILE HEAD RESPONSE

The response of the head of a rigid pile can be found by combined the elastic solutions for the shaft and the base. The system can be idealised as parallel shaft and base springs (Figure 5.27), so the stiffness of the pile head, $V/w_{head}$, is the sum of the two stiffness contributions:

$$\frac{V}{w_{head}} = \frac{Q_b}{w_{base}} + \frac{Q_s}{w}$$
(5.43)

The stiffness of a rigid pile can, therefore, be expressed as:

$$\frac{V}{w_{head}} = \frac{4R_{base}G_{base}}{1-v} + \frac{2\pi L G_{avg}}{\zeta}$$

$$\frac{V}{w_{head}DG_L} = \frac{2}{1-v}\frac{G_{base}}{G_L}\frac{D_{base}}{D} + \frac{2\pi}{\zeta}\frac{G_{avg}}{G_L}\frac{L}{D}$$

$$\frac{V}{w_{head}DG_L} = \frac{2}{1-v}\frac{\eta}{\zeta} + \frac{2\pi}{\zeta}\rho\frac{L}{D}$$
(5.44)

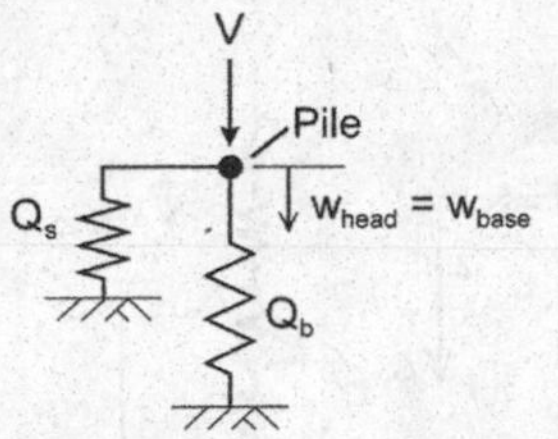

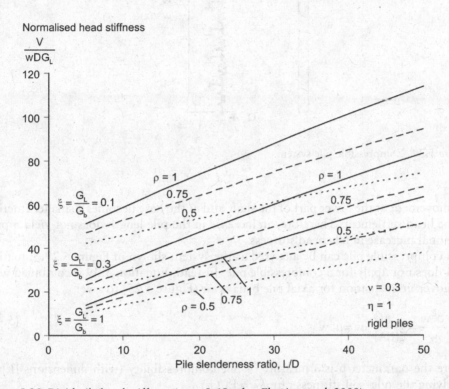

*Figure 5.27* Rigid pile system

*Figure 5.28* Rigid pile head stiffness (eqn. 5.65, after Fleming *et al.* 2009)

These expressions are simplified using dimensionless variables:

$\eta = R_{base}/R = D_{base}/D$      Base enlargement ratio
$\rho = G_{avg}/G_L$      Stiffness gradient ratio
$\xi = G_L/G_{base}$      Base stiffness ratio
$L/D$      Slenderness ratio

Equation 5.44 is plotted for various combinations of input parameters in Figure 5.28.

### 5.8.5 Elastic solutions for compressible pile response

The solution in Section 5.8.4 is for a rigid pile. If the pile is compressible then the rigid solution will overestimate the stiffness at the pile head. As the pile length increases,

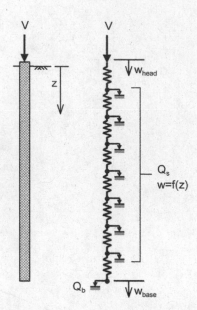

*Figure 5.29* Compressible pile system

the movement at the lower part of the shaft and at the base becomes a smaller fraction of the head settlement. Therefore, an increase in the pile length does not yield a proportional increase in pile head stiffness.

A compressible pile can be analysed as the system shown in Figure 5.29. Equation 5.44 does not apply for a compressible pile. Instead, Equation 5.39 is combined with the governing equation for axial pile behaviour (Figure 5.25) to give:

$$\frac{d^2 w}{dz^2} = \frac{2\pi G}{\zeta (EA)_p} w = \mu^2 w \tag{5.45}$$

where the parameter $\mu$ is a measure of pile compressibility (with dimensions $[L]^{-1}$), involving the pile–soil stiffness ratio, $\lambda = E_p / G_L$:

$$\mu = \frac{\sqrt{8/\zeta\lambda}}{D} \tag{5.46}$$

Equation 5.67 is solved by substituting for the boundary conditions at the pile base to give an expression for dimensionless pile head stiffness:

$$\frac{V}{w_{head} D G_L} = \frac{\dfrac{2\eta}{(1-v)\zeta} + \rho \dfrac{2\pi}{\zeta} \dfrac{\tanh\mu L}{\mu L} \dfrac{L}{D}}{1 + \dfrac{1}{\pi\lambda} \dfrac{8\eta}{(1-v)\zeta} \dfrac{\tanh\mu L}{\mu L} \dfrac{L}{D}} \tag{5.47}$$

The numerator of this expression is made up of the base stiffness (first term) and shaft stiffness (second term). The second term in the denominator is generally small (except for very compressible piles). The influence of pile compressibility on the pile

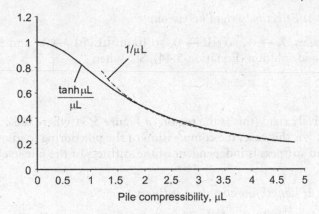

*Figure 5.30* Form of compressibility term in pile stiffness solution

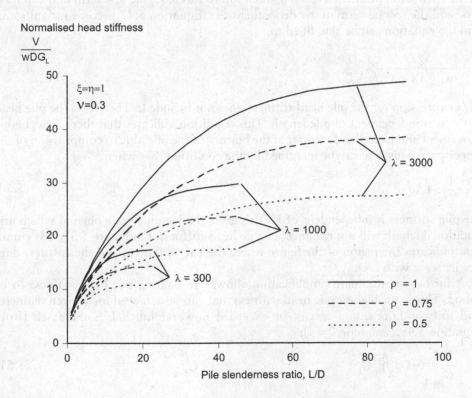

*Figure 5.31* Compressible pile head stiffness (Equation 5.69, after Fleming *et al.* 2009)

head stiffness is controlled by the $(\tanh \mu L)/\mu L$ term, which is plotted in Figure 5.30. Equation 5.47 is evaluated for a range of dimensionless input parameters in Figure 5.31 (for cases with no under-ream, $\eta = 1$, and a base stiffness ratio, $\xi = 1$).

Two limiting regions of behaviour can be identified from Figure 5.30 and Figure 5.31:

*Limiting region 1: Effectively rigid behaviour*

For short stiff piles, $\lambda \to \infty$, so $\mu L \to 0$, so $(\tanh \mu L)/\mu L \to 1$, and Equation 5.47 reduces to the rigid solution (Equation 5.44). So, when

$$\frac{L}{D} < \frac{\sqrt{\lambda}}{4} \tag{5.48}$$

the pile is effectively rigid (this is the region of Figure 5.31 where the curves of different $\lambda$ converge). For this case, the compression of the pile during loading is negligible, and the pile head stiffness is independent of the stiffness of the pile itself, $E_p$.

*Limiting region 2: Length-independent stiffness*

As the pile lengthens or the flexibility increases the parameter $\mu$ increases. For $\mu L = 2$, $\tanh \mu L = 0.96$, and as $\mu L \to \infty$, $\tanh \mu L \to 1$. Therefore, for values of $\mu L$ greater than 2, the term $(\tanh \mu L)/\mu L \sim 1/\mu L$ (Figure 5.30). In this case, the first term in the numerator and the second term in the denominator of Equation 5.47 become insignificant, and the equation can be simplified to:

$$\frac{V}{w_{head}DG_L} = \frac{\pi \rho}{\sqrt{2}} \sqrt{\frac{\lambda}{\zeta}} \tag{5.49}$$

This expression for the pile head stiffness does not include L. Therefore, the pile head stiffness is independent of pile length. This condition indicates that there is negligible movement and hence load transfer at the bottom of the pile, due to compression of the upper part. This case can be identified by the condition that when

$$\frac{L}{D} > 1.5\sqrt{\lambda} \tag{5.50}$$

the pile stiffness is independent of L. This gives the length of pile beyond which any additional length will not enhance the pile head stiffness. On Figure 5.31, this condition indicates the region of the figure where the curves flatten, and the stiffness does not increase with L/D.

A further approximate simplification allows the limiting pile head stiffness to be found. This is the highest pile head stiffness that can be achieved for a given diameter and soil conditions, and cannot be exceeded however much L is increased (from Equation 5.49, assuming $\zeta = 4$):

$$\frac{V}{w_{head}} \approx D\rho\sqrt{E_p G_L} \tag{5.51}$$

For consistency, when using this approximate simplification for limiting pile head stiffness, $G_L$ should be taken as the value at the bottom of the 'active' part of the pile, which is transmitting load. Since the 'active' length is defined as $L/D = 1.5\sqrt{\lambda}$ (Equation 5.50), $G_L$ should be the shear modulus at a depth of $z = 1.5D\sqrt{\lambda}$. Below this depth no significant load transfer occurs.

These two elastic solutions – for a rigid pile and a compressible pile – are idealisations that are not often directly applicable in practice. However, they provide a basis for understanding the general interactions between the t–z load transfer stiffness and the stiffness of the pile itself. They are also important benchmarking tools that can be used to check numerical analyses of axial pile behaviour, before site-specific conditions such as layering and non-linear t–z curves are incorporated.

### 5.8.6 Non-linear t–z curves

The elastic solutions described in the preceding section provide analytical expressions for the initial slope of the t–z curve. Since soil stiffness degrades with strain level as failure is approached, t–z curves also flatten as settlement increases.

The t–z response approaches a gradient of zero in the limit, as $\tau_s \rightarrow \tau_{sf}$, followed by softening to a residual value, $\tau_{s,res}$, in some cases. Softening is commonly observed in clay and can be due to the development of a residual shear surface at the pile-soil interface, with a reduced interface friction angle, or to the generation of excess pore pressure as the interface is sheared. On grouted piles, softening commonly occurs as the grout–soil bond fails and the response decays to only the frictional resistance between the pile and the surrounding material.

Various mathematical forms of t–z curve are used in practice; hyperbolic and parabolic forms are described in the following section, as well as the standard curves provided in the API RP 2A Recommended Practice.

### 5.8.7 t–z curves: API guidelines

API RP 2A (2000) provides load transfer curves that may be used for axial analysis (Figure 5.32). In clay, the response is approximately parabolic, reaching the peak shear stress at a displacement of 1 percent of the pile diameter. Beyond the peak, a degree of strain softening is assumed, with the unit shaft resistance reducing by up to 30 per cent at a displacement of 2 per cent.

In sand, a linear t–z response is assumed, up to a specified settlement, with no strain-softening under monotonic loading. A similar curve is recommended for the base response in both sand and clay (Figure 5.33). This response is likely to be conservative. The ultimate capacity, $q_{bf}$, reached at a pile settlement of 0.1D is assumed to be the maximum value, but it is generally recognised that the resistance will increase further with additional settlement. However, compensating for this is the neglect of residual loads in most load transfer calculations. After installation, a driven pile is unlikely to be unstressed, but will have a component of base load trapped by negative shaft resistance on the lower part of the pile. Load transfer analyses usually neglect residual loads, so underestimate the *additional* shaft resistance available on the lower part of the pile, and overestimate an equal component of base resistance.

### 5.8.8 Other 't–z' models

Forms of t–z curve based on a hyperbolic stress–strain response of the soil, and also parabolic shapes, have been implemented in load transfer programs. Examples from the program RATZ are shown in Figure 5.34. The hyperbolic model uses a stiffness degradation parameter $R_f$ to vary $\zeta$ with settlement and takes the form:

$$w = \frac{\tau_s D}{2 G_i} \zeta_{hyp} \qquad (5.52)$$

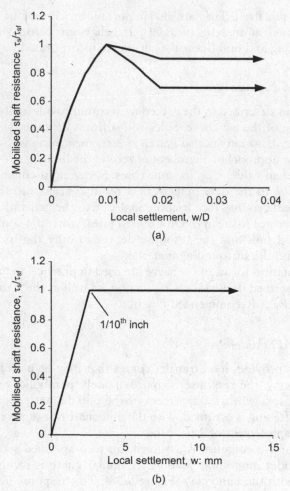

*Figure 5.32* API guidelines for 't–z' curves (a) clay (b) sand (API 2000)

where $G_i$ is the stiffness used to determine the initial slope of the t–z curve (often taken as the small strain initial stiffness $G_0$), and $R_f$ governs the shape of the non-linearity according to:

$$\zeta_{hyp} = \ln\left(\frac{r_m/R - \psi}{1 - \psi}\right) \qquad (5.53)$$

where

$$\psi = \frac{R_f \tau_s}{\tau_{sf}} \qquad (5.54)$$

The hyperbolic t–z curve is limited to $\tau_s < \tau_{sf}$ (see Figure 5.34).

In a parabolic model, the initial slope of the t–z curve is calculated following Equation 5.39. From the geometry of a parabola, the displacement to mobilise $\tau_{sf}$

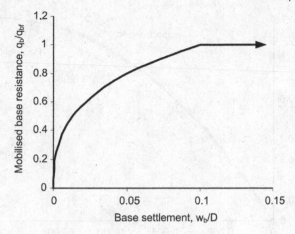

*Figure 5.33* API guidelines for base response (API 2000)

is twice the value for a linear response. This mobilisation displacement, $w_f$, is, therefore:

$$w_f = \frac{2\tau_{sf} R}{G} \zeta \qquad (5.55)$$

The resulting t–z response is (Figure 5.34):

$$\tau_s = \tau_{sf} \left[ 2\frac{w}{w_f} - \left(\frac{w}{w_f}\right)^2 \right] \qquad (5.56)$$

Softening to a residual shaft resistance, $\tau_{s,res}$, over a post-peak displacement of $\Delta w_{res}$ can be modelled using a softening parameter $\eta_{res}$ in the following relationship, where $\Delta w$ is the displacement post-peak (Figure 5.34):

$$\tau_s = \tau_{sf} - 1.1(\tau_{sf} - \tau_{s,res}) \left[ 1 - \exp\left( -2.4 \left(\frac{\Delta w}{\Delta w_{res}}\right)^{\eta_{res}} \right) \right] \qquad (5.57)$$

### 5.8.9 *Progressive failure of long piles*

Strain softening effects can lead to the progressive failure of long, slender piles, such that the actual capacity is less than the ideal (rigid pile) capacity. If a pile is rigid, the peak shaft resistance is mobilised simultaneously along the entire length. However, if the pile is compressible, the top part of the pile will be on the softening part of the t–z response when the bottom part of the pile is reaching the peak. Therefore, the maximum shaft capacity will be only a fraction of the ideal value (Figure 5.35).

The degree of progressive failure depends on the relative magnitudes of the pile compression and the displacement required for the resistance to decay from the

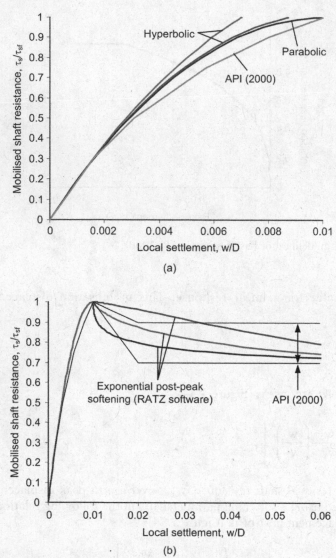

*Figure 5.34* Typical 't–z' models (Randolph 2003b) (a) pre-peak (b) post-peak

peak value to the residual. The resulting reduction in shaft resistance can be expressed as:

$$Q_{sf,actual} = R_{prog} Q_{sf,ideal} \qquad (5.58)$$

A simple design chart can be used to estimate the potential reduction in capacity as a function of the relative compressibility of the pile. The non-dimensional flexibility factor, K, is the ratio of the elastic shortening of the pile under the maximum

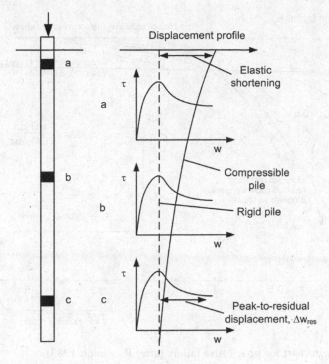

*Figure 5.35* Progressive failure of long piles

shaft capacity, to the displacement needed to strain soften from peak to residual, expressed as:

$$K = \frac{\text{Shortening}}{\text{Displacement from peak to residual}} = \frac{(\pi D \tau_{sf} L)L}{(EA)_p \Delta w_{res}} \quad (5.59)$$

Figure 5.36 shows a relationship between K and $R_{prog}$, based on a particular form of softening response.

The peak-to-residual displacement, $\Delta w_{res}$, is typically 20–40 mm, but can be measured for a particular soil in ring shear tests or CNS tests. The softening strength ratio $\xi_{soft} = \tau_{s,res}/\tau_{sf}$ is the ratio of residual to peak strength in clay. In cemented soils and in particular in carbonate materials, $\xi_{soft}$ can be very low, not due to changes in tan $\delta$, but due to the loss of horizontal effective stress with shearing.

### 5.8.10 Cyclic loading

Cyclic loading can lead to significant degradation of shaft resistance in both sands and clays, with the effects being more severe in carbonate sands, particularly when cemented. In clay, the behaviour is commonly linked to excess pore pressure build-up, whereas in sand the response is due to stress relief due to densification adjacent to the pile. These two mechanisms are manifestations of the same behaviour – a contractile response to shearing at the pile–soil interface. The presence of cyclic horizontal loads may exacerbate the degradation of axial shaft resistance, if the lateral

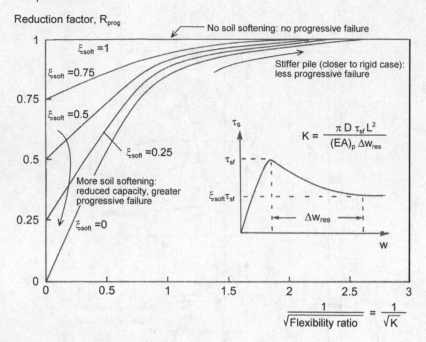

*Figure 5.36* Design chart for progressive failure (after Randolph 1983)

movements cause additional softening or the opening of a gap around the upper part of the pile.

Shaft resistance degrades progressively with cycles and at a higher rate for larger cycles. At the element level (modelling one particular point down the pile), one-way cyclic loading will lead to increasing settlement, followed by failure at a value of shaft resistance that is lower than obtained under monotonic loading. Under two-way cyclic loading, the average settlement may not change much, but the cyclic amplitude will increase as the soil stiffness drops, followed by failure. This failure may be at a reduced level of shaft resistance. Two-way loading causes a more rapid degradation of shaft resistance than one-way loading.

The overall effect of this behaviour is for load transfer to be shed from the upper part of the pile, where the loading and, therefore, the degradation is more severe, towards the lower part of the pile.

Examples of the many studies in the field and in the laboratory that have illustrated cyclic t–z behaviour are, in clay, Bea and Audibert (1979), Karlsrud and Haugen (1985), Bea (1992), and Bogard and Matlock (1990); in uncemented siliceous sands, Jardine and Standing (2000); and in carbonate sands (uncemented and cemented), Poulos (1988) and Randolph *et al.* (1996).

Simple models to capture this behaviour within the t–z response have been described by Matlock and Foo (1980), Randolph (1988), Poulos (1989), Kalrsrud and Nadim (1990) and Bea (1992). The general strategy is to incorporate rules for updating, on a cycle-by-cycle basis, the ultimate shaft resistance (and in some cases the shape of the t–z response). This strategy is illustrated in the following discussion by a cyclic t–z model that was developed to simulate the response of grouted piles in weak calcarenite (Randolph *et al.* 1996).

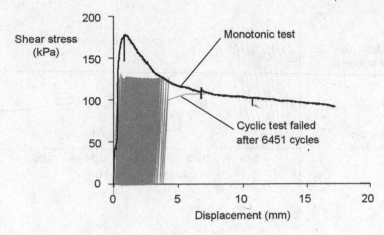

*Figure 5.37* One-way cyclic response of calcarenite (rod shear test data) (Randolph *et al.* 1996)

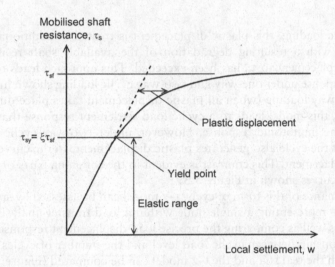

*Figure 5.38* 'Elastic' and 'plastic' parts of a non-linear t–z response

A one-way cyclic loading test on a short model pile (which can be idealised as rigid, thus indicating directly the t–z response) in calcarenite is shown in Figure 5.37, and compared with the response under monotonic loading. In this case, the failure occurs once the cumulative displacement under cyclic loading reaches the monotonic loading envelope. This observation can be used to derive simple algorithms to allow cyclic load transfer response to be modelled.

The axial pile load transfer program, RATZ, has in-built algorithms to model degradation under cyclic loading. The plastic displacement within each cycle is defined as all of the displacement additional to the 'elastic' displacement, which is found by extrapolating the initial slope of the t–z curve (Figure 5.38). In RATZ, an initial elastic range of the t–z curve can be defined, until a yield point, at shear stress $\tau_{sy}$, beyond which the response is hyperbolic or parabolic. In the nomenclature used by RATZ, the yield point is defined as a fraction $\xi$ of $\tau_{sf}$.

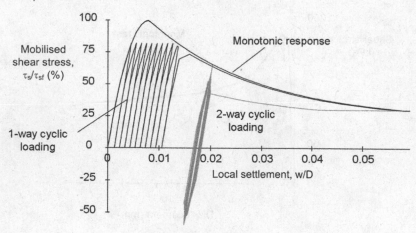

*Figure 5.39* RATZ simulation of cyclic loading (Randolph *et al.* 1996)

Under cyclic loading, this plastic displacement is treated as additional monotonic displacement, with a resulting degradation of the available shaft resistance if the monotonic displacement to $\tau_{sf}$ has been exceeded. This approach leads to the predictions of t–z response under one-way and two-way cyclic loading shown in Figure 5.39.

Under one-way loading (when all plastic displacement takes place during positive displacement), this model leads to a cyclic load settlement response that follows the envelope of the monotonic response. However, under two-way cyclic loading, the reverse part of the cycle also generates plastic displacement, so the curve softens at a lower net displacement. This contrast is evident in the softening parts of the one-way and two-way curves shown in Figure 5.39.

The performance of this form of cyclic t–z model can be assessed by simulating soil element tests – representing a single node within a load transfer analysis – using the t–z element. As well as comparing the precise load-displacement response within each test, the relationship between cyclic load level and the number of cycles required for failure for both the real soil and the t–z model can be compared (Figure 5.40).

### 5.8.11 Cyclic stability diagrams

In the same way that cyclic stability diagrams can be derived from the results of cyclic soil element tests, they can also be constructed for pile shaft resistance. A cyclically applied shaft resistance load can be expressed in terms of the average and cyclic components, normalised by the maximum static capacity (Figure 5.41):

Normalised mean shaft resistance = $Q_{s,cyc,mean}/Q_{sf}$
Normalised cyclic shaft resistance = $Q_{s,cyc,amp}/Q_{sf}$

Stable modes of cyclic loading can be defined using interaction diagrams linking $Q_{s,cyc,mean}$ and $Q_{s,cyc,amp}$. Cyclic load combinations that lie beneath the cyclic stability envelope will not lead to failure. The diagram may be made more sophisticated by the inclusion of contours showing numbers of cycles to cause failure. Three simple forms of interaction diagram are shown in Figure 5.42.

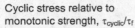

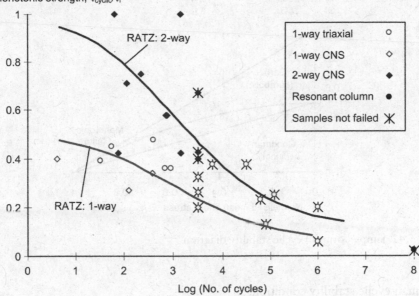

*Figure 5.40* Cyclic stability (S–N) diagram for carbonate soil, and t–z element (Randolph *et al.* 1996)

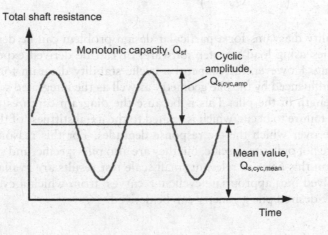

*Figure 5.41* Notation for cyclic shaft resistance

The Gerber relationship is widely used in the fatigue analysis of metals, and the more simple Goodman relationship is slightly conservative by comparison.

Gerber cyclic stability condition:

$$\frac{Q_{s,cyc,amp}}{Q_{sf}} < \frac{1}{2}\left(1 - \frac{Q_{s,cyc,mean}}{Q_{sf}}\right)^2 \tag{5.60}$$

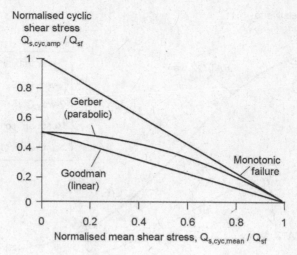

*Figure 5.42* Simple forms of cyclic stability diagram

Goodman cyclic stability condition:

$$\frac{Q_{s,cyc,amp}}{Q_{sf}} < \frac{1}{2}\left(1 - \frac{Q_{s,cyc,mean}}{Q_{sf}}\right)$$    (5.61)

Cyclic stability diagrams for a particular design problem can be derived from parametric studies using load transfer software, or can be derived experimentally from model testing. However, the shape of a cyclic stability diagram for the overall pile capacity is influenced by the pile geometry as well as the integrated soil characteristics along the length of the pile. This is because the diagram captures the tendency for progressive failure to occur, which is related to the axial stiffness of the pile relative to the distance over which the t–z response degrades. For this reason, cyclic stability diagrams are not only site-specific, but they are also pile-specific, and should, therefore, be derived on this basis. If relevant small-scale test results are available, they should be deconvolved into appropriate cyclic t–z curves, from which a cyclic stability diagram for the desired pile geometry can be derived.

### 5.8.12 Examples load transfer analysis

The calibration of a cyclic t–z model to CNS data and the resulting cyclic pile response is illustrated with an example taken from Randolph and White (2008a), which is from the design of a drilled and grouted pile in variably cemented carbonate soils. Figure 5.43 shows the type of load transfer response that has emerged both from CNS testing and from field load tests on grouted section tests (Randolph *et al.* 1996). Under monotonic displacement, cemented carbonate soils exhibit a very brittle reduction in shear resistance, and this appears to degrade gradually over a considerable displacement. Under cyclic displacements, a very low shear resistance is measured, in some cases as low as 1 per cent of the peak shear stress. The low shear resistance extends through the

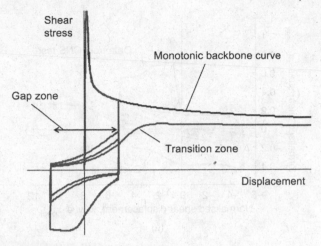

*Figure 5.43* Form of load transfer curve for cyclic pile software (AG 2007)

major portion of the displacement range over which failure has been reached, which is indicated as the 'gap' zone in Figure 5.43 (analogous to the physical gap that might appear around a pile loaded laterally). On emerging from the gap zone, the shear resistance builds back up towards the monotonic backbone curve, but with a lower failure shear stress, reflecting the additional plastic shearing within the cyclic gap zone. This form of load transfer curve has been built into a load transfer program, Cyclops (AG 2007), which has evolved from the software, RATZ (Randolph 2003b).

The brittleness of cemented carbonate sediments such as calcarenite and calcisiltite, and the vulnerability to damage during cyclic shearing, leads to the potential for significant progressive failure. Under monotonic loading, failure will occur near the pile head and propagate downwards, so that the maximum pile load that may be mobilised is much less than the ideal capacity of a rigid pile. The challenge is to quantify the strain softening behaviour from appropriately devised monotonic and cyclic laboratory shear tests.

Figure 5.44a shows the results of a CNS test where a series of post-failure cyclic displacements of ±5 mm have been applied. The data have then been fitted by simulating the element test using the pile analysis software, Cyclops. Figure 5.44b shows the match to the gradual reduction in the shear resistance mobilised with each cycle. After the 25 cycles, the sample is subjected to further monotonic shearing up to the limits of the apparatus. The load transfer fit includes modelling of the transition back to the monotonic shearing curve.

An example of the resulting response of a drilled and grouted pile under monotonic and cyclic loading is shown in Figure 5.45a. The load has been normalised by the ideal 'rigid pile' capacity $Q_{rigid}$ while the displacement is normalised by the diameter, as w/D. It may be seen that the peak monotonic capacity is only 64 per cent of the rigid pile capacity. Following cyclic loading to simulate the design storm, plus a full 'life cycle' of more moderate environmental loading, the pile capacity is further reduced, to just over 50 per cent of the ideal capacity. This compares with the peak design load, which is just under 40 per cent, indicating a material factor of safety of about 1.2.

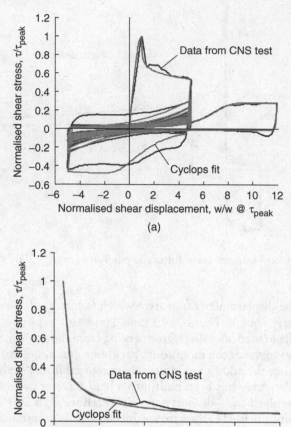

*Figure 5.44* Fitting of data from constant normal stiffness (CNS) test with cyclic pile software, Cyclops (a) normalised shear-displacement response (b) reduction in shear stress mobilised with each cycle (Randolph and White 2008a)

Figure 5.45b shows the pattern of degradation of shaft friction at the end of the cyclic loading stages. The upper part of the pile, down to just over 50 per cent of the grouted length, has been reduced to residual shaft friction, while the bottom 36 per cent has suffered no degradation. The transition zone over which partial degradation has occurred is, therefore, around 14 per cent of the grouted length of the pile.

## 5.9 Lateral response

### 5.9.1 Design considerations

Offshore piles must often withstand significant lateral loads and moments. For anchor piles and monopile foundations – supporting wind turbines or small platforms – the dominant design load generally acts laterally.

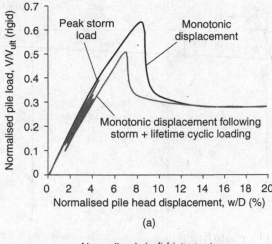

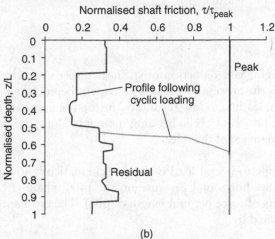

*Figure 5.45* Response of drilled and grouted pile to monotonic and cyclic loading, computed using Cyclops (a) pile head response (b) profiles of peak, residual and mobilised shaft resistance (Randolph and White 2008a)

The capacity and response of laterally loaded piles is usually assessed by means of non-linear load transfer analysis, using load transfer (or 'p–y') curves appropriate for the soil type. This load transfer approach was introduced in Section 5.8.2.

A major difference between axial and lateral loading of piles is that lateral loading effects are confined to the upper ~10–15 diameters of the pile. The design process includes the following considerations:

- Lateral pile stiffness:
  - prevent excessive lateral deflection of the pile
- Lateral pile strength:
  - prevent failure by the pile moving sideways or rotating through the soil
  - prevent bending failure of the pile

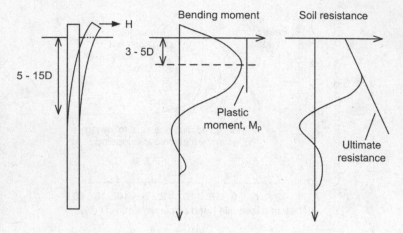

*Figure 5.46* Response of a pile to lateral load

- Cyclic effects:
  - assess any significant reduction in stiffness under cyclic loading due to post-holing; this is the process by which a gap opens behind the pile during horizontal deflection, leading to low resistance during the first part of the return cycle
- Assess the distribution of cyclic bending moments along the pile, which contributes to fatigue damage of the pile.

The response of a pile to lateral load is shown schematically in Figure 5.46. Lateral loading leads to a net horizontal pressure on the pile, which is expressed as a soil resistance, P (in units of force per unit length of pile). The ultimate lateral capacity of a pile, $H_{ult}$, is estimated by:

1. Assuming a distribution of maximum lateral resistance, $P_f$, on the pile.
2. Treating the pile as a beam under distributed load $P_f$ and point load $H_{ult}$. $H_{ult}$ can be found by solving for equilibrium. A check must also be made to ensure that the plastic moment capacity of the pile itself is not exceeded.

'Short' pile failure mechanisms involve failure within the soil only – the pile does not fail in bending. 'Long' pile failure mechanisms involve the formation of a plastic hinge at one or more points within the pile. Since these are collapse mechanisms (like upper bounds) rather than permissible load combinations (like lower bounds), both types of potential failure mechanism must be checked in design.

A full analysis of the lateral stiffness of a pile, and the response to cyclic loading, is usually evaluated by a p–y approach using software such as LPILE (www.ensoftinc.com) or PYGMY (Stewart 2000).

### 5.9.2 Short pile failure mechanisms

When loaded laterally to failure, the net lateral force (per unit length) on the pile is $P_f$, and acts against the direction of movement. For short piles, failure occurs by rigid

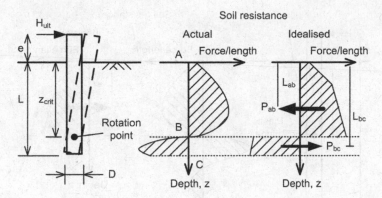

*Figure 5.47* Lateral failure mechanism for short piles

rotation of the entire pile about a centre of rotation at depth $z_{crit}$ below the soil surface. Typically, $z_{crit}$ will be 70–80 per cent of the embedded pile length.

The soil resistance $P_f$ will be positive above the rotation centre, and negative below it. This distribution is idealised, as in retaining wall design, by assuming a sharp transition from full positive to full negative resistance at the point of rotation (Figure 5.47).

In order to calculate the lateral pile capacity $H_{ult}$ when the load acts at an eccentricity, e, above the soil surface, the first step is to evaluate the profile of limiting soil resistance down the pile – this is covered in Section 5.9.5. The second step is to calculate, for an assumed value of $z_{crit}$, the integrated positive and negative resistance forces, $P_{ab}$ and $P_{bc}$, together with their lines of action, at depths $L_{ab}$ and $L_{bc}$ below the soil surface.

By considering the horizontal and moment equilibrium of the pile, two equations for $H_{ult}$ can be written:

$$H_{ult} = P_{ab} - P_{bc} \tag{5.62}$$

$$H_{ult}e = -P_{ab}L_{ab} + P_{bc}L_{bc} \tag{5.63}$$

If the forces and distances $P_{ab}$, $P_{bc}$, $L_{ab}$ and $L_{bc}$ can be expressed analytically in terms of $z_{crit}$, these two equations can be solved simultaneously to find $H_{ult}$ by eliminating $z_{crit}$. Otherwise, it is necessary to iterate the calculation until the correct value of $z_{crit}$ is found.

### 5.9.3 Long pile failure mechanisms

The long pile failure mechanism involves the formation of a plastic hinge at a distance $z_{crit}$ below the ground surface (Figure 5.48). The plastic hinge will form where the bending moment is maximum, and hence the shear force is zero ($dM/dz = S = 0$). This implies that the lateral forces on the pile above the hinge are in horizontal equilibrium, as are those below.

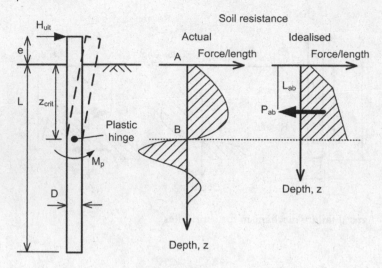

*Figure 5.48* Lateral failure mechanism for long piles

Therefore, separate horizontal equilibrium equations can be written for the pile above and below the hinge. The soil resistance components mobilised below the hinge point are self-equilibrating. Calculation of the lateral capacity need only consider the positive soil resistance above the level of the hinge. Above the hinge point (where the pile is moving in one direction only), the maximum soil resistance must equal the horizontal capacity $H_{ult}$. Therefore, from horizontal equilibrium and by taking moments about the top of the pile:

$$H_{ult} = P_{ab} \tag{5.64}$$

$$M_p = P_{ab}(L_{ab} + e) \tag{5.65}$$

For a given pile $z_{crit}$ can be found from Equation 5.65 (knowing $M_p$) using an iterative process if $L_{ab}$ and $P_{ab}$ cannot be expressed analytically in terms of $z_{crit}$. Once $z_{crit}$, and hence $P_{ab}$ are known, $H_{ult}$ can be found from Equation 5.64.

### 5.9.4 *Fixed head pile failure mechanisms*

In many situations, piles form part of a pile cap and rotation is restrained at this level; these piles are termed 'fixed-head'. This end condition introduces three variants on the 'short' and 'long' pile failure mechanisms. Failure under lateral loading can then occur in three modes shown in Figure 5.49. The failure load corresponding to each of these three variants must be considered when designing a fixed-head pile; the lowest failure load will govern.

- Mode A: no hinges.
  - A short fixed-head pile will fail by rigid body translation. The failure load is the integrated lateral resistance along the length of the pile.

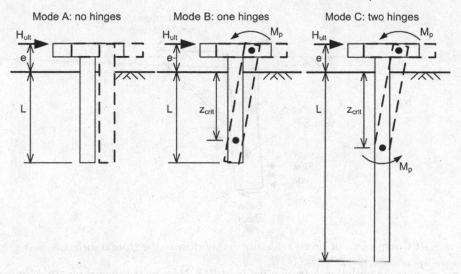

*Figure 5.49* Lateral failure mechanisms for fixed-head piles

- Mode B: one hinge.
  - An intermediate-length fixed head pile will fail by rigid body rotation with a single plastic hinge forming at the pile cap.
  - This failure mechanism is analysed in the same manner as a short (unrestrained) pile, with an additional $M_p$ term in the moment equilibrium equation about the line of action of H (Equation 5.63)
- Mode C: two hinges.
  - A long fixed-head pile will fail by the formation of plastic hinges at the pile cap and at the point of zero shear force.
  - This failure mechanism is analysed in the same manner as a long (unrestrained) pile, with an additional $M_p$ term in the moment equilibrium equation about the line of action of H (Equation 5.65).
  - Therefore, the lateral capacity of a long fixed-head pile may be estimated from that for a long unrestrained pile, by substituting an equivalent moment capacity, $M_p$, equal to twice the actual moment capacity, in the latter solution.

### 5.9.5 Limiting lateral resistance

The lateral resistance $P_f$ in homogeneous clay is related to the undrained shear strength via a failure mechanism. Different failure mechanisms apply close to the ground surface and at depth (Figure 5.50). Close to the ground surface, a conical wedge is lifted, resisted by friction on the pile shaft, shear within the soil and the self-weight of the wedge. A gap may open behind the pile, if the tension cannot be held by negative pore water pressure. Deeper down, soil flows in the horizontal plane around the pile shaft.

UNIFORM CLAY: SHALLOW

The shallow failure mechanism involves the failure of a wedge of soil in front of the pile, with a gap forming behind the pile. This leads to a lower limiting resistance compared to at depth. Traditionally, for clay, the limiting soil resistance has been

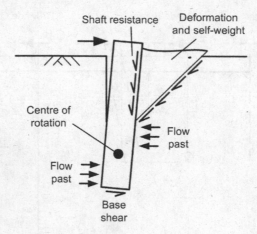

*Figure 5.50* Components of lateral resistance in clay close to the ground surface

taken to increase with depth from a value of $P_f = 2Ds_u$ near the surface, to a limiting value of $P_f = 9Ds_u$ at depths greater than 3 pile diameters (Broms 1964a). The lower limit of $2Ds_u$ corresponds to passive failure in front of the pile with a gap forming behind it. The upper limit of $9Ds_u$ is the value at which horizontal flow of soil around the pile becomes an easier failure mechanism than the conical wedge. This horizontal flow mechanism is described in the following section and governs the deep limiting resistance.

An upper bound solution for the wedge failure mechanism near the soil surface was derived by Murff and Hamilton (1993). Resistance to lateral displacement and rotation of the top part of the pile shaft is calculated from the work associated with:

- Deformation of soil within the wedge
- Lifting the soil within the deforming wedge as it moves upwards
- Shear along the wedge–soil interface
- Shear along the pile–soil interface as the wedge moves upwards.

For very short piles, Murff and Hamilton also derived components of resistance due to the flow of soil around the pile base. Their solutions for weightless soil compare well with the empirical correlations of Matlock (1970) and Broms (1964a). However, their solution suggests that the limiting resistance will be mobilised more rapidly with depth, once the weight of the soil is taken into consideration, which is supported by model test data (Hamilton and Murff 1995). Figure 5.51. shows the variation in $P_f$ with depth for both uniform clay, and clay with linearly increasing strength with depth, ignoring the soil self-weight (which is conservative).

UNIFORM CLAY: DEEP

Below a certain depth, the wedge failure mechanism of Murff and Hamilton is no longer critical. Instead, less resistance is offered by a flow mechanism in the horizontal plane. Broms (1964a) proposed a value of $P_f = 9Ds_u$. Rigorous plasticity solutions for flow around a cylindrical pile, with the limiting resistance, $P_f$, varying slightly with pile roughness, $\alpha = \tau_s/s_u$ were later presented by Randolph and Houlsby (1984) and Martin

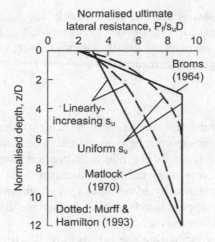

*Figure 5.51* Variation in lateral resistance with depth in clay (weightless one-sided analysis)

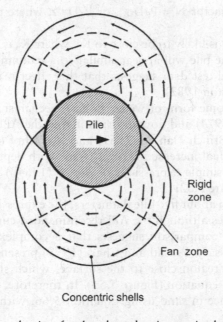

*Figure 5.52* Flow round mechanism for deep lateral resistance in clay

and Randolph (2006). Their upper bound flow mechanism consists of small 'rigid' regions of soil attached to the front and back of the advancing pile, fan zones adjacent to the pile and concentric sliding shells completing the mechanism (Figure 5.52). The resistance, $P_f$, varies from $9.14Ds_u$ to $11.92Ds_u$ between fully smooth and fully rough interface conditions. A design value of $P_f = 9Ds_u$ is conservative, being about 15 per cent lower than the bearing factor of 10.5 that is often adopted for the reverse analysis of deriving soil strength from T-bar penetration resistance.

There is considerable experimental evidence to support values at least as high as $P_f = 10.5Ds_u$, for example, Murff and Hamilton (1993), although the conservative approach of Broms (1964a) is still the most widely adopted, with $P_f = 9Ds_u$. Rate effects, reflecting the high strain rate of wave-induced loading relative to conventional laboratory tests, can be invoked to justify a modest increase in the operative soil strength during lateral loading (Jeanjean 2009).

SAND

The limiting lateral resistance in sand is more difficult to analyse than in clay, since simple plasticity mechanisms cannot be constructed. The combined influence of self-weight and the stresses created by the pile load prevent simple lower bound stress fields being constructed. Any deep failure mechanism must take place at constant volume in order to be kinematically admissible. This requirement for constant volume cannot be satisfied by any upper bound solution derived for sand since normality must be obeyed and hence dilation must occur.

Therefore, methods for estimating the variation of $P_f$ with depth for piles in sand are empirical, and checked against field or model test data. It is usual to non-dimensionalise the lateral resistance by dividing by the *in situ* vertical effective stress to create a dimensionless factor $N = P_f/D\sigma'_v$, or $P_f/D\gamma'z$, where this factor depends on friction angle $\phi$.

At the soil surface, N is likely to be close to $K_p$, where $K_p$ is the passive earth pressure coefficient, since the pile will appear similar to a retaining wall at very shallow depths. However, model test data suggests that this effect may be confined to very close to the surface (Barton 1982).

At shallow depths, some form of wedge failure mechanism will develop, as proposed by Reese *et al.* (1974) and now incorporated in the API (2000) guidelines. To account for this, both Brinch Hansen (1961) and Meyerhof (1995) have developed charts describing a gradual increase in the N factor with depth. These relationships are compared with the simple expressions of Broms (1964b) ($N = 3K_p$) and Barton (1982) ($N = K_p^2$) in Figure 5.53:

At depths of less than about five pile diameters, the expressions all yield relatively similar limiting pressures although the API-recommended curves appear optimistic at greater depths. The comparison suggests that a complex variation of N with depth may not be necessary. Prasad and Chari (1999) presented data of the lateral pressure in the critical region close to the surface, which shows good agreement with Barton's empirical equation (Figure 5.54). It, therefore, appears reasonable to estimate lateral resistance in sand using Equation 5.66, with $P_f$ increasing linearly with depth:

$$P_f = DK_p^2\sigma'_v = DK_p^2\gamma'z \tag{5.66}$$

where $K_p$ is the passive earth pressure coefficient:

$$K_p = \frac{1+\sin\phi}{1-\sin\phi} \tag{5.67}$$

In carbonate sand, there is no distinct limit on the lateral resistance, which increases steadily with increasing lateral deflection. Forms of p–y response for carbonate sands are discussed in Section 5.9.14

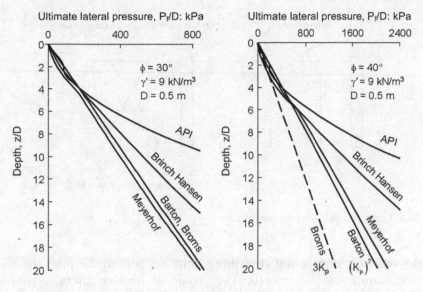

*Figure 5.53* Variation in lateral resistance with depth in sand: proposed approaches

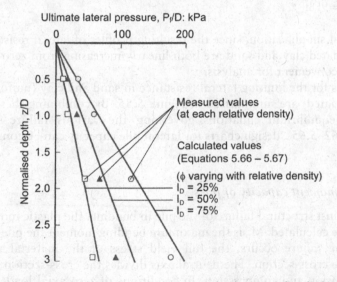

*Figure 5.54* Model test data of lateral resistance in sand (after Prasad and Chari 1999)

### Normally consolidated clay

In normally consolidated clay the strength increases linearly with depth from close to zero at the ground surface. In this case, the undrained strength profile can be simplified as:

$$s_u = k_{su} z \qquad (5.68)$$

In this soil profile, the limiting lateral resistance is reached very close to the surface so the deep flow round mechanism can be assumed to act on the entire pile length.

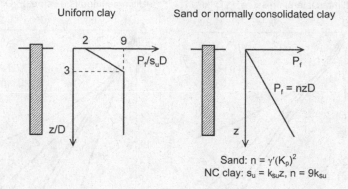

*Figure 5.55* Limiting lateral resistance profiles

Since the very soft surface soil contributes little support to the pile, any shallow wedge mechanism can be ignored. The profile of limiting lateral resistance is, therefore:

$$P_f = 9s_u D = 9Dk_{su}z \qquad (5.69)$$

This is a useful simplification, since the resulting profiles of lateral resistance in normally consolidated clay and sand are both linearly increasing from zero at the mudline, which is convenient for analysis.

The profiles for the limiting lateral resistance in sand and clay (uniform and normally consolidated) are summarised in Figure 5.55. By combining these profiles of $P_f$ with the equilibrium equations governing the various failure mechanisms (Equations 5.62–5.65), design charts for lateral pile capacity can be constructed (see Section 5.9.7).

### 5.9.6 *Plastic moment capacity of piles*

To design against structural failure of the pile in bending, the plastic moment capacity $M_p$ must be calculated. $M_p$ is the maximum bending moment the pile can sustain. When bending failure occurs, the full yield stress of the material is mobilised over the entire cross-section. The neutral axis divides the cross-section into regions that are in tension and compression. In conditions of zero axial load, the area on each side of the neutral axis is equal, but if there is a significant axial force this is not the case.

The tension and compression forces in the cross-section are in equilibrium with the applied bending moment and axial force. Normally the axial force is ignored, although a reduced value of $M_p$ can be derived if significant axial force is present. For a tubular pile, the bending moment about the neutral axis due to a small element at yield stress $\sigma_y$ is (Figure 5.56):

$$dM_p = \sigma_y \, y dA = \sigma_y (R\sin\theta)(Rd\theta)t \qquad (5.70)$$

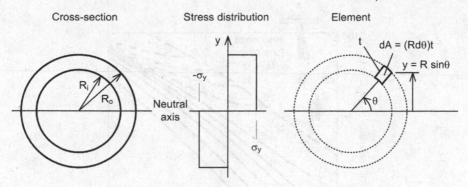

*Figure 5.56* Plastic moment capacity of a tubular pile

Integrating over a single quadrant ($\theta = 0 \rightarrow \pi/2$) and multiplying by 4 leads to:

$$M_p = 4R^2 t\sigma_y = D^2 t\sigma_y \tag{5.71}$$

### 5.9.7 Design charts for lateral pile strength

*Uniform clay*

Using the assumed lateral resistance distributions derived in the foregoing section, combined with equilibrium (Equations 5.62–5.65), design charts for lateral capacity can be derived. This process is carried out iteratively for uniform clay since the variation of $P_f$ with depth cannot be expressed analytically.

Two design charts are derived: one for the short pile failure mechanisms (Figure 5.57, in which normalised $H_{ult}$ is related to slenderness L/D) and one for the long pile failure mechanisms (Figure 5.58, in which normalised $H_{ult}$ is related to normalised $M_p$). Each chart has different failure lines for different values of eccentricity, e, which represents the elevation of the line of action of $H_{ult}$ above the ground surface. If a combined HM load is applied at the ground surface (with H and M acting in the same sense), this can be converted into a single H load, applied at an eccentricity, e, above the ground surface; e = M/H.

The designer should check that the proposed load and pile combination (H, $s_u$, $M_p$, D) lie at a point below the appropriate failure line (i.e. the failure line, which represents $H_{ult}$, corresponds to a higher load than the design value, H). This check should be carried out on both charts, to confirm that neither a short or long failure mechanism will occur.

As noted earlier, results for long piles that are restrained from rotating at the pile head are obtained by substituting an artificial moment capacity for the restrained pile that is twice the actual value, and using the corresponding curve for an unrestrained pile. So, design curves for restrained piles are shifted by a factor of 2 to the left, as shown for zero eccentricity (e/D = 0).

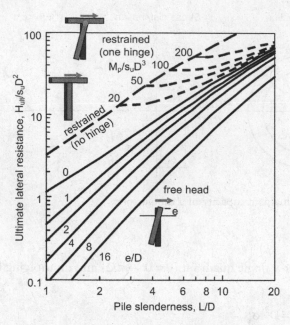

*Figure 5.57* Short pile design charts for lateral capacity in uniform clay (Fleming *et al.* 2009)

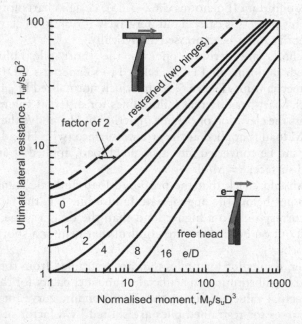

*Figure 5.58* Long pile design charts for lateral capacity in uniform clay (Fleming *et al.* 2009)

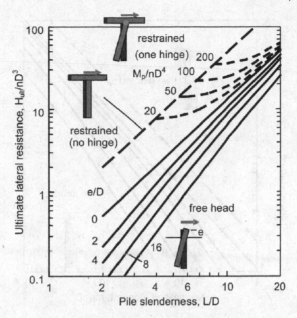

*Figure 5.59* Short pile design charts for lateral capacity with linearly increasing soil resistance with depth (Fleming *et al.* 2009)

## Linearly increasing soil resistance

In sand or normally consolidated clay, for which the limiting lateral resistance is assumed to increase linearly with depth, design charts have been constructed in the same manner, for the short (Figure 5.59) and long (Figure 5.60) failure mechanisms. In these charts, the lateral resistance gradient with depth is denoted n, where:

$$P_f = nzD \tag{5.72}$$

For sand, using Barton's empirical expression (from Equations 5.66–5.67):

$$n = \gamma' K_p^2 = \gamma' \left( \frac{1 + \sin\phi}{1 - \sin\phi} \right)^2 \tag{5.73}$$

For normally consolidated clay, with strength increasing linearly with depth (from Equation 5.69):

$$n = 9k_{su} \tag{5.74}$$

Since the lateral resistance in sand (or n.c. clay) can be expressed analytically as $P_f = nzD$, it is possible to derive some of the lines on the design charts directly. The free body diagrams for two simple cases are shown in Figure 5.61.

The dimensionless horizontal capacity of a short fixed-head pile, $H_{ult}/nD^3$ (no-hinge failure), can be linked to slenderness ratio L/D by considering the horizontal equilibrium of the pile in Figure 5.61a:

$$H_{ult} = \bar{P}_f L = \frac{1}{2} nLDL \tag{5.75}$$

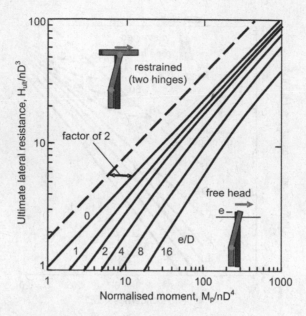

*Figure 5.60* Long pile design charts for lateral capacity with linearly increasing soil resistance with depth (Fleming *et al.* 2009)

so

$$\frac{H_{ult}}{nD^3} = \frac{1}{2}\left(\frac{L}{D}\right)^2 \tag{5.76}$$

for a short pile with a fixed head. This expression is the heavy dotted line on the short pile design chart shown earlier in Figures 5.57 and 5.59.

Similarly, the long pile failure mechanism for the case of e = 0 can be solved directly, for both the fixed-head and free-head (unrestrained) cases. From horizontal equilibrium of the pile in Figure 5.61b:

$$H_{ult} = P_f z_{crit} = \frac{1}{2} n z_{crit} D z_{crit} \tag{5.77}$$

Taking moments about the pile head (where the bracketed 2 applies to fixed-head case):

$$(2) M_p = \frac{1}{2} n z_{crit} D z_{crit} \frac{2}{3} z_{crit} \tag{5.78}$$

Combining these equations to eliminate $z_{crit}$, for the unrestrained case:

$$\frac{H_{ult}}{nD^3} = \frac{3^{2/3}}{2}\left(\frac{M_p}{nD^4}\right)^{2/3} = 1.04\left(\frac{M_p}{nD^4}\right)^{2/3} \tag{5.79}$$

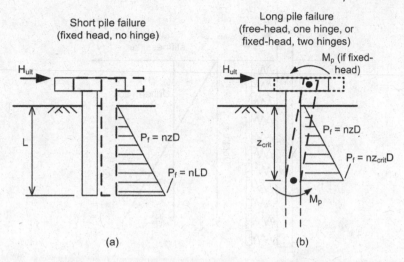

*Figure 5.61* Simple lateral capacity solutions for linearly increasing resistance

This expression is the solid line for $e/D = 0$ on the long pile design chart (Figure 5.60). Combining these equations to eliminate $z_{crit}$, for the fixed-head case:

$$\frac{H_{ult}}{nD^3} = \frac{6^{2/3}}{2}\left(\frac{M_p}{nD^4}\right)^{2/3} = 1.65\left(\frac{M_p}{nD^4}\right)^{2/3} \qquad (5.80)$$

This expression is the thick dotted line on the long pile design chart (Figure 5.60).

Equations 5.79 and 5.80 are related by a factor of 2 on the plastic moment, as indicated on the design chart. This simple analysis allows it to be shown that the effect of restraining the pile cap is to increase the horizontal capacity $H_{ult}$ by a factor of $2^{2/3} = 59$ per cent.

These solutions are idealised, in that they do not account for layering or non-linear profiles of shear strength. In reality, it is common for pile capacity to be assessed using a beam-column load transfer method, which provides both the load-displacement behaviour as well as the ultimate capacity. Alternatively, capacity calculations can be performed using a simple spreadsheet program that checks the short and long piles failure mechanisms by applying the governing equilibrium equations in an iterative manner, searching for the critical depth to either a rotation point or a plastic hinge.

### 5.9.8 Elastic lateral stiffness solutions

To assess the lateral load-displacement response of a pile it is necessary to assess the local lateral p–y stiffness. For an elastic idealisation of the response, there are analytical solutions available to provide the resulting head stiffness, but if a non-linear p–y response is to be modelled then a numerical solution is required.

#### Sub-grade reaction approach

The most simple approach to model the lateral response of a pile is to treat the soil as a series of springs along the length of the pile; this is known as the Winkler idealisation

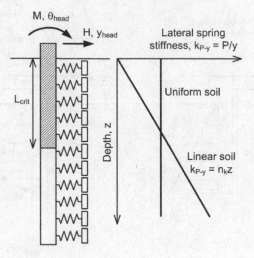

*Figure 5.62* Lateral response: sub-grade reaction approach

(Figure 5.62). The most rudimentary application of the Winkler method is to treat the soil as a linear spring. The stiffness of the Winkler spring $k_{P-y}$ is expressed as the lateral load per unit length of pile divided by the local lateral deflection y. This stiffness has units of modulus, contrasting with the conventional 'coefficient of subgrade reaction', which is expressed in units of stress/displacement.

If $k_{P-y}$ is assumed to be constant or linearly increasing with depth, closed form solutions are available for the deflection $u_{head}$ and rotation $\theta_{head}$ at the pile head due to applied load H and moment M (Matlock and Reese 1960). A simple correlation between the sub-grade reaction modulus $k_{P-y}$ or its gradient with depth, $n_k = dk_{P-y}/dz$, is that $k_{P-y} \sim 4G$, where G is the operative shear modulus of the soil (or $n_k \sim 4(dG/dz)$. Theoretical studies have shown, however, that the exact link between G and k is influenced by the stiffness and deformed shape of the pile (Baguelin *et al.* 1977).

There is a critical length $L_{crit}$ beyond which the pile behaves as if it is infinitely long – this is the depth to which lateral load is transferred. In uniform soil, the critical depth is given by

$$L_{crit} = 4\left(\frac{(EI)_p}{k_{P-y}}\right)^{1/4} \tag{5.81}$$

where $(EI)_p$ denotes the bending rigidity of the pile, where I is the second moment of area about the neutral axis, defined as:

$$I = \int y^2 dA \tag{5.82}$$

Integrating this expression over a thick-walled cylinder of inner and outer radii $R_i$ and $R_o$ (or diameters $D_i$ and $D_o$) gives:

$$I_{thick} = \frac{\pi}{4}(R_o^4 - R_i^4) = \frac{\pi}{64}(D_o^4 - D_i^4) \tag{5.83}$$

For a thin-walled cylinder (or pile) of thickness t, this expression simplifies to:

$$I_{thin} = \pi R^3 t = \frac{\pi}{8} D^3 t \qquad (5.84)$$

In practice, most piles are longer than the critical length. In uniform soil, the sub-grade reaction solutions give the following pile head deflections, if $L > L_{crit}$:

$$y_{head} = \sqrt{2} \frac{H}{k_{P-y}} \left( \frac{L_{crit}}{4} \right)^{-1} + \frac{M}{k_{P-y}} \left( \frac{L_{crit}}{4} \right)^{-2} \qquad (5.85)$$

$$\theta_{head} = \frac{H}{k_{P-y}} \left( \frac{L_{crit}}{4} \right)^{-2} + \sqrt{2} \frac{M}{k_{P-y}} \left( \frac{L_{crit}}{4} \right)^{-3} \qquad (5.86)$$

Similar expressions have been derived for the case of stiffness increasing with depth from zero at the ground surface (Reese and Matlock 1956). In this case, the coefficient of sub-grade reaction, $k_{P-y} = n_k z$, and the critical length is:

$$L_{crit} = 4 \left( \frac{(EI)_p}{n_k} \right)^{1/5} \qquad (5.87)$$

In linear soil, the sub-grade reaction solutions give the following pile head deformations, if $L > L_{crit}$:

$$y_{head} = 2.43 \frac{H}{n_k} \left( \frac{L_{crit}}{5} \right)^{-2} + 1.62 \frac{M}{n_k} \left( \frac{L_{crit}}{4} \right)^{-3} \qquad (5.88)$$

$$\theta_{head} = 1.62 \frac{H}{n_k} \left( \frac{L_{crit}}{4} \right)^{-3} + 1.73 \frac{M}{n_k} \left( \frac{L_{crit}}{4} \right)^{-4} \qquad (5.89)$$

*Elastic continuum approach*

A solution that is similar in nature to the sub-grade reaction approach may be devised using a continuum model for the soil. Elastic continuum solutions have been generated using finite element and boundary element analyses, and the results have been normalised and simple relationships fitted (Randolph 1981). The notation for this analysis is shown in Figure 5.63.

As for the sub-grade reaction approach, a critical pile length is expressed in terms of the pile–soil stiffness ratio $E_p/G_c$ where $E_p$ is the Young's modulus of a solid pile of equivalent bending rigidity, and $G_c$ is a characteristic shear modulus. Since the second moment of area of a solid pile $I_{solid} = \pi R^4/4$ the equivalent Young's modulus $E_p$ is:

$$E_p = \frac{(EI)_p}{\pi R^4/4} = \frac{(EI)_p}{\pi D^4/64} \qquad (5.90)$$

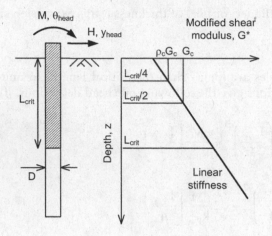

*Figure 5.63* Notation for elastic continuum analysis of lateral response

It was found that a modified shear modulus $G^*$ could be used to normalise results for different values of Poisson's ratio, $v$:

$$G^* = G\left(1 + \frac{3v}{4}\right) \tag{5.91}$$

The characteristic shear modulus $G_c$ is then the average value of $G^*$ over the critical pile length (see Figure 5.63). A parameter $\rho_c$ is also introduced, to quantify the degree of non-homogeneity of the soil:

$$\rho_c = G^*_{z=L_{crit}/4} / G_c \tag{5.92}$$

The critical length for elastic continuum analyses is calculated as:

$$L_{crit} = D\left(\frac{E_p}{G_c}\right)^{2/7} \tag{5.93}$$

Since $L_{crit}$ depends on $G_c$, which is defined as the average modified shear modulus over the critical length, some iteration is needed to establish these parameters. The surface response of the pile is then expressed in terms of the foregoing parameters and the applied loading, $H$ and $M$. The following equations were derived as best fits to parametric finite element studies.

$$y_{head} = \frac{(E_p/G_c)^{1/7}}{\rho_c G_c}\left[0.27\frac{H}{L_{crit}/2} + 0.30\frac{M}{(L_{crit}/2)^2}\right] \tag{5.94}$$

$$\theta_{head} = \frac{(E_p/G_c)^{1/7}}{\rho_c G_c}\left[0.30\frac{H}{(L_{crit}/2)^2} + 0.80\sqrt{\rho_c}\frac{M}{(L_{crit}/2)^3}\right] \tag{5.95}$$

These expressions have a similar form to the sub-grade reaction equations described previously. A modification of these equations can be used for cases where a pile cap prevents rotation of the pile at ground level. For a fixed-head pile $\theta_{head} = 0$ and the resulting fixing moment at the pile head $M_{fix}$ is (from Equation 5.95):

$$M_{fix} = -\frac{0.375}{\sqrt{\rho_c}}\frac{HL_{crit}}{2} \tag{5.96}$$

The ground level deflection for the fixed-head pile $y_{head,fix}$ may then be obtained, substituting for $M_{fix}$ into Equation 5.94, as:

$$y_{head,fix} = \frac{(E_p/G_c)^{1/7}}{\rho_c G_c}\left[0.27 - \frac{0.11}{\sqrt{\rho_c}}\right]\frac{H}{L_{crit}/2} \tag{5.97}$$

The algebraic expressions for surface displacement and rotation can be extended to give profiles of displacement down the length of the pile, and also profiles of bending moment. The depth scale is normalised by the critical pile length $L_{crit}$ while the displacement axis is normalised as:

$$\bar{y} = \frac{yDG_c}{H}\left[\frac{E_p}{G_c}\right]^{1/7} \tag{5.98}$$

The results in Figures 5.64 and 5.65 are for the case of a free-head pile, with $M = 0$ at the soil surface. Cases indicated cover three different profiles of soil modulus. For the bending moment profiles, the moment is normalised by $HL_{crit}$. The maximum bending moment occurs at a depth of between 0.3 and 0.4 $L_{crit}$. This is where a plastic hinge would be expected to form as the pile is loaded to failure. The value of the maximum bending moment is largest for the case of soil with stiffness proportional to depth ($\rho_c = 0.5$).

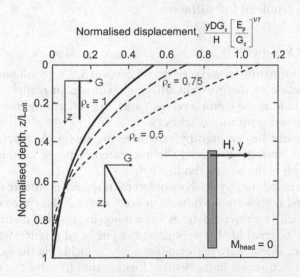

*Figure 5.64* Displacement profiles from lateral continuum solutions (free-head piles) (Randolph 1981)

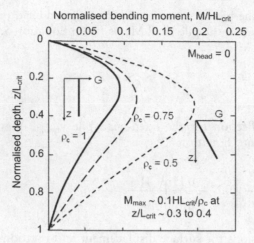

*Figure 5.65* Bending moment profiles from lateral continuum solutions (free-headed piles) (Randolph 1981)

For fixed-head piles, the condition of $\theta_{head} = 0$ leads to the maximum bending moment occurring at the point of fixity (the pile head). The maximum bending moment is typically 30–80 per cent greater than the maximum value that occurs under the same load for a free-head pile (Figure 5.66). The maximum displacement is reduced by a factor of about 2 in comparison with the free-head case (Figure 5.67).

In general, for design calculations, it is recommended that low values of soil modulus are assumed at shallow depths, since the large displacements and hence strains in this region will lead to a lower operative soil stiffness. As such, the results for soil modulus proportional to depth ($\rho_c = 0.5$) are generally more relevant than those for homogeneous soil ($\rho_c = 1$). Usually the maximum bending moment is more critical in design than the head displacement. The calculated value of $M_{max}$ is relatively insensitive to the chosen profile of soil stiffness.

### 5.9.9 Non-linear p–y analysis

The methods for estimating lateral response in Section 5.9.8 are all based on elasticity, with a constant value of the pile–soil stiffness resulting. In reality, the initially stiff response will degrade as the strain level increases, the soil around the pile approaches failure, and the lateral resistance reaches $P_f$.

This behaviour can be captured by extending the sub-grade reaction approach to include non-linear 'springs' to model the integrated response of the deforming soil around the pile; this is the p–y method.

The solution method for p–y analysis of lateral response is similar to t–z analysis of axial response, and both were introduced in Section 5.8.2. The pile is modelled as an elastic beam, which is prevented from deforming by springs distributed along its length, whilst the external load is applied at the pile head (Figure 5.68). Many computer programs are available to perform this analysis including the program PYGMY (Stewart 2000), which uses a finite element formulation to solve the governing equation in incremental steps. At each step, the current displacement at each node is used to find the current tangent stiffness based on the chosen p–y curve.

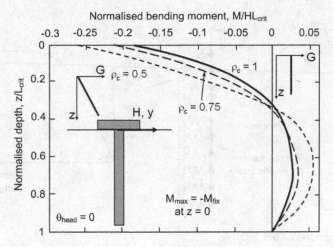

*Figure 5.66* Bending moment profiles from lateral continuum solutions (fixed-head piles) (Randolph 1981)

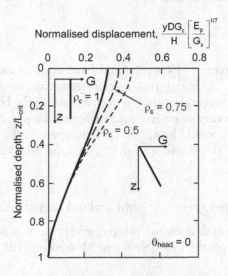

*Figure 5.67* Displacement profiles from lateral continuum solutions (fixed-head piles) (Randolph 1981)

Different sources use different notation for the 'p' in p–y. The net pressure acting on the pile is denoted here by p, and has units of stress (e.g. kPa). The soil resistance acting per unit length of pile is denoted by P, where

$$P = pD \qquad (5.99)$$

To clarify the distinction between these conventions, the initial gradient of an (upper case) P–y curve, $k_{P–y}$, has units of modulus, and is typically 4G for an elastic

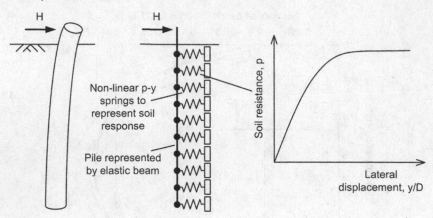

*Figure 5.68* Schematic of p–y analysis

analysis (Section 5.9.8). The initial gradient of a P–y curve can be estimated based on $G = G_{max}$, with the gradient defined as

$$\frac{P}{y} = k_{P-y} \sim 4G \tag{5.100}$$

The gradient of a lowercase p–y curve would have units of modulus/distance, but is also often denoted by k. Although this is consistent with the original concept of the coefficient of sub-grade reaction, it should not be used for the lateral pile–soil stiffness as the value would then vary in inverse proportion to the pile diameter. For example, in an elastic analysis, p/y would be about 4G/D where G is the shear modulus. In the following, the 'correct' definition, as for $k_{P-y}$ in Equation 5.100, is used for all k values.

### 5.9.10 Recommended p–y curves for sand and soft clay

The API RP2A recommendations for the p–y response in sand follow a hyperbolic tangent curve, given originally by O'Neill and Murchison (1983):

$$p = Ap_f \tanh\left(\frac{n_{ki}zy}{Ap_f D}\right) \tag{5.101}$$

The parameter A depends on whether the loading is cyclic or static and varies with depth for static loading. For cyclic loading, it is assumed that:

$$A = 0.9 \tag{5.102}$$

whereas under static loading

$$A = \left(3.0 - 0.8\frac{z}{D}\right) \geq 0.9 \tag{5.103}$$

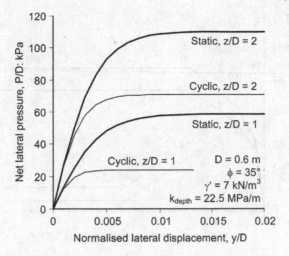

*Figure 5.69* Typical p–y curves for sand using API model

The effect of this adjustment is for cyclic loading to significantly reduce the resistance near the surface but to have no effect below a depth of 2.6D.

The parameter $n_{ki}$ is the gradient with depth of the initial stiffness of the P–y response (i.e. P, force per unit length versus y), $k_{P-y}$ at $y = 0$, and is taken as 45 MPa/m in dense soil (nominally $\phi \sim 40°$), and halving with each 5° decrease in friction angle.

The API recommendations for $P_f$ differ from the Barton approach described earlier (Equation 5.73). Instead, the API guidelines suggest that $P_f$ is found as the lesser of:

$$P_f = Dp_f = D\left[\left(C_1\frac{z}{D}+C_2\right)\gamma'z\right]$$ (5.104)

and

$$P_f = Dp_f = DC_3\gamma'z$$ (5.105)

The coefficients $C_1$, $C_2$ and $C_3$ are shown graphically in the API guidelines and a profile for $p_f$ for typical conditions is shown in Figure 5.53. Typical p–y curves for sand using the API model are shown in Figure 5.69. A lateral movement of less than 1 per cent of the pile diameter is required to mobilise most of the lateral resistance.

### 5.9.11 *Recommended p–y curves for soft clay*

For soft clay, API RP2A provides a p–y response that can be expressed as:

$$\frac{p}{p_f}=0.5\left(\frac{y}{y_r}\right)^{1/3} \leq 1$$ (5.106)

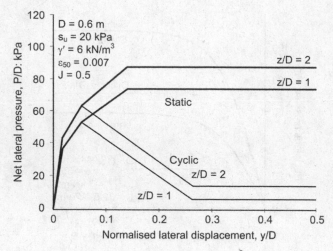

Figure 5.70 Typical p–y curves for soft clay using API model

as proposed by Matlock (1970). The limiting soil resistance $p_f$ is calculated as:

$$p_f = \left(3 + 0.5\frac{z}{D}\right)s_u + \sigma'_{v0} \leq 9s_u \tag{5.107}$$

The depth at which the limiting resistance is reached is several diameters, which is conservative. The plasticity solutions derived by Murff and Hamilton (1993) suggest a more rapid increase in lateral resistance with depth, once soil self-weight is accounted for. Under cyclic loading, which is the usual design condition, the maximum resistance is reduced to $0.72p_f$ (reached when $y = 3y_c$) at depths below $z_r$ (where $p_f$ reaches $9s_u$). At shallower depths, the response is taken to soften at displacements greater than $3y_c$, reducing to $0.72z/z_r$ at deflections of $15y_c$; example responses are shown in Figure 5.70. The resulting pattern is for the cyclic lateral resistance to decay to zero at the mudline, reflecting the opening of a gap around the pile or extreme softening of the soil.

The characteristic distance for mobilisation, $y_c$, is defined as:

$$y_c = 2.5\varepsilon_{50}D \tag{5.108}$$

where $\varepsilon_{50}$ is the characteristic strain in an undrained laboratory compression test at which 50 per cent of the failure stress is mobilised. This approach links the stiffness of the p–y response to the soil stiffness. Typically, $\varepsilon_{50}$ increases as soil strength reduces, with values in the range $\varepsilon_{50} = 1$–2 per cent spanning the range $s_u = 50$–100 kPa (Reese and van Impe 2001).

### 5.9.12 Comparison of p–y curves for sand and soft clay

Stepping back from the details of the correlations, some key points should be noted.

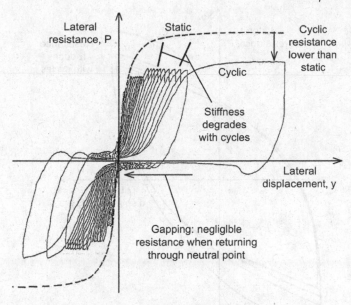

*Figure 5.71* Typical cyclic p–y response with near-surface gapping (after Bea and Audibert 1979)

Sand is stiffer than clay. Reflecting the relative stiffness of the soil types, the sand p–y curve mobilises the full resistance at a pile head movement of ~1 per cent D, whilst in clay a movement of ~10 per cent D is required. The rapid mobilisation of limiting resistance for sand appears unrealistic in light of the expected ultimate conditions.

Sand suffers little degradation during cyclic loading, whereas the recommendations include a ~30 per cent softening in clay at depth, increasing towards the surface.

### 5.9.13 Observations of cyclic p–y behaviour

The API guidance on p–y modelling is based on recommendations that are more than 40 years old, and is highly idealised. Modern p–y analysis software is capable of capturing more faithfully the detailed shape of the lateral load-displacement response at a given soil horizon, so there have been significant efforts to simulate the full cyclic load response of laterally loaded piles in model tests in order to refine and revise these recommendations. Some salient observations are outlined in the following section.

Under cyclic loading, the lateral load transfer in clay can degrade significantly, by more than suggested by the current API guidelines, particularly if the loading is two-way. At shallow depths, lateral movement of the pile remoulds the local soil and causes water to become entrained within the soil. A pronounced 'gapping' is observed, with very low resistance as the pile passes through its neutral position, before the resistance recovers as the pile moves back into intact soil. This form of behaviour, as identified in model tests, is illustrated in Figure 5.71. Despite the apparent complexity of this behaviour, some relatively simple mechanical analogues comprising non-linear springs and sliders have been implemented into

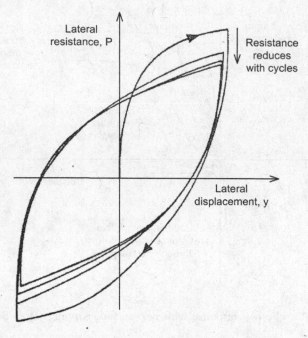

*Figure 5.72* Typical cyclic p–y response below depth of gapping (after Bea and Audibert 1979)

numerical programs to mimic the key features of this behaviour (Boulanger *et al.* 1999).

At greater depths, where the overburden is sufficient to prevent a gap remaining open, the response is more stable, but softens as the soil becomes remoulded (Figure 5.72).

The softening behaviour is exacerbated if piles are located at close spacing, since the zones of remoulding overlap. Doyle *et al.* (2004) report the results of centrifuge model tests undertaken in relation to the Ursa TLP, where the piles were spaced at 3.08D centres. They found that under static loading the trailing pile gave lateral resistances of about 2/3 that of the leading pile. Under cyclic loading, the lateral soil resistance for a single pile reduced to about 0.6 of the static resistance (less than Matlock's value of 0.72 adopted in the API code). More significant, however, was the observation that the cyclic soil resistance on the 2-pile group decayed to only 29 per cent of the static value. In this particular case, it appears that the cyclic loading was sufficient for the soil to be fully remoulded, much as in a cyclic T-bar test, with the resistance reducing by a factor $1/S_t$, where $S_t$ is the soil sensitivity.

Similar observations are reported by Zhang *et al.* (2010), from cyclic lateral load tests on short rigid piles in soft clay. They linked the softening of the p–y response during an initial packet of cyclic displacements directly with the progressive remoulding observed in a cyclic T-bar test. However, they also found that periods of reconsolidation between the cyclic episodes led to an increase in the p–y stiffness. In this case, the harmful influence of cyclic loading was compensated for by an intervening period of consolidation. The steady p–y stiffness approached after many episodes of

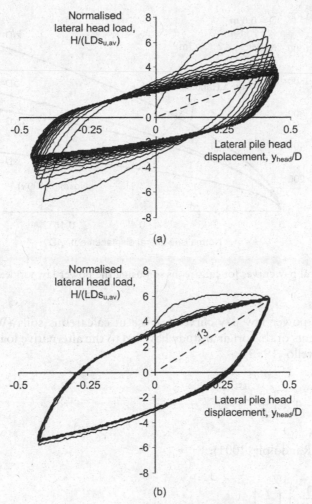

*Figure 5.73* Episodes of cyclic lateral loading with intervening periods of reconsolidation (a) sequence 1 (20 cycles) (b) sequence 5 (20 cycles) (Zhang *et al.* 2010)

cyclic movement followed by reconsolidation was comparable to the stiffness in the first cycle of the first episode (Figure 5.73).

### 5.9.14 p–y curves for calcareous sands and silts

Lateral load transfer curves for piles in calcareous soil have a markedly different format to those described earlier for sand and clay. This unusual response was identified in centrifuge modelling undertaken in the late 1980s at Cambridge University (Wesselink *et al.* 1988), as part of the design of the retrofit strut strengthening system for the first generation Bass Strait platforms.

Model testing showed that p–y curves in calcareous soils have a lower initial stiffness than in silica sand. There is a gentle degradation of stiffness with increasing displacement but no distinct limiting lateral resistance. These observations led to the

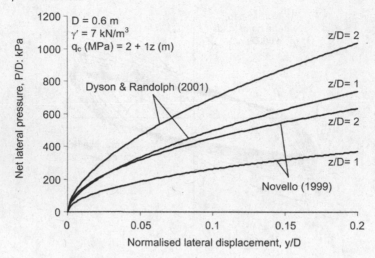

*Figure 5.74* Typical p–y curves for calcareous soil using the power law models

development of power law p–y curves for use in calcareous soils (Wesselink *et al.* 1988). Refinements of this original study have led to the alternative load transfer relationships of Novello (1999):

$$\frac{P}{D} = 2\left(\gamma'z\right)^{0.33} q_c^{0.67} \left(\frac{y}{D}\right)^{0.5} \tag{5.109}$$

and Dyson and Randolph (2001):

$$\frac{P}{\gamma'D^2} = R\left(\frac{q_c}{\gamma'D}\right)^n \left(\frac{y}{D}\right)^m \tag{5.110}$$

where R, n and m were optimised as 2.7, 0.72 and 0.58 (Figure 5.74). These forms of load transfer curve are not bounded as y increases, and also have an infinite initial gradient (set to a finite value for numerical implementation). For practical values of deflection, with y ≪ D, the net lateral pressure on the pile remains as a small fraction of the cone resistance. Thus for $q_c$ = 10 MPa, and $\gamma'D$ = 20 kPa, the latter expression gives a net pressure, p = P/D, of 1.86 MPa for y = 0.2D, and remains below $0.5q_c$ even for y ∼ D. While this form of load transfer curve has only been proposed for carbonate sediments at this stage, it is attractive to link the load transfer curve to the cone resistance for silica sands as well, avoiding the need to estimate a friction angle.

A cemented cap rock is often found in carbonate sediments in the upper few metres of the seabed. At shallow depths, such cemented material is generally quite brittle, which will lead to reduction in resistance once a failure surface is formed. Abbs (1983) suggested a strain-softening p–y model to simulate this. However, a more rational approach was developed (AG 2003; Erbrich 2004) for application in the carbonate deposits on the North-West Shelf of Australia. The basis of the model (Figure 5.75) is

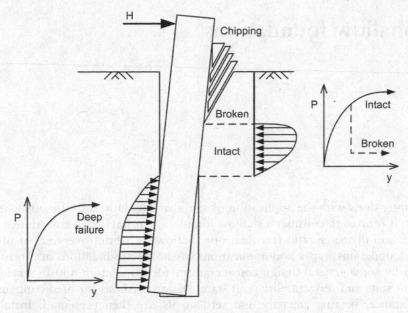

*Figure 5.75* Modelling of lateral pile behaviour in cemented soil: CHIPPER (AG 2003)

the concept of wedges of 'chipped' material forming near the surface. This concept was combined with the kinematic mechanism of Murff and Hamilton (1993) in order to evaluate the net resistance once a chip has occurred, and to develop a criterion for the maximum depth to which chips, as opposed to deformation of intact rock (which offers significantly higher capacity than chipped material), will occur.

The 'chipper' model proposed by Erbrich (2004) was calibrated by means of 3D finite element analyses, and also through comparisons with centrifuge model tests. It was found to yield much greater lateral pile capacity than the Abbs (1983) model, but provided a sound physical basis on which many anchor piles have been designed off-shore Australia and in other regions of carbonate sediments.

# 6 Shallow foundations

This chapter deals with the application of shallow foundations in the offshore environment. It is usual to consider a shallow foundation as having an embedment depth to foundation diameter ratio less than one. Following a brief overview of offshore shallow foundations, types and applications of shallow foundations are introduced followed by some general design considerations for installation and in-service ultimate limit state and serviceability limit state. Design methods for predicting installation resistance, bearing capacity and settlements are then presented. Installation resistance based on basic soil mechanics principles is outlined. Bearing capacity predictions using classical methods are presented along with advanced solution methods based on explicitly derived uniaxial limit states and load interaction through three-dimensional failure envelopes. Serviceability and settlements are addressed using classical elastic theory and recent, advanced solutions, including consideration of cyclic loading effects.

## 6.1 Introduction

### 6.1.1 Overview

Shallow foundations have become an economic, and sometimes the only practical, solution, as an alternative to deep piled foundations. Offshore development originated in the Gulf of Mexico with steel jacket (template) structures founded on piles that were well suited to the soft clays. With the development of the North Sea, deep-piled foundations were not required or practical for the harder overconsolidated clays and dense sands initially encountered and the concrete gravity-based structure (GBS) was developed as an alternative to deep-piled steel jacket platforms.

Historically, offshore shallow foundations either comprised large concrete gravity bases, supporting large fixed sub-structures, or steel mudmats used as temporary support for conventional piled jackets before the piles are is installed. In recent times, shallow foundations have become more diverse, and now include concrete or steel bucket foundations used as anchors for floating platforms or as permanent supports for jacket structures instead of piles, or as foundations for a variety of (usually) small sea bottom structures. A number of different applications for offshore shallow foundation systems are shown in Figure 6.1. Spudcans are a type of temporary shallow foundation solution used for mobile drilling rigs (commonly called jack-ups) and are dealt with separately in Chapter 8.

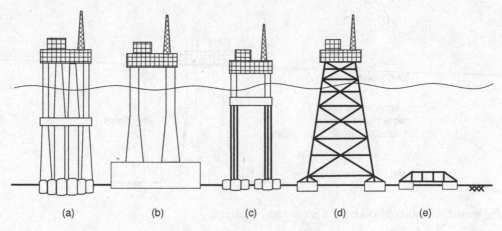

*Figure 6.1* Applications of offshore shallow foundations (a) Condeep gravity-based structure (GBS), (b) GBS, (c) Tension-leg platform (TLP), (d) Jacket and (e) Subsea frame

Gravity-based foundations rest on the surface of competent seabeds, but where softer surficial deposits exist, skirts are provided to confine soft surface soil and transmit foundation loads to deeper, stronger soil. Skirts are provided around the periphery of the foundation, penetrating vertically into the seabed, confining a soil plug. If the foundation area is large, interior skirts are also provided to form skirt compartments under the foundation base. This construction approach contrasts with onshore shallow foundation systems, since onshore soft surficial soils would more often be removed (or treated) prior to construction.

If a structure is relatively heavy and the soil relatively soft, skirted gravity bases and bucket foundation systems can be installed under self-weight alone. However, for light jackets, dense materials or deep skirts, penetration is assisted with suction (e.g. Tjelta *et al.* 1986, Bye *et al.* 1995; Andenaes *et al.* 1996). Suction-assisted installation is discussed in more detail in Section 6.2.

Skirts are also useful to help compensate for irregularities in the seabed and improve erosion resistance under the foundation periphery. The presence of skirts will, in most cases, increase the capacity of the foundation to resist vertical, horizontal and over-turning actions and decrease vertical and horizontal displacements and rotations. Skirts can also significantly increase the ability of shallow foundations to sustain tension loads. Skirts enable the development of transient tensile capacity to withstand moments or uplift arising from environmental loading, which cannot be sustained with conventional shallow foundations. Uplift capacity is well documented by model tests and field observation on clays (e.g. Dyvik *et al.* 1993, Anderson *et al.* 1993, Gourvenec *et al.* 2007, 2008a) as well as sand (e.g. Tjelta and Haaland 1993, Bye *et al.* 1995). The duration of the tension load must be relatively short compared to the time required for dissipation of the under pressure developed in the pore water under the foundation plate. Cyclic uplift loads from ocean waves may be safely resisted even on relatively pervious sand deposits and uplift loads may be carried over much longer durations by skirted foundations on clays with lower permeability. The reliance on uplift capacity provided by passive suction developed within foundation skirt compartments is only recently being accepted by industry for design purposes.

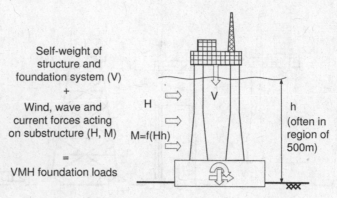

*Figure 6.2* Combined loading of a gravity base platform

Offshore shallow foundations are larger than those typically required onshore due to the size of the structures they support and the harsher environmental conditions. Even small gravity base structures are often 70 m high (equivalent to an 18 or 20 storey high-rise) with a footprint 50 m by 50 m; larger structures can be over 400 m tall supported by foundations with a plan area in excess of 15,000 m². Even a single bucket foundation may have a diameter of 15 m. Apart from the sheer size of offshore structures, their foundations are required to resist severe environmental forces from wind, waves and currents (and in some cases ice, e.g. offshore Canada or the Baltics), which impart significant horizontal and moment loads that are not experienced onshore. Loading conditions for a gravity base platform are shown schematically in Figure 6.2.

Design loads for an offshore gravity base structure and an onshore high-rise of similar height are compared in Figure 6.3. The self-weight of the two structures is moderately comparable with the design vertical load of the offshore structure approximately 30 per cent smaller than the onshore high-rise, despite being 10 per cent taller. It is worth noting that the self-weight of an offshore structure does not increase proportionately with increasing height as the foundation and topsides provide the majority of vertical load such that additional height provides little additional (relative) vertical load. In contrast, each additional storey of an onshore building contributes a similar increment of load. The environmental loads offshore are approximately 500 per cent greater than onshore, which is reflected in the increased foundation area. More important than the absolute magnitude of the environmental loads is the lever arm of the horizontal load and hence the ratio of moment to horizontal load. Overturning becomes dominant for offshore structures as water depth increases. In shallow waters (< 200 m), M/HD typically lies in the range 0.35–0.7, indicating sliding will govern failure.

Limit load, whether sliding, overturning or vertical bearing capacity, is generally the key design criterion for offshore shallow foundations with less emphasis on displacements than would be typical in onshore foundation design. Allowable settlements of offshore shallow foundations are usually constrained by tolerances of allowable deformation the superstructure and to maintain the integrity of oil wells

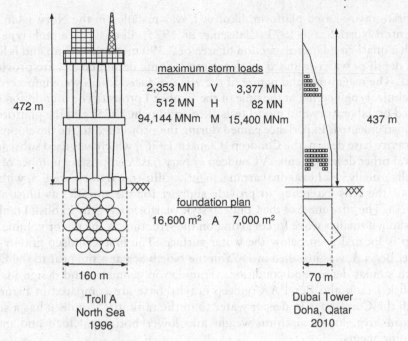

maximum storm loads

| | | |
|---|---|---|
| 2,353 MN | V | 3,377 MN |
| 512 MN | H | 82 MN |
| 94,144 MNm | M | 15,400 MNm |

foundation plan

16,600 m²     A     7,000 m²

472 m

437 m

160 m

70 m

Troll A
North Sea
1996

Dubai Tower
Doha, Qatar
2010

*Figure 6.3* Comparison of offshore and onshore design conditions

and pipelines, and it is feasible to design offshore shallow foundations for settlements of the order of a metre or so.

Due to the dominance of environmental loading offshore, accounting for the influence of cyclic loading on the soil response is much more important than for a typical onshore structure. Environmental forces impart significant cyclic horizontal, vertical and moment loads to offshore foundation systems, and generate excess pore water pressures in the vicinity of the foundation, reducing the effective stresses in the seabed. Foundation stability is ultimately compromised by accumulated residual strain and degradation of cyclic shear strength (depending on the combination of average and cyclic stress). If drainage can take place, some pore water pressure dissipation may occur between storm events, although cyclically induced pore pressures may still accumulate from one storm to another. The cyclic response of soils is considered in detail in Chapter 4.

### 6.1.2 Types of offshore shallow foundations

#### Gravity bases

Gravity-based structures (GBS) rely mostly on their weight and the size of their footprint on the seabed to withstand the environmental lateral and moment loading, although foundation skirts are helpful in improving lateral resistance and provide some short-term tensile capacity. Gravity-based foundations are usually equipped with skirts that may range in length from 0.5 m in competent strong clays and dense sands to greater than 30 m in softer sediments.

The first gravity-based platform, Ekofisk I, was installed in the Norwegian sector of the central North Sea in 1973 (Clausen *et al.* 1975). Ekofisk I is a tank-type structure with a quasi-circular cross-section of area of 7,390 m² ($D_{eq}$, = 97 m) and is located in 70 m depth of water founded on dense sand. Some design details are provided in Table 6.1. The magnitude and nature of the wave-induced forces were unprecedented in geotechnical engineering at the time of the Ekofisk I project. The foundation design was based on laboratory and model tests (Lee and Focht 1975) and the platform was heavily instrumented. Experience gained during the project led to the development of a new gravity base design, the Condeep (Clausen 1976), which was used subsequently for several other developments. A Condeep gravity base comprises a number of cylindrical cells usually in a hexagonal arrangement, as illustrated in Figure 6.3, with three or four of the cells extending to provide support for the topside, as illustrated in Figure 6.1a. The advantage of the Condeep style platform over the Ekofisk I tank style design is much smaller wave forces acting on the structure as the major volume of the Condeep is located deep below the water surface. The first Condeep gravity-based structure, Beryl A, was installed in 1975 in the North Sea at a site near to the Ekofisk field with similar dense sand conditions. Foundation geometry and design loads for the Ekofisk I tank and Beryl A Condeep gravity base are compared in Figure 6.2. Although the Condeep is in deeper water than the tank, the Condeep has a smaller foundation area, lower platform weight and lower horizontal load and moment design components.

As exploration of the North Sea oil and gas reserves moved into deeper water, softer normally consolidated clays were encountered and the Condeep gravity base design was adapted to include the provision of deep skirts to transmit the foundation loads to the deeper, stronger soils. To achieve the deeper penetration, active suction is required to draw the skirted base into the seabed. Gullfaks C, installed in 1989 at a soft clay site in the Norwegian North Sea, was the first deep-skirted gravity base to use suction to assist installation (Tjelta 1993). Active suction was also maintained for some time after installation to accelerate consolidation for ground improvement (see Section 6.1.3 for more details). Gullfaks C was installed in 220 m depth of water with a foundation plan area of 16,000 m² and skirts penetrating 22 m below the seabed. At the time of construction, Gullfaks C was the largest and heaviest offshore structure ever built (Tjelta 1993). Some details of the project are summarised in Table 6.1.

The Troll A deep-skirted Condeep platform in the Norwegian section of the North Sea, is the world's largest concrete platform, standing 472 m tall with a foundation area of 16,600 m² ($D_{eq}$ = 145 m) and skirts penetrating 36 m into the seabed (Table 6.1). Permanent monitoring of Gullfaks C and Troll A has enabled verification of the designs and improved understanding of the response of deep-skirted Condeeps. The experience from these projects led to application of the technology to individual suction installed 'bucket' foundations (discussed later).

In other offshore regions where gravity-based structures have been installed, different seabed conditions have necessitated different shallow foundation solutions. Wandoo B, installed on the North-West Shelf of Australia on a thin layer of sand overlying competent calcareous rock comprises a ballasted rectangular concrete gravity base resting on the seabed (Figure 6.4). The shallow rock provides sufficient vertical bearing and overturning capacity, leaving lateral sliding as the key design issue.

*Table 6.1.* Details of shallow foundation case studies

| Project Year | Type | Location | Water depth m | Soil conditions | Foundation dimensions m, m² | Skirts m (d/D) | V MN | H MN | M MN | M/HD |
|---|---|---|---|---|---|---|---|---|---|---|
| Ekofisk I Tank 1973 [1,2] | GBS | N. Sea Norway | 70 | Dense sand | $A = 7{,}390$ $D_{eqiv} = 97$ | 0.4 (0.004) | 1,900 | 786 | 28,000 | 0.37 |
| Beryl A 1975 [3] | Condeep | N. Sea Norway | 120 | As at Ekofisk | $A = 6{,}360$ $D_{eqiv} = 90$ | 4 (0.04) | 1,500 | 450 | 15,000 | 0.37 |
| Brent B 1975 [2] | Condeep | N. Sea Norway | 140 | Stiff to hard clays with thin layers of dense sand | $A = 6{,}360$ $D_{eqiv} = 90$ | 4 (0.04) | 2,000 | 500 | 20,000 | 0.44 |
| Gullfaks C 1989 [4,5] | Deep skirted Condeep | N. Sea Norway | 220 | Soft nc silty clays and silty clayey sands | $A = 16{,}000$ $D_{eqiv} = 143$ | 22 (0.13) | 5,000 | 712 | 65,440 | 0.64 |
| Snorre A 1991 [6,7] | TLP with concrete buckets | N. Sea Norway | 310 | Soft nc clays | $A_{total} = 2{,}724$ $D_{cell} = 17$ | 12 (0.7, CFT 0.4) | 142 per CFT | 21 per CFT | 126 per CFT | 0.20 |
| Draupner E (Europipe) 1994 [8] | Jacket with steel buckets | N. Sea Norway | 70 | Dense to very dense fine sand over stiff clay | $A_{total} = 452$ $D = 12$ | 6 (0.5) | 57 per bucket | 10 | 30 | 0.25 |
| Sleipner SLT 1995 [8] | Jacket with steel buckets | N. Sea Norway | 70 | As at Draupner E | $A_{total} = 616$ $D = 14$ | 5 (0.35) | 134 per bucket | 22 | - | - |
| Troll A 1996 [9,10] | Deep skirted Condeep | N. Sea Norway | 305 | Soft nc clays | $A = 16{,}596$ $D_{eqiv} = 145$ | 36 (0.25) | 2,353 | 512 | 94,144 | 1.27 |
| Wandoo 1997 [11] | CGBS | NW Shelf Australia | 54 | Dense calc. sand over calcarenite | $A = 7{,}866$ 114x69x17 | 0.3 (0.003) | 755 | 165 | 7,420 | 0.45 |
| Bayu-Undan 2003 [12] | Jacket with steel plates | Timor Sea Australia | 80 | Soft calc. sandy silt over cemented calcarenite and limestone | $A_{total} = 480$ $A_{plate} = 120$ | 0.5 (0.04) | 125 per plate | 10 | - | - |
| Yolla 2004 [13] | Skirted hybrid | Bass Strait Australia | 80 | Firm calc. sandy silt with soft clay and sand layers | $A = 2500$ 50 x 50 | 5.5 (0.1) | - | - | - | - |

1. Clausen (1976), 2. O'Reilly & Brown (1991), 3. Clausen (1976), 4. Tjelta *et al.* (1990), 5. Tjelta (1998) 6. Christophersen (1993), 7. Støve *et al.* (1992), 8. Bye *et al.* (1995), 9. Andenaes *et al.* (1996), 10. Hansen *et al.* (1992), 11. Humpheson (1998), 12. Neubecker & Erbrich (2004), 13. Watson & Humpheson (2005)

*Table 6.2.* Comparison of tank and Condeep gravity bases

| Design details | Ekofisk I Tank | Beryl A Condeep |
|---|---|---|
| Water depth, $h_w$: m | 70 | 120 |
| Foundation footprint area, A: m$^2$ | 97 | 90 |
| Equivalent footprint diameter, $D_{eq}$: m$^2$ | 7,360 | 6,360 |
| Vertical load, V: MN | 1,900 | 1,500 |
| Horizontal load, H: MN | 786 | 450 |
| Moment, M: MNm | 28,000 | 15,000 |

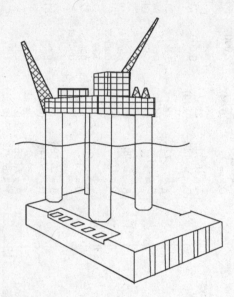

*Figure 6.4* Schematic of Wandoo gravity base structure

An active drainage layer was provided to prevent accumulation of excess pore water pressures in order to maintain capacity against sliding.

Two unusual gravity base platforms were installed at Bayu-Undan, in the Timor Sea, offshore North Australia. These comprise a steel jacket with steel plate foundations provided at each corner, with the weight of the topsides, rather than the substructure, providing the dead load required to resist environmental loading (Neubecker and Erbrich 2004). Conditions at Bayu-Undan are relatively unusual: the central processing platforms have particularly heavy topsides, water depth is moderate and environmental loading is mild (see Table 6.1). Another unusual aspect of the Bayu-Undan project was that seabed preparation works were carried out prior to installation of the jackets in order to remove the surficial silty sand. This was necessary due to the high bearing stresses imposed from the foundation. The silt and sand layer was blasted away and the steel plate foundations were grouted directly onto the underlying caprock (Sims *et al.* 2004).

A novel gravity-based hybrid structure, the Yolla A platform, operates in the Bass Strait, offshore South Australia founded in (mostly) carbonate sandy silt and silty sand (Table 6.1). The platform comprises a steel-skirted gravity base foundation supporting a steel jacket sub-structure (see Figure 6.5) and relies on transient suction developed inside

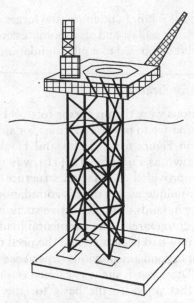

*Figure 6.5* Schematic of Yolla A hybrid gravity base structure

the skirts to resist the environmental lateral and overturning loads rather than relying on the weight of the foundation (Watson and Humpheson 2005).

## Concrete bucket foundations for tension-leg platforms (TLPs)

A progression from the development of suction-installed deep-skirted concrete gravity base foundations (such as for Gullfaks C and Troll A) was the use of individual or clusters of small concrete cells or 'bucket' foundations (Figure 6.1c). The first application of the new technique was for the Snorre A tension-leg platform (TLP), installed in 1991 in the Norwegian North Sea (details in Table 6.1). At the time of construction, Snorre A was the deepest North Sea installation at 310 m. The foundation system involved four concrete foundation templates (CFTs) constructed from three cylindrical concrete cells located beneath each of the four corners of the floating pontoon. Resistance to the tether tension was provided by a combination of the self-weight of the CFT and the interaction between the structure and seabed. Following Snorre A, a similar foundation system was employed for the Heidrun TLP, installed in the North Sea in 350 m depth of water (Table 6.1).

At Snorre A and Heidrun, the dead weight of the CFT and ballast counteracts the tension in the tethers under normal, calm weather, working conditions and only during storm conditions is skirt friction and suction under the top cap relied on to provide resistance to the moment incurred as the platform offsets over the CFT. For concrete bucket foundations to provide a cost-effective foundation solution in deeper waters, tensile resistance under normal working loads would need to be relied on. Bucket foundations have certain advantages over piles as anchors for deeper water moorings if they can provide sufficient tensile capacity. For example, using pumps for installation of skirted foundations is less costly and with fewer technical challenges than operating piling hammers in very deep water, although new pile driving systems are being developed to operate in water depths of 3 km. In addition, the larger diameter of skirted foundations provides a larger area for ballast and also mobilises greater reverse end bearing or passive suction during uplift compared to a pile foundation (Clukey *et al.* 1995).

### Steel bucket foundations for jackets

Concrete bucket foundations were the precursor to steel buckets (also known as suction cans) used as an alternative to pile foundations for steel braced jacket structures, as shown schematically in Figure 6.1d (Tjelta and Haaland 1993). The Draupner E platform (formerly known as Europipe 16/11E) was the first occasion that steel bucket foundations were provided on a jacket structure. Some platform details are provided in Table 6.1. A unique aspect of this foundation system is the reliance on mobilising tensile capacity in sands through passive suction under the base-plate when the foundations are subject to extreme environmental loads. Draupner E was the first time that bucket foundations had been used and the need to penetrate steel skirts into very dense sand also went beyond any previous experience (Tjelta 1995). An extensive field investigation at the Draupner E site was carried out in 1992, including penetration and capacity tests that provided the basis for the bucket design (Tjelta and Haaland 1993, Bye *et al.* 1995). Following the successful use of the suction installed steel bucket foundations on Draupner E, a similar foundation solution was employed the following year on the nearby Sleipner Vest SLT jacket. Ground conditions were similar but the jacket was considerably larger with a maximum compression load of 134 MN per leg compared with 57 MN per leg for Draupner E and slightly larger tensile loads (17 MN compared to 13.9 MN). To maintain an economical and practical foundation size a greater percentage of the available capacity had to be relied on than for the Draupner E bucket design. As a result, a comprehensive programme of model testing and numerical analysis was undertaken as part of the detailed design for the Sleipner SLT buckets (Erbrich 1994, Bye *et al.* 1995).

There has been a recent upsurge in interest to employ wind as a clean and renewable energy resource throughout the world, and offshore wind turbines are an attractive option where land is scarce. Compared to oil and gas installations, the vertical weight of offshore wind turbines is very low while the horizontal and moment loads are still very significant. The challenge is to engineer an economic foundation solution that can resist the wave and wind induced forces without unacceptable deflections and with relatively little self-weight. Currently, offshore wind turbines are founded on large diameter monopiles (Figure 6.6a). As the water depth increases, it may prove more economical to use skirted foundations, either as a single foundation or in a tripod or quadruped arrangement (Figure 6.6b and c).

### 6.1.3 Design considerations

The basis of design for a shallow foundation system can be considered with respect to the stages during the life of the structure:

- Installation (including site levelling, skirt penetration and base grouting)
- Capacity (under in-service general VHM loading including uplift and cyclic loading, undrained and drained conditions)
- Serviceability (short and long-term displacements and the influence of cyclic loading, and the magnitude of settlement required to mobilise capacity).

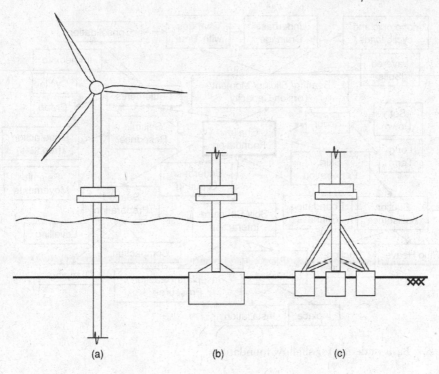

*Figure 6.6* Foundation options for offshore wind farms (a) monopile, (b) single-skirted foundation and (c) skirted foundation tripod (Byrne and Houlsby 2003)

Installation may provide the key design challenge, particularly in dense soils or deposits with cemented lenses. Horizontal sliding or overturning may be more critical than vertical bearing capacity in competent soils while long-term settlements may be more critical than capacity in soft sediments. A selection of issues that may need consideration during the design of an offshore shallow foundation are illustrated in Figure 6.7. Design considerations for installation and in-service operation are discussed in the following section in the context of the various foundation systems introduced in Section 6.1.2.

## Installation

The method of installation of a shallow foundation system may result in particular geotechnical challenges. In the following discussion, gravity-based structures are considered separately from foundation systems involving multiple individual foundation units as the installation procedures differ, therefore, giving rise to specific geotechnical issues.

### INSTALLATION OF GRAVITY BASE STRUCTURES

Concrete gravity-based structures require a deep-water site for construction and a relatively deep tow path to site, which made the Norwegian fjords an ideal location. The construction procedure is illustrated schematically in Figure 6.8. A concrete gravity base structure begins construction in a dry dock. When the base and a

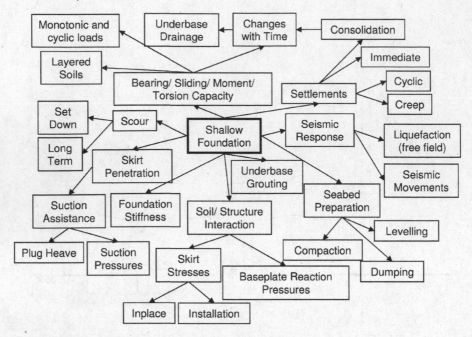

*Figure 6.7* Basis of design for shallow foundations

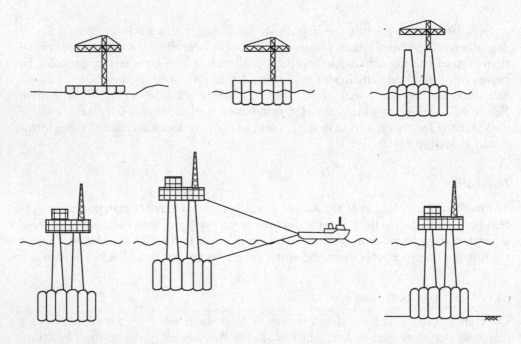

*Figure 6.8* Construction sequence for concrete gravity-based structures (Mo 1976)

portion of the upright have been completed the structure is moved to a wet dock where it is partially submerged by filling it with, typically, a combination of water and iron ore. Construction of the concrete pillars continues in a piecemeal fashion, building and submerging in increments. When the structure reaches its design height, the topside is floated out and mounted on the top of the concrete pillars. Some of the ballast water is then pumped out of the base and pillars to partially raise the structure above the water line. The platform is then towed to the final location. By increasing the volume of water in the base by controlled flooding, the platform gradually sinks through the water column and embeds itself in the seabed. Suction assistance may be required to achieve the design embedment depth if deep skirts are provided. Once the structure reaches its required position, the base may be weighted with concrete or iron ore to achieve a permanent pre-determined on-bottom force. Once ballasting is completed, sub-base grouting is carried out to complete the construction process.

Sub-base grouting is carried out to avoid any further penetration of the foundation and to keep the platform level, to obtain uniform soil stresses across the foundation and to avoid piping from water pockets below the base during environmental loading. Grouting is particularly important with uneven or sloping seabeds. Key uncertainties during installation include whether the design depth of penetration can be achieved (even with suction) and the magnitude and distribution of contact pressures on the platform base. This latter uncertainty is complicated as conditions may vary across the foundation base area due to a sloping seabed profile, varying soil strength or a combination of the two. Other considerations include how much suction is required for installation (and if this can be achieved) and whether active drainage is required after installation, either for ground improvement or to limit development of excess pore pressures due to cyclic loading. The risk of immediate bearing capacity failure in the short-term following installation of a gravity-based structure also requires consideration as the soil strength will not have had time to improve by means of consolidation from the weight of the structure, while some degree of consolidation may have been assumed for the permanent design. Active drainage or staged ballasting may be options to reduce the risk of short-term bearing capacity failure.

A programme of active drainage (i.e. sustained under pressure or suction) was adopted following installation of the Gullfaks C deep-skirted Condeep platform (Tjelta 1993). The combination of the heavy structure and soft soils necessitated a deep-skirted foundation and an active drainage system for seabed improvement. It was the intention that the majority of settlement was completed before wells and pipelines were in operation (as these can only tolerate limited settlement). The initial three months of high suction coincided with the summer months and it was important that consolidation of the seabed, and the associated strength improvement, was achieved before the winter storms arrived. Tjelta (1993) commented that for such a heavy structure on relatively soft soils, the reduced settlement rate only 15 months after installation was remarkable and illustrated the benefit of deep-skirted foundation systems. Earth pressure measurements showed that immediately after platform installation the entire platform weight was carried as contact pressure between the foundation base and the soil. By the end of the active drainage period, pore water pressures had dissipated sufficiently that 100 per cent of the submerged weight of the platform was carried as skirt wall friction. The platform had changed from

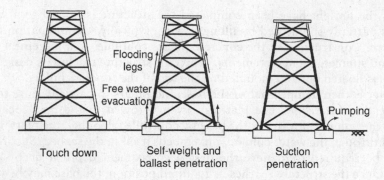

Figure 6.9 Stages of installation of bucket foundations on a jacket structure

being a traditional gravity base structure with all the load taken as base contact pressure to a 'piled' type structure where all load was carried as wall friction and tip resistance.

The Troll A deep-skirted Condeep platform relied on a different design approach to that for Gullfaks C. The low permeability clays at the Troll A site, in conjunction with the long drainage paths (due to the large plan area of the foundation base and the deep skirts), were relied on to limit excess pore pressure dissipation. The time for primary consolidation was estimated at approximately 1,000 years so only a fraction of the consolidation would take place during the lifetime of the structure. A basic design criterion was a watertight base, ensuring an undrained response and acceptable settlements (dictated mostly by constraints related to the integrity of the oil wells). Reliance on the undrained response of the foundation soils was in contrast to the approach adopted for Gullfaks C where active drainage was used to increase the strength of the soil such that foundation performance improved continuously after installation due to consolidation.

INSTALLATION OF SUCTION CANS OR BUCKET FOUNDATIONS

Figure 6.9 shows schematically the various stages of installation of a suction can foundation on a jacket structure, although the general procedure is applicable to concrete buckets for TLPs or other buoyant facilities. Initially the jacket is lowered to the seabed. The foundations penetrate under the self-weight of the jacket structure, enhanced by flooding the legs and other members. If full skirt penetration cannot be obtained under the self weight alone, as would be the case for relatively light jackets in stiff materials, the required penetration is obtained by pumping water from within the bucket (after sealing vent valves) generating a differential pressure, i.e. suction relative to the high ambient water pressure, across the foundation base plate. Installation of a suction can in sand leads to steady state seepage flows in the soil when suction is applied, due to the high permeability. Seepage flows can be represented by a flow net as shown schematically in Figure 6.10. The effect of these seepage flows is to increase the effective stresses in the soil outside the skirts but to decrease the effective stresses within the skirts. Since the external and internal friction are related to the effective stresses, the external friction is increased whilst the internal friction is decreased. More importantly though, the skirt tip end bearing is also

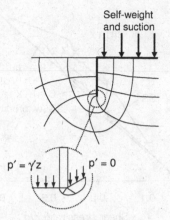

*Figure 6.10* **Seepage pressures set up during suction installation of bucket foundation**

decreased (see inset in Figure 6.10) and this is the most important contribution facilitating skirt penetration. An additional driving force is also obtained through the differential water pressure across the base plate but this is a relatively small factor compared to degradation of the skirt tip resistance.

The most obvious and serious problem when suction is applied is the potential for liquefaction of the internal soil plug and the potential formation of localised flow paths outside the caisson, which could stop further penetration and severely impair the in-place performance of the foundation (Senpere and Auvergne 1982). The model testing and analytical work carried out for the Sleipner SLT bucket design showed that localised liquefaction and piping failure can be avoided with carefully controlled pumping, attributed to the increased soil permeability as the initially dense sand loosens within the bucket (Erbrich and Tjelta 1999). More recent studies have shown that it is necessary to establish a critical hydraulic gradient in the soil plug adjacent to the skirts, reducing effective stresses at the skirt tip to close to zero, for suction installation in dense sand (Tran and Randolph 2008). Other geotechnical concerns during installation of steel bucket foundations include the potential for buckling of the very thin walled skirts during installation due to the applied suction pressures and lateral soil resistance (Barbour and Erbrich 1995).

Concrete buckets have a larger relative wall thickness compared to steel buckets so that buckling is not an issue but other issues are relevant. For example, it is important to ensure a minimum self-weight penetration is achieved to provide a sufficient water tight seal to withstand the differential pressures applied during suction (and which must be achieved without the additional weight of the jacket if for a buoyant structure).

*Capacity*

GENERAL LOADING

The loading applied to a shallow foundation under working conditions generally comprises monotonic and cyclic vertical load V (in either compression or tension), horizontal load H and moment M (as illustrated in Figure 6.2). The ratio of the

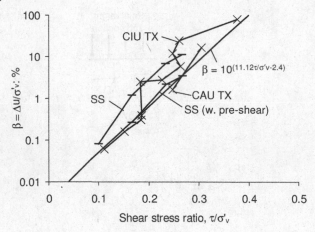

*Figure 6.11* Anticipated excess pore water pressures in sand layer beneath Wandoo GBS (Humpheson 1998)

moment to horizontal component indicates whether a structure is prone to a sliding or overturning failure.

The Troll A deep-skirted Condeep, the largest offshore concrete gravity-based structure standing 472 m above the seabed, is a relatively light structure for its size and the design overturning moment was the dominating load factor (Hansen *et al.* 1992). The relatively light weight of Troll A in conjunction with the deep-water site led to dynamic effects of the platform to be a governing criterion for the foundation design. The consequent sensitivity to and requirement of foundation stiffness necessitated a larger foundation than required to fulfil stability requirements. From Table 6.1, it can be seen that the moment to horizontal load ratio for Troll A, M/HD = 1.27 or 0.64 for Gullfaks C, also a deep-skirted Condeep but in shallower water, indicating overturning is more dominant than sliding. By contrast, the moment to horizontal load ratio for skirtless condeeps Beryl A and Brent B are lower, M/HD = 0.37 and 0.44 respectively, indicating the structures are more prone to sliding.

Sliding resistance was a critical aspect in the design of the ballasted gravity-based platform Wandoo B (Figure 6.4, Table 6.1), where the seabed comprised a thin surface layer of dense carbonate sand (0.5–1.4 m), overlying strong calcarenite (Humpheson 1998). The competent calcarenite near the surface ensured sufficient vertical bearing and overturning capacity but horizontal shear resistance relied on the surface sand layer. Skirts were not feasible due to the shallow depth of sand so friction between the base slab and the surficial sand was critical. Sliding stability is affected by the level of excess pore pressure in the sand, generated by wave induced cyclic shear loading during storm events. Cyclic shear test results, carried out to predict the accumulation of excess pore water pressures within the sand layer, are shown in terms of average excess pore water pressure per load cycle, $\beta$ (Bjerrum 1973), against shear stress ratio $\tau/\sigma'_v$ in Figure 6.11. A passive drainage system was provided in the base slab of Wandoo B to facilitate dissipation of cyclically induced excess pore pressures. Finite element dissipation analyses were used for the design to calculate the amount of ballast required to resist the severe environmental loads and ensure against sliding failure (Humpheson 1998).

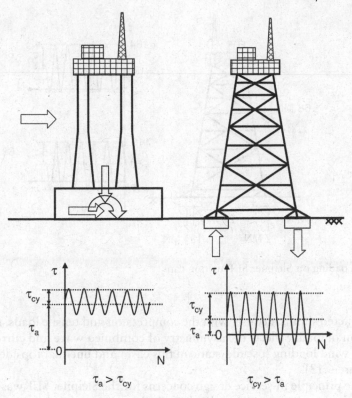

*Figure 6.12* Cyclic loading of fixed-bottom structures

CYCLIC LOADING AND UPLIFT

For a fixed-bottom structure, the vertical component of foundation load mainly comprises the weight of the platform with an additional (small) cyclic component representing the variation in water surface elevation as the wave passes. The horizontal load and moment components of a foundation load are (typically) entirely due to the environmental forces and include a cyclic component about a steady asymmetric offset. (If the sub-structure is mounted off-centre on the foundation, usually to facilitate drilling access, or the topsides impart an uneven pressure distribution, a permanent moment is carried by the foundation.) Gravity base structures may incur tensile loads over part of the foundation due to strong overturning moments although the cyclic loading is likely to be one-way and compressive in nature. Jacket structures with bucket foundations are more likely to sustain two-way cyclic loads, i.e. undergoing cycles of tension and compression, due to the relatively light structure compared to a gravity base structure (Figure 6.12).

The 100-year return maximum compression load on an individual suction can of the Sleipner SLT jacket, was 134 MN (factored), whilst the maximum tension load was 17 MN. For both the maximum compression and tension cases, the double amplitude cyclic vertical load was approximately 68 MN, which implies substantially different mean loads on each leg; 66 MN for the maximum compression case and 51 MN for the maximum tension case. In addition, horizontal shear loads of 22 MN

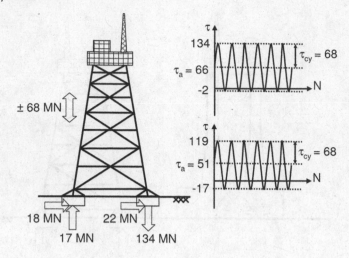

*Figure 6.13* Loading on Sleipner SLT suction cans

and 18 MN acted simultaneously with the compression and tensile loads, respectively. The variation of loads reflects the asymmetry of combined wave and current loading, the effect of wind loading (pseudo-static in this case) and unequal topside load distribution (Figure 6.13).

One of the principle in-service design concerns for the Sleipner SLT was the reliance on tensile resistance during moment loading. Resistance to substantial tensile forces in sand had previously been considered possible only with the use of active suction, maintained by continuous pumping during the life of a platform, while the bucket foundation is a totally passive system in which suction is generated in response to short-term tension load.

Previous consideration of tensile resistance had been based on the assumption of idealised elastic and non-dilatant plastic soil, from which it follows that tensile loading would cause a decrease in mean effective stress and, therefore, a monotonic decrease in strength with time, starting from the *in situ* frictional strength. In this case, once pore water pressures have dissipated, tensile resistance is very small, resulting only from skirt friction and the self-weight of the system. In reality, the tendency for volumetric expansion of dense sands during shear is resisted in the short term by suctions generated due to the incompressibility of the pore water; expansion of the soil matrix can only occur if water is allowed to flow freely into the pore spaces. If water ingress is inhibited the mean effective stresses will increase, increasing the shear strength. While this mechanism predicts very high monotonic pullout capacity for bucket foundations, cavitation of the pore water, which is a function of the water depth at the site, may limit the ultimate achievable pullout capacity (Bye *et al.* 1995).

Sustained uplift resistance is not currently relied on in design. For example, the dead weight of the foundation and ballast of the Snorre A TLP counteracts the tension in the tethers under normal, calm weather, working conditions. During a storm, the TLP will offset from the foundations resulting in a moment on each CFT, which gives the most critical load situation (Christophersen 1993). A series of static and dynamic scale model tests were carried out for the Snorre foundation design (Dyvik *et al.* 1993, Andersen *et al.* 1993). Recent centrifuge tests have shown that uniaxial tensile loads of around 30 per cent of undrained uplift capacity may be resisted for several years by shallow-skirted foundations with an embedment ratio as low as 0.3, indicating potential efficiencies in skirted foundation design (Gourvenec *et al.* 2007, 2008a). Irrespective of the potential for sustained uplift resistance, reverse end bearing is only mobilised if a competent seal is achieved, which may be compromised by non-vertical or cyclic loading.

A significant component of environmental loading applied to an offshore foundation is cyclic and, for such conditions, pore pressure accumulation must also be considered. Cyclic loads give rise to pore pressure accumulation for most soils. In clays, this can give rise to softening of the stress–strain response and some reduction in strength (usually modest except in quick clays). In sands, liquefaction can occur in which either the soil loses most of its stiffness (initial liquefaction) or most of its strength (leading to flow failures). Initial liquefaction is most significant for offshore conditions, because sands are rarely loose enough for flow failure liquefaction to occur. Although a dense sand will tend to dilate under monotonic shearing, contraction always occurs under cyclic loading. Ultimately, a state of initial liquefaction may be reached when the effective stresses in the soil reduce to zero.

Observations from model tests indicate that once such a state is triggered, a bucket foundation will continuously sink into the soil even if the magnitude of the applied cyclic loads is subsequently reduced. For very dense sands, it was found that this condition could only be initiated when the foundation was cycled from compression to tension, and it was this mechanism that limited the design tensile capacity to only a very small component of the monotonic pullout resistance (Bye *et al.* 1995). In less dense sands, it remains a moot point as to the magnitude of cyclic tensile loads that may be safely resisted.

In addition to ongoing cyclic loading during calm weather conditions, offshore structures are also subject to extreme cyclic loading during storms. Extreme loading events build up over a length of time, due to winds generated over a long fetch (for instance in the North Sea) or by hurricane events (e.g. in the Gulf of Mexico and along the North-West Shelf of Australia). Structures and foundations are not hit by just one large wave but are subjected to continual loading over a period. Most storms are considered to be three-hour events, with approximately 1,000 waves, but the build up and build down can last up to 72 hours. Ocean waves vary with time and the loading they apply to the foundations will vary according to their size. Bearing capacity analyses must account for the effect of cyclic loads, but this is complicated since a real cyclic loading comprises many cycles of varying amplitude while laboratory tests are limited to uniform cycling. The effect of cyclic loading on soil response is considered in Chapter 4.

SHEAR STRENGTH FOR STABILITY ANALYSES

The relevance of stability calculations depends on the choice of appropriate shear strength parameters. This is not straightforward due to the variety of stress paths under a foundation. A simplified picture of the shear stresses in a few typical elements along a potential failure surface is shown in Figure 6.14. At different locations, the relevant shear strengths will approximate more to those measured in triaxial

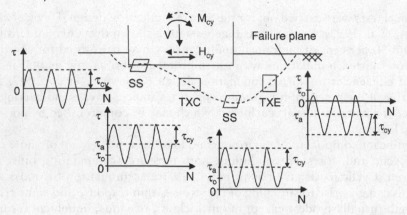

*Figure 6.14* Simplified stress conditions along a potential failure surface beneath a shallow foundation (O'Reilly and Brown 1991)

compression, simple shear or triaxial extension, which may differ by a factor of two (see Chapter 4). In addition, the relative magnitude of average and cyclic shear stresses to which the soil is subjected will vary through the potential failure mechanism, again leading to variations in available shear strength or accumulation of shear strain.

Detailed analysis of the foundation stability would ideally require sophisticated numerical analysis, with a soil constitutive model that allows for anisotropy of shear strength and also the effects of cyclic loading. In practice, most design adopts a simpler approach, either assessing appropriate shear strengths to insert at different locations in limit equilibrium (or upper bound plasticity) analyses or estimating a single 'average' shear strength to use in standard design approaches. The average shear strength should reflect both the range of stress paths, for example, taking an average of shear strengths in triaxial compression, simple shear and triaxial extension, and also the effects of cyclic loading. This is discussed further in Chapter 4.

DRAINAGE

Prediction of foundation bearing capacity requires assessment of the drainage conditions beneath the foundation under both static and cyclic foundation loads. If soil or loading conditions are such that dissipation of excess pore pressures does not take place within the period of interest then an undrained bearing capacity analysis is sufficient. Undrained conditions would be assumed to prevail, for example, in the short term after application of a load to a fine grained (i.e. clayey soil), or in a storm event when the period of loading is very short. It should also be recognised that for a large GBS with a long drainage path, undrained conditions may prevail for months or years even on relatively pervious deposits. If drainage is allowed to take place after application of a load then a drained bearing capacity analysis should be used.

It is relatively straightforward to assess whether consolidation will occur sufficiently quickly under vertical static loads to enable soil strengths to increase before design wave loads are applied. If pore pressures will not dissipate, then the *in situ* undrained strength should be used for all bearing capacity calculations. Generally, this case applies for most fine-grained soils (clays and many silts). It is also possible to determine the drainage that occurs during a single wave cycle. The governing load condition acting on most offshore foundations involves a significant magnitude of cyclic loading applied over short durations (i.e. wave periods of generally less than about 15 s). Most materials exhibit an undrained response during a single wave cycle, even fairly coarse materials, such as most sands (which is the basis for the tension capacity of bucket foundations discussed earlier). This is currently not reflected in design codes (e.g. API 2000, ISO 2000, DNV 1992). For sands and silts, it is, therefore, usually necessary to assess the consolidated undrained bearing capacity, that is where the peak loads are applied undrained but the soil has consolidated and strengthened under the weight of the platform before the peak loads are applied. Consequently, even where the soils have fully drained under the static loads, an undrained analysis of the wave-loading event should be performed but using soil strengths determined allowing for strength increases due to consolidation under the vertical static load. For consolidated undrained conditions, the distribution of undrained shear strength beneath the foundation is complex, and there are no simple solutions available in the literature that address such behaviour. Inevitably, the design engineer must resort to either limit equilibrium methods (e.g. Svano 1981, Randolph and Erbrich 2000) or finite element methods (e.g. Bye *et al.* 1995) to address this issue.

FACTOR OF SAFETY

Stability calculations for design must incorporate a factor of safety. Generally, there are three approaches:

1. The overall safety factor approach, also called the working stress design approach (WSD), in which a single factor of safety is applied to the ultimate limit load. The American Petroleum Institute (API) RP 2A-WSD (American Petroleum Institute 2000) recommends a factor of safety of two for bearing failure and 1.5 for sliding failure. The magnitude of the factor of safety is intended to account for uncertainties in soil conditions and design values of dead and live loading.
2. The partial safety factor approach, also called the load and resistance factor design approach (LRFD), uses separate factors for soil strength and applied loads. The International Standardization Organization (ISO) (International Standardization Organization 2000) recommends reducing the shear strength by a material coefficient $\gamma_m = 1.25$, and increasing loads by a factor of 1.1 for dead loads and 1.35 for live loads.
3. A probabilistic approach, in which the uncertainties in soil strength and applied loads are quantified and used to determine a probability of failure, which should be less than a pre-determined value.

The partial safety factor approach is now generally preferred to an overall safety factor approach, since it allows the overall reliability of a design to be quantified better. Probabilistic approaches are yet to find their way into mainstream design, although they form the basis of modern geohazard assessment and are gradually spreading into other areas of offshore geotechnical design.

*Serviceability*

Capacity may be the dominant issue in many offshore shallow foundation designs, but serviceability is also significant, including prediction of displacements required to mobilise capacity. Serviceability issues are pertinent to sandy seabeds, particularly for compressible deposits such as loose silica sand or most carbonate sands and silts. In clays, volume change is restricted by the lack of drainage whereas in sands the bulk compressibility can have a significant effect on practical limits of bearing capacity. Depending on the size of a foundation, design settlement limits of a fraction of the foundation diameter or width may lead to reduced foundation capacity compared to ultimate bearing capacity predictions.

SETTLEMENT

Allowable settlement of an offshore shallow foundation is usually constrained by the displacement tolerance of oil wells and pipelines connected to the foundation and, so long as settlements take place before wells and pipelines are connected, it is feasible to design offshore shallow foundation for settlements of the order of a metre or so. Spatial variability of the seabed deposits can lead to differential settlements, which can be more harmful to a structure than total settlements particularly if the structure is relatively rigid, such as a gravity-based structure. Differential settlements of gravity-based structures can be managed over the lifetime of the structure by controlled ballasting if settlement monitoring is undertaken. Eccentricity of the structure on the foundation (sometimes necessary due to drilling constraints from the adjacent jack-up) causes a permanent moment to be transferred to the foundation, leading to the potential for differential consolidation settlement over the life of the platform.

Active drainage by applying an under pressure can be employed following installation to accelerate soil consolidation and hence settlements in fine-grained materials. This can be beneficial to encourage most of the settlement to take place before oil wells and risers are connected to the main structure. An active drainage programme was adopted following installation of the deep-skirted Condeep platform Gullfaks C in soft clay in the North Sea. Most of the settlement (approximately 0.5 m) took place in the first three months during which the highest under pressures were applied and the observed settlements were similar to predicted values. When active under pressure ceased, settlements reduced to 15–20 mm year. Conversely, a watertight base, ensuring an undrained response and acceptable settlements, was a basic design criterion for the deep-skirted Condeep platform Troll A, which is located in similar soil conditions to Gullfaks C.

Accumulated pore pressures during cyclic loading can lead to ongoing foundation displacements over time. Some pore pressure dissipation may occur between storm events while the remaining cyclically induced pore pressures accumulate from one storm to another. Pore pressure measurements from instrumentation on the first gravity-based platform, Ekofisk I, showed that repeated loading generated excess pore pressures even in dense sands, which was accompanied by large settlements of the tank (Clausen *et al.* 1975).

Particular seabed conditions in some areas of the world can lead to specific serviceability challenges. For example, calcareous soil deposits, which predominate on the seabed offshore Australia, Brazil and many parts of the Middle-East, are characterised by spatial variability in terms of particle size and cementation and high angularity of individual particles. In many regions, this leads to high void ratios and compressibility. Particular problems for shallow foundation design in calcareous deposits include liquefaction and volume collapse under the action of cyclic loading (Randolph and Erbrich 2000). Stratigraphies where low strength material underlies stronger layers may lead to punch-through type failures or large settlements.

SUBSIDENCE

Subsidence is a serviceability issue that can arise through long-term extraction from reservoirs beneath a platform. Considerable subsidence beneath the Ekofisk field was reported in 1984 when facilities on the now highly developed hub had subsided up to 3 m in 13 years of production. In 1989, six installations were jacked up and a concrete breakwater was constructed around the original Ekofisk tank. The seabed continued to subside by up to 0.5 m per year, and the operator resolved in 1994, under pressure from the authorities, to redevelop the field. The Ekofisk II project came on stream in 1998, designed to cope with 20 m of subsidence. By 2005, the seabed had subsided by 8 m and is continuing at a rate of about 0.1 m per year (Holhjem 1998, Etterdal and Grigorian 2001).

## 6.2 Installation resistance

Penetration analyses should include calculation of skirt penetration resistance under self-weight, the magnitude of under pressure (if any) needed to achieve target penetration depth and the allowable under pressure (i.e. not exceeding the limits for either large soil heave inside the skirts, soil plug liquefaction or cavitation in the water). Comparing the suction pressure needed to advance the foundation into the soil, with the pressure required to fail the soil plug, provides an assessment of the relative stability of the soil plug. The limiting depth to diameter embedment ratio d/D that can be installed is obtained when the two forces become equal. Penetration resistance and the required and allowable under pressure can be predicted by simple soils mechanics principles as outlined in the following sections.

### 6.2.1 Installation in fine grained sediments

PENETRATION RESISTANCE

A schematic of the forces acting during installation of a skirted foundation is shown in Figure 6.15. For a simple case where no protuberances on the inner and outer walls of the skirt are considered, the total penetration resistance of a skirted foundation Q is provided by the side shear along the skirt walls and any internal stiffeners, the bearing capacity at the skirt and stiffener tip and the overburden pressure, expressed as

$$Q = A_s \alpha \overline{s}_u + A_{tip}(N_c s_u + \gamma' z) \qquad (6.1)$$

where

$A_s$ = skirt wall surface area (sum of internal and external)
$A_{tip}$ = skirt tip bearing area
$\alpha$ = adhesion factor (assumed equal to the inverse of the sensitivity)
$\overline{s}_u$ = average shear strength over penetration depth (simple shear strength $s_{uss}$)

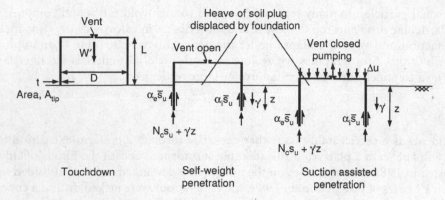

Figure 6.15  Forces during installation of a skirted foundation (Randolph *et al.* 1998)

$s_u$ = undrained shear strength at skirt tip level (average of $s_{uc}$, $s_{ue}$, $s_{uss}$)
$\gamma'$ = effective unit weight of soil
 $z$ = skirt penetration depth
$N_c$ = bearing capacity factor for plane strain conditions (= 7.5).

The overburden term $\gamma'z$ is required due to the increasing penetration resistance arising from the increasing vertical stresses in the soil (essentially this is identical to a buoyancy force due to the displaced effective weight of soil). In reality, the effect of internal stiffeners must also be considered in terms of bearing and frictional resistance.

NECESSARY UNDER PRESSURE

Penetration under self-weight will continue until the penetration resistance (as calculated in Equation 6.1) equals the submerged self weight W' of the foundation. To achieve further penetration, suction assistance is required. The soil resistance during penetration under suction can also be calculated using Equation 6.1. The under pressure (i.e. suction) $\Delta u_{req}$ needed within the skirt compartment in order to penetrate the skirts is then given by

$$\Delta u_{req} = \frac{Q - W'}{A_i} \qquad (6.2)$$

where

 $Q$ = total penetration resistance of a foundation (from Equation 6.1)
$W'$ = submerged weight of foundation
 $A_i$ = internal cross-sectional area of the foundation (i.e. area over which suction is applied).

ALLOWABLE UNDER PRESSURE

The allowable under pressure $\Delta u_a$ is the maximum pressure that can be applied without causing large soil heave within the cylinder due to bottom heave at skirt tip level (i.e. clay sucked into the foundation). The weight of the soil plug, bearing resistance across the base of the plug and the internal shaft friction mobilised along the skirt walls resist soil plug failure. The gravity forces of the soil column adjacent to the foundation assist failure of the soil plug and, therefore, should be subtracted. The allowable under pressure is then given by,

$$\Delta u_a = \frac{A_i N_c s_u + A_{si}\alpha \overline{s}_u + W'_{plug} - \gamma' d A_{plug}}{A_i} \qquad (6.3)$$

where

$A_i$ = internal cross-sectional area of foundation
$N_c$ = bearing capacity factor varying from 6.2 to 9 depending on depth/diameter ratio during penetration.
$s_u$ = undrained shear strength at skirt tip level (average of $s_{uc}$, $s_{ue}$, $s_{uss}$)
$A_{si}$ = internal skirt wall surface area
$\alpha$ = adhesion factor (assumed equal to the inverse of the sensitivity)
$\overline{s}_u$ = average shear strength over penetration depth (using simple shear strength $s_{uss}$)
$W'_{plug}$ = weight of the soil plug inside the skirts (= $\gamma' d A_{plug}$)
$d$ = embedment depth
$\gamma'$ = effective unit weight of soil
$A_{plug}$ = cross-sectional area of soil plug.

$\gamma' d A_{plug}$ = weight of soil column adjacent to foundation and since $W'_{plug} = \gamma' d A_{plug}$, the terms for plug weight and overburden cancel. The allowable under pressure can thus be defined as the sum of the plug base resistance and internal shaft resistance,

$$\Delta u_a = \frac{A_i N_c s_u + A_{si}\alpha \overline{s}_u}{A_i} \qquad (6.4)$$

In shallow water, it should be checked that the allowable pressure does not exceed the cavitation pressure.

The factor of safety F against plug failure is commonly expressed as the ratio of the suction pressure to cause failure of the soil plug $\Delta u_a$ and the suction pressure required to advance the foundation into the soil $\Delta u_{req}$:

$$F = \frac{\Delta u_a}{\Delta u_{req}} \qquad (6.5)$$

A more consistent definition of the factor of safety against plug failure is to rearrange Equations 6.1–6.3 to provide a ratio of the plug end-bearing resistance, $A_i N_c s_u$, to the net plug uplift force at the tip under the required suction (Andersen *et al.* 2005). If material factors are used, then the external shear strength should be *increased* and the shear strength for plug end-bearing resistance *decreased* since that leads to the most adverse situation.

VERIFICATION OF PENETRATION ANALYSES

Skirt penetration resistance is normally measured during installation to control the operation. Predicted and measured skirt penetration resistances are compared for the

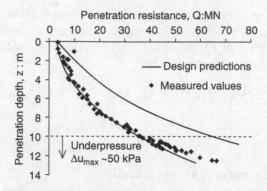

*Figure 6.16* Predicted and measured penetration resistance for the Snorre bucket foundations (Andersen and Jostad 1999)

Snorre TLP concrete bucket foundations in Figure 6.16. The measured data fall close to the lower bound predictions, indicating that the prediction of penetration resistance in clay is reliable.

### 6.2.2 *Installation in permeable sediments*

Installation of suction caissons in sand or other permeable seabeds results in the development of a seepage flow field into the caisson. This is not only inevitable (due to the high permeability and hence rapid decay of any transient response due to consolidation), but also essential in order to reduce the effective stresses, and hence penetration resistance, at the skirt tip (Erbrich and Tjelta 1996). As for suction caissons in clay, the total installation force comprises the submerged weight of the caisson and the differential pressure acting on the lid of the caisson. However, for sand, the dominant resistance to installation is the tip penetration and the principal effect of suction is to reduce that installation resistance rather than to significantly increase the installation force (which is a secondary effect).

Alternative methods for estimating the required level of suction have been proposed. One such method follows a classical approach based on fundamental soil parameters (friction angle, $\phi$, soil effective unit weight, $\gamma'$, and *in situ* stress conditions), with shaft friction calculated as $K\sigma'_v\tan\delta$ and tip resistance estimated as $N_q\sigma'_v$ (Houlsby and Byrne 2005). The seepage field arising from the applied suction is used as the basis to modify $\sigma'_v$ from the *in situ* conditions, which leads to some increase in the external shaft friction, but significant reductions in the tip resistance and internal shaft friction, particularly at levels of suction where a hydraulic piping condition is approached within the soil plug.

An alternative method uses field data directly in the form of a profile of cone penetration resistance $q_c$ (Senders and Randolph 2009). External and internal shaft friction and tip resistance, are estimated from the cone resistance following similar methodologies to those for pile foundations (see Chapter 5). During self-weight penetration the full shaft and tip resistance is assumed to apply; as the suction is increased towards the limiting value that leads to a critical hydraulic condition within the caisson, the internal shaft friction and tip resistance are both

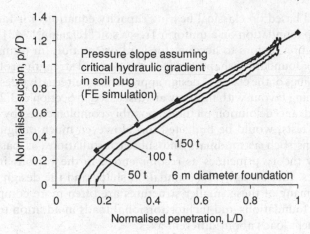

*Figure 6.17* Model test data from installation of suction caisson in dense sand (Tran and Randolph 2008)

reduced (linearly) with the increase in suction. The vertical equilibrium is thus expressed as,

$$W' + 0.25\pi D_i^2 p = F_o + (F + Q_{tip})\left(1 - \frac{p}{p_{crit}}\right) \quad \text{for} \quad p \le p_{crit} \tag{6.6}$$

where $W'$ and $D_i$ are the submerged weight and internal diameter of the suction caisson, p is the applied suction (with $p_{crit}$ the critical value), $F_o$ and $F_i$ are the external in internal shaft resistance and $Q_{tip}$ is the tip resistance.

Experience with both model and field tests has shown that it is possible to install caissons to depths in excess of one diameter using suction, even though current applications have been limited to shallower embedment. In order to reduce the tip resistance sufficiently, especially in the glaciated dense sands of the North Sea (Erbrich Tjelta 1996), values of suction close to the critical value at which the internal soil plug becomes buoyant are required (Tran and Randolph 2008, Senders and Randolph 2009). Figure 6.17 shows example data from centrifuge tests modelling a 6 m diameter prototype, with a wall thickness of 30 mm (0.5 per cent of the diameter), for three different submerged masses. It is clear that for most of the installation process, the suction (normalised as $p/\gamma' D$) follows the line corresponding to a critical hydraulic gradient.

## 6.3 Bearing capacity

### 6.3.1 Overview

Design guidance for calculation of bearing capacity of offshore shallow foundations is set out by the ISO (2000), DNV (1992) and the API (2000) amongst others. Despite the clear differences between offshore and onshore shallow foundation systems and loading conditions the roots of the design methods presented in the recommended practices are the same as adopted for onshore design (e.g. Eurocode 7 1997). These are ultimately all based on classical bearing capacity equations for failure of a vertically loaded strip foundation on a uniform Tresca soil (Terzaghi 1943) combined with various modification factors to account for load orientation (in terms of inclination and eccentricity), foundation shape, embedment and soil strength profile.

The shortcomings of the classical design approach and alternative design approaches increasingly gaining favour with industry are discussed in Section 6.3.2 and 6.3.3. For novel designs, advanced solution methods, possibly complemented by project specific field and model tests, would be budgeted for. However, much design, especially for small foundations such as mudmats and subsea installations, is based on classical bearing capacity theory principles as recommended in the various industry recommended practices. Interestingly, the foundation shape and the design load combinations acting on many of these smaller structures are often more complex than those acting on larger foundations and include torsional loads in addition to vertical, horizontal and moment loads about different axes.

In the following sections, the classical bearing capacity theory approach to predicting ultimate limit states under general loading and an introduction to advanced solution methods for bearing capacity prediction are presented.

### 6.3.2 Classical bearing capacity theory approach

Traditional bearing capacity theory, as well as many advanced recent solutions, is based on plasticity theory. Plasticity solutions are based on the bound theorems to identify an upper and lower bound to the range within which the exact solution lies. For an answer to be exact the upper bound (mechanism) and lower bound (stress field) approaches must give the same result. Limit analysis explores different ways in which these two bounds can be satisfied until identical collapse loads are obtained with each, i.e. the true collapse load is obtained. Plasticity solutions are based on perfect plasticity prevailing at failure, i.e. no hardening or softening behaviour is considered. Although these are simple soil models, they have been widely used historically as the basis of bearing capacity solutions for both onshore and offshore applications.

### Undrained bearing capacity

The classical recommended approach for predicting undrained bearing capacity of a shallow foundation is given by

$$V_{ult} = A' \left( s_{u0} (N_c + kB'/4) \frac{F K_c}{\gamma_m} + p_0' \right) \tag{6.7}$$

where

$V_{ult}$ = ultimate vertical load
$A'$ = effective bearing area of the foundation
$s_{u0}$ = undrained shear strength of the soil at foundation level
$N_c$ = bearing capacity factor for vertical loading of a strip foundation on a homogeneous deposit, i.e. 5.14 (Prandtl 1921)
$k$ = gradient of the undrained shear strength profile (equal to zero for homogeneous deposits).

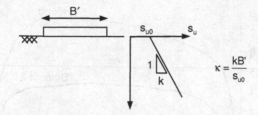

*Figure 6.18* Definition of degree of heterogeneity for linearly increasing shear strength with depth

B′ = effective width of the foundation (Meyerhof 1953)
F = correction factor to account for the degree of strength heterogeneity
$\gamma_m$ = material factor on shear strength
$K_c$ = modification factor to account for load orientation, foundation shape and embedment, given by

$$K_c = 1 - i_c + s_c + d_c \tag{6.8}$$

where

$$i_c = 0.5\left(1 - \sqrt{1 - H/A's_{u0}}\right) \tag{6.9}$$

$$s_c = s_{cv}(1 - 2i_c)B'/L \tag{6.10}$$

$$d_c = 0.3e^{-0.5kB'/s_{u0}} \arctan(d/B') \tag{6.11}$$

$p'_0$ = effective overburden (resisting failure) at foundation level; for foundations with skirts this is usually taken at skirt tip level.

The correction factor F is given as a function of a dimensionless heterogeneity factor $\kappa = kB'/s_{u0}$, as illustrated in Figure 6.18. Values of F as a function of $\kappa$ are shown in Figure 6.19 (after Davis and Booker 1973).

The inclination factor $i_c$ and shape factor $s_c$ are taken directly from Brinch Hansen (1970). The inclination factor is based on the exact solution for loading of a strip foundation under a centrally applied inclined load (Green 1954) in conjunction with the effective width principle to account for moment (Meyerhof 1953). The shape factor $s_c$ depends on a coefficient $s_{cv}$, which is a function of the degree of heterogeneity $\kappa = kB'/s_{u0}$ (Salençon and Matar 1982) (Table 6.3), the inclination factor and foundation aspect ratio. For uniform deposits $s_{cv} = 0.2$ (Brinch Hansen 1970). Generally, guidelines do not make explicit provision for specific three-dimensional foundation geometry but suggest equivalence to a rectangle with the same area and areal moment of inertia. The recommended depth factor in Equation 6.11 is slightly more conservative than the conventional Brinch Hansen (1970) factor. Further, it is often recommended that $d_c = 0$ be used if the installation procedure or other foundation aspects, such as scour, prevent mobilisation of shear stresses in the soil above skirt tip level, or

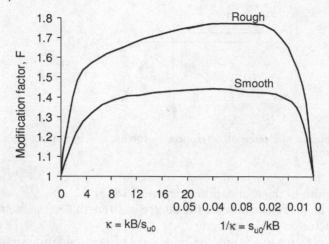

*Figure 6.19* Bearing capacity modification factor for linearly increasing soil strength with depth (Davis and Booker 1973)

*Table 6.3* Shape factor coefficient for heterogeneous soil strength (Salençon and Matar 1982)

| $\kappa = kB'/s_{u0}$ | $s_{cv}$ |
|---|---|
| 0 | 0.20 |
| 2 | 0.00 |
| 4 | −0.05 |
| 6 | −0.07 |
| 8 | −0.09 |
| 10 | −0.10 |

if the in-service horizontal loads lead to large passive earth pressures over the depth of the skirts. The $kB'/4$ term essentially normalises the shear strength profile with respect to the plan size of the foundation.

The undrained bearing capacity under uniaxial vertical loading of a strip foundation resting on the surface of a homogeneous deposit calculated by Equation 6.7 gives $V_{ult} = 5.14As_{u0}$ in agreement with the exact solution (Prandtl 1921). Incorporation of the shape factor gives $V_{ult} = 6.17As_{u0}$, slightly over predicting the exact solution for a rough circular foundation of $6.05As_{u0}$ (Cox *et al.* 1961). The horizontal failure criterion gives $H_{ult}/As_{u0} = 1$ and is independent of foundation geometry or undrained shear strength gradient implying failure by sliding when the applied force is in equilibrium with the undrained shear strength at ground level. For conditions of no vertical load eccentricity and $H/As_{u0} = 1$, the inclination factor $i_c = 0.5$ and the shape factor $s_c$, if applicable, reduces to zero such that substitution in Equation 6.7 indicates lateral failure will prevail for vertical loads $V \le 2.57As_{u0}$, i.e. $V/V_{ult} \le 0.5$. The ultimate moment $M_{ult}$ is predicted in conjunction with a vertical load $V = 0.5V_{ult}$ (Meyerhof 1953). At lower vertical loads, application of moment is assumed to result in separation at the foundation–soil interface leading to a reduced bearing area and hence bearing capacity. Solution of the bearing capacity equation (Equation 6.7) gives

$M_{ult} = 0.64ABs_{u0}$ for a strip foundation and $0.61ADs_{u0}$ for a circular foundation, which compare well with plasticity solutions (Houlsby and Puzrin 1999, Randolph and Puzrin 2003) and finite element analyses (Taiebat and Carter 2002, Gourvenec 2007a).

## Drained bearing capacity

Drained bearing capacity under compression is generally larger than the undrained capacity as the increase in soil stresses due to the foundation load leads to an increase in frictional shear strength. An exception is highly dilatant sands, where the undrained capacity may be much greater due to dilation-induced negative pore pressures. In any soil type, if tensile loads are to be withstood and suctions are to be relied on then undrained conditions are advantageous.

The classical recommended approach for predicting drained bearing capacity of a shallow foundation is given by

$$V_{ult} = A'(0.5\gamma'B'N_yK_y + (p'_0 + a)N_qK_q - a) \tag{6.12}$$

where

$V_{ult}$ = ultimate vertical load
$A'$ = effective bearing area of the foundation
$\gamma'$ = effective unit weight of the soil
$B'$ = effective width of the foundation
$N_y$, $N_q$ = bearing capacity factors for self-weight and surcharge
$K_y$, $K_q$ = modification factors to account for foundation shape, embedment and load inclination
$p'_0$ = effective overburden acting to either side of the foundation
$a$ = soil attraction factor, which accounts for cementation, equal to the point of interception of the tangent to the Mohr Circle and the normal stress axis.

$N_q$ and $N_y$ should be modified to incorporate the material factor on strength, and are given by

$$N_q = \tan^2\left(\frac{\pi}{4} + 0.5\tan^{-1}\left(\frac{\tan\phi}{\gamma_m}\right)\right)e^{\pi\tan\phi}/\gamma_m \tag{6.13}$$

and

$$N_y = 1.5(N_q - 1)\tan\left(\frac{\tan\phi}{\gamma_m}\right) \tag{6.14}$$

where

$\phi$ = effective internal friction angle of the soil
$\gamma_m$ = material factor on shear strength.

The 0.5 coefficient for $\gamma'B$ in the bearing capacity equation arises from the convention to express the contribution due to soil self-weight in terms of the effective overburden stress at a depth of half the foundation width below the foundation.

The solution for $N_q$ (i.e. for a weightless soil with a surcharge) under uniaxial vertical load given in Equation 6.13 is exact, identified as a lower bound by Prandtl (1921) and subsequently formalised by similar upper bounds. Note that because $\phi$ exists as an exponent in the expression for $N_q$, the magnitude of $N_q$ and $V_{ult}$ are sensitive to a small change in $\phi$. There is no exact solution for $N_\gamma$ (i.e. for a heavy soil with no surcharge) and the established solution is based on a lower bound by Lundgren and Mortensen (1953). Davis and Booker (1971) presented rigorous solutions for $N_\gamma$ using the method of characteristics. Curve fits to Davis and Booker's solutions are given by

$$N_\gamma = 0.1054 e^{9.6\phi} \quad \text{for a rough foundation} \tag{6.15}$$

$$N_\gamma = 0.0663 e^{9.3\phi} \quad \text{for a smooth foundation} \tag{6.16}$$

The approximate relationship between $N_\gamma$ and $N_q$ (as given in Equation 6.14) is widely used.

It is prudent to note that the bearing capacity equation assumes that the effect of soil self-weight and surcharge are superposable (i.e. the sum of the effect of either acting independently is equivalent to the two acting simultaneously). This was shown to be conservative even for uniaxial vertical load by Davis and Booker (1971).

The modification factors $K_q$ and $K_\gamma$ may be written as

$$K_q = s_q d_q i_q \tag{6.17}$$

$$K_\gamma = s_\gamma d_\gamma i_\gamma \tag{6.18}$$

where

$$s_q = 1 + i_q \frac{B'}{L} \sin\left(\tan^{-1}\left(\frac{\tan\phi}{\gamma_m}\right)\right) \tag{6.19}$$

$$d_q = 1 + 2\frac{d}{B'}\left(\frac{\tan\phi}{\gamma_m}\right)\left\{1 - \sin\left(\tan^{-1}\left(\frac{\tan\phi}{\gamma_m}\right)\right)\right\}^2 \tag{6.20}$$

$$i_q = \left\{1 - 0.5\left(\frac{H}{V + A'a}\right)\right\}^5 \tag{6.21}$$

$$s_\gamma = 1 - 0.4 i_\gamma \frac{B'}{L} \tag{6.22}$$

$$d_\gamma = 1 \tag{6.23}$$

$$i_\gamma = \left\{1 - 0.7\left(\frac{H}{V + A'a}\right)\right\}^5 \tag{6.24}$$

where $A'$ is the effective area of the foundation $= B'L$.

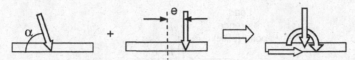

*Figure 6.20* Representation of general loading in classical bearing capacity theory

## Summary

The classical bearing capacity approach uses modification factors to extend the basic solution for the simple conditions for which it was derived, i.e. failure under predominantly vertical load of a strip foundation resting on the surface of a uniform deposit, to the more complex conditions more likely to be encountered in reality, of a three-dimensional, embedded foundation under general loading in heterogeneous material.

The interaction of horizontal and moment load components with a vertical load are treated separately in the classical bearing capacity approach. It is then supposed that coupling of the two solutions for VH and VM will represent a fully combined vertical, horizontal and moment load (Figure 6.20). For conditions of soil strength homogeneity, classical theory adequately predicts bearing capacity under either centrally applied inclined loads *or* vertical eccentric loads, but is less accurate (but conservative) under superposition of the solutions for inclination and eccentricity (i.e. VHM loading). The validity of the superposition of separate solutions for inclination and eccentricity to represent general loading was questioned in the light of bound solutions reported by Ukritchon *et al.* (1998) for strip foundations and similar shortcomings for circular foundations were confirmed by Gourvenec and Randolph (2003a). For non-uniform shear strength profiles classical bearing capacity theory will under predict the capacity even for simple eccentricity with no lateral load. The breakdown of classical bearing capacity theory under combined loading is particularly significant in offshore shallow foundation design due to the large components of horizontal load and moment from the harsh environmental conditions (i.e. wind, wave and current forces, see Figure 6.2) and the often normally consolidated seabed deposits.

The applicability of classical bearing capacity theory to offshore design is also questionable as the approach neglects tensile capacity, which in reality can be mobilised by some offshore shallow foundations provided they are equipped with skirts and sealed base plates.

The embedment provided by foundation skirts will in reality lead to coupling of the horizontal and moment degrees of freedom, enhancing the available horizontal and moment capacity, the interaction becoming more pronounced with increasing embedment (this is discussed in more detail in Section 6.3.3). This coupling of the horizontal and moment degrees of freedom for embedded foundations is not accounted for in the classical bearing capacity approach by the incorporation of depth factors or an overburden term.

Finally, the classical bearing capacity approach results in an expression for an adjusted ultimate vertical limit load, rather than discrete components of vertical load, horizontal load and moment that will cause failure. In this way, the method is not straightforward to apply in a manner that indicates the consequences of changes in the individual components of load on the real factor of safety against collapse. These

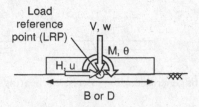

*Figure 6.21* Sign convention for general planar loads and displacements

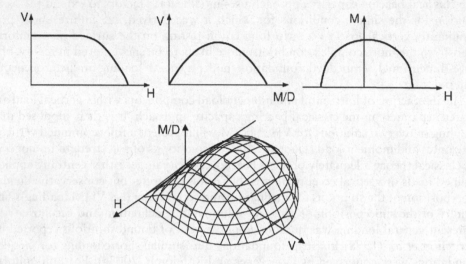

Figure 6.22 Representation of failure envelopes in two- and three-dimensional load space

issues are discussed further in the context of alternative solution methods in the following section.

### 6.3.3 Advanced solution methods

#### Failure envelopes

General loading of a shallow foundation results in a complex state of stress in the underlying soil. It is becoming increasingly accepted that when horizontal and moment loads are applied as well as vertical loads, the interaction of these different components should be explicitly considered when determining the bearing capacity (as opposed to the coupling of solutions using inclination and eccentricity factors as adopted in the classical bearing capacity approach). The most convenient way to represent ultimate limit states under combined loading is through interaction diagrams or failure envelopes. Failure envelopes can be expressed in planes of constant vertical, moment or horizontal load or as a three-dimensional surface in vertical, horizontal, moment (V, H, M) load space. Adopting the sign convention for loads and displacements illustrated in Figure 6.21, schematic representations of failure envelopes are illustrated in Figure 6.22. Any combination of loads within the failure envelope is regarded to be safe for the foundation while any load combination on the outside of the locus will violate the failure criterion.

Figure 6.22 indicates the general form of a failure envelope giving ultimate limit states for a rough surface foundation and undrained conditions showing sliding resistance, but no moment capacity available, at zero vertical load. The specific shape of a failure envelope depends on a variety of conditions, including whether failure takes place undrained or drained, the degree of shear strength heterogeneity, foundation–soil interface roughness and tensile resistance, foundation shape and embedment. An inherent assumption in the classical bearing capacity theory approach is that the incorporation of shape, embedment or heterogeneity factors simply scales the failure envelope in accordance with the relevant ultimate limit state $V_{ult}$ with no change in shape of the failure envelope. It is becoming increasingly recognised that the interaction of vertical, horizontal and moment loads is often more complex than represented by traditional bearing capacity theory; in many situations, particularly to represent embedment, it is not accurate to simply scale failure envelopes by their apex points.

Representation of ultimate limit states as a failure envelope (as opposed to an adjusted ultimate uniaxial vertical load as with classical bearing capacity theory) is particularly useful, as it allows the designer a more direct evaluation of the factor of safety against failure with respect to changes in the individual load components.

### Determination of failure envelopes

There are three main methods to determine a failure envelope in three-dimensional (V, H, M) load space: experimental, analytical and numerical. In their initial development, the most common approach was experimental (e.g. Nova and Montrasio 1991, Martin 1994, Gottardi *et al.* 1999). Just as for purely vertical loading, analytical plasticity solutions for the combined loading problems can be investigated (e.g. Ukritchon *et al.* 1998, Bransby and Randolph 1998, Randolph and Puzrin 2003). However, for combined loading, it is much more difficult to get the lower and upper bounds to approach. Numerical finite element studies are becoming a more common method to define a failure envelope (Bransby and Randolph 1998, Taiebat and Carter 2000, 2002, Gourvenec and Randolph 2003, Gourvenec 2007a and b, 2008).

(1) Experimental
Figure 6.23 shows an experimental rig developed at the University of Oxford, United Kingdom, to investigate shallow foundation response under general loading. Independent movement (or loading) in the vertical, rotational and horizontal planes can be applied. The vertical, moment and horizontal loads are read using a load cell placed just above the foundation and the displacements measured using transducers in each direction. The results from many tests can be compiled and interpolated to construct a continuous three-dimensional representation of a failure envelope.

(2) Analytical determination; plasticity theory
Analytical upper and lower bound plasticity solutions for foundation bearing capacity may be developed using, respectively, a postulated kinematic collapse mechanism or appropriate stress fields within the soil body. Apart from relatively simple cases, however, numerical procedures are required, even if only to optimise

*Figure 6.23* Experimental apparatus for investigating shallow foundations under general loading (Martin 1994)

a particular collapse mechanism. Alternatively, finite element-based limit analysis solutions may be pursued (Sloan 1988, 1989).

Generally, upper bound solutions are more straightforward to pursue, as it is more intuitive to postulate failure mechanisms than stress fields. Typical collapse mechanisms for the collapse of a strip foundation under combined loading are shown in Figure 6.24. These mechanisms may also be extended to three-dimension, in particular for circular foundations (Randolph and Puzrin 2003).

(3) Numerical determination

Failure envelopes can be determined through various numerical techniques such as the finite element and finite difference methods. Numerical approaches are based on discretising an area into a number of elements and/or nodes to which material properties and boundary conditions can be applied. This enables the analyst to realistically represent the foundation geometry, soil conditions and the loading regime and to calculate the failure load for the modelled conditions and the kinematic mechanism that accompanies failure. (This is in contrast to the upper bound approach that requires the analyst to postulate a failure mechanism *a priori*.) Figure 6.25 shows a finite element mesh with an example of a displacement-controlled load path from which the failure load of the foundation can be defined, along with the kinematic mechanism accompanying failure. A number of analyses need to be carried out to provide sufficient data points to construct a failure surface.

Historically, interaction diagrams for drained conditions (frictional soils) have been derived from experimental studies while those for undrained conditions ('cohesive' soils) have been based on analytical and numerical studies, owing to the relative complexity of the alternative approaches for the respective soil conditions.

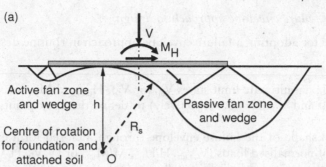

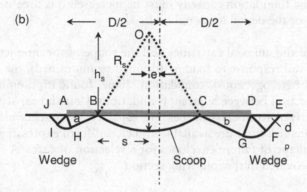

*Figure 6.24* Plane strain mechanisms for strip foundation under combined loading (Randolph and Puzrin 2003) (a) Brinch Hansen mechanism and (b) Bransby-Randolph wedge-scoop-wedge mechanism

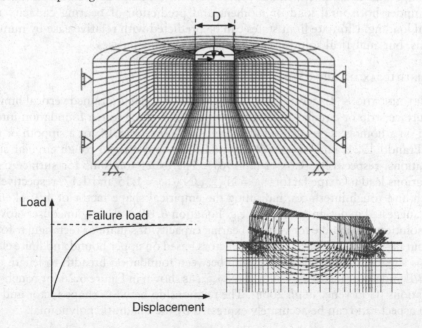

*Figure 6.25* Finite element mesh, load-displacement response and kinematic failure mechanism of a shallow foundations under general loading

*Application of failure envelope approach to design*

The procedure for adopting a failure envelope approach in routine design is straight-forward:

1. Define uniaxial ultimate limit states $V_{ult}$ (or $V_0$), $H_{ult}$ and $M_{ult}$ (i.e. at $H = M = 0$, $V = M = 0$ and $V = H = 0$, respectively) to define the apex points of the failure envelope.
2. Define the shape of the failure envelope through a closed-form expression as a function of normalised loads $(V/V_{ult}, H/H_{ult}, M/M_{ult})$ or $(V/V_0, H/V_0, M/V_0)$.
3. Identify if the design load condition falls within the failure envelope; if so the design is 'safe' if not the foundation capacity must be increased (i.e. area or embedment ratio increased) or the design loads reduced.

The magnitude of the uniaxial capacities and the shape of the interaction diagram will depend on the soil response to loading (drained or undrained), the soil strength profile (uniform or heterogeneous), foundation shape, foundation embedment and any structural connection between adjacent foundations. Uniaxial capacities, $V_{ult}$, $H_{ult}$ and $M_{ult}$ can be derived from classical bearing capacity approaches although more accurate and rigorous solutions are available and closed-form expressions are available to describe the shape of failure envelopes for a selection of cases. A review of relevant literature is provided in the following section.

*Available advanced solutions*

Advanced solutions have enabled refinement of bearing capacity factors for foundation shape and embedment under purely vertical loading, as well as ultimate limit states under horizontal load or moment and prediction of bearing capacity under general loading. Ultimate limit states can be predicted with relative ease by numerical analysis, but analytical solutions are available for limited cases.

UNDRAINED UNIAXIAL LIMIT STATES

SURFACE FOUNDATIONS   Exact solutions exist for the uniaxial undrained vertical limit load of a surface strip or circular foundation with a smooth or rough foundation interface resting on a homogeneous deposit giving $N_{cV} = V_{ult}/As_{u0} = 5.14$ for a smooth or rough strip (Prandtl 1921) and $N_{cV} = 5.91$ and $6.05$ for smooth and rough circular surface foundations, respectively (Cox *et al.* 1961). The exact solutions for surface circular foundations lead to shape factors $s_{cV} = N_{cVcircle}/N_{cVstrip} = 1.15$ and $1.17$, respectively, for smooth and rough interfaces, indicating the empirical shape factor of $1.2$ (Skempton 1951) suggested in design guidelines (e.g. Equation 6.10) is slightly unconservative. No exact solution is available for vertical bearing capacity of square or rectangular foundations but best estimate bearing capacity factors based on upper bound and finite element analysis suggest a quadratic relationship between foundation breadth to length aspect ratio B/L and shape factor $s_{cV} = N_{cV(B/L)}/N_{cVstrip}$, as shown in Figure 6.26 for rough-based foundations (Gourvenec *et al.* 2006). The relationship between shape factor and foundation aspect ratio can be accurately expressed by the quadratic polynomial.

$$s_{cV} = 1 + 0.214\frac{B}{L} - 0.067\left(\frac{B}{L}\right)^2 \tag{6.25}$$

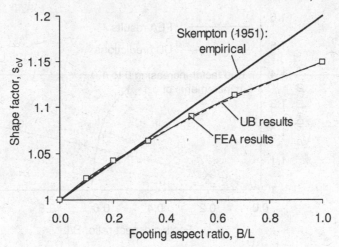

*Figure 6.26* Vertical bearing capacity shape factors for rectangular surface foundations (Gourvenec *et al.* 2006)

Figure 6.26 shows generally lower shape factors than Skempton's original recommendation of a linear variation with a coefficient of 0.2, and as recommended in design guidelines (i.e. indicating the traditional approach is unconservative), with a maximum difference of 7.8 per cent for a square foundation. The difference will be more marked for a smooth foundation interface although no analytical or numerical results are as yet available to quantify the disparity.

Undrained ultimate horizontal load for a surface foundation is trivial since $N_{cH}$ (= $H_{ult}/As_{u0}$) is equal to zero for a smooth surface foundation or unity for a rough surface foundation, independent of foundation plan geometry, as failure occurs by sliding when the applied load mobilises the limiting interface shear strength.

Undrained ultimate moment capacity of a surface strip or circular foundation with zero-tension capacity along the foundation–soil interface is given by the effective width (or effective area) principle (Meyerhof 1953). The ultimate moment of a surface foundation given by the effective width method, effectively a lower bound, is given by $N_{cM}$ (= $M_{ult}/ABs_{u0}$ or $M_{ult}/ADs_{u0}$) = 0.64 and 0.61 for strip and circular foundation geometry, respectively, and is mobilised in conjunction with a vertical load of $0.5V_{ult}$. Analytical solutions are not available for the moment bearing capacity factor for other foundation geometries, but finite element analyses of rectangular surface foundations showed the moment capacity shape factor $s_{cM} = N_{cM(B/L)}/N_{cMstrip}$ was related to the square of the foundation breadth to length aspect ratio, $B/L$ (Gourvenec 2007a):

$$s_{cM} = 1 + 0.075\frac{B}{L} - 0.005\left(\frac{B}{L}\right)^2 \tag{6.26}$$

Undrained ultimate moment capacity of a surface foundation with a full-tension interface is mobilised under zero vertical load and is given by an upper bound solution based on a circular or spherical scoop mechanism giving $N_{cM}$ = 0.69 and 0.67 for a strip and circular foundation, respectively (Murff and Hamilton 1993, Randolph and Puzrin 2003). Undrained ultimate moment capacity of surface rectangular foundations with an unlimited tension interface are given by an upper bound solution based

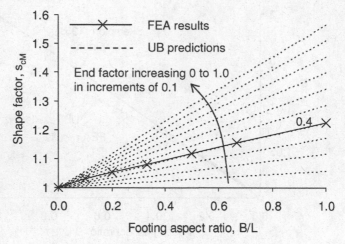

*Figure 6.27* Moment capacity shape factors for rectangular surface foundations (Gourvenec 2007a)

on a cylindrical scoop with a reduction factor to the work done on the out-of-plane section to account for edge effects (Gourvenec 2007a). Calibration against finite element analyses showed a reduction factor of 40 per cent (i.e. using 40 per cent of the work done on the out-of-plane section of a cylindrical scoop) gave good agreement. The relationship between moment capacity shape factor $s_{cM}$ and foundation aspect ratio, B/L, are shown in Figure 6.27 and expressed by the quadratic function

$$s_{cM} = 1 + 0.250\frac{B}{L} - 0.026\left(\frac{B}{L}\right)^2 \tag{6.27}$$

The above discussion of bearing capacity and shape factors relates to homogeneous soil conditions. The effect of shear strength heterogeneity on the vertical bearing capacity of surface foundations for linearly increasing strength with depth is established for strip and circular foundations based on rigorous solutions with the method of characteristics (Davis and Booker 1973, Houlsby and Wroth 1983) with correction factors adopted in current design guidelines (as previously shown in Figure 6.19 and Table 6.3). The effect of soil shear strength heterogeneity on ultimate horizontal capacity of a surface foundation is trivial; as for the homogeneous case, failure will occur by sliding when the applied horizontal force mobilises the interface shear strength, which is independent of the strength gradient. The effect of soil shear strength heterogeneity on the undrained ultimate moment capacity for surface strip and circular foundations with full-tension capacity has been determined from upper bound solutions of circular and spherical scoop mechanisms and finite element analyses (Gourvenec and Randolph 2003). Figure 6.28 shows moment capacity factors as a function of heterogeneity factor $\kappa = kB/s_{u0}$, $kD/s_{u0}$ for strip and circular surface foundations with unlimited tension interface, which are described by the quadratic expressions given by

$$F_M = 1 + 0.197\kappa - 0.003\kappa^2 \text{ for strip foundations} \tag{6.28a}$$

$$F_M = 1 + 0.156\kappa - 0.002\kappa^2 \text{ for circular foundations} \tag{6.28b}$$

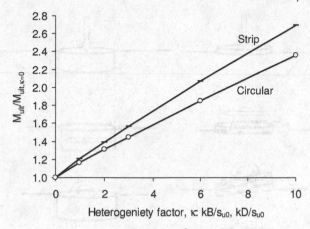

*Figure 6.28* Moment capacity heterogeneity modification factors (Gourvenec and Randolph 2003)

*Figure 6.29* Scoop mechanisms for surface circular foundations on soils with different degrees of shear strength heterogeneity

The effect of increasing shear strength with depth forces the failure mechanism into the shallower, weaker material, as shown by the displacement vectors from finite element analyses in Figure 6.29.

EMBEDDED FOUNDATIONS   Embedment increases vertical, horizontal and moment capacity as failure mechanisms are forced deeper within the soil mass, illustrated schematically in Figure 6.30. For soils with increasing shear strength with depth, capacity is further enhanced as failure takes place in deeper, stronger soil. Classical bearing capacity theory uses depth factors proposed by Skempton (1951) and Brinch Hansen (1970) to modify the uniaxial vertical bearing capacity, with industry recommended practice based on these factors (e.g. Equation 6.11). Classical depth factors were originally derived for smooth-sided circular foundations but are widely adopted for rough and smooth-sided, strip and three-dimensional foundations. Recent work has challenged the use of these conventional modification factors and explored the extent to which capacity is enhanced and the mode of failure changed by increasing embedment ratio.

Rigorous bearing capacity factors have been derived from the method of characteristics for embedded circular foundations with smooth sides, with varying interface base roughness and varying degree of shear strength heterogeneity, and are summarised in Table 6.4 (Houlsby and Martin 2003). Note that the undrained shear strength

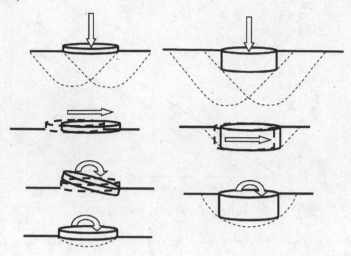

*Figure 6.30* Effect of embedment on capacity

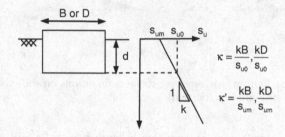

*Figure 6.31* Definition of degree of soil strength heterogeneity for embedded foundations

profile of the site is characterised by a heterogeneity factor $\kappa' = kD/s_{um}$, where $s_{um}$ is the undrained shear strength at the mudline (since typically the embedment ratio will not be known at the time of site characterisation), while the bearing capacity factors are based on the undrained shear strength at foundation level, $N_{cV} = q_{ult}/s_{u0}$, where $s_{u0}$ is the undrained shear strength at foundation level (Figure 6.31). The bearing capacity factors for smooth and rough interfaces given in Table 6.4 are represented graphically as depth factors $d_{cV} = N_{cV}/N_{cV,strip}$ in Figure 6.32 compared with depth factors predicted by the classical approach (Equation 6.11). Figure 6.32 illustrates the difference in predictions by the two approaches and the effect of overlooking interface roughness. The high depth factors at low embedment ratios arise as increasing $kD/s_{um}$ implies low shear strength at foundation level $s_{u0}$ leading to high $N_{cV} = q_{ult}/s_{u0}$. After an initial drop in depth, and hence bearing capacity, factor, continued increases are observed with increasing embedment ratio. The trend of increasing capacity with increasing embedment does not continue indefinitely owing to a change in failure mode (Figure 6.33).

*Table 6.4* Bearing capacity factors $N_{c0V} = q_{ult}/s_{u0}$ for embedded circular smooth-sided foundations (Houlsby and Martin 2003)

| $\kappa'$ ($kD/s_{um}$) | $d/D$ | *Base roughness factor* $\alpha$ | | | | | |
|---|---|---|---|---|---|---|---|
| | | 0.0 | 0.2 | 0.4 | 0.6 | 0.8 | 1.0 |
| 0.0 | 0.0 | 5.690 | 5.855 | 5.974 | 6.034 | 6.052 | 6.052 |
| | 0.1 | 5.967 | 6.127 | 6.238 | 6.290 | 6.298 | 6.298 |
| | 0.25 | 6.314 | 6.467 | 6.570 | 6.611 | 6.613 | 6.611 |
| | 0.5 | 6.785 | 6.927 | 7.020 | 7.048 | 7.047 | 7.048 |
| | 1.0 | 7.492 | 7.627 | 7.703 | 7.709 | 7.714 | 7.714 |
| | 2.5 | 8.824 | 8.944 | 8.991 | 8.993 | 8.987 | 8.990 |
| 1.0 | 0.0 | 6.249 | 6.469 | 6.651 | 6.794 | 6.895 | 6.946 |
| | 0.1 | 6.482 | 6.692 | 6.867 | 7.003 | 7.095 | 7.138 |
| | 0.25 | 6.741 | 6.940 | 7.106 | 7.234 | 7.317 | 7.350 |
| | 0.5 | 7.048 | 7.237 | 7.393 | 7.509 | 7.577 | 7.599 |
| | 1.0 | 7.469 | 7.644 | 7.787 | 7.884 | 7.933 | 7.942 |
| | 2.5 | 8.264 | 8.319 | 8.525 | 8.595 | 8.608 | 8.615 |
| 2.0 | 0.0 | 6.725 | 6.983 | 7.203 | 7.385 | 7.529 | 7.632 |
| | 0.1 | 6.852 | 7.084 | 7.295 | 7.463 | 7.593 | 7.676 |
| | 0.25 | 6.979 | 7.203 | 7.394 | 7.547 | 7.660 | 7.725 |
| | 0.5 | 7.148 | 7.357 | 7.532 | 7.667 | 7.760 | 7.804 |
| | 1.0 | 7.447 | 7.628 | 7.782 | 7.897 | 7.963 | 7.984 |
| | 2.5 | 8.157 | 8.266 | 8.427 | 8.503 | 8.527 | 8.527 |
| 3.0 | 0.0 | 7.156 | 7.445 | 7.694 | 7.906 | 8.080 | 8.210 |
| | 0.1 | 7.132 | 7.395 | 7.622 | 7.813 | 7.965 | 8.072 |
| | 0.25 | 7.147 | 7.375 | 7.581 | 7.750 | 7.880 | 7.962 |
| | 0.5 | 7.211 | 7.422 | 7.605 | 7.751 | 7.856 | 7.912 |
| | 1.0 | 7.433 | 7.617 | 7.777 | 7.896 | 7.968 | 7.992 |
| | 2.5 | 8.134 | 8.227 | 8.385 | 8.462 | 8.491 | 8.493 |
| 4.0 | 0.0 | 7.560 | 7.872 | 8.145 | 8.382 | 8.583 | 8.734 |
| | 0.1 | 7.375 | 7.642 | 7.885 | 8.091 | 8.260 | 8.385 |
| | 0.25 | 7.258 | 7.497 | 7.714 | 7.892 | 8.033 | 8.127 |
| | 0.5 | 7.245 | 7.462 | 7.651 | 7.803 | 7.919 | 7.983 |
| | 1.0 | 7.442 | 7.609 | 7.772 | 7.894 | 7.971 | 7.995 |
| | 2.5 | 8.086 | 8.194 | 8.361 | 8.440 | 8.470 | 8.471 |
| 5.0 | 0.0 | 7.943 | 8.274 | 8.572 | 8.828 | 9.051 | 9.228 |
| | 0.1 | 7.555 | 7.847 | 8.103 | 8.321 | 8.504 | 8.641 |
| | 0.25 | 7.341 | 7.590 | 7.812 | 7.998 | 8.147 | 8.249 |
| | 0.5 | 7.269 | 7.490 | 7.683 | 7.839 | 7.956 | 8.025 |
| | 1.0 | 7.435 | 7.604 | 7.768 | 7.892 | 7.973 | 8.003 |
| | 2.5 | 8.069 | 8.180 | 8.346 | 8.428 | 8.456 | 8.461 |

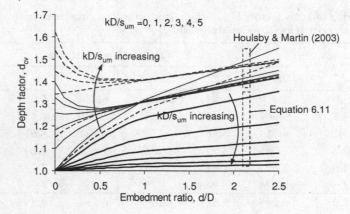

*Figure 6.32* Comparison of rigorously defined and classical, empirically derived, depth factors for shallow foundations

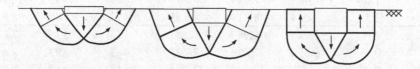

*Figure 6.33* Modes of vertical bearing failure of embedded foundations

Ultimate horizontal capacity of an embedded foundation is governed by a translational scoop mechanism, as opposed to pure sliding, due to coupling of the horizontal and moment degrees of freedom (Figure 6.34). Ultimate horizontal capacity depth factors for rough embedded strip foundations have been shown to be related to the square of the embedment ratio, expressed as (Yun and Bransby 2007b, Gourvenec 2008)

$$d_{cH_{ult}} = 1 + 4.46\frac{d}{B} - 1.52\left(\frac{d}{B}\right)^2 \quad \text{for uniform } s_u \tag{6.29a}$$

$$d_{cH_{ult}} = 1 + 3.01\frac{d}{B} - 1.12\left(\frac{d}{B}\right)^2 \quad \text{for nc } s_u \text{ profile or uniform } s_u \text{ with crack} \tag{6.29b}$$

Less horizontal resistance can be mobilised in a normally consolidated soil compared with that in a soil with uniform shear strength (for the same shear strength at foundation level), as would be expected. The effect of soil strength heterogeneity on horizontal capacity in a broader sense is yet to be addressed. Horizontal capacity depends on the degree of contact that is maintained between the foundation and soil during displacement, which in turn depends on the soil strength profile. A soil with uniform undrained shear strength is likely to be able to support a crack, whereas a softer, normally consolidated soil is less likely to be able to. Interestingly, horizontal capacity depth factors for an embedded foundation, in a soil with a uniform shear strength with a zero-tension foundation–soil interface (i.e. permitting crack development under tensile contact stress), is well described by Equation 6.29b, for an embedded foundation in a normally consolidated deposit.

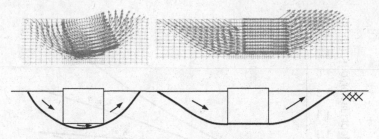

*Figure 6.34* Modes of horizontal failure for embedded foundations

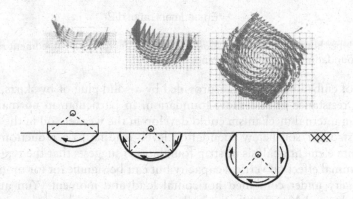

*Figure 6.35* Modes of failure of embedded foundations under pure moment, full tension interface

The ultimate moment capacity of an embedded foundation is governed by a scoop mechanism, similar to that for a surface foundation, with the scoop intersecting the edges of the base of the foundation and the centre of rotation moving towards foundation level with increasing embedment (Figure 6.35). Moment capacity depth factors $d_{cM}$ calculated from an upper bound solution based on a embedded scoop mechanism are shown graphically for a range of linearly increasing soil strength profiles in Figure 6.36 (Bransby and Randolph 1999). From the graph it can be seen, for example, that embedment to a depth equal to half a diameter leads to an 85 per cent increase in moment capacity in a material with a uniform shear strength compared to a 30 per cent increase in a normally consolidated deposit. The moment capacity depth factor reduces with increasing heterogeneity as the mechanism tends to take place in the weaker, shallower material, rather than penetrating to depth.

For uniform shear strength with depth, ultimate moment capacity of a strip foundation can be described by a quadratic expression for the depth factor (Gourvenec 2008)

$$d_{cM_{ult}} = 1 + 1.27\frac{d}{B} + 1.27\left(\frac{d}{B}\right)^2 \tag{6.30}$$

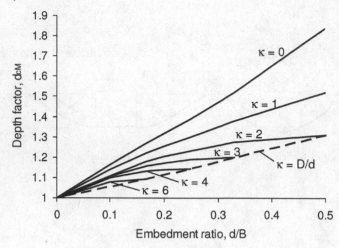

*Figure 6.36* Upper bound prediction of moment capacity factor with embedment ratio for strip foundations (Bransby and Randolph 1999)

The type of embedment, whether provided by a solid plug or by skirts, may affect the ultimate resistance of a shallow foundation. In particular, in normally consolidated soils, an internal mechanism could develop in the weaker soil in the plug rather than in the stronger soil below foundation level, leading to a reduction in bearing capacity. Finite element analysis of strip foundations suggests that the type of embedment has minimal effect on vertical capacity but can be significant for moment capacity and capacity under combined horizontal load and moment (Yun and Bransby 2007a, Bransby and Yun 2009). For soils with a uniform shear strength profile, a Prandtl-type mechanism governs failure and the soil plug tends to move as a rigid block under uniaxial vertical load such that the type of embedment has little effect on bearing capacity. In soils with significant soil strength heterogeneity, such as normally consolidated deposits, a Hill-type mechanism governs failure and a mechanism will tend to develop within the soil plug leading to reduced bearing capacity mobilised across the base of the foundation, albeit limited to a small percentage compared to a solid foundation (Figure 6.37). (Note that only base resistance is shown in Figure 6.37.) Conversely, capacity under pure moment or combined horizontal load and moment can be dramatically reduced due to an internal 'inverted' scoop mechanism developing within the soil plug, the reduction in capacity becoming less significant with increasing embedment ratio (Figure 6.38, Figure 6.39). The reduction in capacity is likely to be less significant for circular rather than strip foundations, since a hemispherical rather than hemi-cylindrical inverted scoop will develop. In practice, internal skirts should be provided to ensure that the soil plug deforms rigidly.

FOUNDATIONS ACTING IN CONSORT    Various offshore structures are founded on multiple shallow foundations such that the kinematic constraint of the structure leads to the foundation elements to act in consort. Examples include jackets permanently supported on suction cans, piled jackets temporarily supported by mudmats, and a variety of subsea infrastructure.

The vertical and horizontal capacity of structurally connected shallow foundation systems is relatively unaffected compared to a single foundation of equivalent bearing

d/D = 0.2

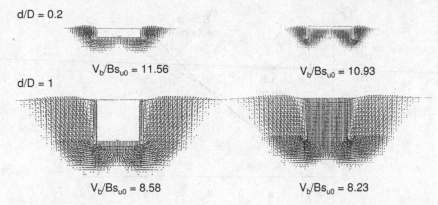

$V_b/Bs_{u0}$ = 11.56          $V_b/Bs_{u0}$ = 10.93

d/D = 1

$V_b/Bs_{u0}$ = 8.58          $V_b/Bs_{u0}$ = 8.23

*Figure 6.37* Base resistance $V_b$ and kinematic mechanisms accompanying failure of solid and skirted shallow foundations in normally consolidated soil (Yun and Bransby 2007a)

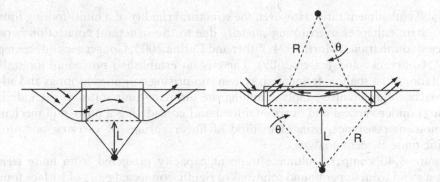

*Figure 6.38* Internal kinematic mechanisms under general loading (Bransby and Yun 2009)

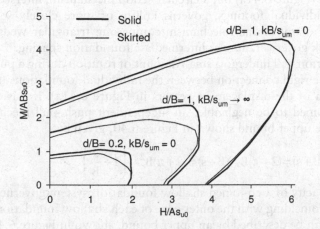

*Figure 6.39* Comparison of failure envelopes for solid and skirted embedded foundations (after Bransby and Yun 2009)

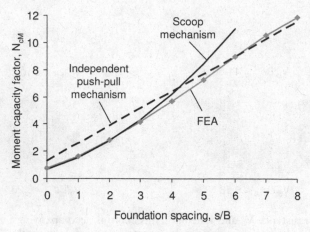

*Figure 6.40* Ultimate moment capacity of co-joined foundations (Gourvenec and Steinepreis 2007)

area and embedment ratio. However, the structural rigidity of a multi-footing foundation system enhances overturning capacity due to the structural connection between adjacent foundations (Murff 1994, Fisher and Cathie 2003, Gourvenec and Steinepreis 2007, Gourvenec and Jensen 2009). There is no established procedure for stability calculations of a shallow foundation system comprising co-joined footings and ad-hoc approaches may involve simply summing the ultimate limit states of the individual footings under vertical and horizontal load and considering a push–pull mechanism for moment resistance, using a method of linear springs or carrying out project-specific finite element analysis.

Figure 6.40 compares ultimate moment capacity predicted from finite element analyses and from upper bound solutions of rigidly connected pairs of surface foundations, with full-tensile foundation-soil interface, as a function of the normalised separation between foundations s/B. Under pure moment, opposing foundations of a co-joined system fail in uniaxial compression and tension when separated beyond a critical distance (Figure 6.41a) but a circular scoop mechanism, intersecting the outer edges of the individual footings, governs failure of more closely spaced footings (Figure 6.41b). A transitional mechanism comprising triangular wedges flanking a translating block governs failure at intermediate foundation spacing.

Each foundation will undergo a small amount of rotation during a push–pull mechanism due to the rigid connection between the individual foundations (as can be seen by the direction of the displacement vectors in Figure 6.41a). If this component of rotation is assumed to be negligible, an independent push–pull mechanism can be described by the upper bound shown in Figure 6.40, given by

$$M_{ult} = V_{ult}(B+s) = (2+\pi)Bs_u(B+s) = (2+\pi)B^2s_u\left(1+\frac{s}{B}\right) \tag{6.31}$$

Moment capacity of a co-joined shallow foundation system governed by a perfect circular scoop coinciding with the outer edge of each shallow foundation (as shown in Figure 6.41b) can be described by an upper bound, shown in Figure 6.40, given by

$$M_{ult} = 0.69(2B+s)^2 s_u = 0.69B^2\left(4+\frac{4s}{B}+\left(\frac{s}{B}\right)^2\right)s_u \tag{6.32}$$

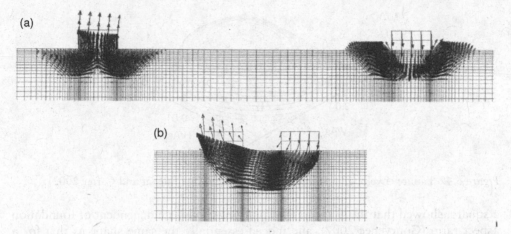

*Figure 6.41* Failure mechanisms of rigidly connected foundations under pure moment (Gourvenec and Steinepreis 2007) (a) independent push–pull mechanism and (b) scoop mechanism

Figure 6.40 indicates that failure of rigidly connected pairs of surface strip foundations is governed by an independent push–pull mechanism for separations s/B = 4 – 6 potentially an upper limit of foundation spacing encountered in reality. It should be borne in mind that circular foundation geometry would tend to reduce the critical separation (due to a less extensive failure mechanism in axi-symmetry compared to plane strain) while foundation embedment will increase the zone of interaction of failure mechanisms and, therefore, the critical distance for an independent push–pull mechanism to govern failure. The independent push–pull mechanism predicts higher moment capacity than that predicted by the finite element analyses for low spacing, such that the assumption of push–pull governing failure may be unconservative (i.e. unsafe).

Shallow foundations of subsea frames and small tripods or quadrupeds may be assumed to act in a quasi-rigid manner. An effectively rigid response is less likely for shallow foundations systems of larger structures, for example, suction cans on a jacket structure, where a certain flexibility will be inherent in the system. Nonetheless, some degree of structural connectivity will enhance the overturning capacity of even these larger structures. When assessing the stability of structurally connected shallow foundation systems, specific consideration should be given to the degree of structural connection, foundation shape, embedment ratio and foundation arrangement.

FAILURE ENVELOPES FOR UNDRAINED CONDITIONS

SURFACE FOUNDATION WITH A ZERO-TENSION INTERFACE Figure 6.42 shows a three-dimensional failure envelope for undrained limit states under general loading of a surface circular foundation on a deposit with uniform undrained shear strength with depth. The envelope was derived from finite element analyses that explicitly represented individual components of vertical load, horizontal load and moment and the results are presented non-dimensionalised by the foundation area and undrained shear strength. Finite element analyses of rectangular foundations with breadth to length aspect ratios 0 ≤ B/L ≤ 1, i.e. covering the geometric spectrum from an infinite strip to

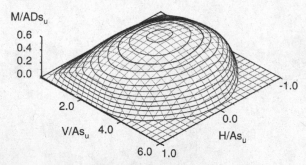

*Figure 6.42* Failure envelope for surface circular foundation (Taiebat and Carter 2002)

a square, showed that the shape of the failure envelope is independent of foundation aspect ratio (Gourvenec 2007), and indeed essentially the same shape as that for a circular foundation (Taiebat and Carter 2010), such that a unique normalised failure envelope scaled by the appropriate ultimate vertical, horizontal and moment limits ($V_{ult}$, $H_{ult}$ and $M_{ult}$) can be used to determine the limit state under general loading for a surface circular or rectangular foundation of any aspect ratio. To avoid having to perform involved calculations on a case by case basis, it is desirable to identify a simple closed-form function to approximate the shape of the normalised failure envelope. An envelope of the form shown in Figure 6.42 can be described by a circular ellipse in terms of general loads normalised by their respective ultimate limit states, i.e. $v = V/V_{ult}$, $h = H/H_{ult}$ and $m = M/M_{ult}$ (Gourvenec 2007)

$$\left(\frac{h}{h*}\right)^2 + \left(\frac{m}{m*}\right)^2 = 1 \tag{6.33}$$

where h* and m* are functions of the normalised vertical load v and are derived from the shape of the failure envelopes in normalised vh and vm space, respectively.

$$h* = 4(v - v^2) \quad \text{for} \quad 0.5 < v < 1. \quad \text{For} \quad 0 < v < 0.5, h^* = 1 \tag{6.34a}$$

$$m* = 4(v - v^2) \tag{6.34b}$$

Equation 6.33 was derived for a uniform shear strength profile and further work would be required to quantify the effect of soil strength heterogeneity on the shape of the normalised failure envelope.

A simple zero-tension interface condition allows comparison of limit states predicted by explicitly modelling individual load components with those predicted by the classical approach (as set out in Equation 6.7). Failure envelopes can be derived from classical theory by solving for ultimate moment at specified intervals of horizontal load and constant ratio of vertical load to ultimate vertical load. Figure 6.43 shows a comparison of limit states in a plane of horizontal load and moment at a constant vertical load of $0.5V_{ult}$. The failure envelope based on classical bearing capacity theory exhibits a quasi-linear relationship between horizontal load and moment in contrast to the curvilinear relationship observed when horizontal load and moment are explicitly considered. The quasi-linear profile arises from the superposition of the separate solutions for load inclination and load eccentricity. The potential conservatism in bearing capacity predicted by the classical approach is clearly apparent in Figure 6.43.

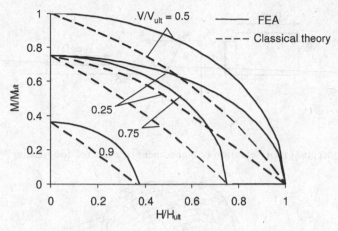

*Figure 6.43* Comparison of ultimate limit states under general VHM loading from classical bearing capacity theory and finite element analysis

SURFACE FOUNDATION WITH FULL-TENSION INTERFACE   The tensile capacity provided by foundation skirts has often been represented by modelling a surface foundation with a full-tension interface (e.g. Tani and Craig 1995, Bransby and Randolph 1998, Taiebat and Carter 2000, Gourvenec and Randolph 2003). This representation is appropriate for shallowly skirted foundations where the effect of the physical embedment on the uniaxial ultimate limit states $V_{ult}$, $H_{ult}$ and $M_{ult}$ is to be overlooked (e.g. due to disturbance during installation or scour) but the benefit of the passive suction developed within the skirted compartment and any increase in shear strength with depth is to be taken advantage of in design.

The conceptual representation of embedment is illustrated in Figure 6.44. The embedded foundation is represented as a surface foundation with a full-tension interface resting on a reduced seabed, set at foundation level of the embedded foundation. A three-dimensional failure envelope for a surface foundation with a full-tension interface is illustrated in Figure 6.45a. The full-tension interface allows moment capacity to be mobilised at low vertical loads; indeed the maximum moment capacity is mobilised in conjunction with zero vertical load (as opposed to peak moment capacity being mobilised in conjunction with a vertical load of $0.5V_{ult}$ with a zero-tension interface, as seen in Figure 6.42). The envelope is asymmetric, seen more clearly in Figure 6.46, due to the different modes of failure when horizontal load and moment act in the same direction or in opposition to each other. Additional moment capacity is available when horizontal load and moment act in the same direction with a maximum moment (in excess of the ultimate moment capacity) mobilised in conjunction with a horizontal load. A closed-form expression is available to describe the failure envelope for a circular foundation with a full-tension interface resting on a deposit with uniform shear strength as shown in Figure 6.45 (Taiebat and Carter 2000):

$$f = \left(\frac{V}{V_{ult}}\right)^2 + \left[\left(\frac{M^*}{M_{ult}}\right)\left(1 - 0.3\frac{HM}{H_{ult}M}\right)^2\right] + \left(\frac{H}{H_{ult}}\right)^3 - 1 \qquad (6.35)$$

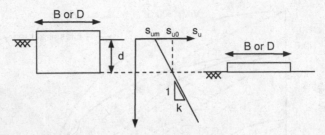

*Figure 6.44* Conceptual representation of embedment of a skirted foundation

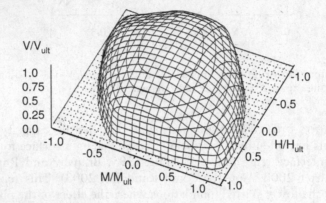

*Figure 6.45* Failure envelope for surface circular foundation with full-tension interface (Taiebat and Carter 2000)

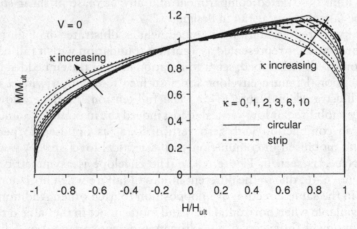

*Figure 6.46* Effect of foundation shape and soil strength heterogeneity on failure envelopes (Gourvenec and Randolph 2003)

A closed-form expression is also available to describe the failure envelope for a strip foundation with a full-tension interface resting on a deposit with a linearly increasing undrained shear strength with $\kappa = 6$ (Bransby and Randolph 1998).

$$f = \left(\frac{V}{V_{ult}}\right)^{2.5} - \left(1 - \frac{H}{H_{ult}}\right)^{\frac{1}{3}}\left(1 - \frac{M^*}{M_{ult}}\right) + \frac{1}{2}\left(\frac{M^*}{M_{ult}}\right)\left(\frac{H}{H_{ult}}\right)^5 \tag{6.36}$$

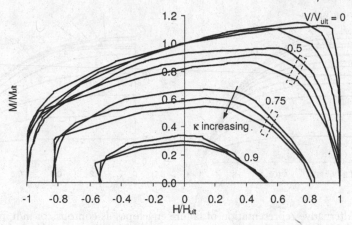

*Figure 6.47* Effect of soil strength heterogeneity on failure envelopes for circular foundations with full-tension interface

where M* is a modified moment parameter given by the expression

$$\frac{M^*}{ABs_{u(tip)}} = \frac{M}{ABs_{u(tip)}} - \frac{h_s}{B}\frac{H}{As_{u(tip)}}$$ (6.37)

where $h_s$ is the height of the rotation of the scoop mechanism above foundation level. Equation 6.36 was derived for a specific degree of heterogeneity ($\kappa = 6$) and re-optimisation of the exponents is required for application of the expression to other degrees of heterogeneity.

It is important to choose a failure envelope appropriate to the field conditions being designed for as the shape of the failure envelope depends on the mode of horizontal and moment loading, the level of vertical load, foundation geometry and the degree of soil strength heterogeneity (Gourvenec and Randolph 2003). Figure 6.46 shows failure envelopes of normalised horizontal and moment capacity for strip and circular foundations with a full-tension interface for degrees of heterogeneity 0 (homogeneous) $\leq \kappa \leq 10$. The normalised size of the failure envelope generally reduces in size with increasing degree of heterogeneity for the most common combination of horizontal load and moment acting in the same direction (H, M of the same sign in the graph) although the normalised failure envelope for a uniform soil strength profile (shown in bold in Figure 6.46) is slightly smaller than those for low heterogeneities. It would, therefore, be unconservative to use a failure envelope derived for homogeneous conditions scaled by ultimate limit states for heterogeneous materials with $\kappa > 2$.

The shape of the normalised failure envelope also depends on the level of vertical load as seen in Figure 6.47 for circular foundations on deposits with heterogeneity factors $\kappa = 0$, 2 and 6, further increasing the challenge to identify a versatile closed-form expression to describe the failure envelope as a function of v, h and m. An alternative approach is to represent combinations of load to cause failure as contours of normalised moment to horizontal load m/h in normalised vertical and moment load space (Figure 6.48). These curves can then be used to identify ultimate limit states under combinations of vertical, horizontal and moment load to construct

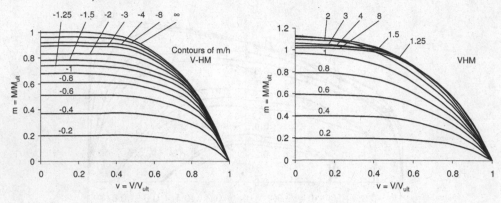

*Figure 6.48* Alternative representation of failure envelopes as contours of m/h in (v,m) space. Example shown for a circular foundation with full-tension interface on a uniform deposit (Gourvenec 2007b)

any plane or part of a failure envelope containing the range of loading relevant to design or to reconstruct a complete three-dimensional failure envelope with a minimal compromise on accuracy. The benefit of expressing the ultimate limit states in terms of constant m/h is that the horizontal load H, due to environmental loading, is a known quantity, as is the height of the line of action of H above the load reference point $L_H$. Given $M = HL_H$ and the theoretical solutions for $M_{ult}$ and $H_{ult}$, the normalised m/h ratio can be determined. The curves in vm space can be described by simple polynomial functions, cubic for combinations of VHM and quartic for combinations of V-HM. The polynomial coefficients for heterogeneity coefficients $\kappa = 0$, 2 and 6 (corresponding to the failure envelopes shown in Figure 6.47) are given in Table 6.5.

EMBEDDED FOUNDATION WITH A FULL-TENSION INTERFACE    When a pure horizontal load is applied at the base of an embedded foundation, the resulting displacement will involve rotation as well as translation, as opposed to a surface foundation, which will simply translate without rotation. Likewise, if a pure moment is applied at the base of an embedded foundation the resulting displacement will involve translation as well as rotation. If translation or rotation can be constrained, additional moment and horizontal capacity can be mobilised. The coupling of the horizontal and moment degrees of freedom of an embedded foundation is reflected in the shape of the failure envelope compared to that for a surface foundation. The coupling leads to an oblique failure envelope, the effect becoming more pronounced with increasing embedment ratio as illustrated in Figure 6.49. The shape of the envelope depends on the magnitude of vertical load and the embedment ratio of the foundation, making fitting a closed-form expression a considerable challenge. One possible solution is a rotational transformation to reduce the eccentricity of the failure envelope (Yun and Bransby 2007b) such that

$$f = \left(\frac{H}{H_{max}}\right)^2 + \left(\frac{M^*}{M_{ult}}\right)^2 - 1 \tag{6.38}$$

*Table 6.5* Polynomial coefficients for reconstruction of failure envelopes in (v, h, m) space for (a) $\kappa = 0$ (b) $\kappa = 2$ (c) $\kappa = 6$. VHM, $m = c_1 v^3 + c_2 v^2 + c_3 v + c_4$; V-HM $m = c_1 v^4 + c_2 v^3 + c_4 v + c_5$ (Gourvenec 2007b)

| | | *(a1)* $\kappa = 0$, *VHM* | | |
|---|---|---|---|---|
| m/h | $c_1$ | $c_2$ | $c_3$ | $c_4$ |
| 0.2 | −0.6771 | 0.6196 | −0.1399 | 0.2014 |
| 0.4 | −0.9013 | 0.6257 | −0.1210 | 0.4011 |
| 0.6 | −1.0205 | 0.5429 | −0.1182 | 0.6004 |
| 0.8 | −0.8119 | 0.0420 | −0.0155 | 0.7909 |
| 1 | −0.7330 | −0.1992 | −0.0351 | 0.9706 |
| 1.25 | −0.7196 | −0.2893 | −0.1193 | 1.1319 |
| 1.5 | −0.9266 | −0.2022 | 0.0065 | 1.1190 |
| 2 | −1.3989 | 0.4425 | −0.1465 | 1.1001 |
| 3 | −1.6475 | 0.7689 | −0.1781 | 1.0605 |
| 4 | −1.6645 | 0.7688 | −0.1398 | 1.0407 |
| 8 | −1.7847 | 0.9597 | −0.1896 | 1.0212 |

| | | *(a2)* $\kappa = 0$, *V-HM* | | | |
|---|---|---|---|---|---|
| m/h | $c_1$ | $c_2$ | $c_3$ | $c_4$ | $c_5$ |
| ∞ | −1.1320 | 0.5228 | −0.5062 | 0.1179 | 0.9998 |
| −8 | −0.9186 | −0.0257 | −0.0496 | 0.0374 | 0.9597 |
| −4 | −0.9861 | 0.0128 | 0.0457 | 0.0104 | 0.9197 |
| −3 | −0.6575 | −0.5830 | 0.4003 | −0.0477 | 0.8898 |
| −2 | −1.3377 | 0.5988 | −0.0990 | 0.0076 | 0.8300 |
| −1.5 | −2.4101 | 2.7125 | −1.2635 | 0.1793 | 0.7802 |
| −1.25 | −2.5806 | 3.1482 | −1.5328 | 0.2243 | 0.7401 |
| −1 | −3.0500 | 4.1755 | −2.1236 | 0.3184 | 0.6800 |
| −0.8 | −2.7243 | 3.6245 | −1.7779 | 0.2722 | 0.6096 |
| −0.6 | −2.6204 | 3.5688 | −1.7126 | 0.2591 | 0.5095 |
| −0.4 | −2.3486 | 3.2960 | −1.5411 | 0.2283 | 0.3695 |
| −0.2 | −1.0971 | 1.5069 | −0.7175 | 0.1109 | 0.1997 |

| | | *(b1)* $\kappa = 2$, *VHM* | | |
|---|---|---|---|---|
| m/h | $c_1$ | $c_2$ | $c_3$ | $c_4$ |
| 0.2 | −0.6339 | 0.5616 | −0.1244 | 0.2013 |
| 0.4 | −0.8313 | 0.5589 | −0.1254 | 0.4009 |
| 0.6 | −0.8804 | 0.4093 | −0.1270 | 0.6000 |
| 0.8 | −0.8164 | 0.1578 | −0.1165 | 0.7806 |
| 1 | −0.6093 | −0.2665 | −0.0589 | 0.9403 |
| 1.25 | −0.5721 | −0.3835 | −0.1218 | 1.0808 |
| 1.5 | −0.4782 | −0.6514 | −0.0136 | 1.1444 |

*(continued)*

### (b1) κ = 2, VHM (Continued)

| m/h | $c_1$ | $c_2$ | $c_3$ | $c_4$ |
|---|---|---|---|---|
| 2 | −0.7836 | −0.2721 | −0.0618 | 1.1210 |
| 3 | −1.0845 | 0.1916 | −0.1989 | 1.0970 |
| 4 | −1.1300 | 0.2618 | −0.2074 | 1.0828 |
| 8 | −1.1345 | 0.2347 | −0.1355 | 1.0425 |

### (b2) κ = 2, V-HM

| m/h | $c_1$ | $c_2$ | $c_3$ | $c_4$ | $c_5$ |
|---|---|---|---|---|---|
| ∞ | −1.5647 | 1.9743 | −1.6461 | 0.239 | 0.9998 |
| −8 | −0.9457 | 0.5598 | −0.6396 | 0.0691 | 0.9596 |
| −4 | −0.844 | 0.272 | −0.3566 | 0.0107 | 0.9198 |
| −3 | −0.7176 | −0.0328 | −0.1013 | −0.0363 | 0.8898 |
| −2 | −0.9789 | 0.349 | −0.1661 | −0.0434 | 0.8399 |
| −1.5 | −1.1477 | 0.4913 | −0.0131 | −0.1312 | 0.8001 |
| −1.25 | −1.5675 | 1.484 | −0.7439 | 0.0773 | 0.75 |
| −1 | −1.6634 | 1.684 | −0.8402 | 0.1289 | 0.6901 |
| −0.8 | −1.7405 | 1.7795 | −0.7659 | 0.1065 | 0.62 |
| −0.6 | −1.6893 | 1.719 | −0.6391 | 0.0799 | 0.53 |
| −0.4 | −1.567 | 1.742 | −0.6284 | 0.0736 | 0.38 |
| −0.2 | −1.2662 | 1.8333 | −0.9053 | 0.1423 | 0.1996 |

### (c1) κ = 6, VHM

| m/h | $c_1$ | $c_2$ | $c_3$ | $c_4$ |
|---|---|---|---|---|
| 0.2 | −0.5929 | 0.5186 | −0.1216 | 0.2016 |
| 0.4 | −1.016 | 0.8557 | −0.2358 | 0.4004 |
| 0.6 | −0.8939 | 0.4423 | −0.1256 | 0.5796 |
| 0.8 | −0.7449 | 0.0657 | −0.0535 | 0.7404 |
| 1 | −0.6168 | −0.1974 | −0.0601 | 0.8796 |
| 1.25 | −0.4171 | −0.4953 | −0.0808 | 1.0017 |
| 1.5 | −0.5557 | −0.3469 | −0.1421 | 1.0513 |
| 2 | −0.8909 | 0.1175 | −0.324 | 1.0996 |
| 3 | −0.6622 | −0.2672 | −0.1554 | 1.0913 |
| 4 | −0.6361 | −0.3148 | −0.1019 | 1.0632 |
| 8 | −0.8819 | 0.0517 | −0.194 | 1.032 |

### (c2) κ = 6, V-HM

| m/h | $c_1$ | $c_2$ | $c_3$ | $c_4$ | $c_5$ |
|---|---|---|---|---|---|
| ∞ | −1.1634 | 1.5282 | −1.5111 | 0.1499 | 0.9996 |
| −8 | −1.2194 | 1.4839 | −1.3283 | 0.1072 | 0.9597 |
| −4 | −1.1375 | 1.2265 | −1.0722 | 0.0555 | 0.9298 |
| −3 | −0.7842 | 0.3928 | −0.4327 | −0.0788 | 0.9048 |

| -2 | -1.0758 | 0.7417 | -0.3587 | -0.1757 | 0.8698 |
| -1.5 | -0.8427 | 0.1627 | 0.1249 | -0.2629 | 0.81 98 |
| -1.25 | -0.9138 | 0.3458 | 0.0009 | -0.1999 | 0.7697 |
| -1 | -0.8282 | 0.2114 | 0.1207 | -0.2198 | 0.7196 |
| -0.8 | -0.9241 | 0.4114 | 0.0244 | -0.1683 | 0.6597 |
| -0.6 | -0.9132 | 0.6038 | -0.2836 | 0.0564 | 0.5397 |
| -0.4 | -1.3731 | 1.3987 | -0.4684 | 0.0537 | 0.3899 |
| -0.2 | -1.6045 | 2.4861 | -1.2808 | 0.205 | 0.1994 |

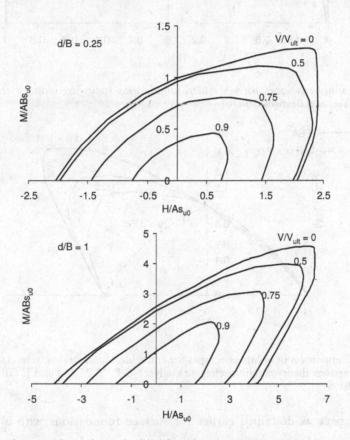

*Figure 6.49* Failure envelopes for embedded foundations (Gourvenec 2008)

where

$$M^* = M - d^* H \tag{6.39}$$

where $d^*$ is the translation reference point, taken as half the depth of embedment.

The result of this transformation is shown for failure envelopes for a deposit with uniform shear strength profile in Figure 6.50. The fit is not exact, and is less so for a normally consolidated shear strength profile. Nonetheless, the approach provides a practical lower limit to capacity. An alternative for generalising the failure envelopes for embedded foundations is to re-express the limit states as polynomials of

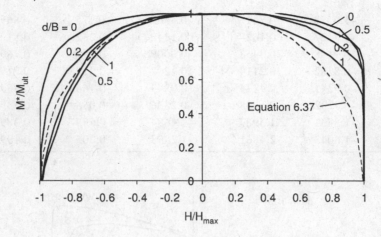

*Figure 6.50* Failure envelopes for embedded foundations following moment transformation (Yun and Bransby 2007b)

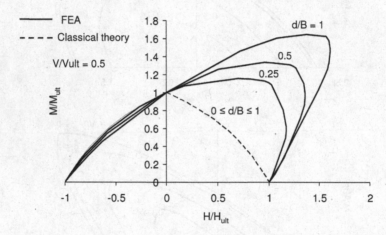

*Figure 6.51* Comparison of failure envelopes for embedded foundations from classical bearing capacity theory and finite element analyses (d/B = 0.25, 0.5 and 1) (after Gourvenec 2008)

m/h in vm space as described earlier for surface foundations with a full-tension interface.

When considering the apparent lack-of-fit of some of the generalising expressions summarised in the foregoing section, it is reassuring to bear in mind the conservatism in estimated bearing capacity incurred by adopting classical bearing capacity theory, as illustrated in Figure 6.51. For combinations of horizontal load and moment acting in opposition (a less common field condition) classical bearing capacity provides an adequate prediction of limit states. However, for the common case of horizontal load and moment acting in the same direction, the coupling of the horizontal and moment degrees of freedom is not represented in classical theory and considerable bearing capacity is overlooked. It should also be borne in mind that classical bearing capacity theory does not account for tension capacity, in reality provided by passive suctions developed in the skirt compartment during undrained loading, and hence is increasingly conservative in respect of moment capacity as vertical load diminishes below $0.5V_{ult}$.

FOUNDATIONS ACTING IN CONSORT    Prediction of ultimate limit states under general loading for foundations acting in consort can be made considering the failure envelope of the individual foundations. The approach takes advantage of normality (and is therefore, limited to consideration of undrained failure). In the case of a foundation, normality means that the direction of movement of the foundation at failure is perpendicular to the failure surface defined in load space when using conjugate displacement axes. If the structure joining multiple foundations is considered rigid, then any assumed failure mechanism for the structure defines the trajectories of each foundation in the system and, therefore, using normality, the failure load for that mechanism. All failure mechanisms for a rigid structure can be specified via a point of rotation – the special case of pure translation involves rotation about a point at infinity. The critical failure mechanism, i.e. the minimum collapse load, can then be assessed via a search routine for the critical point of rotation. An elegant iterative solution for this task is presented by Murff (1994).

The method outlined in the preceding section essentially stitches together multiple failure envelopes for individual shallow foundations, of the kind described in this chapter, in order to create a single failure envelope for a rigid structure supported by multiple shallow foundations. A limitation of this approach is that interaction between the individual footings is not considered and, therefore, this type of approach would be unconservative for conditions where the individual failure mechanisms of the foundations interacted. Finite element analysis explicitly modelling two connected foundations overcomes this issue. Figure 6.52 shows failure envelopes for a single strip foundation and for a rigidly connected two-footing system at a spacing of 3B, indicating the variation in capacity due to the interaction of the individual foundations.

DRAINED UNIAXIAL LIMIT STATES    Bearing capacity factors for limit states of strip and circular foundations under vertical loading on sand may be obtained to a very high accuracy, for a given friction angle of the soil, using numerical implementations of the method of characteristics, such as the freeware ABC (Martin 2003, http://www-civil.eng.ox.ac.uk/people/cmm/software/abc). The greatest source of uncertainty lies in the appropriate choice of friction angle (discussed further in Chapter 4). Numerical implementations of the method of characteristics: (i) obviate the need to rely on tabulated or curve-fitted values of the bearing capacity factor $N_\gamma$, therefore, eliminating inaccuracies incurred by interpolation; (ii) explicitly account for self-weight and overburden as opposed to assuming simple superposition; and (iii) explicitly account for foundation geometry as opposed to relying on multiplicative shape factors to modify plane strain solutions.

Limit analyses, such as the method of characteristics and other theoretical plasticity solutions, are based on an associated flow rule, i.e. dilation angle $\psi$ equal to effective friction angle $\phi$. In reality, dilation angle affects the ultimate bearing capacity and the settlement required to mobilise bearing capacity. Studies have shown that dilation angle has little effect for low friction angles but starts to become significant for friction angles of 35° or more (Frydman and Burd 1997, Erickson and Drescher 2002) with values of $N_\gamma$ for realistic dilation angles being lower than those predicted by rigorous plasticity solutions based on associated flow. In reality, the effect may

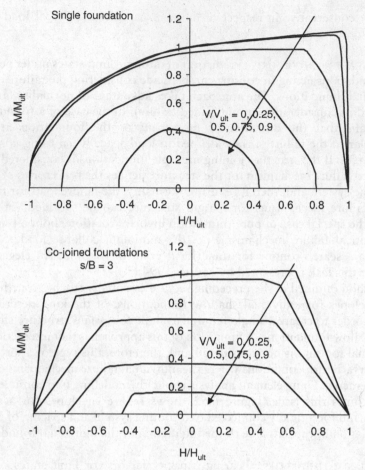

*Figure 6.52* Failure envelopes for a single foundation and a rigid two-footing system with full tension interface (Gourvenec and Steinepreis 2007)

not be very significant as the effect of soil compressibility starts to dominate at high friction angles and theoretical values of ultimate bearing capacity for friction angles in excess of 35° are unlikely to be achievable in practice. For a moderate friction angle $\phi = 25°$, the settlement to mobilise bearing capacity is more than doubled for a dilation angle of zero compared with the case of associated flow (Potts and Zdravkovic 2001).

YIELD ENVELOPE FOR DRAINED CONDITIONS

The term yield envelope rather than failure envelope is commonly adopted for drained bearing capacity. Yield and failure are synonymous for the undrained limit state of a rigid plastic material while the stress dependency of shear strength for a frictional material leads to gradual yield and hardening with increased displacement. The general form of yield envelope for drained bearing capacity of surface foundations comprises a system of inclined ellipses in (H, M) space and intersecting parabolas in (V, H) and (V, M) space. The three-dimensional yield envelope in (V, H, M) space is shown

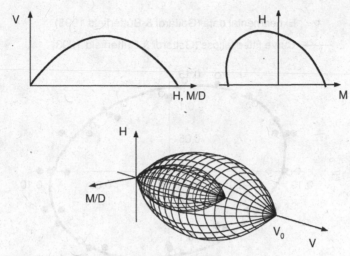

*Figure 6.53* Hardening yield envelope for drained failure under general loading

in Figure 6.53 and can be described by a closed-form expression in terms of normalised loads given by (Gottardi *et al.* 1999)

$$f = \left(\frac{m_n}{m_0}\right)^2 + \left(\frac{h_n}{h_0}\right)^2 - 2a\left(\frac{h_n m_n}{h_0 m_0}\right) - 1 \tag{6.40}$$

where

$$m_n = \frac{M/DV_0}{4v(1-v)} \tag{6.41a}$$

$$h_n = \frac{H/V_0}{4v(1-v)} \tag{6.41b}$$

$$v = \frac{V}{V_0} \tag{6.41c}$$

where $V_0$ is the uniaxial vertical yield load. The subscript '0' is used, rather than 'ult', to emphasise the work-hardening aspect of foundations in sand, with $m_0 V_0$ and $h_0 V_0$ representing the moment and horizontal load to cause yield for the current (uniaxial) yield load $V_0$. Experimental studies have led to parameter values of $m_0 = 0.09$, $h_0 = 0.12$ and $a = -0.22$ (Gottardi and Butterfield 1993). A hardening rule defines the increase in $V_0$ with increasing penetration, which can be determined from the vertical load–displacement relationship.

Other experimental studies have confirmed the general shape of the yield envelope shown in Figure 6.53 and described by Equation 6.40 for surface foundations for different plan geometry. Most experimental studies have been carried out at 1*g* (Nova and Monstrasio 1991, Gottardi and Butterfield 1993, 1995, Byrne and Houlsby 2001, Bienen *et al.* 2006) although some centrifuge tests have been carried out (Tan 1990,

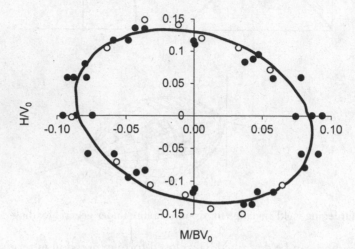

*Figure 6.54* Comparison of finite element analysis results with experimentally derived yield envelope (Zdravkovic *et al.* 2002)

Dean *et al.* 1992, Govoni *et al.* 2006, Cassidy 2007, Govoni *et al.* 2010). The effect of embedment on drained bearing capacity under general loading has not yet been comprehensively addressed.

Historically, it has been assumed that small-scale 1g tests produce a response, that is generally similar to that of the prototype. The applicability of small-scale 1g model tests on frictional materials to replicate prototype conditions is questionable owing to scale effects (since angles of friction and soil stiffness are stress-level dependent). For a shallow foundation on sand, the vertical bearing resistance and the form of the vertical load–displacement response are significantly affected by scale (Ovesen 1975, Kimura *et al.* 1985, Zhu *et al.* 2001) and similar scale effects may influence the response to general loading. Scale effects are potentially more significant if foundations are embedded in a frictional material as opposed to resting on the soil surface. Thus, when considering embedded foundations a properly scaled self-weight stress distribution is crucial. Experimental determination of a three-dimensional yield surface is a time-consuming and formidable task. The issue is further complicated when experiments are carried out in the centrifuge as limited space inhibits mounting a loading apparatus that can independently control vertical load, horizontal load and moment. Numerical determination of a yield surface provides an attractive alternative if it can be shown to accurately replicate experimental results.

Small strain finite element analysis results of surface foundations on a weightless soil with friction angle $\phi = 25°$ and dilation angle $\psi = 0°$ (Zdravkovic *et al.* 2002) show remarkably close agreement with Equation 6.40, given the numerical analysis modelled a weightless, non-dilant, non-hardening soil, in stark contrast to the physical characteristics of the dense sand used in the experiments on which Equation 6.40 is based (Figure 6.54). Figure 6.55 shows good agreement between large strain finite

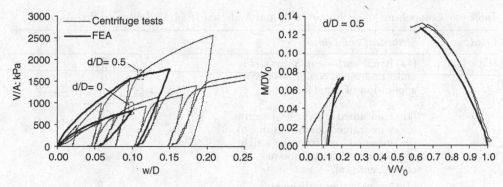

*Figure 6.55* Comparison of large-strain finite element analysis results with centrifuge test data (Gourvenec *et al.* 2008b)

element analyses, representing the hardening behaviour of real sand, compared with centrifuge model test results of vertical penetration and swipe tests for surface and embedded foundations (Gourvenec *et al.* 2008b).

The application of a work-hardening yield envelope within a complete plasticity model is discussed in more detail in Chapter 8 with respect to design and performance of spudcan foundations for mobile drilling rigs. Pre-loading of foundations for mobile drilling rigs leads to larger displacements than typically encountered for shallow foundations for permanent structures.

## 6.4 Serviceability

### 6.4.1 *Introduction*

Serviceability assessment requires prediction of vertical, horizontal and rotational movements under static and cyclic loads over the lifetime of the structure. Simple methods of predicting displacements based on elastic theory can be used to provide a rough estimate of settlement while numerical analysis, perhaps complemented by centrifuge testing, is required for more accurate representation of the soil profile, soil behaviour, foundation geometry and loading conditions.

In this section, an overview of different types of settlement is presented followed by a summary of simple, established elastic solutions. The simple solutions presented here underpin advanced analyses, providing a framework for assessing their validity. The theory behind the approaches to settlement prediction presented in this section is presented in more detail in Chapter 4.

### 6.4.2 *Types of deformation*

Settlement, i.e. vertical displacement, of an offshore shallow foundations will generally comprise an immediate, undrained component, a component due to consolidation and an accumulated, permanent component due to cyclic loading. Table 6.6 shows a detailed sub-division of these components. The immediate component of settlement is that portion of the settlement that occurs essentially with load application, primarily as a consequence of distortion in the foundation soils. The distortion

*Table 6.6* Components of settlement (Eide and Andersen 1984; O'Reilly and Brown, 1991)

| Load | Settlement component | |
|------|---------------------|---|
| Static load | (1a) Initial settlement: Shear strain under undrained conditions due to application of static load) | |
| | (1b) Undrained creep: Shear strains under undrained conditions due to the sustained load from the weight of platform. Continuation of component (1a). | $\Delta$ vol = 0 |
| | (2) Consolidation settlement: Volumetric strain due to pore pressure dissipation under weight of platform. | $\Delta$ vol > 0 |
| | (3) Secondary settlement: Volumetric and shear strains under drained conditions and constant effective stresses. | $\Delta$ vol > 0 |
| Cyclic load | (4a) Local plastic yielding and redistribution of stresses during cyclic loading (undrained). | $\Delta$ vol = 0 |
| | (4b) Shear strains caused by cyclically induced excess pore pressure and the corresponding reductions in effective stress and soil stiffness (undrained). | $\Delta u$ > 0 $\Delta$vol = 0 |
| | (5) Volumetric strains due to dissipation of the cyclically induced excess pore pressures. | $\Delta u$ > 0 $\Delta$vol = 0 |

settlement is generally not elastic, although it is often calculated using elastic theory. Consolidation settlement results from the gradual expulsion of water from the voids and the concurrent compression of the soil skeleton. The distinction between primary and secondary consolidation settlements is made on the basis of the physical processes that control the time rate of settlement. In primary consolidation, the time rate of settlement is controlled by the rate at which water can be expelled from the void spaces in the soil. During secondary consolidation, or creep, the rate of settlement is controlled largely by the rate at which the soil skeleton itself yields. Horizontal and rotational movements will generally only involve an immediate component, although some

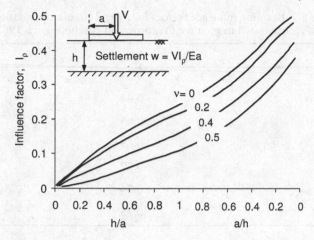

*Figure 6.56* Settlement influence factors for a rigid circular foundation under vertical load (Poulos and Davis 1974)

additional component due to consolidation may also develop in permeable soils and permanent horizontal displacements may accumulate due to preferred wind, current and wave directions (Poulos 1988).

### 6.4.3 *Estimation of deformation*

*Elastic solutions*

Preliminary estimates of movements of a shallow foundation are typically calculated based on elastic theory. A range of elastic solutions are available for the deformation of a rigid circular foundation (Poulos and Davis 1974) and a selection are summarised in the following section.

For a homogeneous elastic soil layer, the vertical displacement w of a rigid circular foundation due to a vertical load V, the rotation $\theta$ due to an applied moment M and the horizontal displacement u due to horizontal loading H are given by

$$w = \frac{VI_\rho}{Ea} \tag{6.42}$$

where

$I_\rho$ = settlement influence factor defined as a function of the normalised layer depth h/a and Poisson's ratio v (Figure 6.56)
E = Young's modulus of the soil
a = radius of the foundation

$$\theta = \frac{M(1-v^2)I_\theta}{E(a^3)} \tag{6.43}$$

*Table 6.7* Rotation influence factors for a rigid circular foundation subjected to moment (Yegorov and Nitchporovich 1961)

| $h/a$ | $I_\theta$ |
|-------|------------|
| 0.25  | 0.27 |
| 0.5   | 0.44 |
| 1.0   | 0.63 |
| 1.5   | 0.69 |
| 2.0   | 0.72 |
| 3.0   | 0.74 |
| $\geq 5.0$ | 0.75 |

where

$I_\theta$ = rotation influence factor defined as a function of the normalised layer depth $h/a$ (Table 6.7)

$$u = \frac{H(7-8v)(1+v)}{16(1-v)Ea} \tag{6.44}$$

The solution for the horizontal displacement u due to horizontal loading H is derived for an infinitely deep layer. However, the depth of the soil layer has limited influence on horizontal displacement, considerably less so than for vertical displacement.

For a heterogeneous elastic soil layer, the soil modulus can be taken to vary according to a power law with depth governed by the exponent $\alpha$

$$G_{(z)} = G_D \left( \frac{z}{D} \right)^\alpha \tag{6.45}$$

where $G_{(z)}$ = shear modulus at depth z

$G_D$ = shear modulus at depth of one diameter (z = D)

$\alpha$ = power law exponent

$\alpha = 0$ defines a homogeneous deposit and $\alpha = 1$ defines a linearly increasing modulus with depth - a so-called 'Gibson soil'. The configurations of stiffness profile described by the power law are illustrated in Figure 6.57.

Solutions for surface and embedded circular foundations of diameter D have been provided by Doherty and Deeks (2003a, b), in terms of stiffness factors, $k_v$, $k_h$ etc, defined as:

$$\begin{Bmatrix} V \\ H \\ M/D \\ T/D \end{Bmatrix} = G_D D \begin{bmatrix} k_v & 0 & 0 & 0 \\ 0 & k_h & k_{mh} & 0 \\ 0 & k_{hm} & k_m & 0 \\ 0 & 0 & 0 & k_t \end{bmatrix} \begin{Bmatrix} w \\ u \\ \theta D \\ \phi D \end{Bmatrix} \tag{6.46}$$

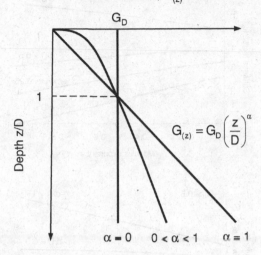

*Figure 6.57* Definition of heterogeneous stiffness profile

Example plots of $k_v$ for a surface foundation and skirted foundations with skirt depths of 0.25, 0.5 and 1 times the diameter are shown in Figure 6.58 for Poisson's ratios of 0.2 and 0.5. Other solutions are provided by Doherty and Deeks (2003a, b) for different types of embedded foundations. Note that the horizontal stiffness will be zero, for a surface foundation, in a soil with modulus varying according to Equation 6.46.

Immediate deformations should be computed using the undrained Young's modulus $E_u$ and Poisson's ratio $\nu_u$ while total deformations should be predicted using the drained Young's modulus E' and Poisson's ratio ν'. The reliability of predictions of deformation using these expressions is limited not only by the general caveats of applying elastic theory to soil but also by the selection of appropriate elastic parameters. In reality, the stress–strain response of soil is non-linear, yet 'equivalent linear' stiffness parameters are required for elastic calculations. Appropriate stiffness parameters must be selected carefully since they are a function of both the confining stress (which varies throughout the soil) and applied load level. Soil parameters for use in these analyses must be selected from site specific soil data, ideally benchmarked against other known similar soils. When the available data is inadequate for this task (due to insufficient site investigation), correlations from other 'similar' soils have to be used. However, 'similar' can in practice mean 'quite different' since soil is a highly variable medium and the closest match may be sufficiently different to change the result substantially. Selection of appropriate parameters, therefore, requires experience and good judgement.

*Rate of consolidation*

It is important to know not only how much consolidation will take place, but also how long consolidation will take to occur, i.e. the rate of consolidation. One-dimensional consolidation theory (Terzaghi 1923) may be used for cases where a shallow foundation sits on a relatively shallow layer of compressible soil overlying

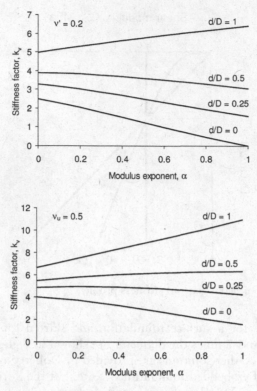

*Figure 6.58* Elastic stiffness factors for surface and skirted foundations

competent strata, or for consideration of the soil response within a soil plug confined by foundation skirts. However, it should generally be avoided for field conditions where three-dimensional flow and strain will govern consolidation. A theory for three-dimensional consolidation was first formulated by Biot (1935, 1956) since when analytical solutions of the time-settlement behaviour of shallow foundations have been presented for a range of permeable to impermeable and flexible to rigid rafts, typically assuming a smooth interface and an underlying elastic half space (e.g. McNamee and Gibson 1960, Gibson *et al.* 1970, Booker 1974, Chiarella and Booker 1975, Booker and Small 1986). Figure 6.59 shows the displacement time history for a smooth, rigid, impermeable, circular foundation resting on an elastic half-space for a range of drained Poisson's ratio $\leq v' \leq 0.1$–$0.3$.

Little attention has been directed towards quantifying the effect of foundation embedment on consolidation response. Embedment will generally reduce the magnitude and rate of consolidation settlements of shallow foundations due to the resistance of the soil above foundation level and longer drainage paths. An analytical solution to the problem of an embedded foundation would involve considerable computational difficulty. The problem is complicated for offshore shallow foundations since embedment generally takes the form of foundation skirts and additional compression of the confined soil plug will lead to a faster rate of consolidation and larger displacements compared with an equivalent buried plate or solid embedded foundation, particularly for smooth-sided foundations. Figure 6.60 shows elastic

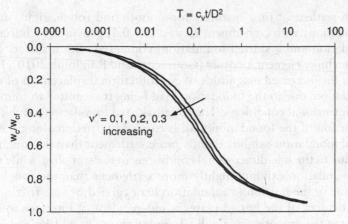

*Figure 6.59* Settlement time history of smooth, rigid, impermeable circular foundation

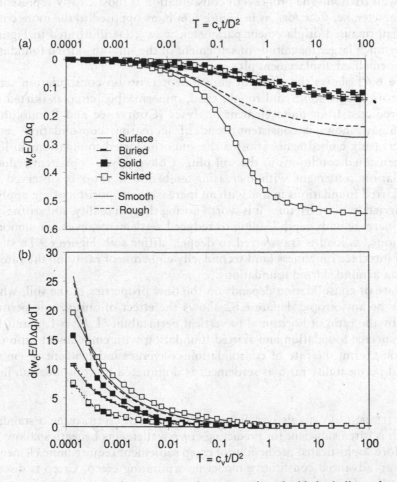

*Figure 6.60* Consolidation settlement time histories of embedded shallow foundations, $d/D = 0.5$, $v' = 0.2$, (a) magnitude and (b) rate (Gourvenec and Randolph 2010)

consolidation settlement time histories for smooth and rough, rigid, impermeable, circular foundations with embedment ratio $d/D = 0.5$, provided by burial of a plate, a solid foundation and a skirted foundation, compared with a surface foundation, predicted from finite element analyses (Gourvenec and Randolph 2010). Figure 6.60a shows clearly the increased magnitude of consolidation displacement of the smooth-skirted foundation, due to the foundation load being transmitted to foundation level under one-dimensional conditions. The rough-skirted foundation exhibits less settlement as a portion of the foundation load is carried by friction along the skirts. The rough-skirted foundation exhibits slightly more settlement than the rough-sided solid foundation due to the one-dimensional conditions in the soil plug, while the smooth-sided solid foundation exhibits slightly more settlement than the rough-sided solid foundation due to the portion of foundation load carried by side friction. The magnitude of settlement of the buried plates is independent of interface roughness and similar that for the smooth-sided solid foundation, as would be expected. Figure 6.60b shows the time histories of the rate of consolidation settlement, clearly indicating the opposing effects of one-dimensional conditions within the soil plug and skirt or side-wall friction. The progress of consolidation is most clearly represented by a rate parameter, i.e. $f(dw_c/dt)$, as in Figure 6.60b, as opposed to the more commonly adopted normalised displacement parameter, $w_c/w_{cf}$, (as illustrated in Figure 6.59) since the much larger magnitude of settlement of the smooth-skirted foundation distorts a normalised displacement plot.

Figure 6.61 shows the effect of embedment ratio on consolidation settlement time histories for smooth and rough, rigid, impermeable, circular-skirted foundations, predicted from finite element analyses (Gourvenec and Randolph 2009). Figure 6.61a shows a consistent trend of increasing consolidation settlement with increasing embedment ratio for the smooth-skirted foundations in line with one-dimensional conditions in the soil plug. Conversely, a consistent reduction in consolidation settlement with increasing embedment ratio is observed for the rough-skirted foundations in line with an increasing proportion of the applied load being carried by skirt friction. It is worth noting that, in reality, soil stiffness would usually increase with depth leading to reduced settlements, even for smooth sided foundations, as load is transferred to deeper, stiffer soil. Figure 6.61b shows the effect of interface roughness (and secondarily embedment ratio) on the rate of consolidation around skirted foundations.

The rate of consolidation depends on the flow properties of the soil, which will generally be anisotropic. Figure 6.62 shows the effect of anisotropic permeability defined by the ratio of horizontal to vertical permeability $k_h/k_v = 1$, $3$ and $10$ for a smooth surface foundation and skirted foundation with embedment ratio $d/D = 1$. In the long term, the rate of consolidation converges independent of embedment ratio and permeability ratio as settlement is dominated by consolidation in the far field.

CREEP SETTLEMENT   A simple one-dimensional model of creep, using a standard '$C_\alpha$' approach is often sufficient for predicting creep settlements beneath shallow foundations. More sophisticated predictions of creep settlement require finite element analysis with an advanced constitutive model incorporating creep. Creep is discussed in more detail in Chapter 4.

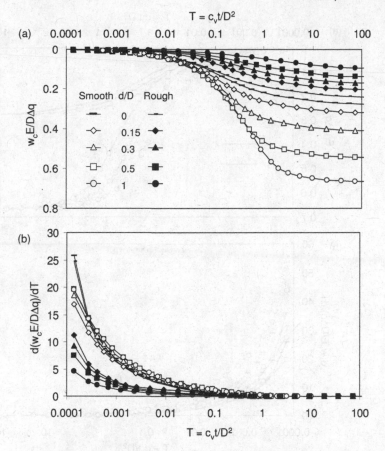

*Figure 6.61* Consolidation settlement time histories of skirted foundations, $v' = 0.2$, (a) magnitude and (b) rate (Gourvenec and Randolph 2009)

## Cyclic settlement

The effect of cyclic loading on settlements must be addressed for the foundations of offshore structures. To calculate cyclically induced displacements, it is necessary to determine the stress–strain behaviour of soil under cyclic loading, as discussed in Chapter 4.

An approximate prediction of shear strain induced settlement may be calculated using similar methods as those for calculating immediate settlement but with 'equivalent' shear moduli accounting for accumulated cyclic shear strain over the design number of cycles N. Since it is important to assess the accumulation of settlement over the lifetime of the foundation, it is usual to account for a lifetime history of cycling in the analysis, not just a single storm as used for bearing capacity assessment. Calculating the resulting settlements is a difficult procedure, since different equivalent shear moduli will be obtained throughout the soil; judicious weighting procedures can be used as the basis for approximate calculations.

An approximate calculation of cyclically induced volumetric strain settlement can be made using similar methods as for consolidation settlement. Pore pressure

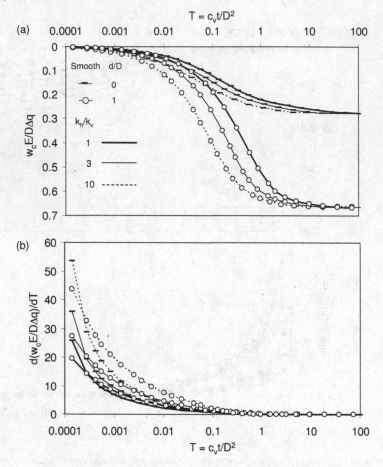

*Figure 6.62* Effect of anisotropic permeability on consolidation response of surface and skirted foundations with smooth interface, $v' = 0.2$, (a) magnitude and (b) rate (Gourvenec and Randolph 2009)

generated by undrained cyclic loading leads to a reduction in the effective stresses in the soil that can be regained as drainage occurs and the excess pore pressure dissipates. The reduction in effective stress due to excess pore pressure generation is indicated by A–B in Figure 6.63 and the increase in effective stress due to dissipation of excess pore pressures is indicated by B–C. At point C, the effective stress has reached the same value as before the start of cyclic loading, but some vertical compression due to the dissipation of cyclically induced pore pressure has taken place.

The vertical compression $\varepsilon_1$ due to the dissipation of cyclically induced pressure can be calculated by

$$\varepsilon_1 = \frac{u_c}{M_r} = \frac{C_r}{1+e_0} \log \left[ \frac{\sigma'_{vc}}{\sigma'_{vc} - u_c} \right]$$

(6.47)

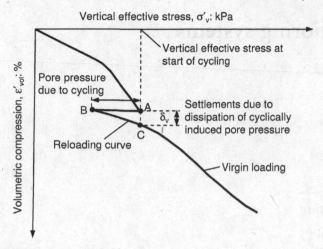

*Figure 6.63* Stress-volume changes due to pore pressure changes (Yasuhara and Andersen 1991)

where

$\varepsilon_1$ = vertical strain
$u_c$ = cumulative pore pressure generated by cyclic loading
$M_r$ = secant recompression modulus over the range B–C
$C_r$ = recompression index in the range B–C
$e_0$ = void ratio at the start of cyclic loading
$\sigma'_{vc}$ = vertical effective stress at the start of cyclic loading (after consolidation).

The settlement can then be found by integrating $\varepsilon_1$ over the soil depth beneath the platform base. The recompression modulus can be determined from a cyclic simple shear test with subsequent drainage of the cyclically induced pore pressure or can be taken as 2/3 of the recompression modulus from a conventional oedometer test (Yasuhara and Andersen 1991). The recompression index $C_r$ is similar to the swelling index $C_s$, but calculated with the change in void ratio during recompression rather than unloading in $\Delta e/\Delta \log \sigma'_v$.

Generally, reasonable agreement has been reported between cyclic and permanent displacements observed in laboratory and field model tests (and indeed observed platform displacements in a few cases) and predictions based on approximate methods outlined here and in Chapter 4 (Andersen *et al.* 1989, Aas and Andersen 1992, Andersen *et al.* 1993, Keaveny *et al.* 1994). Nonetheless, non-linear finite element methods are the best way forward, if an accurate result is required.

# 7  Anchoring systems

## 7.1 Introduction

### 7.1.1 Overview

Anchoring systems are needed to moor buoyant facilities in position and sometimes to provide extra stability to fixed or flexible structures. Floating facilities are a practical option in deep water where it is not possible to install fixed structures (due to the water depth or the hostile environmental conditions). Floating production platforms with storage facilities may also be used in moderate water depths if there is no pipeline to shore.

In this Chapter, an introduction to buoyant platforms, mooring systems and types of anchors is presented followed by consideration of design principles for anchors and anchor chains.

### 7.1.2 Buoyant platforms

Floating facilities come in many shapes and sizes with various degrees of functionality. In all cases, what sets them apart from fixed systems (e.g. gravity-based structures or jackets) is the buoyancy of displaced water that holds them up, as opposed to a concrete or steel sub-structure. A selection of buoyant platforms is shown in Figure 7.1.

Floating production, storage and offloading platforms (FPSOs) are the most widely used floating facility. FPSOs are ship-shaped processing facilities, often converted oil tankers, that sit on station and weathervane about a rotating turret moored to the seabed. Shuttle tankers periodically relieve the FPSO of its oil and take it to shore. Since the first FPSO was installed for the Castellon field in the Mediterranean Sea in 1977, FPSOs have been widely used, mostly in marginal fields, remote or hostile environments or where no pipeline infrastructure exists. Currently, there are around 120 FPSOs in production worldwide. Floating production systems (FPSs) have their origins in drilling platforms and typically comprise a hull of four circular steel columns connected to a ring pontoon, although they can be ship-shaped like FPSOs. FPSs are moored by catenary or taut-lines to anchors at the seabed. FPSs can serve several fields processing oil or gas from subsea wet trees. For example, the Na Kika development in the Gulf of Mexico is designed to handle six oil and gas fields, some several kilometres away, with tie backs to the main semi-submersible FPS.

SPAR platforms comprise a long cylindrical hull oriented vertically in the water that is ballasted (typically with iron ore) to ensure sufficient mass to provide stability to the platform and maintain a centre of gravity below the centre of buoyancy. The SPAR is

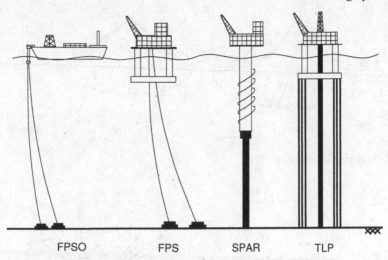

FPSO      FPS      SPAR      TLP

*Figure 7.1* Buoyant platforms

kept in position with a catenary or taut-line mooring anchored to the seabed. The Genesis SPAR in the Gulf of Mexico has a hull 40 m in diameter and 235 m long, weighing 26,700 tonnes unballasted. Ballasted, the hull weighs nearly eight times its self-weight.

Tension-leg platforms (TLPs) typically comprise a pontoon-and-column style hull like FPSs. Unlike for an FPS, the TLP hull is held down under tension by vertical steel cables anchored into the seabed. The Ursa TLP, installed in 1997 in the Gulf of Mexico, has hull columns 28 m in diameter and 60 m high connected to a ring pontoon 12.5 m wide and 10 m high, the whole assembly weighing approximately 28,600 tonnes. The hull was fabricated in Italy and was transported across the Atlantic to the Gulf of Mexico by Heerema's TT-851 barge, the longest transport barge in the world.

### 7.1.3 Mooring systems

Mooring systems provide the connections to the seabed that keep the floating facility in place and include catenary, taut or semi-taut line and vertical configurations as illustrated in Figure 7.2. Mooring systems combine steel wire or synthetic rope with chain while steel tendons are used to moor TLPs.

### Catenary moorings

Floating facilities have historically been secured by catenary moorings. A catenary is a mathematical definition of the curve assumed by a perfectly flexible uniform inextensible string when suspended from its ends. A catenary mooring is, therefore, one that takes up a catenary curve between the floating facility and the seabed. A catenary mooring touches down on the seabed in advance of the anchor, such that the angle of the anchor chain at the mudline is close to zero and the anchor is only subjected to horizontal forces. In a catenary mooring, most of the restoring forces are generated by

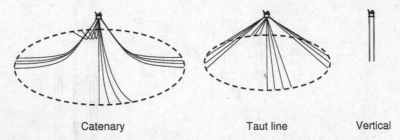

*Figure 7.2* Mooring configurations

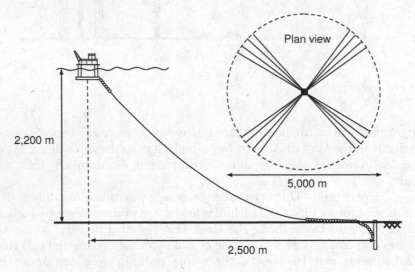

*Figure 7.3* Catenary mooring spread for the Na Kika FPS

the weight of the mooring line. Catenary moorings derive their compliance from the change in suspended line weight and thus line tension as the grounded line lifts off and is replaced on the seabed due to motion of the floating system. A typical mooring configuration has 100 mm diameter rope, steel wire or polyester, stretching down to heavy chain links that can weigh more than 200 kg each. The chains connect at the seabed to an anchor system. Mooring spreads have at least eight separate lines coming from the floating system and some have up to 16. The Na Kika FPS, located in 2,200 m depth of water, used 3,200 m of wire and 580 m chain for each of 16 separate lines (Figure 7.3). The anchors to which the mooring chains are attached sit 2.5 km away from the FPS.

## Taut or semi-taut line moorings

In deep and ultra-deep water, the weight of the mooring line of a catenary becomes a limiting factor in the design of the platform. To overcome this problem synthetic rope (which is lighter than conventional mooring chain or cables) and taut (or semi-taut) line moorings were developed. The major difference between a catenary mooring and

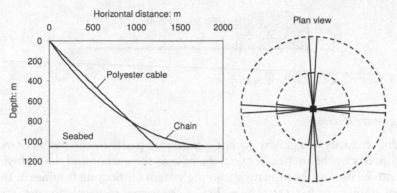

*Figure 7.4* Comparison of taut line and catenary moorings

a taut line mooring is that, where a catenary mooring arrives at the seabed horizontally, a taut line mooring arrives at the seabed at an angle. This means that the anchor point of a taut line mooring has to be capable of resisting both horizontal and vertical forces (while in a catenary the anchor point is only subjected to horizontal forces). The restoring forces in a taut line mooring are provided by the elasticity of the mooring line, compared to a catenary where most of the restoring forces are generated by the weight of the mooring line. A taut line mooring will usually make an angle of 30°–45° to the horizontal between the floating facility and the anchor attachment point, with little change in angle over the length of the line, as a result of its low mass per unit length. Semi-taut line moorings will show a wider operating range, but reach similar maximum angles under critical design conditions. There are various advantages of adopting mooring lines that make an angle with the seabed, as opposed to a catenary mooring arrangement. Offsets under steady conditions can be controlled better and tension variations due to vessel motions are a smaller proportion of the mean line tensions, with better load sharing achieved between adjacent lines so improving the overall efficiency of the system. Also, shorter lines are required giving a more compact seabed footprint, with typical diameters for the anchor spread (normalised by the water depth) ranging from four for a catenary system, to three for a semi-taut line mooring to two for a taut line mooring (Figure 7.4).

### Vertical moorings

Vertical moorings are used to anchor tension-leg platforms and involve taut steel cables or pipe applying tension between a seabed template and the floating platform. The Ursa TLP is held on-station by 16 tendons, four per corner, with a diameter of 80 mm and a wall thickness of 38 mm. Each tendon is approximately 1,266 m long and the total weight for the 16 tendons is approximately 16,000 tonnes. The lower end of each tendon is attached to an anchor pile 2.4 m in diameter and 147 m long, weighing approximately 380 tonnes each.

### 7.1.4 Types of anchors

A variety of anchoring systems are used to secure mooring lines to the seabed, but they can be divided conveniently into surface or 'gravity' anchors and embedded anchors.

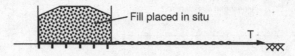

*Figure 7.5* Schematic of gravity box anchor

### Surface gravity anchors

The holding capacity of a gravity anchor is generated partly by the weight of the anchor itself and partly by the friction between the base of the anchor and the seabed. Gravity anchors can be used as the primary anchoring system for floating facilities or to provide additional stability to fixed structures, but are restricted to developments in relatively shallow waters because of their limited practical size and, therefore, holding capacity.

#### BOX ANCHORS

The simplest type of gravity anchor is a dead weight placed on the seabed. In order to minimise crane capacity for the installation of gravity anchors, they are generally designed as a structural component, for example, an empty box, together with bulk granular fill (either rock-fill, or heavier material such as iron ore). The box is typically equipped with ribs that penetrate into the seabed such that shearing is developed in the seabed material, rather than along the anchor–soil interface. The structural element is placed first, and then the bulk fill is added (deposited down tubes extending from the vessel to close to the seabed, guided by ROVs – remotely operated vehicles). A schematic of a box anchor is shown in Figure 7.5. Box anchors provide additional stability to the guyed flare support tower of the North Rankin A platform on the North-West Shelf of Australia. Located in 125 m of water, each of the four anchor boxes are 18 m by 19 m in plan and 6 m deep, weighing 4,000 tonnes each when ballasted (see Figure 1.7).

#### GRILLAGE AND BERM ANCHORS

A novel type of gravity anchor involves a buried grillage, beneath a rock-fill or iron ore berm, illustrated schematically in Figure 7.6. The grillage is placed towards the back of the berm, but such that the complete berm must be moved if the grillage starts to fail. The grillage and berm anchor is considerably more efficient in terms of quantity of steel for a given holding capacity than a simple box anchor (and hence can be installed with smaller crane vessels) but is much less efficient in terms of the quantity of ballast required. Design of this type of anchor is also more complex since a variety of failure modes must be considered, ranging from sliding of the complete berm, pulling out of the grillage, or combinations involving asymmetric mechanisms. Grillage and berm anchors were adopted for the CALM buoy at the Apache Stag field on the North-West Shelf (Erbrich and Neubecker 1999). Sited in 50 m water depth, the berm is approximately 27 m square and 3.35 m high overlying a 20 m by 10 m grillage (see Figure 1.8).

### Embedded anchors

Embedded anchors are required for situations in which more holding capacity is required than can be provided by gravity anchors.

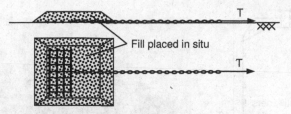

*Figure 7.6* Schematic of grillage and berm anchor

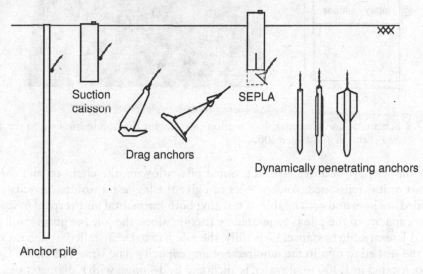

*Figure 7.7* Types of embedded anchors

Historically, three main types of embedded anchor have been used in practice:

- Driven or drilled and grouted piles
- Suction caissons
- Drag anchors (traditional fixed fluke or plate).

Two other types of embedded anchor have been introduced over the last decade: the SEPLA (suction embedded plate anchor) and dynamically penetrating anchors. The latter include the torpedo anchor, developed by Petrobras for commercial applications in Brazilian waters, the DPA (deep penetrating anchor) currently under development in the North Sea and the proprietary OMNI-Max anchor first used in 2007 in the Gulf of Mexico. Schematics of different types of embedded anchors are illustrated in Figure 7.7.

ANCHOR PILES

Anchor piles are hollow steel pipes that are either driven or drilled and grouted into the seabed, in the same manner as piles used for foundations for fixed bottom structures. Different types of piles and their installation method are discussed in detail in Chapter 5. Different arrangements of the anchor chain are possible, either attaching it partway down the pile (particularly for a driven pile) or grouting the anchor chain

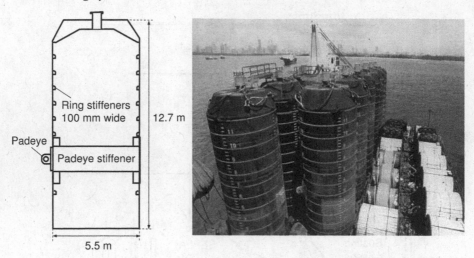

*Figure 7.8* Schematic and photograph of suction caissons for the Laminaria field in the Timor Sea (Erbrich and Hefer 2002)

into the upper part of a drilled and grouted pile, allowing the chain to pull through the grout as it is tensioned. Anchor piles can give the highest absolute capacity of the embedded anchors and are capable of resisting both horizontal and vertical loads. The holding capacity of the pile is generated by friction along the pile (or grout)–soil interface and lateral soil resistance. Generally, the pile has to be installed at a great depth below the seabed to obtain the required holding capacity. The Ursa TLP in the Gulf of Mexico, located in 1,300 m of water, is anchored by 16 piles, with a diameter of 2.4 m, driven to a depth of 130 m. Most pile driving hammers have a practical operation depth limit of around 1,500 m although specialist hammers with reduced rated energy have been developed capable of operating in water depths up to 3,000 m. The lower energy and complexity of pile driving operations mean that pile driving is still somewhat unattractive in very deep water.

SUCTION CAISSON ANCHORS

Suction caissons comprise large diameter cylinders, typically in the range of 3–8 m, open at the bottom and closed at the top, and generally with a length to diameter ratio L/D in the range three to six, considerably less than offshore piles, which have slenderness ratios up to 60. The Na Kika semi-submersible FPS in the Gulf of Mexico (shown previously in Figure 7.3) is moored in 2,200 m depth of water by 16 suction caisson anchors 4.7 m diameter and 26 m long. Figure 7.8 shows the relatively stocky suction caissons used for the Laminaria FPSO in the Timor Sea, which are 5.5 m in diameter and 12.7 m in length, each weighing 50 tonnes.

A suction caisson anchor is installed in the same way as a skirted foundation (described in Chapter 6). Initially, penetration into the seabed is achieved by self-weight, with the top cap vented, with the remaining penetration achieved by suction, via a pump connected to a valve in the top cap of the caisson. Water is pumped out from inside the caisson causing the pressure inside the caisson to fall below that outside. This causes a net downward pressure on the top of the caisson forcing it into the

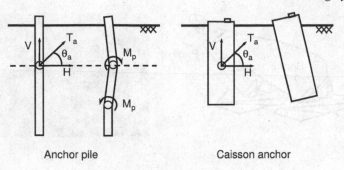

*Figure 7.9* Failure modes of anchor pile and suction caisson

seabed. The holding capacity of a suction caisson anchor, assuming a sealed cap, is derived from bearing resistance between the soil and the projected area of the caisson (on a vertical plane for horizontal resistance, and the cross-sectional area for reverse end-bearing), aided by frictional resistance along the outside of the caisson shaft. Reverse end-bearing relies on passive suctions developed within the soil plug, and so consideration needs to be given to the time over which these can be sustained.

Although concrete caissons have been used, the vast majority of suction caissons are fabricated from steel with ratios of diameter to wall thickness $D/t \sim 100$–$250$. Internal stiffeners are provided to prevent structural buckling during installation and due to the mooring loads and soil resistance during operation. The low embedment ratio and relatively high stiffness of suction caissons leads to failure by rigid body motion compared with an anchor pile, which forms plastic hinges during failure (Figure 7.9). Mooring loads are applied by an anchor line to the padeye attached to the side of the caisson at a depth that optimises the holding capacity. Generally, this requires the line of action of the load to pass through a point on the axis at a depth of 60–70 per cent of the embedded depth, which is at or near the centroid of the lateral soil resistance for the typical normally to lightly overconsolidated fine-grained sediments found in deep water. The padeye location is calculated by moment equilibrium so that the suction caisson is designed to translate horizontally and not rotate under lateral loading to achieve the maximum lateral load carrying capacity.

DRAG ANCHORS

High capacity drag anchors evolved from conventional ship anchors. Traditional drag anchors comprise a broad fluke rigidly connected to a shank, as shown in Figure 7.10. The angle between shank and fluke is pre-determined, though may be adjusted prior to anchor placement on the seabed. This angle is typically around 50° for clay conditions (provided soft material occurs near the seabed) and 30° in sand or where clay of high strength occurs at the seabed. For installation, the anchor is placed on the seabed in the correct orientation (with the aid of ROVs) and embedded by pre-tensioning the chain to an appropriate proof load. Traditional fixed fluke anchors come in standard sizes defined by weight, up to 65 tonnes with fluke lengths of up to around 6.3 m. Depending on soil conditions, penetration depths of between one and five fluke lengths are typically achieved through dragging a distance of 10–20 times the fluke length, mobilising holding capacities of 20–50 times the anchor weight. The holding capacity

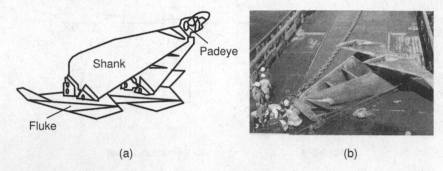

*Figure 7.10* Fixed fluke plate anchor (a) schematic and (b) 32 tonne Vryhof Stevpris Mk5 (Courtesy of Vryhof Anchors)

*Figure 7.11* Vertically loaded anchors (VLAs) (a) Vryhof Stevmanta and (b) Bruce DENNLA (Courtesy of Vryhof Anchors and Bruce Anchors)

of the traditional fixed fluke anchor is developed from the soil in front of the anchor and may exceed 10 MN. Traditional fixed fluke anchors are not designed to take large vertical loads, indeed these anchors are withdrawn by applying vertical load to the anchor chain. They are, therefore, suitable for catenary moorings but not for deepwater applications using taut or semi-taut line moorings.

Vertically loaded anchors (VLAs), also called drag-in plate anchors, were developed to overcome the limitations of the traditional fixed fluke anchor. VLAs are based on the traditional fixed fluke drag anchor except the broad shank is replaced by either a thinner shank or a wire bridle (Figure 7.11). VLAs are installed like a traditional fixed fluke anchor by applying a horizontal load at the mudline to achieve embedment. Deeper embedment, to depths of up to 7–10 fluke lengths, can be achieved with

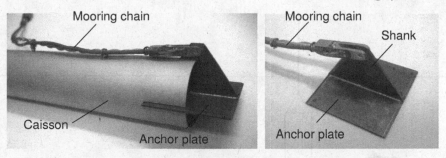

*Figure 7.12* Component of a suction embedded plate anchor (Gaudin *et al.* 2006)

plate anchors than with a traditional fixed fluke anchor, because their improved slim-line geometry provides less resistance. When embedded, the anchor is rotated until the applied load acts normal to the fluke (plate), mobilising the maximum possible soil resistance and enabling the anchor to withstand both horizontal and vertical loading. Plate anchors are typically smaller than fixed fluke plate anchors, with fluke areas up to 20 m² and fluke lengths up to 6 m.

Drag anchors were initially used predominantly for semi-permanent moorings, for example, for MODUs (mobile offshore drilling units) but are also used to moor permanent floating facilities. For example, the Roncador FPSO in the Campos Basin, Brazil is located in 1,600 m depth of water with a taut-line mooring secured to the seabed by nine 14 m² Vryhof Stevmanta VLAs.

SEPLAS

Suction embedded plate anchors (SEPLA) are a form of plate anchor slotted into the toe end of a suction caisson (Figure 7.12) and were developed to combine the economic benefits of plate anchors with the installation certainty of suction caissons, allowing holding capacity to be more accurately assessed than if a plate anchor is installed by dragging. The suction caisson is embedded, then withdrawn (using reverse pumping) leaving the plate anchor embedded. The plate anchor is then keyed by pre-loading the anchor chain leading to rotation of the plate normal to the anchor line load (Figure 7.13). Plate anchors of up to 4.5 m by 10 m are anticipated for permanent installations with smaller plates for temporary installations. Small-scale field trials and a full-scale field trial in the Gulf of Mexico, in water depths up to 1,300 m with penetration depths up to 25 m, are reported by Wilde *et al.* (2001). They have since been used extensively in the Gulf of Mexico and West African waters to anchor short-term facilities such as MODUs.

DYNAMICALLY PENETRATING ANCHORS

The cost of installing anchors in deep water has led to the development of anchors that embed themselves into the seabed under free-fall, such as the deep penetrating anchor (DPA) (Lieng *et al.* 1999, 2000). This comprises a rocket-shaped anchor, 1–1.2 m in diameter, with a dry weight of 500–1,000 kN and a height of 10–15 m, and is designed to be released from a height of 20–50 m above the seabed (Figure 7.14a).

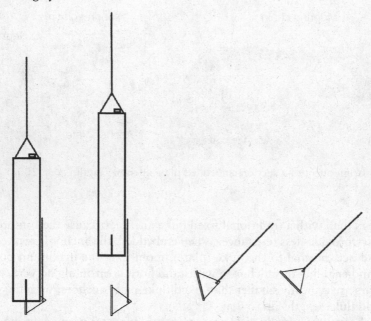

*Figure 7.13* Installation and deployment of a suction embedded plate anchor (O'Loughlin *et al.* 2006)

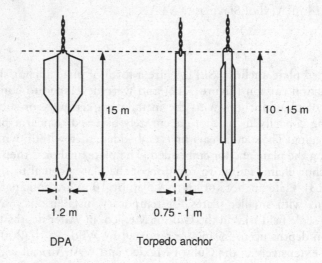

*Figure 7.14* Dynamically penetrating anchors

Field trials with the anchor are proceeding although it has yet to be used in practice. A less sophisticated torpedo anchor has been used by Petrobras in the Campos Basin (Medeiros 2001, 2002). The dimensions of the anchor range from 0.76 to 1.1 m in diameter by 12–15 m long, with a weight of 250–1,000 kN. Some versions of the anchor have been fitted with four flukes at the trailing edge, ranging in width from 0.45 to 0.9 m, and 9–10 m long (Figure 7.14b). Both of the aforementioned anchors

*Table 7.1* Advantages and disadvantages of different deep-water anchor types (Ehlers *et al.* 2004)

| Anchor | Advantages | Disadvantages |
|---|---|---|
| Suction caisson anchor | Simple to install accurately with respect to location, orientation and penetration.<br>Leverage design experience with driven piles.<br>Well-developed design and installation procedures.<br>Anchor with the most experience in deep water for mooring MODUs and permanent facilities. | Heavy: derrick barges may be required.<br>Large: more trips to shore to deploy full anchor spread.<br>Requires ROV for installation.<br>Requires soil data from advanced laboratory testing for design.<br>Concern with holding capacity in layered soils.<br>Lack of formal design guidelines.<br>Limited data on set-up time for uplift. |
| Drag embedment vertically loaded anchor (VLA) | Lower weight.<br>Smaller: fewer trips to transport the full anchor spread to a site.<br>Well-developed design and installation procedures. | Requires drag installation, keying and proof testing; limited to bollard pull of installation vessel.<br>Requires 2 or 3 vessels and ROV.<br>No experience with permanent floating facilities outside of Brazil.<br>Difficult to assure installation to, and orientation at, design penetration. |
| Suction embedded plate anchor (SEPLA) | Uses proven suction caisson installation methods.<br>Cost of anchor element is the lowest of all the deep-water anchors.<br>Provides accurate measure of penetration and positioning of anchor plate.<br>Design based on well developed design procedure for plate anchors. | Proprietary or patented installation.<br>Installation time about 30 % greater than for suction caissons and may require derrick barge.<br>Requires keying and proof testing: limited to bollard pull of installation vessel; also requires ROV.<br>Limited field load tests and applications limited in numbers to MODU only. |
| Dynamically penetrating anchor | Simple to design: conventional API RP 2A pile design procedures used for prediction of capacity; thus capacity calculations are likely to be readily acceptable for verification agencies.<br>Simple and economical to fabricate.<br>Robust and compact design makes handling and installation simple and economical using 1 vessel and no ROV.<br>Accurate positioning with no requirements for specific orientation and proof testing during installation. | Proprietary or patented design.<br>No experience outside Brazil.<br>Lack of documented installation and design methods with verification agencies.<br>Verification of verticality. |

are referred to here as dynamically penetrating anchors. They are designed to reach velocities of 25–35 m/s at the seabed, allowing tip penetrations of two to three times the anchor length and holding capacities after consolidation that are anticipated in the range three to six times the weight of the anchor. While such efficiencies are lower than would be obtained with other sorts of anchor, this is compensated by the low cost of fabrication and ease of installation.

A selection of advantages and disadvantages of different embedded anchor types are summarised in Table 7.1.

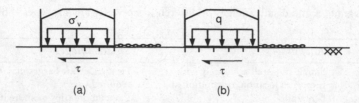

*Figure 7.15* Failure mode for gravity anchors in (a) free-draining material and (b) slow draining material

## 7.2 Design principles for gravity anchors

The design of gravity anchors follows similar principles to those for shallow foundations (as discussed in Chapter 6). A brief overview is presented here for design of simple box anchors on relatively free-draining and slow-draining materials. Numerical analyses should be employed for more complex conditions, such as for design of a grillage and berm gravity anchor.

### 7.2.1 Box anchor in sand (relatively free-draining)

Failure of a gravity box in sand will occur by horizontal sliding, with a mechanism similar to that in a direct shear test, as illustrated in Figure 7.15a. Under monotonic loading, failure shear stress ratios $\tau_f/\sigma_v'$ may be as high as 0.8, although design is generally based on more conservative values. Cyclic loading is less damaging for an anchor than a fixed bottom structure, as the loading is entirely unidirectional. This limits the extent of pore pressure development and (partial) liquefaction, although significant incremental displacements can accumulate. Monotonic capacity is governed by peak strength, with shear stress ratios typically in the region 0.5–0.7. Cyclic capacity is governed more by yield point, although the one-way nature of cyclic loading allows more optimistic design than for platforms, with typical design cyclic shear stress ratios in the region of 0.3.

### 7.2.2 Box anchor in silt or clay (relatively slow draining)

For a gravity anchor on less permeable material, such as clays and silts, the limiting friction is equal to the current undrained shear strength of the underlying material (as measured in a simple shear test), as illustrated in Figure 7.15b. Bearing failure will occur for pressures in the range 6–12 times the mudline shear strength with no horizontal shear, reducing by a factor of two during sliding shear. Thus, the maximum shear stress ratio $\tau/q$ is limited to about one-third, or lower in soil with an increasing strength profile with depth. Cyclic loading effects may further reduce the design shear stress ratio such that site specific laboratory test data are essential.

### 7.2.3 Grillage and berm anchor

Grillage and berm anchors pose a more varied array of design issues in comparison to a simple box anchor and with consideration of a range of failure modes. Bearing and sliding failure of the grillage and berm under combined static and cyclic loads must be assured against, as for box anchors. Slope stability failure of the berm and pullout of the grillage from the berm due to the chain loads and hydrodynamic forces

(a)

(b)

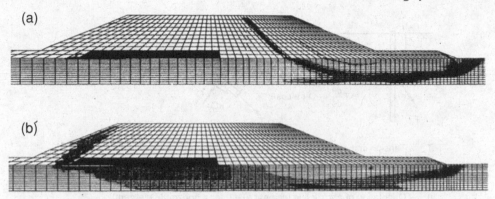

*Figure 7.16* Numerical analysis results showing global failure mechanisms of a grillage and berm anchor (a) slope stability failure and (b) combined sliding and bearing failure (Erbrich and Neubecker 1999)

must also be considered. In addition to resisting mooring line and hydrodynamic forces, the berms must resist scour or erosion. Sufficient particle size of the armour material is required to prevent erosion of the berm due to seabed shear stresses caused by currents and waves that may undermine the stability of the berm slopes. Erosion of the seabed due to current induced shear stresses must also be protected against. Since the particle size of the seabed material is unlikely to be sufficient to resist erosion, a scour mat is likely to be required, extending sufficient distance that should the berm be undermined the resulting inclination would not result in slope instability of the berm face. The various issues requiring consideration during design rely on a combination of analysis techniques that may include limit equilibrium methods, plasticity methods and numerical methods. Figure 7.16 shows contours of maximum shear strain from finite difference analyses, indicating the global failure mechanisms for the grillage and berm anchors for the CALM buoy at the Apache Stag field that were illustrated in Figure 1.8 (Erbrich and Neubecker 1999).

## 7.3 Anchor line response for embedded anchors

For vertical tethering of a tension-leg platform the anchor chain will be attached to the pile or suction caisson top cap and there will be no interaction between the chain and the soil. For catenary and taut-line moorings using anchor piles, suction caissons or drag anchors the optimal attachment point of the anchor chain is below the mudline. It is, therefore, essential to consider the performance of the anchor chain and its interaction with the soil as this will determine the angle of loading on the anchor. For anchor chains below the mudline, horizontal tensioning will result in cutting and sliding of the chain through the soil, which in turn results in large soil resistive forces acting on the chain. The chain will form an inverse catenary shape from the mudline to the anchorage leading to two important conditions:

- Friction developed along the chain causes the tension in the chain at the anchor attachment point (called the 'padeye') to be smaller than at the mudline, allowing optimisation of anchor size.
- The angle of the chain at the padeye is greater than at the mudline, resulting in a component of uplift being applied to the anchor.

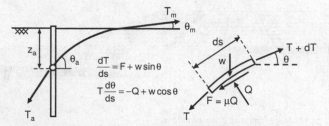

$T_m$ = Tension in the anchor line at mudline
$T_a$ = Tension in the anchor line at the anchor attachment point (padeye)
$\theta_a$ = Inclination of anchor line at attachment point (measured from the horizontal)
$\theta_m$ = Inclination of anchor line at mudline (measured from the horizontal)
$\theta$ = Orientation of anchor line element ($\theta$ = 0 for a horizontal element)
$Q$ = Normal force transmitted to the anchor line from the soil
$F$ = Frictional force acting on the anchor line (parallel to the line)
$\mu$ = F/Q = friction coefficient between chain and soil
$z_a$ = Depth to the padeye (i.e. chain attachment point)
$w$ = Self weight of the anchor line

*Figure 7.17* Configuration of loads on an anchor line (Neubecker and Randolph 1995)

This information is critical to the anchor design, since it will determine the failure mode for anchor piles and suction caissons and the embedment performance of drag anchors. Design principles for embedded anchors are set out in the following sections. The depth of water in which embedded anchors are employed results in applications that are mainly in fine-grained soils, therefore the following discussion concentrates on design principles for undrained conditions.

### 7.3.1 Equilibrium equations of an embedded anchor line

The anchor line configuration and the loads carried by an anchor line are illustrated in Figure 7.17. The illustration shows the case of an anchor pile although the same conditions apply to a suction caisson or drag anchor since only the position of the padeye influences the line behaviour. The governing equations for chain profile and load development consist of a pair of simultaneous partial differential equations, describing the change in line tension dT over an element length ds and the corresponding angular change dθ:

$$\frac{dt}{ds} = F + w\sin\theta$$
$$\frac{d\theta}{ds} = \frac{-Q + w\cos\theta}{T}$$

(7.1)

where

$F$ = local soil friction (per unit length, parallel to line) acting on the anchor line
$w$ = self weight of the anchor line
$\theta$ = orientation of anchor line element to the horizontal (in radians)
$Q$ = local normal force (per unit length) transmitted to the anchor line from the soil
$T$ = anchor line tension.

The soil friction acting on the anchor line (parallel to the line) may be expressed as:

$$F = A_s \alpha s_u$$

(7.2)

where

$A_s$ = effective surface area of the anchor line per unit length

$\alpha$ = interface friction coefficient (typical values of $\alpha$ in clay lie in the range 0.2–0.6 depending on whether wire rope or chain is used)

$s_u$ = local undrained soil strength at that position (average, or as measured in simple shear $s_{uss}$).

The limiting normal force transmitted to the anchor line from the soil is evaluated as

$$Q = A_b N_c \bar{s}_u \tag{7.3}$$

where

$A_b$ = effective bearing area of the anchor line per unit length

$N_c$ = bearing capacity factor (typically taken as 7.6, based on a buried strip)

For wire or polyester rope, the unit bearing area is equal to the diameter of the rope, $A_b = d$, and the surface area $A_s = \pi d$. For a standard link chain an effective width b is given in terms of the bar diameter d as b = 2.5d, giving normal resistance $A_b = b = 2.5d$ and for sliding $A_s = 8$–$11d$, depending on the chain geometry. Generally, it is found that $\mu$ for chains will lie in the range 0.1–0.3 for clay soils, and up to 0.5 for sands.

A full solution for predicting the resulting anchor chain profile and load development requires numerical integration of the governing differential equations (Equation 7.1), together with iteration of one of the unknown boundary conditions in order to match the known position of the padeye (Vivatrat *et al.* 1982). The main complication in the solution of the governing differential equations arises from the self-weight of the chain w. Chain weight is generally only significant at shallow embedment depths and can be omitted from the governing equations in their original form, instead allowing for it by a simple adjustment of the profile of bearing resistance with depth. This assumption allows an analytical solution to be developed as outlined in the following section (Neubecker Randolph 1995).

### 7.3.2 Simplified analytical solution of an embedded anchor line

The force equilibrium equations of the anchor line are simplified to produce closed form expressions for both the geometric profile of the anchor line and the force distribution. The analytical solution obviates the need for unwieldy numerical computations. The solutions also give direct insight into the key variables that dictate anchor line performance and provide a theoretical basis for the design of anchor piles, suction caisson anchors, drag anchors and other embedded anchoring systems.

First, the change in anchor line angle between the mudline $\theta_m$ and padeye $\theta_a$ is related to the anchor line tension at the padeye $T_a$ and the average bearing resistance $Q_{av}$ over the depth range $0 \leq z \leq z_a$, where $z_a$ is the depth to the padeye. More rigorous relationships are given by Neubecker and Randolph (1995), but they show that the approximation given here is relatively accurate, with $T_a$ and the changes in $\theta$ related by

$$\frac{T_a}{2}(\theta_a^2 - \theta_m^2) = z_a Q_{av} \tag{7.4}$$

where the bearing resistance

$$z_a Q_{av} = b N_c \int_0^{z_a} s_u \, dz \tag{7.5}$$

For the situation where the anchor line angle at the mudline is zero (i.e. for catenary moorings), the anchor line angle at the padeye is

$$\theta_a = \sqrt{\frac{2Z_a Q_{av}}{T_a}} = \sqrt{\frac{2}{T^*}} \tag{7.6}$$

where $T^*$ is a normalised anchor line tension given by

$$T^* = \sqrt{\frac{T_a}{z_a Q_{av}}} \tag{7.7}$$

Secondly, the anchor line tension at the mudline $T_m$ is related to that at the padeye $T_a$ by

$$\frac{T_m}{T_a} = e^{\mu(\theta_a - \theta_m)} \tag{7.8}$$

For the situation where the anchor line angle at the mudline is zero, Equation 7.8 reduces to

$$\frac{T_m}{T_a} = e^{\mu\theta_a} \tag{7.9}$$

The anchor line profile may be expressed in terms of a normalised depth $z^* = z/z_a$ and normalised distance from the anchor $x^* = x/z_a$. For a soil with a uniform strength with depth, the normalised profile of the anchor line is given by

$$z^* = \left(1 - \frac{x^*}{\sqrt{2T^*}}\right)^2 \tag{7.10}$$

For soils with strength increasing linearly with depth, the normalised anchor line profile is given by

$$z^* = e^{-x^*\theta_a} \tag{7.11}$$

In order to allow for the effect of the self-weight of the anchor line w, an effective bearing resistance is defined as

$$Q_{eff} = Q - w \tag{7.12}$$

Similarly, the depth of the padeye is adjusted by an amount $\delta$, given by

$$\delta = \frac{w}{k} \tag{7.13}$$

where k is the gradient of the bearing resistance with depth. Typically, the adjustment to the padeye depth is less than 0.5 m and makes little difference to the computed chain response. The bearing resistance and padeye depth adjustments are illustrated in Figure 7.18.

The pivotal analytical equations (Equations 7.4 and 7.8) are solved simultaneously to determine the anchor line angle at the padeye for given padeye depth, mudline tension and angle, soil shear strength profile and chain properties. The anchor line tension at the padeye can then be determined from either of the pivotal equations. The simplified analytical approach is very useful for mooring design, as it allows an instant appraisal of the length of submerged anchor chain, the load dissipated by the chain, and the tension and inclination of the chain at the anchorage.

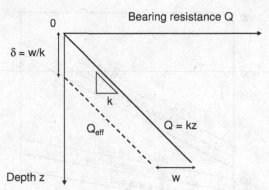

*Figure 7.18* Soil strength adjustment to account for chain weight (Neubecker and Randolph 1995)

## 7.4 Design principles for drag anchors

### 7.4.1 Overview

The holding capacity of a drag anchor, either a traditional fixed fluke or VLA, depends on the ability to penetrate and to reach the target installation depth. The most challenging aspect of a drag anchor's performance at site is the uncertainty with installation and final location of the anchor. Prediction of the load carrying or holding capacity of a drag anchor at a given depth is relatively straightforward compared with the prediction of the anchor trajectory. Prediction of the ultimate holding capacity for an anchor at a single location and orientation lends itself well to conventional bearing capacity theory while prediction of installation is complex. The penetration path and ultimate penetration depth of a drag anchor is a function of:

- Soil conditions (soil layering and variation in $s_u$)
- Type and size of anchor
- Anchor fluke-shank angle
- Type and size of the anchor forerunner (wire or chain)
- Line uplift angle at the seabed.

Anchor trackers, comprising a small reactive propeller mounted on the anchor to measure distance travelled, combined with tilt sensors etc, can reduce the uncertainty in final anchor penetration depth. Different systems are available depending on the application. For anchor tests, a system in which data can be downloaded from the tracker following retrieval of the anchor is adequate, while for permanent field applications a wireless modem must transmit real-time data back to the surface. Anchor trackers are still a relatively new development and can be easily damaged during installation; traditional and analytical design methods are routinely relied on to predict anchor trajectory.

In the following sections, the traditional approach to drag anchor design is set out followed by a discussion of some of the increasingly favoured analytical approaches.

### 7.4.2 Traditional approach to drag embedment design

The traditional approach to predicting anchor penetration, drag distance and holding capacity relies on empirically derived design charts based on anchor dry weight W for

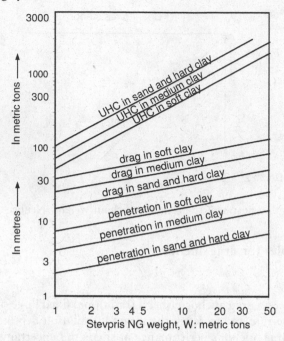

*Figure 7.19* Manufacturer's design chart for traditional fixed fluke drag anchor, Stevpris New Generation (Vryhof Anchors 2008)

broad categories of soil type. Design charts are developed and supplied by anchor manufacturers and also provided by independent bodies such as the American Petroleum Institute (API RP-2SK 2005). Figure 7.19 shows a design chart for penetration depth, drag distance and holding capacity over a range of soil conditions published by Vryhof Anchors for their Stevpris New Generation (NG) anchor (Vryhof 2008).

Although design charts are traditionally based on anchor dry weight W, drag anchor performance is determined more by size. However, since historically all anchors have been made from similar materials (steel) there is a close correlation between fluke size and anchor weight. The distinction has become increasingly important with the advent of VLAs due to their low weight to area ratio. Generic design curves are derived from a database of model tests and suffer from the limitations in the database and inaccuracies involved in the simple extrapolation of the ultimate holding capacity measured in small-scale tests to larger anchors. In addition, most of the anchor tests in the database were carried out with chain forerunners, which cause the charts to under predict anchor depth and hence holding capacity for anchors with wire forerunners. The effect of a wire forerunner, therefore, needs to be estimated and incorporated into the calculation. A further limitation arises from the broad soil types, such as very soft clay, medium clay, and stiff clay or sand, defined by the generic design curves. Specific soil properties, for example, friction and dilation angles for sand, shear strength profiles in clay, and soil layering can cause significant variations in behaviour that are not captured by the broad categories of soil type in design charts. Further, the anchor resistance resulting from these charts relates to ultimate penetration of the anchor and, therefore, represents a factor of safety of

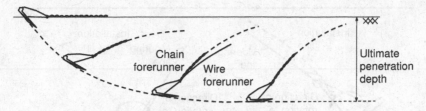

*Figure 7.20* Deployment and operational mode of a fixed fluke drag anchor

unity. As anchors are seldom installed to their ultimate depth, anchor resistance derived from theses charts must be corrected for depth of penetration, or degree of mobilisation.

The limitations mentioned in the foregoing section, among others, justify using a design procedure based on geotechnical principles. A comprehensive review of advanced design methods for VLAs has been prepared by Murff *et al.* (2005) summarising the results from an industry sponsored study. Det Norske Veritas (DNV) RP-E301 and E02 (2000a, b) provide recommended practice for design of fixed fluke anchors and VLAs in clay, recommending analytical methods over empirically derived design charts, although emphasise that it is still necessary to calibrate analytical methods against high-quality model tests, with due consideration of scale effects.

### 7.4.3 Analytical approach to predicting installation of a drag anchor

Installation of a drag embedment anchor is illustrated in Figure 7.20 for the case of a traditional fixed fluke anchor. The anchor is embedded into the seabed by a horizontal force acting at the seabed. From a starting point on the seabed with the anchor shank horizontal, the anchor follows a path where the anchor line makes an almost constant angle to the fluke. A small variation of angle takes place due to the gradually lessening effect of the anchor weight as the anchor embeds and the soil resistance increases. Ultimate penetration depth is achieved when the flukes are horizontal.

The embedment process for VLAs is similar to that for traditional fixed fluke anchors. Once the anchor has embedded sufficiently, the anchor is rotated ('keyed') such that the mooring line applies a force acting approximately normal to the fluke. Different methods are employed to key-in different types of VLAs and, in some cases, different methods of keying are available for a given type of VLA. Figure 7.21 shows the deployment and operational modes of a Vryhof Stevmanta VLA and a Bruce Dennla VLA. The Vryhof VLA is keyed continuing in the same direction as installation, causing forward rotation of the anchor, such that the fluke tip points downwards. This can be achieved by either pulling on a second line attached to the back of the upper shackle or by failing a shear pin. The Bruce VLA is keyed by rotating the anchor backwards, such that the fluke tip points upwards. In either case, in the final orientation, the anchor line force acts essentially perpendicular to the fluke.

The holding capacity of a drag anchor, either traditional fixed fluke or VLA, is a direct function of the depth to which the anchor can be installed. The issue of final penetration depth is further complicated for VLAs as rotation of the anchor to its operational mode causes some upward movement, reducing the embedment, moving

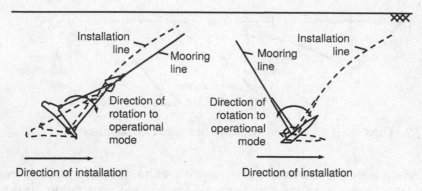

*Figure 7.21* Deployment and operational modes of VLAs

into softer soil and remoulding the soil in the immediate vicinity of the anchor. The reduction in soil strength due to the latter effect may be recovered in due course by consolidation, but the loss of embedment is crucial. Wilde *et al.* (2001) report upward movements during keying of SEPLAs that range between 0.5 and 1.7 times the plate height, which is a disconcertingly wide range. Predicted embedment depths and capacities for drag anchors show a significant degree of scatter while there is much closer agreement on the anchor capacity for a given embedment depth (Murff *et al.* 2005). Thus, accurate prediction of anchor performance relies on being able to simulate the installation process accurately. Anchor trajectory and final embedment, for traditional fixed fluke or VLAs, can be modelled using anchor simulation programs or with simplified analytical models based on theoretical principles.

Once embedded, and keyed so that the mooring line is normal to the anchor fluke, the holding capacity of a VLA, or other plate anchor, may be calculated from the plasticity solution for an embedded plate loaded normal to its surface (Martin and Randolph 2001), given by

$$T_a = A_f N_c s_u \tag{7.14}$$

where

$T_a$ = tension in the anchor line at the padeye
$A_f$ = fluke bearing area
$N_c$ = bearing capacity factor for deep 'flow-around' bearing failure, typically taken in the range 12–13, with slight dependence on the anchor roughness
$s_u$ = undrained shear strength at fluke level.

### Drag anchor simulation programs

Drag anchor simulation programs currently operate by incrementally advancing the anchor through the soil, assuming that it travels parallel to its flukes. The anchor characteristics for kinematic analysis are indicated in Figure 7.22. (An example of a system of soil reaction that may be used to model force and moment equilibrium of an anchor is shown later in Figure 7.28.) At each step, the soil pressures acting on the various elements of the anchor are calculated using standard bearing capacity type equations. The orientation of the anchor is updated with each step, calculated by moment equilibrium considerations assuming centres of pressure for each of the soil

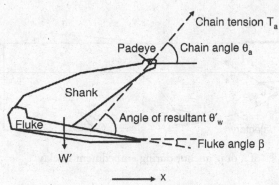

*Figure* 7.22 Characteristics of a drag anchor for kinematic analysis

reactions. At each incremental advance of the anchor, the chain equations need to be solved to evaluate the chain inclination at the shackle (which affects moment equilibrium) and the chain tension at the mudline.

Simulation of anchor embedment can be summarised as follows:

1. From a given starting point (e.g. anchor shank horizontal, embedment depth zero), advance the anchor through a horizontal increment $\Delta x$.
2. Calculate the new embedment depth assuming motion parallel to the previous fluke orientation or at a small (given) offset angle.
3. Calculate the anchor resistance $T_a$ and its orientation relative to the anchor $\theta'_w$ from the anchor characteristics and local soil strength.
4. Calculate the chain angle $\theta_a$ taking account of the average chain resistance through the superficial soil.
5. Calculate the anchor tension at the mudline $T_m$.
6. Adjust the anchor angle $\beta$ in order to satisfy equilibrium (see comment below).
7. Increment the anchor displacement by a further $\Delta x$ and repeat from Step 2.

The anchor will follow a path as indicated in Figure 7.23 where the anchor line makes an almost constant angle to the fluke (the variation being due to the gradually lessening effect of the anchor weight, as the anchor embeds and capacity increases). At ultimate penetration depth, the anchor flukes reach a horizontal attitude (although in practice ultimate penetration depth is often not achieved). Step 6 in the foregoing summary is critical to the process, and may be achieved either by considering moment equilibrium of the various forces acting on the anchor, or by the simple assumption that the angle $\theta'_w$ (= $\theta_a + \beta$) stays constant during the embedment process, as confirmed through detailed analysis (Aubeny and Chi 2010).

*Simplified analytical procedures*

The full procedure for simulating the trajectory of a drag anchor requires considerable computational effort. To overcome this, simplified procedures for predicting the trajectory and anchor line force of a drag anchor can be derived based on limit equilibrium or plasticity methods (e.g. Neubecker and Randolph 1996, Thorne 1998, Dahlberg 1998, Bransby and O'Neill 1999, O'Neill *et al*. 2003). Neubecker and

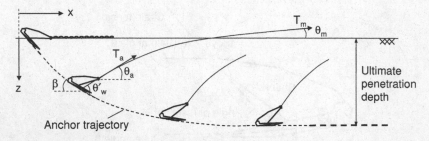

*Figure 7.23* Kinematics of a drag anchor during embedment in clay

Randolph (1996) proposed a limit equilibrium approach where the anchor resistance is taken as a function of the anchor area projected at right angles to the direction of travel (assumed approximately parallel to the flukes), using a conventional bearing capacity approach. As the anchor is dragged through the soil, the resisting force is assumed to maintain a fixed angle with the flukes, while the magnitude varies according to the local average shear strength. By coupling this approach with the chain solution (Section 7.3.2), the complete embedment response of the anchor can be simulated. Bransby and O'Neill (1999) and O'Neill *et al.* (2003) proposed an alternative approach based on macro plasticity concepts using interaction diagrams (yield envelopes in combined load and moment space) as plastic potential surfaces from which to derive the relative motion of the anchor at each stage. These two simplified approaches are presented in the following sections.

LIMIT EQUILIBRIUM APPROACH

In order to simplify the full anchor installation procedure as outlined preceding section, it is rational to express the geotechnical resisting force acting on the anchor parallel to the direction of travel $T_p$ as a product of the frontal projected area in the direction of travel $A_p$ and the local bearing capacity of the soil

$$T_p = (fA_p)N_c s_u \tag{7.15}$$

where f is a form factor for the anchor, which can be seen as a factor of the effective (projected) area of the anchor or as a modification to the bearing capacity factor $N_c$ (normally taken as 9). Figure 7.24 illustrates the forces acting on a traditional fixed fluke anchor during installation, although the procedure is equally applicable to VLAs.

It is judicious to use the undrained shear strength at fluke depth for the prediction of the anchor resistance $T_p$ as the theory for the anchor is based on the soil resistance parallel to the fluke motion of the anchor, with the bulk of the resistance assumed to act at fluke level. In contrast, the chain solution depends on the average bearing resistance over the depth down to the padeye and use of an average value of undrained shear strength is more appropriate. In reality, some form of weighted undrained shear strength for the anchor might be more appropriate, but probably 80 or 90 per cent for $s_u$ at the fluke level and 20 or 10 per cent at the padeye level (the projected area is dominated by the fluke, with the shank being relatively low profile).

It is evident from moment equilibrium that for a weightless anchor, there will be geotechnical forces normal to the fluke $T_n$, such that the resultant resisting force of a

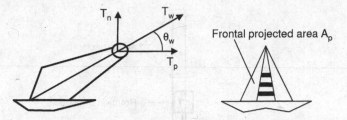

*Figure 7.24* Forces acting on a weightless anchor during installation

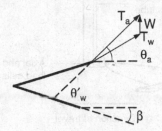

*Figure 7.25* Rotation of a weighty anchor during embedment

weightless anchor $T_w$ will make a unique angle $\theta_w$ with the fluke. Therefore, the anchor capacity at any embedment can be taken as proportional to the local shear strength, given by

$$T_w = \frac{T_p}{\cos\theta_w} = \frac{fA_pN_cs_u}{\cos\theta_w} \tag{7.16}$$

The angle of the resultant to the fluke $\theta_w$ may be taken as a geometric characteristic of the anchor. A minor variation will arise due to the differential soil strength over the depth occupied by the anchor. However, considering the deep embedment achieved by drag anchors in soft clay, this differential is assumed to be small and hence $\theta_w$ is taken as a constant.

During embedment, the anchor fluke will be at some angle $\beta$ to the horizontal and the anchor weight will lead to modified resistance components parallel and normal to the anchor fluke, leading to a modified resultant angle $\theta'_w$ (Figure 7.25). The chain angle at the padeye $\theta_a$ is then given by

$$\theta_a = \theta'_w - \beta \tag{7.17}$$

At the ultimate penetration depth, the flukes tend to the horizontal, i.e. $\beta \to 0$, and $\theta'_w \to \theta_a$. The resultant force in the anchor chain at the anchor attachment point $T_a$ can be expressed as

$$T_a = \frac{T_p}{\cos\theta'_w} = \frac{fA_pN_cs_u}{\cos\theta'_w} \tag{7.18}$$

The angle of the resultant anchor line tension $T_a$ to the fluke for a weighty anchor $\theta'_w$ can be determined (from simple trigonometry) as

$$\theta'_w = \tan^{-1}\frac{W + T_p\tan\theta_w}{T_p} \tag{7.19}$$

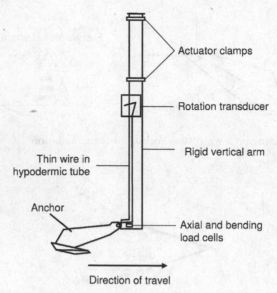

*Figure 7.26* Anchor loading arm (O'Neill *et al.* 1997)

FORM FACTOR f AND ANGLE OF RESULTANT FORCE $\theta_w$

Every drag anchor can be considered to have inherent properties f and $\theta_w$ that are unique to an anchor and are capable of describing its behaviour in any fine-grained soil profile. These anchor properties could be evaluated by experimental investigation, calibration against field data, results from analyses using the currently available drag anchor simulation procedures or detailed finite element analyses.

The apparatus illustrated in Figure 7.26 was developed to determine the form factor and angle of the resultant to the fluke $\theta_w$ by forcing the anchor to travel horizontally at a fixed embedment but allowing gradual rotation of the anchor. At any given stage, the anchor angle and the vertical and horizontal components of resistance are measured. The anchor considered is a 1:80 scale model of a 32 tonne Vryhof Stevpris with fluke-shank angle 50°. The resulting variation of the angle $\theta_w$ is shown in Figure 7.27a for different drags, in one of which the anchor was pre-buried at an orientation corresponding to the end of the previous drag (hence the rising curve for $\theta_w$). The angle stabilises rapidly towards a limiting value with an average from several tests of 28°. The resultant force at the anchor padeye $T_a$ also reaches a steady value within about one fluke length of drag giving a resulting form factor, averaged from several tests, of ~1.4 (Figure 7.27b).

INTERACTION DIAGRAM APPROACH

Figure 7.28 illustrates the various forces acting on an anchor as it embeds. The first step, using conventional soil mechanics approaches to estimate the resisting forces along the shank, is to transfer the tension in the anchor line at the padeye $T_a$ and the overall shank forces $F_{bn}$ and $F_{bs}$ (acting at point b) to equivalent normal ($F_{dn}$), parallel (sliding) ($F_{ds}$) and moment ($M_d$) resultants acting at the mid-point of the fluke (point d). A failure envelope derived from finite element analysis (generally for a simplified fluke shape) is then used to assess (a) conditions for plastic motion of the fluke, as determined by the size of the failure envelope and (b) the relative normal, parallel

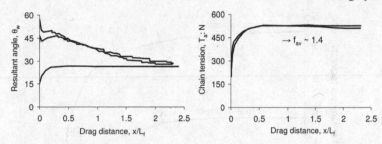

*Figure 7.27* Variation of resultant angle $\theta_w$ and anchor force $T_a$ with drag distance for 32 tonne Vryhof Stevpris (O'Neill *et al.* 1997)

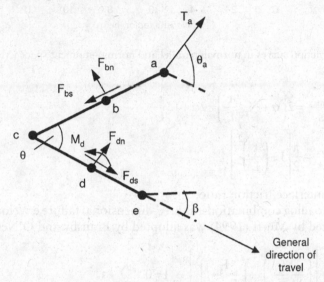

*Figure 7.28* Schematic of anchor for transfer of shackle force to fluke (Murff *et al.* 2005)

and rotational motion of the plate, determined from the local gradients of the failure envelope (Bransby and O'Neill 1999). Typical failure envelopes for an anchor fluke, modelled as a two-dimensional strip with length (across the width of the strip) to thickness ratio of $L/t = 7$ are shown in Figure 7.29 (Murff *et al.* 2005). The normal, parallel and moment load factors (per unit length) are defined, respectively, by

$$N_n = \frac{F_n}{Ls_u} \tag{7.20a}$$

$$N_s = \frac{F_s}{Ls_u} \tag{7.20b}$$

$$N_m = \frac{M}{L^2 s_u} \tag{7.20c}$$

Maximum values of these factors for uniaxial loading may be estimated from upper-bound analyses (Bransby and O'Neill 1999, O'Neill *et al.* 2003), giving

$$N_{n\,max} = \frac{F_{n\,max}}{Ls_u} = 3\pi\,2 + \frac{t}{L}\left(\alpha + \frac{1+\alpha}{\sqrt{2}}\right) \tag{7.21}$$

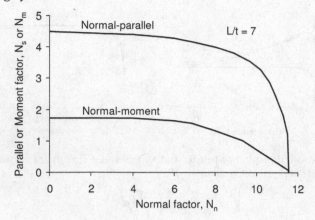

*Figure 7.29* Interaction curves in normal–parallel and normal–moment space (Murff *et al.* 2005)

$$N_{s\,max} = \frac{F_{s\,max}}{Ls_u} = 2\left(\alpha + N_{tip}\frac{t}{L}\right) \approx 2\alpha + 15\frac{t}{L} \tag{7.21b}$$

$$N_m = \frac{M_{max}}{L^2 s_u} = \frac{\pi}{2}\left[1+\left(\frac{t}{L}\right)^2\right] \tag{7.21c}$$

where $\alpha$ is the interface friction ratio.

For general loading combinations, a three-dimensional failure envelope is required. A form suggested by Murff (1994) was adopted by Bransby and O'Neill (1999)

$$\left(\frac{F_n}{F_{n\,max}}\right)^q + \left[\left(\frac{M}{M_{max}}\right)^r + \left(\frac{F_s}{F_{s\,max}}\right)^s\right]^{\frac{1}{p}} - 1 = 0 \tag{7.22}$$

with coefficient values optimised to fit the results of finite element computations. Coefficients deduced from three dimensional finite element analysis (Elkhatib 2006) are summarised in Table 7.2.

The procedure for determination of an anchor's trajectory using the failure envelope approach is summarised in the following steps (Bransby and O'Neill 1999):

1. Assume an initial orientation and anchor depth.
2. Calculate the normal and sliding shank resistances $F_{bn}$ and $F_{bs}$.
3. Assume a padeye chain tension $T_a$ and calculate the padeye chain angle $\theta_a$ using the simplified chain solution (Neubecker and Randolph 1995).
4. Calculate $F_n$, $F_s$ and M using static equilibrium and check if they lie on the yield locus. If not then adjust $T_a$ until they do.

*Table 7.2* Exponents for the Bransby-O'Neill failure envelope (Equation 7.22)

| Parameter | p | q | r | s |
|-----------|-----|-----|-----|-----|
| Value | 1.1 | 4.0 | 1.1 | 4.2 |

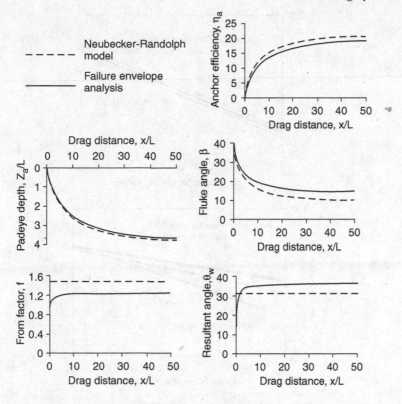

*Figure 7.30* Comparison of kinematic failure envelope analysis with Neubecker Randolph method; 32 tonne Vryhof Stevpris anchor with 50° fluke angle and 5 m fluke length, soft clay with $s_u = 1.5z$ kPa (O'Neill *et al.* 2003)

5. Use normality to determine the direction of travel and relative rotation of the anchor.
6. Assume an incremental displacement $\delta h$ parallel to the fluke face and calculate $\delta v$ and $\delta\beta$.
7. Move the anchor to the next position and repeat the procedure.

COMPARISON OF DIFFERENT SIMPLIFIED ANALYTICAL PROCEDURES

An example kinematic analysis using the simplified limit equilibrium and failure envelope approaches described earlier is shown in Figure 7.30. The analysis is based loosely on a 32 tonne Vryhof Stevpris anchor, with 50° fluke angle, fluke length of 5 m and a total fluke area of about 25 m², penetrating soft clay with $s_u = 1.5z$ kPa. Similar predictions of anchor efficiency $\eta = UHC/W$ and anchor trajectory are achieved with both methods. A small disparity is observed in the prediction of fluke angle during installation, with the Neubecker and Randolph method assuming travel parallel to the fluke while the failure envelope analysis indicates the anchor travels at a slightly deeper angle, reaching a maximum of 4° below the fluke angle at ultimate penetration depth. Predictions of form factor f and the angle of resultant chain load to the fluke $\theta_w$ also differ between the different types of analyses, with the failure envelope approach predicting lower form factor and higher resultant angle than the Neubecker and Randolph approach. In other words, the different

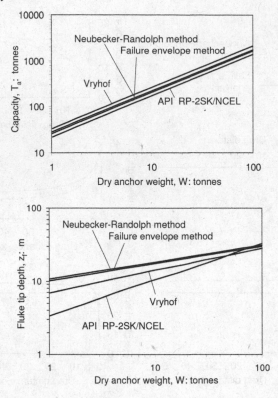

*Figure 7.31* Comparison of generic design charts with simplified analytical methods (O'Neill *et al.* 2003)

methods predict slightly different contributions of the fluke to anchor capacity. Given the idealisation of the shape of the failure envelope in the one approach and the assumptions of the geotechnical forces in the other, the slight differences in predictions are to be expected.

COMPARISON OF SIMPLIFIED ANALYTICAL METHODS WITH DESIGN CHARTS

Figure 7.31 compares predictions of holding capacity and final penetration depth as a function of dry anchor weight W for 50° Vryhof Stevpris anchors in soft clay determined by generic design charts (NCEL 1987, Vryhof Anchors 1990, API 2005) and simplified analytical approaches (Neubecker and Randolph 1996, Bransby and O'Neill 1999). The simplified analytical approaches predict similar responses to each other. The analytical methods predict higher capacity than the API/NCEL curve but lower than the Vryhof curve, with a similar disparity irrespective of anchor size. The analytical methods predict deeper embedment than both the generic design curves, particularly for small and mid-size anchors.

EFFECT OF INTERFACE FRICTION

Figure 7.29 showed results for anchors with a fully rough foundation–soil interface. Figure 7.32 shows installation trajectories and anchor line loads for anchors with a

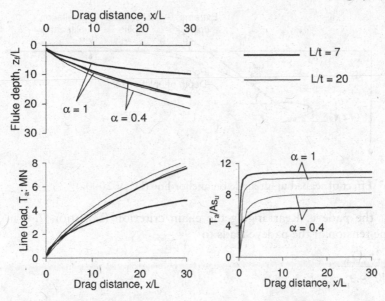

*Figure 7.32* Effect of interface friction on drag anchor performance (Elkhatib and Randolph 2005)

rough foundation soil interface ($\alpha = 1$) and with an interface friction ratio $\alpha = 0.4$, predicted by finite element analyses. The anchor geometry in this study was idealised as a strip, with fluke length 3.5 m, but representing a square fluke of area 10.5 m², with a fixed shank, 4.2 m long, at an angle of 55° to the fluke. The soil was modelled as a soft clay with an undrained shear strength profile $s_u = 1.2z$ kPa, where z is depth in metres, and results are shown for fluke length to thickness ratios L/t = 7 and 20 (Elkhatib and Randolph 2005).

Final anchor embedment increases by 75 per cent for the anchor with L/t = 7 when $\alpha$ is reduced to 0.4. The effect is not so significant for the thinner anchor with L/t = 20, with a resulting increase in embedment of 20 per cent. Anchor line loads also increase with reducing interface roughness, more so for the thicker fluke anchor. In all cases, the normalised load $T_a/As_u$ reaches a plateau following a similar drag distance, of around 10 fluke lengths, indicating the anchor has reached a steady state and the chain angle relative to the fluke (i.e. $\theta_a + \beta$) remains constant as the anchor embeds.

### 7.4.4 Design considerations for drag anchors

Different approaches are required to estimate the holding capacity of traditional fixed fluke anchors and VLAs. Traditional fixed fluke anchors mobilise holding capacity equal to the load required for installation, such that the ultimate penetration depth and, therefore, holding capacity is determined by combining the chain solution (outlined in Section 7.3.2) with the analytical solution for installation of a drag anchor for a given form factor f and angle of the resultant anchor force $\theta_w$. Ultimate embedment, or the final depth of the flukes $z_f$, is identified by equating the tension in the anchor line at the padeye from the anchor criterion with the chain criterion for anchor line

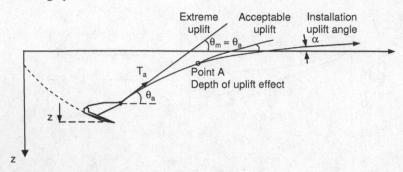

*Figure 7.33* Effect of seabed uplift angle on anchor line (DNV 2000a)

tension at the padeye. Rearranging the chain criterion (Equation 7.4) in terms of anchor line tension at the padeye leads to

$$T_a = \frac{2z_a Q_{av}}{(\theta_a^2 - \theta_m^2)} \tag{7.23}$$

For equilibrium at final embedment depth, the anchor line tension at the padeye $T_a$ from the anchor solution (Equation 7.18) must equal that from the chain solution (Equation 7.23). Solving for $T_a$ gives the final depth to flukes $z_f$. The tension in the anchor chain at the mudline, i.e. the holding capacity, is related to the tension at the padeye by the chain criterion (Equation 7.8). The chain angle at the mudline $\theta_m$ depends on the mooring configuration and will be a given parameter of the design. This relationship can be used to provide a direct estimate of the ultimate capacity of a drag anchor in soft sediments (Neubecker and Randolph 1996).

Traditional fixed fluke anchors are intended for essentially catenary moorings, with zero uplift angle at the seabed, i.e. $\theta_m = 0$ although non-zero seabed uplift angles may be incurred during storm loading, even with the use of a tensioner. Modest seabed uplift angles can be sustained without adversely affecting an anchor's holding capacity but must be significantly less than the anchor line angle at the padeye $\theta_a$. Significant uplift angle at the seabed affects the anchor force and moment equilibrium leading to reduced anchor capacity. Figure 7.33 illustrates the effect of non-zero seabed uplift angles on the shape of the embedded anchor line. Acceptable uplift angles at the seabed only affect the inverse catenary shape of the anchor line to a limited depth (point A in Figure 7.33) such that the anchor is unaffected. Non-zero seabed anchor line angles should be avoided during installation, especially at the start of drag when uplift will prevent the anchor from penetrating the seabed. With increased embedment, small uplift angles can be tolerated without significantly affecting the penetration trajectory and hence final penetration depth and holding capacity. DnV RP-E302 (2000) recommends a minimum depth of 2.5 fluke lengths should be achieved before any uplift can be tolerated during installation and even then, the mudline chain angle should not exceed 10°. Higher uplift angles can be tolerated in operational conditions during storm loading than during installation, but should not typically exceed 15°–20°.

As mentioned earlier in this section, the embedment process for VLAs is similar to that for traditional fixed fluke anchors. Installation depth of VLAs can, therefore, be predicted using similar methods to those outlined before. The holding capacity of a

VLA, however, can be predicted by simple bearing capacity theory for an embedded plate loaded normal to its surface (Martin and Randolph 2001).

$$T_a = A_f N_c s_u \tag{7.24}$$

where

$T_a$ = tension in the anchor line at the padeye
$A_f$ = fluke bearing area
$N_c$ = bearing capacity factor for deep 'flow-around' bearing failure, typically taken as $12 < N_c < 13$.
$s_u$ = undrained shear strength at fluke level

The ratio of holding capacity to installation load is called the performance ratio and typically lies in the range 1.5–2.

## Other considerations

### REMOULDING AND CONSOLIDATION

Installation of a drag anchor leads to remoulding of the soil around the anchor, as it penetrates though the soil, such that frictional resistance is governed by the remoulded shear strength $s_{ur}$ in a narrow zone close to the anchor. Remoulded shear strength is typically accounted for through a friction coefficient $\alpha$, based on soil sensitivity $S_t$ where

$$S_t = \frac{s_u}{s_{ur}} \tag{7.25}$$

$$\alpha \approx \frac{1}{S_t} \tag{7.26}$$

Reconsolidation, either following installation or during an interruption during installation, will lead to a time-dependent increase in shear strength, commonly referred to as set-up or soaking (as for piles or shallow foundations). Temporary stoppages during installation can adversely affect the achievable final penetration depth, and therefore, holding capacity, as increased penetration resistance develops during reconsolidation during the stoppage. Conversely, delays between installation and hook-up to the floating facility has a beneficial effect on holding capacity. Rotation of a VLA to its operational mode following installation means that the zone of soil remoulded during installation has little effect on the holding capacity of the anchor although the rotation itself leads to remoulding. The amount of strength recovery following remoulding depends on the elapsed time, soil sensitivity, anchor geometry and the depth and orientation of the anchor. DNV RP-E301 recommend a consolidation factor $U_{con}$ in the range of 1.25–1.55 for fixed fluke drag anchors for typical soil sensitivities $S_t$ between 2 and 2.5.

### CYCLIC LOADING

Loads transmitted to an anchor through an anchor line will always remain in tension, i.e. anchors will experience one-way tensile cycling, leading to failure through accumulation of residual strains (hence displacements) but limited detrimental effect on shear strength compared with cyclic loading involving stress reversal, i.e. two-way cycling (see Chapter 4 for further details on cyclic loading of soil). In practice, the potential negative effects of cyclic loading are likely to be compensated by increased shear strength of the soil in front of the anchor, due to consolidation under the sustained mooring loads. Overall, and given that mooring systems have low sensitivity to (moderate) movement of the anchor, the effects of cyclic loading are unlikely to represent a critical design consideration for anchors.

## 7.5 Design principles for suction caissons

### 7.5.1 Overview

A summary of applications of suction caisson anchors for both catenary and taut-wire moorings is shown in Table 7.3. To date, suction caissons have not been used for deepwater TLPs. Many of the developments listed are taut and semi-taut line moorings that have been installed since 2000 indicating the relatively limited experience to date of the suction caisson anchor.

Design issues for suction caissons can be divided into those associated with installation and those concerned with operational conditions, principally capacity. In both areas, structural integrity of the caisson must be considered although that is not considered further here. Until recently, no formal recommended practice or design guidelines were available for suction caisson design. A comprehensive report on suction caisson design methods was presented by Andersen *et al.* (2005), the culmination of a collaborative project sponsored jointly by the American Petroleum Institute (API) and Deepstar that was completed in 2003, and has now been incorporated into API RP-2SK (2005).

### 7.5.2 Installation of suction caissons

#### Prediction procedure

The installation resistance during penetration of caissons comprises external and internal friction along the shaft and along any extended plate stiffeners within the caisson, and end-bearing resistance on the tip of the caisson and any external or internal protuberances (e.g. the padeye or internal stiffeners). This resistance is estimated using conventional soil mechanics principles, generally taking an end-bearing factor of $N_c \sim 7.5$ for the caisson tip (since it is closer to a deeply embedded strip than a circular foundation), and shaft friction based on the remoulded shear strength of the soil. An effective stress approach may be used to estimate the shaft friction during installation and is particularly relevant where painted sections render the caisson surface smooth (Colliat and Dendani 2002, Dendani and Colliat 2002) or in unusual soil conditions (Erbrich and Hefer 2002).

Equations for predicting penetration resistance, required and allowable underpressure and the factor of safety of the soil plug are presented in Chapter 6 with respect to shallow skirted foundations. The same principles apply for deeper-skirted suction caissons although the effect of internal stiffeners and the external padeye must also be considered in terms of bearing and frictional resistance, taking into consideration the effect of soil flow into or outside the caisson.

#### Calculated and observed penetration resistance

Figure 7.34 shows a comparison of measured installation pressures required for suction caissons in two different soft clay sites, drawn from six case histories described in

*Table 7.3* **Applications of suction caisson anchors (Andersen *et al.* 2005)**

| Name (Operator) | Year | Location | Facility | Depth (m) | No. | D (m) | L (m) | L/D |
|---|---|---|---|---|---|---|---|---|
| Gorm (Shell) | 1981 | North Sea | FPS | 40 | 12 | 3.5 | 8.5 | 2.4 |
| Nkossa (Elf) | 1995 | Gulf of Guinea | Barge | 200 | 8 | 4.0 | 11.8 | 3.0 |
|  |  |  |  |  | 4 | 4.5 | 12.5 | 2.8 |
| YME (Statoil) | 1995 | North Sea | Buoy | 100 | 8 | 5.0 | 7.0 | 1.4 |
| Harding (BP) | 1995 | North Sea | Buoy | 110 | 8 | 5.0 | 10.0 | 2.0 |
| Norne (Statoil) | 1996 | North Sea | FPSO | 375 | 12 | 4.9 | 10.0 | 2.0 |
| Aquila (Agip) | 1997 | Adriatic |  | 850 | 8 | 4.5 | 16.2 | 3.6 |
|  |  |  |  |  |  | 5.0 |  | 3.2 |
| Njord (Norsk Hydro) | 1997 | North Sea | FPS | 330 | 12 | 5.0 | 8.0–10 | 1.4– |
|  |  |  |  |  | 8 | 5.0 | 7.0–10 | 2.0 |
| Marlim P19, 26 | 1997 | Offshore Brazil | FPS | 830 | 32 | 4.7 | 13.1 | 2.8 |
| Schiehallion (BP) | 1997 | Shetlands | FPSO | 400 | 14 | 6.5 | 12.0 | 1.8 |
| Curlew (Shell) | 1997 | North Sea | FPSO | 90 | 9 | 5.0–7.0 | 9.0–12.0 | 1.7–1.8 |
| Visund (Norsk Hydro) | 1997 | North Sea | FPS | 345 | 16 | 5.0 | 11.0 | 2.2 |
| Aasgard A (Statoil) | 1998 | North Sea | FPSO | 350 | 12 | 5.0 | 11.0 | 2.2 |
| Laminaria (Woodside) | 1998 | Timor Sea | FPSO | 400 | 9 | 5.0 | 12.0 | 2.4 |
| Siri (Statoil) | 1998 | North Sea | Loading | 60 | 1 | 4.3 | 4.6 | 1.1 |
| Marlim P18 (Petrobras) | 1998 | Offshore Brazil | Riser | 900 | 2 | 18.0 | 16.2 | 0.9 |
| Marlim P33 (Petrobras) | 1998 | Offshore Brazil | FPSO | 790 | 6 | 4.7 | 20.0 | 4.3 |
| Aasgard B (Statoil) | 1999 | North Sea | FPS | 350 | 16 | 5.0 | 10.0 | 2.0 |
| Aasgard C (Statoil) | 1999 | North Sea | FPSO | 350 | 9 | 5.0 | 12.0 | 2.4 |
| Marlim P35 (Petrobras) | 1999 | Offshore Brazil | FPSO | 860 | 6 | 4.8 | 17.0 | 3.5 |
| Troll C (Norsk Hydro) | 1999 | North Sea | FPS | 350 | 12 | 5.0 | 15.0 | 3.0 |
| Kuito (Chevron) | 1999 | West Africa | FPSO | 400 | 12 | 3.5 | 11.0–14.0 | 3.1–4.2 |
| Diana (ExxonMobil) | 2000 | Gulf of Mexico | SPAR | 1,500 | 12 | 6.5 | 30.0 | 4.6 |
| Girassol (TFE) | 2001 | West Africa | Riser | 1,350 | 3 | 8.0 | 20.0 | 2.5 |
|  |  |  | FPSO |  | 16 | 4.5 | 17.0 | 3.8 |
|  |  |  | Buoy |  | 6 | 5.0 | 18.0 | 3.6 |
|  |  |  | Buoy |  | 3 | 5.0 | 16.1 | 3.2 |
| Horn Mountain (BP) | 2002 | Gulf of Mexico | SPAR | 1,650 | 9 | 5.5 | 27.4–29.0 | 5.0–5.3 |
| Na Kika (Shell/BP) | 2002 | Gulf of Mexico | FPS | 1,920 | 16 | 4.3 | 23.8 | 5.5 |
| Wen-chang (CNOOC) | 2002 | South China Sea | FPSO | 120 | 9 | 5.5 | 12.1 | 2.2 |
|  |  |  |  |  |  |  | 12.8 | 2.3 |
| Barracuda (Petrobras) | 2003 | Offshore Brazil | FPSO | 825 | 18 | 5.0 | 16.5 | 3.3 |
| Caratinga (Petrobras) | 2003 | Offshore Brazil | FPSO | 1,030 | 18 | 5.0 | 16.5 | 3.3 |
| Bonga (Shell Nigeria) | 2003 | Offshore Nigeria | FPSO | 980 | 12 | 5.0 | 17.5 | 3.5 |
|  |  |  |  |  |  | 5.0 | 16.0 | 3.2 |

*Table 7.3* (**Continued**)

| | | | | | | | | |
|---|---|---|---|---|---|---|---|---|
| Red Hawk (Kerr McGee) | 2003 | Gulf of Mexico | SPAR | 1,600 | 8 | 5.5 | 22.9 | 4.2 |
| Devils Tower (Dominion) | 2003 | Gulf of Mexico | SPAR | 1,700 | 9 | 5.8 | 34.8 | 6.0 |
| Holstein (BP) | 2003 | Gulf of Mexico | SPAR | 1,280 | 16 | 5.5 | 36.3 | 6.6 |
| Panyu (CNOOC) | 2003 | South China Sea | FPSO | 105 | 9 | 5.0 | 11.7 | 2.3 |
| | | | | | | 6.0 | 12.7 | 2.1 |
| Thunder Horse (BP) | 2004 | Gulf of Mexico | FPS | 1,830 | 16 | 5.5 | 27.5 | 5.0 |
| | | | Manifold | | 4 | 6.4 | 23.8 | 3.7 |
| | | | PLET | | 4 | 5.5 | 26.0 | 4.7 |
| | | | Injection | | 3 | 3.4 | 19.0 | 5.6 |
| | | | Injection | | 2 | 3.4 | 20.0 | 5.9 |
| Mad Dog (BP) | 2004 | Gulf of Mexico | SPAR | 1,600 | 11 | 5.5 | 11.0 | 2.0 |

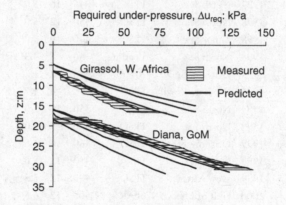

*Figure 7.34* Measured and (Class C) predictions of penetration resistance for suction caissons at two soft clay sites (Andersen *et al.* 2005)

detail by Andersen *et al.* (2005). The predictions were provided by four internationally experienced practitioners, and the range of results is indicative of current uncertainties in estimating parameters (even for Class C predictions!). As an example, the lowest prediction for the Diana site was based on an effective stress approach using an interface friction angle $\delta$ of 12° based on correlations with plasticity index (in the absence of direct measurements). Increasing $\delta$ to 17° below a depth of 20 m, where the plasticity index of the soil reduces to less than 35 per cent, would provide a much improved fit and is consistent with other data from the Gulf of Mexico (Jardine and Saldivar 1999). At the Diana site, there was a significant delay between self-weight penetration and the application of suction, during which consolidation led to an increase in capacity, as noted by the step increase in suction before further penetration occurred. By contrast, without such delay, only a small amount of suction is necessary, as in the Girassol example. In this example, and by contrast with the Diana predictions, the measured data tend to fall below any of the predictors' estimates.

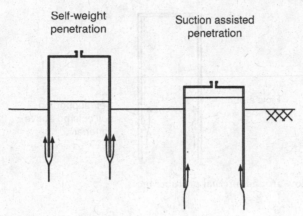

*Figure 7.35* Soil flow around caisson tips

Andersen *et al.* (2005) identified two main areas of variation in practice for estimating installation performance, other than estimating key soil parameters. These concern the treatment of ring stiffeners in regard to soil flow, the choice of end-bearing coefficient $N_c$ for internal stiffeners and the degree to which this coefficient may be affected by friction transferred above the stiffener.

## Soil flow and plug heave

Penetration of the caisson will give rise to heave of the inner soil plug to accommodate (partly) the wall thickness of the caisson and also the full volume of any internal stiffeners. The plug heave will be enhanced by using suction, as opposed to self-weight or other external force, to install the caisson (Figure 7.35). There has been debate over the proportion of the caisson wall that is accommodated by flow of soil inwards into the caisson, or outwards. It is customary to assume a 50:50 split during self-weight penetration, but up to 100:0 split in favour of inward flow once suction is applied (Andersen and Jostad 2002, 2004). Measurements of radial stress changes and long-term axial capacity suggest that the difference between suction and jacked installation is minimal (Chen and Randolph 2007). Field data reported by Newlin (2003) reported measurements of soil plug heave consistent with less than 25 percent (on average) of the total caisson steel volume being accommodated by inward soil flow; these data may have been influenced by an external chamfer on the thickened shoe of the caisson. In practice, the amount of soil flow into the caisson is likely to be strongly affected by the proximity to internal plug failure. During initial penetration by suction, the factor of safety against plug failure is high, and soil flow at the caisson tip is likely to be similar to that during self-weight penetration. Towards the end of penetration, however, the soil plug will be closer to failure, with end-bearing resistance required for stability and a greater proportion of the soil displaced at the caisson tip is likely to flow inwards. Detailed consideration of the soil flow into suction caissons, and the effects on the eventual shaft capacity, has been presented by Chen *et al.* (2009).

The soil that flows into the caisson may not flow fully around ring stiffeners, but may tend to be extruded as a self-supporting inner plug (Figure 7.36), until such point

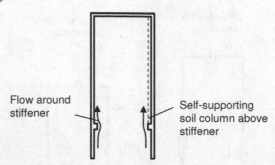

*Figure 7.36* Soil flow around internal protuberances

as the self-weight stresses in the plug cause collapse (Erbrich and Hefer 2002). By that point, however (and particularly for closely spaced stiffeners), trapped water will prevent the soil from relaxing against the inner caisson wall. For such cases, the internal frictional resistance may be extremely low, with a mixture of softened soil and water providing lubrication adjacent to the caisson wall.

### Soil plug stability

The resistance of the internal soil plug against failure during suction installation comprises internal friction (as used to evaluate the caisson penetration resistance) together with the end-bearing resistance of the soil plug. Andersen *et al.* (2005) suggest that, since the internal friction contributes equally to penetration resistance and to soil plug stability, an appropriate design procedure is to adopt a soil strength material factor greater than unity for external caisson resistance (representing a worst case of required suction) and less than unity for the plug base resistance (again representing a worst case estimate). A consensus appeared to be for a minimum material factor of ~1.5 against soil plug failure during suction installation.

For a given situation, design curves can be developed showing suction pressure and soil plug stability ratio, as a function of the installed depth of the caisson (Figure 7.37). The soil plug stability ratio is the ratio of pressure to cause plug failure (at any given penetration of the caisson) to the pressure required to penetrate that caisson to that depth. The ratio is influenced by the weight of the caisson, the soil heterogeneity ratio $s_{um}/kD$, the external and internal friction coefficients, $\alpha_e$ and $\alpha_i$, and the ratio of effective unit weight of the soil to the strength gradient $\gamma/k$. For normally consolidated soil, where the undrained shear strength profile is given by $s_u = kz$, the limiting aspect ratio may be approximated by

$$\left(\frac{L}{D}\right)_{limit} \approx \frac{1}{4\alpha_e}\left[N_c + \left(N_c^2 + \frac{32W\alpha_e}{\pi kD^3}\right)^{\frac{1}{2}}\right] \tag{7.27}$$

For typical normally consolidated clay soils in deep water, plug failure is not an installation design issue for suction caissons with an L/D ratio up to around eight.

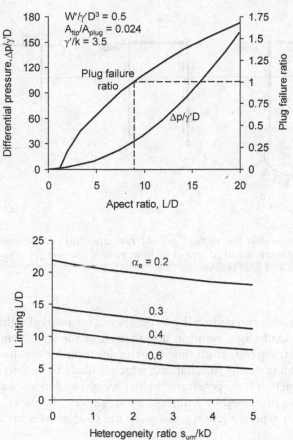

*Figure 7.37* Design curves for installation of a suction caisson for suction pressure and soil plug
stability ratio

### 7.5.3 Capacity of suction caisson anchors

Various predictive methods based on limit equilibrium are employed in the design of
suction caissons, and sometimes these analyses may be augmented by numerical analyses using, for example, the finite element method.

*Vertical capacity*

The vertical uplift capacity of a suction caisson anchor comprises three components:

- Weight of the caisson
- Frictional resistance over the caisson shafts (particularly the external shaft, assuming the caisson has a sealed cap)
- Upwards or reverse end bearing (provided the suction caisson is designed with a
  sealed cap and is able to develop an under pressure during uplift)

The weight of the caisson is the only component that is considered to have a deterministic value with no uncertainty. Calculation of the frictional resistance is not as

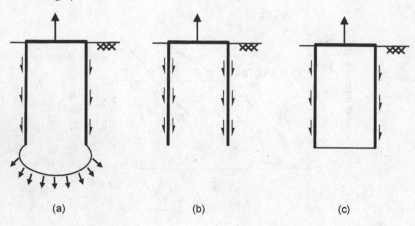

*Figure 7.38* Failure models for vertical pullout resistance (a) reverse end bearing, i.e. with passive suction and (b and c) without passive suction (b) caisson pull out and (c) caisson and plug pull out

straightforward as for conventional pile design as, for example, the cyclic nature of the mooring line load might result in degradation of the soil strength. There is also uncertainty concerning the set-up time (i.e. the duration for regaining shear strength following remoulding during installation), which is likely to be different over the portions of the caisson that were penetrated by self-weight and those that were penetrated with suction. Data presented by Jeanjean (2006) suggest, however, that times for 90 per cent set-up for typical suction caissons in the Gulf of Mexico were in the range 30–90 days.

Reverse end bearing relies on passive suction and can be relied upon if the top cap is sealed. Reverse end bearing is a function of depth to diameter ratio of the caisson and is often assumed to be lower than for end bearing in compression. For pure vertical pullout, centrifuge tests suggested that reverse end bearing is about 80 per cent of the theoretical bearing capacity in compression, (Clukey and Morrison 1993), although other model tests have indicated similar end-bearing in uplift and compression. A bearing capacity factor of $N_c = 9$ is typically used for pile foundations with a depth to diameter ratio greater than 2.5 while Clukey and Morrison suggest a factor as low as 7 might be appropriate for suction caissons. By contrast, Jeanjean *et al.* (2006) noted $N_c$ values of 12 mobilised at large displacements, although values of around 9 were mobilised at the displacement where peak external shaft friction was achieved. If passive suction can be mobilised, axial pullout resistance $V_{ult}$ consists of outer skirt friction and reverse end bearing, hence the use of $N_c = 9$ is appropriate.

If the top plate is not sealed, or sustained load is applied, reverse end bearing will either not be relevant, or will be reduced in value. In the extreme, the vertical capacity $V_{ult}$ consists of outer shaft friction plus the lesser of inner shaft friction and the soil plug self-weight, in addition to the caisson submerged weight. Figure 7.38 illustrates the various components of pullout resistance for sealed and vented conditions.

Vertical pullout resistance $V_{ult}$ is equal to one of the following:

WITH PASSIVE SUCTION

(a) Caisson weight + external shaft friction + reverse end bearing

$$V_{ult} = W' + A_{se}\alpha_e \bar{s}_{u(t)} + N_c s_u A_e \tag{7.28}$$

The overburden from the soil column outside the caisson above tip level, and the weight of the soil plug within the caisson are equal and opposite and so their effects cancel out.

WITHOUT PASSIVE SUCTION

(b) Caisson weight + external shaft friction + internal shaft friction

$$V_{ult} = W' + A_{se}\alpha_o \bar{s}_{u(t)} + A_{si}\alpha_i \bar{s}_{u(t)} \tag{7.29}$$

or

(c) Caisson weight + external wall friction + soil plug weight

$$V_{ult} = W' + A_{se}\alpha_e \bar{s}_{u(t)} + W'_{plug} \tag{7.30}$$

where

$A_{se}$ = external shaft surface area
$A_{si}$ = internal shaft surface area
$A_e$ = external cross-sectional area
$\alpha_e$ = coefficient of external shaft friction (i.e. steel to soil)
$\alpha_i$ = coefficient of internal shaft friction (i.e. steel to soil)
$N_c$ = reverse end bearing factor (~9)
$s_u$ = representative undrained soil shear strength at tip level
$\bar{s}_{u(t)}$= average undrained soil shear strength over penetrated depth at time t after installation
$W'_{plug}$ = effective weight of the soil plug
$W'$ = submerged caisson weight.

There has been debate over whether it is appropriate to adopt a higher friction coefficient $\alpha$ for external shaft friction over the zone where self-weight penetration occurred, compared to the lower zone where suction installation took place. However, recent work suggests negligible difference in $\alpha$ during jacking or suction installation (Chen and Randolph 2007, Chen *et al.* 2009). After full consolidation, the value of $\alpha$ for suction caissons is generally lower than those recommended by the American Petroleum Institute (API) for driven piles in normally consolidated clays, typically by 15–20 per cent. This may be partly attributable to the much higher diameter to wall thickness ratios for suction caissons, which results in lower external excess pore pressures (since reduced volume of steel forced into the soil) and lower final effective stresses compared with a driven pipe pile. It should also be noted that lower values of friction coefficient for internal shaft friction, compared with external friction, were reported from model tests conducted on a double-walled caisson (Jeanjean *et al.* 2006).

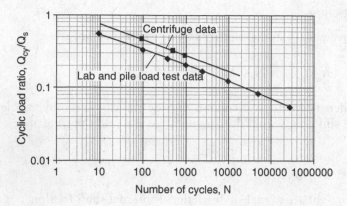

*Figure 7.39* Cyclic loading of suction caissons under uplift (Clukey *et al.* 1995)

EFFECT OF CYCLIC AND SUSTAINED LOADING ON UPLIFT CAPACITY

The combined effects of cyclic loading and an oscillating horizontal component of load can be damaging to the uplift capacity of a suction caisson. Figure 7.39 shows the fatigue strengths measured in centrifuge model tests of suction caissons subjected to quasi-vertical tension loading (Clukey *et al.* 1995). The test results show a consistent trend with the laboratory element data, although somewhat higher strengths. A reduction in capacity of about 50 per cent was observed for 100 cycles of loading for tests where the loading direction was varied by up to 6° from vertical. Other published data for purely vertical loading have shown essentially no degradation in capacity for cyclic loading levels up to 75–80 per cent of the monotonic capacity.

The ability of (sealed cap) caissons to withstand sustained loading, such as from loop currents in the Gulf of Mexico, was examined by Clukey *et al.* (2004). Comparison of data from physical and numerical modelling suggested that no appreciable reduction in capacity occurred over likely design periods (less than eight weeks) although, as for cyclic loading, a reduced capacity of 75–80 per cent of the monotonic capacity may be prudent for loading sustained over a year or more (Chen and Randolph 2007).

*Horizontal capacity*

Early suction caissons were designed for catenary moorings, where the chain load was close to horizontal at the seabed, increasing to angles typically of 10°–20° at the padeye, positioned a depth $z_a$ down the caisson. Horizontal capacity of a suction caisson is maximised by positioning the padeye such that the caisson translates at failure without rotating. Essentially, this is dictated by the centre-line intersection of the line of action of the load $z^*$, as indicated in Figure 7.40. For normally or lightly overconsolidated clay, where the strength gradient is significant, the optimal depth is for $z^*/L \sim 0.65$–$0.7$, with the optimal depth decreasing slightly with increasing loading angle (Andersen *et al.* 2005). Andersen and Jostad (1999) suggest positioning the padeye just below the optimal depth in order to ensure backward rotation at failure, thus reducing the potential for a crack to open on the trailing edge of the caisson.

The failure modes of suction caissons loaded at the seabed are similar to those for laterally loaded piles (Murff and Hamilton 1993). The main features of this mechanism are a conical wedge extending from the edge of the caisson at depth, $z_o$, to a maximum

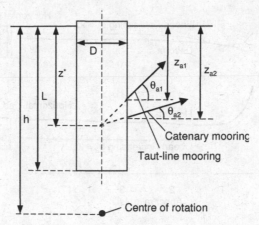

*Figure 7.40* Variation of padeye depth with loading angle for given centre of rotation (Randolph and House 2002)

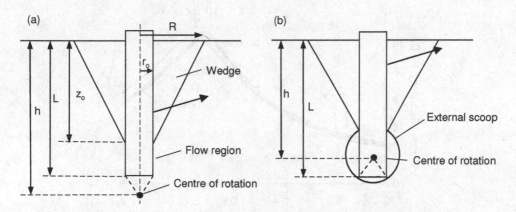

*Figure 7.41* Elements of soil failure mechanism for suction caisson (a) conical wedge and flow region and (b) external base rotational scoop (Randolph and House 2002)

radius of R at the seabed (Figure 7.41a). The radial soil velocity within the wedge may be represented by

$$v_r = v_o \left(\frac{r_o}{r}\right)^\mu \left(\frac{h-z}{h-z^*}\right) \cos \psi \qquad (7.31)$$

where $v_o$ is the horizontal velocity at $z = z^*$, h is the depth to the centre of rotation and $\psi$ is the circumferential angle in plan view.

Below the conical wedge, a confined flow mechanism is assumed with the net resistance given by the plasticity solution derived for laterally loaded piles (Randolph and Houlsby 1984, Martin and Randolph 2006). For caissons loaded above the optimal depth, where the centre of rotation falls within the caisson, an external scoop mechanism can replace the flow region, as shown in Figure 7.41b. Figure 7.42 shows how caisson capacity varies with embedment ratio for normally consolidated and uniform soil strength profiles for the two extreme cases of optimal loading (no rotation) and horizontal loading at seabed level with free rotation. The capacity allowing free

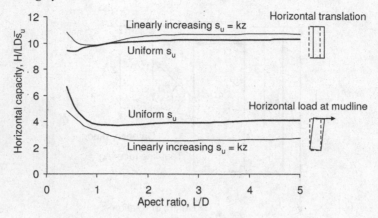

*Figure 7.42* Capacity of suction caissons in soil with uniform strength and with strength proportional to depth (Randolph *et al.* 1998b)

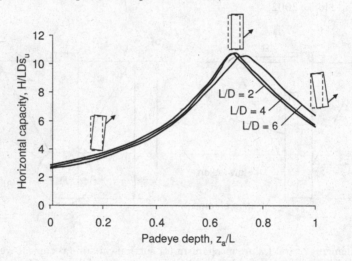

*Figure 7.43* Effect of padeye depth on lateral capacity, results for a constant load angle of 30°, normally consolidated strength profile (Supachawarote *et al.* 2004)

rotation is only about 25 per cent of the optimal capacity for normally consolidated soil, i.e. where the shear strength is proportional to depth.

Figure 7.43 shows the variation of horizontal capacity with padeye depth for suction caissons of varying embedment ratios for a particular case of a constant load angle of 30° and a normally consolidated shear strength profile. The optimum position of the chain attachment (or padeye) is generally at a depth $z_a$ of 60–70 per cent of the embedment depth, leading to horizontal translation of the caisson, rather than forward or backward rotation. Where the chain load is applied at an angle to the horizontal, the critical consideration is the depth at which the load resultant crosses the central axis of the caisson. That depth should correspond to the optimum depth of about 70 per cent of the caisson embedment in order to minimise rotation of the caisson as it fails (thus giving maximum horizontal capacity).

It is generally assumed that gapping behind the caisson does not occur in normally consolidated clays and, therefore, both active and passive wedges occur during failure.

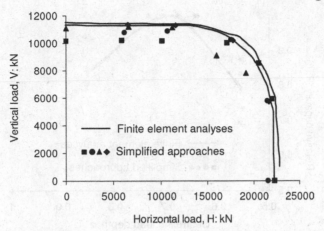

*Figure 7.44* Failure envelope for suction caisson under inclined loading, L/D = 5, normally consolidated soil (Andersen *et al.* 2005)

For failure to occur as pure translation, i.e. with no rotation of the caisson, the padeye must be placed at the optimum level. Assuming this is the case, the maximum horizontal resistance

$$H_{max} \approx LD_e N_p \bar{s}_u \qquad (7.32)$$

where

L = embedded length of caisson

$D_e$ = external diameter of caisson

$N_p$ = lateral bearing capacity factor–varying slightly with embedment ratio L/D (see Figure 7.42)

$\bar{s}_u$ = undrained shear strength averaged over penetration depth

### Inclined loading

For general loading conditions with combinations of vertical and horizontal loading, the pure vertical and pure horizontal capacities will be reduced by the presence of loading in the orthogonal direction. Interaction at the base of the caisson will also lead to a reduction in the vertical reverse end bearing capacity as the caisson is simultaneously displaced laterally or rotated. Interaction between vertical and horizontal loading can be modelled by developing a failure envelope in (V, H) load space (as introduced in Chapter 6 for shallow foundations).

Figure 7.44 shows a failure envelope for a 5 m diameter suction caisson with L/D = 5 under inclined loading applied at the optimal depth; the soil was normally consolidated with $s_{uSS} = 1.25z$ kPa, $\gamma' = 6$ kN/m³ and interface friction taken as $\alpha s_{uSS}$, with $\alpha = 0.65$. The lines represent three-dimensional finite element analyses and the individual points are predictions made prior to the finite element predictions using simplified approaches such as HCMCap (Norwegian Geotechnical Institute 2000), AGSPANC (AG 2001) and profiles of lateral resistance based on Murff and Hamilton (1993). The failure envelope lies significantly within the rectangular boundary defined by the uniaxial vertical and horizontal capacities.

The sensitivity of capacity to the loading depth is illustrated for a 30° loading angle in Figure 7.45. In practice, the mooring line will be attached to a padeye on the caisson at a fixed depth, but the loading angle will vary for different design cases,

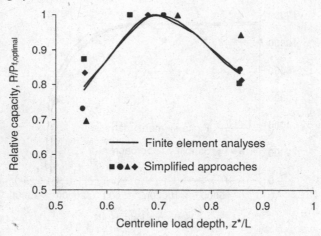

*Figure 7.45* Effect of padeye depth on capacity (Andersen *et al.* 2005)

resulting in a variation in the centre-line intercept for the applied load. As may be seen in Figure 7.45, while the optimum capacity is attained for a centre-line depth of 0.7L, significant reduction in capacity, by 20 per cent, occurs as the loading depth is varied by ±0.15L. A further point is that for typical loading angles applied by taut and semi-taut mooring systems, which are generally in excess of 30° giving V > 0.58H, the caisson capacity is essentially governed by the vertical capacity of the caisson.

The shape of the failure envelopes for suction caissons has been modelled by an elliptical relationship

$$\left(\frac{H}{H_{ult}}\right)^a + \left(\frac{V}{V_{ult}}\right)^b = 1 \tag{7.33}$$

where $H_{ult}$ and $V_{ult}$ are the uniaxial horizontal and vertical capacities, respectively, and the exponents a and b vary with caisson aspect ratio L/D according to (Supachawarote *et al.* 2004)

$$a = \frac{L}{D} + 0.5$$

$$b = \frac{L}{3D} + 4.5 \tag{7.34}$$

The high value of the power b implies that the vertical capacity is less affected by the horizontal load than vice versa.

The inclined load capacity of a suction caisson depends on whether a crack develops along the trailing edge of the caisson. A crack is unlikely to form in a normally consolidated soil while in overconsolidated soils a crack may be sustained. Figure 7.46 compares results obtained using three-dimensional finite element analysis of 5 m diameter suction caissons with and without a crack allowed to form for suction caissons with aspect ratios L/D = 3 and 5 in a lightly overconsolidated soil represented with a strength gradient as $s_u = 10 + 1.5z$ kPa, with $\gamma' = 7.2$ kN/m³ and $K_0 = 0.65$ (Supachawarote *et al.* 2005). The finite element results are compared with results obtained using AGSPANC with either a two-sided (uncracked), or a one-sided (cracked) failure mechanism. The finite element results show a reduction of between 20 and 22 per cent in the pure horizontal capacity, which is maintained for loading

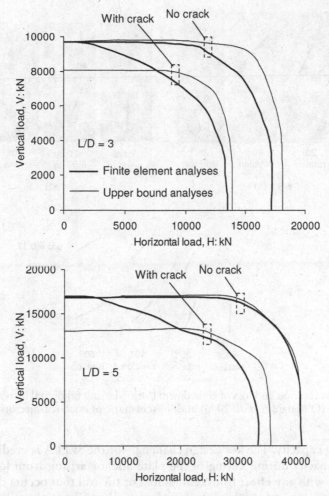

*Figure 7.46* Comparison of failure envelopes for suction caissons with and without crack formation along the caisson–soil interface, lightly overconsolidated deposit (Supachawarote *et al.* 2005)

angles of up to about 45° for L/D = 3 and 30° for L/D = 5. While loading close to vertical produces no reduction due to the formation of a gap, in any practical case the loading direction would range over several degrees from vertical. It, therefore, seems prudent to design suction caissons with allowance for a gap to form and the potential for a reduction in capacity by around 20 per cent for any loading angle. Comparing Figure 7.45 and Figure 7.46, the reduction in capacity due to locating the padeye below optimal level in order to force reverse rotation of the caisson, thus minimising potential gapping, is similar in magnitude to the loss in capacity due to gapping itself.

## 7.6 Design considerations for recent anchor types

### 7.6.1 Suction embedded plate anchor (SEPLA)

SEPLA design can rely on conventional suction caisson design principles for installation and retrieval of the suction caisson and the same principles as applied to VLAs to

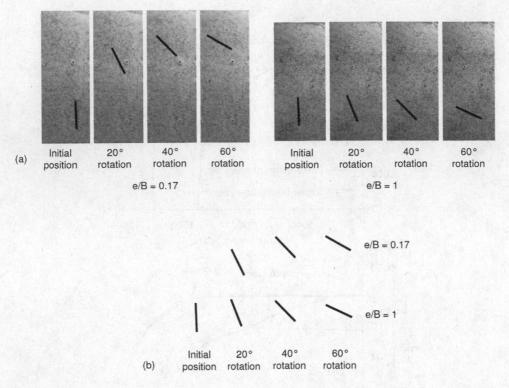

*Figure 7.47* Plate rotation and loss of embedment during keying (a) digital images taken during test (O'Loughlin *et al.* 2006) and (b) schematic of anchor trajectories

predict holding capacity. The key design challenge for the SEPLA is predicting the loss in potential capacity during keying of the plate anchor arising from loss of embedment, together with any effect from remoulding of the soil that occurs during extraction of the caisson and keying of the plate anchor. Strength reduction during remoulding may be partially regained due to set-up, while loss of embedment can be critical to the stability of the anchor, especially where the strength gradient is high. The issues surrounding keying are relevant for VLAs, although the situation is more extreme for SEPLAs as the initial orientation of the plate is vertical, as opposed to near-horizontal for VLAs.

Loss of embedment, and the uncertainty over the expected magnitude, is particularly significant for typical soft clay conditions with a linearly increasing shear strength profile with depth, as loss of embedment is proportional to loss of holding capacity. For example, at an installation depth of five times the plate breadth B, a loss of embedment of 0.5B corresponds to a loss of 10 per cent holding capacity, increasing to 50 per cent loss of holding capacity for a loss of embedment of 2.5B. High vertical translation during keying also leads to a higher potential for reduced anchor capacity due to shallow failure mechanism developing.

Figure 7.47 shows displacement paths for plane strain plate anchors during keying under vertical loading at two padeye eccentricities, as observed in centrifuge tests carried out against a plexiglass window. Figure 7.48 presents loss of embedment observed in the same suite of centrifuge tests during keying at a range of padeye eccentricity,

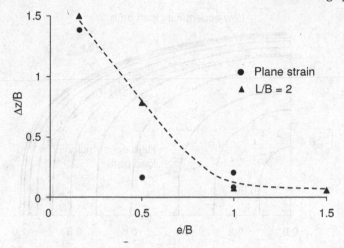

*Figure 7.48* Effect of padeye eccentricity on loss of plate anchor embedment during keying (O'Loughlin *et al.* 2006)

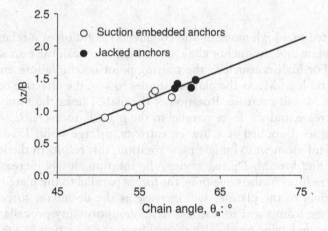

*Figure 7.49* Effect of padeye chain inclination on loss of plate anchor embedment during keying (Gaudin *et al.* 2006)

confirming the reduction in loss of embedment with increasing eccentricity of the keying load, but a rapidly diminishing effect with eccentricity e/B > 1. Loss of embedment also reduces with reducing inclination of the chain at the padeye during keying, as seen in Figure 7.49. Considerable uncertainty surrounds prediction of loss of embedment during keying with field trials and centrifuge tests indicating a range of 0.5–2.5 times the plate breadth B (Wilde *et al.* 2001, Foray *et al.* 2005, O'Loughlin *et al.* 2006, Gaudin *et al.* 2006).

During keying, the plate anchor is subjected to a combination of shear ($F_s$), normal ($F_n$) and moment (M) loading. The combined loading problem lends itself well to a failure envelope approach based on plasticity concepts, such as introduced with respect to installation of drag anchors (Section 7.4.3). Figure 7.50 shows typical load paths followed during keying of plate anchors for vertical keying loads applied at low and high eccentricity to the fluke. Considering keying with a highly eccentric load, the

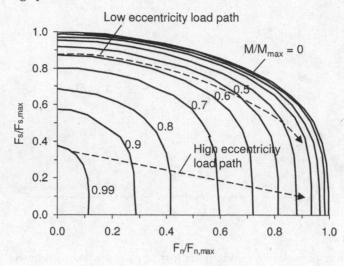

*Figure 7.50* Combined load paths during keying of plate anchors (O'Loughlin *et al.* 2006)

plate is subjected to a high moment and commences rotation at a relatively low-load uplift. As the plate and the anchor chain are initially vertical, the normal force $F_n$ will start at zero. For high eccentricity the starting point on the failure envelope is relatively low $F_s$ and high M. As the plate continues to key, the effective eccentricity and hence moment M will decrease. Rotation of the plate causes the force normal to the plate $F_n$ to increase and the force parallel to the plate $F_s$ to decrease. For the case of vertical keying load applied at a low eccentricity, a large uplift load is required to develop sufficient moment to initiate plate rotation; this results in the load path starting at high $F_s$ and low M. During keying, the rotation slowly increases causing the moment M to reduce. At the same time, the forces parallel to the plate $F_s$ will decrease and those normal to the plate $F_n$ will increase as the dominant forces on the plate gradually change from shear to normal. Assuming normality prevails, the trajectory of the plate during keying can be determined from the direction of the normal to the failure envelope at the intersection with the load path.

Loss of embedment during keying depends upon many factors such as the plate geometry, anchor weight, loading inclination and loading eccentricity (i.e. the perpendicular distance of the padeye from the anchor plate). Large deformation finite element analysis is proving a powerful tool to conduct parametric studies in order to develop design approaches. In a recent study (Wang *et al.* 2010), the keying response of rectangular anchors was studied using a three-dimensional, large-deformation, finite element strategy coupled with the general-purpose FE package ABAQUS. The rectangular anchor was pre-installed vertically in normally consolidated soil and then forced to rotate by loading through a (notional) chain. The theoretical solution for 'anchor chain–soil' interaction was included, so the changing loading angle at the anchor padeye was simulated correctly. The numerical results of embedment loss and pullout resistance during anchor keying were validated by comparison with existing experimental data. A series of numerical analyses were then conducted to evaluate the factors influencing the embedment loss, including anchor geometry, soil properties and pulling angle.

The results were interpreted using dimensional analysis to identify the main non-dimensional groups that affect the embedment loss. The embedment loss $\Delta z_u/L$ was found to be a function of certain non-dimensional groups, according to:

$$\frac{\Delta z_u}{L} = f\left(\frac{e}{l}, \frac{B}{L}, \frac{t}{L}, \frac{kL}{s_{u0}}, \frac{\gamma'_\alpha t}{s_{u0}}\right) \tag{7.35}$$

where $s_{u0}$ represents the local soil strength at the initial embedment depth of the anchor, k is the shear strength gradient (assuming a linear strength profile), t is the anchor fluke thickness and $\gamma'_a$ is the effective unit weight of the anchor when submerged in soil. The last group affects the loss of embedment, since the greater the anchor weight, the higher is the applied moment at the stage when the net vertical force on the anchor becomes neutral (i.e. anchor weight is just balanced by the vertical component of the applied padeye force). Essentially, this starts the load path at a higher moment (such as the higher eccentricity path in Figure 7.50), reducing the loss of embedment for a given amount of rotation.

For high shear strength ratios, the weight of the anchor became irrelevant (since the moment applied at the point of neutral vertical force became small in relation to the moment capacity). The loss of embedment, therefore, reached a maximum value, which was expressed as

$$\frac{\Delta z_{max}}{L} \approx a\left(\left(\frac{e}{L}\right)\left(\frac{t}{L}\right)^p\right)^q \tag{7.36}$$

where the three coefficients were fitted as a = 0.144, p = 0.2 and q = –1.15.

It was also found that the trend for the ultimate embedment to gradually reach the maximum value could be expressed through a combined dimensionless factor $s_{u0}/\gamma'_a\sqrt{te}$, according to:

$$\frac{\Delta z_u}{L} = \frac{\Delta z_{max}}{L}\tanh\left(b\left(\frac{s_{u0}}{\gamma'_a\sqrt{te}}\right)^r\right) \tag{7.37}$$

with best fit coefficients b = 5 and r = 0.85 (see Figure 7.51).

A final part of the study explored the effect of different mudline loading angles, $\theta_m$. Typical chain properties were adopted, and the chain was assumed to start vertical but gradually cut through the normally consolidated soil under a constant mudline angle. The results showed a proportional increase in the ultimate embedment loss as $\theta_m$ increased from zero to 90°, so that for a typical mudline loading angle of 40°, the embedment loss would be only around 45 per cent of that for a vertical mudline loading angle.

## 7.6.2 Dynamically penetrating anchors

A key aspect of predicting the penetration depth, and thus the eventual holding capacity for dynamically penetrating anchors (DPAs), is understanding the penetration resistance at such high velocities. The resistance will be dominated by fluid mechanics drag resistance at shallow depths, and viscous-enhanced shearing resistance as the anchor penetrates further. Relationships have been developed between impact velocity, penetration depths and holding capacity from centrifuge tests with DPAs with various fluke configurations deployed in soft, normally consolidated clay (O'Loughlin *et al.* 2004).

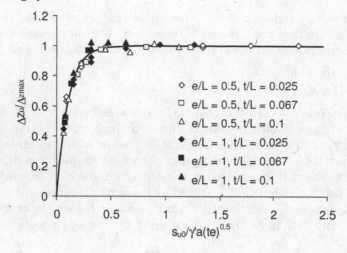

*Figure 7.51* Effect of anchor weight on ultimate embedment loss of square anchors during vertical pullout (Wang *et al.* 2010)

Figure 7.52a shows the relationship between impact velocity and embedment depth measured in the centrifuge together with theoretical predictions, for model anchors simulating a prototype of 1.2 m diameter, 15 m long and 100 tonne mass in air. The data indicate a linear increase in embedment depth with impact velocity, provided a minimum threshold velocity is achieved, resulting in embedment depths of between 1.5 and 3 anchor lengths. Increased embedment depth is also observed with decreasing surface area (i.e. fewer flukes), to be expected as greater surface area results in greater frictional resistance during penetration. Comparison of the 'blunt' and 'sharp' tipped 'flukeless' anchors indicates tip geometry also affects embedment depth.

The predicted values of embedment are based on an approach combining fluid mechanics drag resistance (which dominates at shallow penetrations) and viscous enhanced shearing resistance. It is rational to account for the velocity dependence of the soil resistance in terms of a drag term (True 1976) with the anchor acceleration deduced from the force-balance equation:

$$m\frac{d^2z}{dt} = W' - R_f(N_c s_{u,tip} A_{tip} + \alpha s_{u,side} A_{side}) - 0.5 C_d \rho_s A_{tip} v^2 \qquad (7.38)$$

where

$W'$ = submerged weight of the anchor
$R_f$ = rate or velocity dependent term
$C_d$ = drag coefficient, estimated as 0.24
$\rho_s$ = soil density.

The rate dependency $R_f$ can be expressed using logarithmic or power law functions, as

$$R_f = \left(1 + \lambda \log \frac{v/D}{\dot{\gamma}_{ref}}\right) \quad \text{or} \quad R_f = \left(\frac{v/D}{\dot{\gamma}_{ref}}\right)^\beta \qquad (7.39)$$

where $\lambda$ and $\beta$ are constants and $\tilde{a}_{ref}$ is a reference shear strain rate. Typical values of $\lambda$ were found to range between 0.2 and 0.3 at the very high strain rates, v/D, with corresponding $\beta$ values of 0.06–0.09 (O'Loughlin *et al.* 2009).

The vertical components of holding capacity during inclined pullout at 45° measured during the centrifuge tests are presented in Figure 7.52b along with predictions

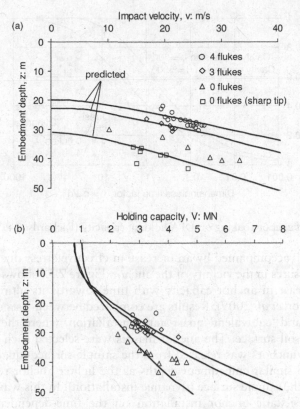

*Figure* 7.52 Relationships between DPA impact velocity, embedment and holding capacity (O'Loughlin *et al.* 2004)

of holding capacity based on basic bearing capacity principles and some available field data. The centrifuge data was found to predict higher holding capacity than obtained from field trials (O'Loughlin *et al.* 2004) although this may be attributed to the higher prototype mass of the model anchors compared to the field anchors. Adopting bearing capacity factors of 9 and 7.5 for the shaft and the flukes, respectively, and an adhesion factor $\alpha = 0.8$, the predicted vertical capacity may be derived from:

$$F_v = W' + F_f + F_b \qquad (7.40)$$

where

W′ = submerged weight of the anchor
$F_f$ = frictional resistance along the anchor shank and fluke faces
$F_b$ = bearing resistance at the top and bottom of the anchor shank and along the top and bottom of each fluke.

The theoretical predictions of holding capacity agree with the measured values (see Figure 7.52b), indicating that simple bearing capacity calculations based on end-bearing resistance and shaft capacity are appropriate.

Installation of a DPA leads to considerable disturbance and remoulding of the soil in the vicinity of the anchor. Anchor soaking, or set-up, will result in an increase in holding capacity with time as consolidation leads to the gradual recovery of shear

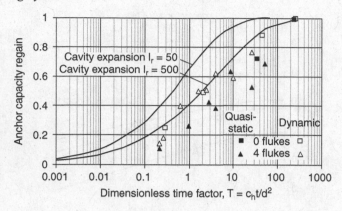

*Figure 7.53* Effect of anchor soaking on DPA holding capacity (Richardson *et al.* 2009)

strength of the soil, accompanied by an increase in effective stress due to dissipation of excess pore pressures in the vicinity of the anchor. Figure 7.53 shows centrifuge test results of the increase in anchor capacity with time towards its ultimate long-term capacity (Richardson *et al.* 2009). Results are compared between tests of dynamically installed anchors and 'equivalent' quasi-static installation, where the anchors were released from the soil surface. The anchor masses were selected such that when the heavier of the two anchors was released from the sample surface (quasi-static installation), it achieved similar embedment depths as the lighter anchor released from a height well above the sample surface (dynamic installation). In this way, the effects of dynamic and quasi-static anchor installation on the time-dependent capacity of dynamic anchors could be compared objectively. The progression of the degree of consolidation with time after installation was similar for all the tests, although slightly quicker for the dynamically installed anchors.

Figure 7.53 also shows predictions of increase in anchor capacity from cavity expansion solutions for radial consolidation, based on the analysis of consolidation following installation of a pile (Randolph and Wroth 1979). The method assumes that the initial pore pressure distribution is a function of the rigidity index $I_r$ of the soil, with typical values for $I_r$ ranging from 50 to 500 (Randolph 2003). Figure 7.53 shows the degree of consolidation predicted by the cavity expansion solutions for $I_r = 50$ and 500. The theoretical solution for the upper bound value of $I_r = 500$ provides a relatively accurate representation of the measured increase in capacity with time for the dynamically installed anchors. Somewhat unexpectedly, the cavity expansion solutions predict shorter consolidation times than observed in the quasi-static tests. Nonetheless, the extent of conformity between the analytical solution and the experimental data, particularly for the dynamic installation tests, suggests that cavity expansion techniques for determining consolidation following the installation of DPAs are suitable to predict the capacity regain after installation. Based on the consolidation data presented in Figure 7.53, for typical values of coefficient of radial consolidation $c_h = 3\text{--}30$ m²/yr, the time required for 50 per cent consolidation, $t_{50}$ of a prototype dynamic anchor with a shaft diameter of 1.2 m ranges from approximately 35–350 days. Similarly, the time required to achieve 90 per cent consolidation $t_{90}$ ranges from approximately 2.4–24 years.

# 8 Mobile jack-up platforms

## 8.1 Introduction

### 8.1.1 Overview of jack-up platforms and spudcans

Most offshore drilling in shallow to moderate water depths is performed from self-elevating mobile jack-up units. They are also used for fixed platform work-overs and increasingly for production support.

Typical units consist of a buoyant triangular platform resting on three independent truss-work legs (Figure 8.1), with the weight of the deck and equipment more or less equally distributed. A rack and pinion system is used to jack the legs up and down through the deck.

The foundations of independent-leg jack-up platforms approximate large inverted cones and are commonly known as 'spudcans'. Roughly circular in plan, spudcans typically have a shallow conical underside (in the order of 15–30° to the horizontal) with a sharp protruding spigot. In the larger jack-ups in use today, the spudcans can be in excess of 20 m in diameter, with shapes varying with manufacturer and rig. Figure 8.2 shows some typical spudcan shapes.

As an alternative, some jack-ups use a mat-support that connect all of the legs together and is of a size comparable to the jack-up hull. These have applicability in very soft sediments where increasing the bearing area of the foundations is a priority.

Jack-ups vary in size for deployment in different water depths. The larger units today have leg lengths in the order of 100–170 m and hull lengths of 50–70 m. Dimensions of jack-ups in operation are provided in Table 8.1.

### Installation method

Jack-ups play a vital role in the offshore industry with proven flexibility and cost-effectiveness in field development and operation. This is mainly due to their self-installation capacity. Jack-ups are towed to site floating on the hull with the legs elevated out of the water (Figure 8.3a). On location, the legs are lowered to the seabed, where they continue to be jacked until adequate bearing capacity exists for the hull to be lifted clear of the water (Figure 8.3b). The spudcan foundations are then pre-loaded by pumping seawater into ballast tanks in the hull. This 'proof tests' the foundations by exposing them to a larger vertical load than would be expected during service. The ballast tanks are emptied before operations on the jack-up begin (Figure 8.3c). It is usual for the total

---

Primary author of this chapter was Mark Cassidy.

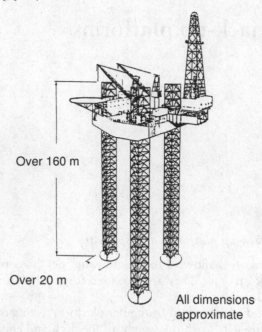

*Figure 8.1* Typical jack-up unit used today (Cassidy *et al.* 2009, after Reardon 1986)

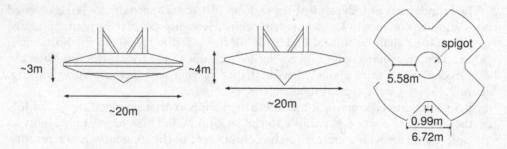

*Figure 8.2* Some example spudcan shapes and plan

combined pre-load (*i.e.* jack-up and seawater) to be between 1.3 and 2 times the weight of the jack-up.

## History of jack-up platforms

Although the earliest reference to a jack-up platform is in the description of a US patent application filed by Samuel Lewis in 1869 (Veldman and Lagers 1997), it was not until 85 years later in 1954 that Delong McDermott No. 1 became the first unit to utilise the jack-up principle for offshore drilling. Delong McDermott No. 1 was a conversion of one of the successful 'Delong Docks': a mobile wharf with a number of tubular legs, which could be moved up and down through cut-outs in the pontoon.

However, it was only two years later that former entrepreneur in earthmoving equipment R.G. LeTourneau revolutionised the design of jack-ups by reducing the number of legs to three and introducing an innovative electrically driven rack and pinion jacking system. Both revolutionary features are common on today's rigs.

Table 8.1 Typical jack-up dimensions (Bienen 2007)

| Rig | Hull length (m) | Hull width (m) | Hull depth (m) | Fwd. leg to of aft legs (centre to centreline) (m) | Aft legs (centre to centre) (m) | Leg length (m) | Spudcan diameter D (m) |
|---|---|---|---|---|---|---|---|
| ENSCO (57, 86, 94) [1] | 54.9–63.2 | 53.0–53.6 | 6.1–7.6 | 36.6–37.8 | 35.1–37.1 | 109.7–113.7 | 12.2–15.2 |
| F&G (Alpha 350, JU-2000, Universal M class) [2] | 67.1–70.4 | 71.8–76.2 | 8.2–9.5 | 39.6–47.2 | 43.3–54.6 | 140.5–166.9 | 17.0–18.3 |
| GSF (High Island I, Main Pass I, Rig 103, Rig 127) [3] | 54.9–63.1 | 51.2–53.6 | 6.1–7.6 | Not known | Not known | 106.7–126.8 | 12–14.0 |
| Noble (Carl Norberg, Charles Copeland, Dick Favor, Ed Noble, George McLeod) [4] | 53.0–63.1 | 49.4–53.6 | 5.5–7.6 | Not known | Not known | 76.3–127.1 | 11.5–14.0 |
| Average, as a proportion of the spudcan diameter | 4.51D | 4.18D | 0.52D | 2.61D | 2.71D | 8.74D | 1.00D |

[1] http://www.enscous.com/ [2] http://www.fng.com/ [3] http://www.globalsantafe.com/ [4] http://www.noblecorp.com

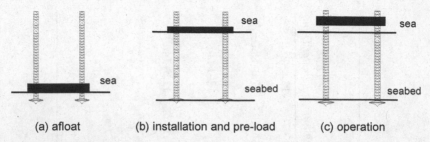

*Figure* 8.3 Jack-up installation procedure (a) afloat, (b) installation and pre-load and (c) operation

Zepata's 'Scorpion', used in water depths up to 25m in the Gulf of Mexico, was the first of many operated by the company Marathon LeTourneau.

Since their first employment, jack-up use has continued to be extended to deeper waters and in harsher environments. This development is continuing with some of the largest units being deployed in about 120m of water. In 2009, there were 485 jack-ups operating worldwide (Breuer and Rousseau 2009). Jack-ups are also now operating for extended periods at one location, often in the role of a production unit (Bennett and Sharples 1987, Scot Kobus *et al.* 1989). Examples of jack-ups being used as mobile offshore productions units include the Legendre oil field in the Carnarvon Basin off Western Australia and the Siri field in Denmark.

### 8.1.2 *Specific considerations required for jack-ups*

#### Site specific assessments

Before a jack-up can operate at a given location, a site-specific assessment of its operation must be performed. This continual assessment is what differentiates jack-up analysis from that of conventional fixed platforms and most onshore operations.

For any new site, assessment must be provided for three distinct phases: installation of the jack-up, capacity under storm loading during operation and removal of the jack-up. In an attempt to standardise jack-up assessment procedures and provide analysis guidance the offshore industry has published the 'Guidelines for the Site Specific Assessment of Mobile Jack-Up Units' (SNAME 2002) and more recently, the International Standard Organisation has begun drafting assessment standards. However, even with these guidelines, jack-ups are still perceived to have lower reliability than fixed offshore platforms, with many of the accidents attributed to geotechnical 'failures'. Hunt and Marsh (2004) provide historical accident statistics from 1955, with Morandi (2007) and Morandi *et al.* (2009) adding details for recent jack-up losses in the Gulf of Mexico's Hurricane Katrina, Rita and Ike.

#### Geotechnical considerations

Table 8.2 details the main considerations facing a geotechnical engineer in a site specific assessment of the installation, operation and removal of a jack-up. Many of these challenges are unique to jack-up and spudcan behaviour. Aspects such as the significant non-linearities due to cyclic and dynamic response and the high level of uncertainty in all of the modelling inputs further differentiate jack-up analysis from conventional onshore situations.

Table 8.2 Geotechnical considerations in jack-up site assessment

| Installation | Operation | Extraction |
|---|---|---|
| On-bottom hazards | Estimation of pre-loading to be applied to allow safe combined loading on foundations | Retrieval of deeply embedded spudcans in soft sediments (e.g. possible jetting procedure to be followed) |
| Prediction of spudcan penetration (and air gap between design wave and bottom of the hull) | Prediction of jack-up movements and capacity under storm loading (dynamic analysis accounting for spudcan fixity and damping; integrated wave–structure–soil interaction) | |
| Risk of 'punch through' failure or rapid leg penetration, where the spudcan uncontrollably pushes a layer of hard soil into the underlying soft layer (e.g. sand over clay; stiff over soft clay; possible consolidation due to a delay) | Prediction of spudcan failure modes: leg sliding or bearing capacity | |
| Offshore operations to mitigate punch-through, including possible perforation drilling | Repetitive cyclic loading on spudcans due to environmental wind and wave loads | |
| Penetration in multi-layered soils | Scour around the spudcans | |
| Differential leg penetrations | | |
| Eccentric loading on spudcan and legs due to existing footprints on the seabed or a sloped surface | | |
| Impact loading on hard seabed | | |
| Influence of penetrating spudcan on surrounding infrastructure (pipelines, fixed-platform piles) | | |

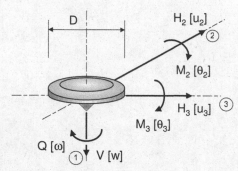

*Figure 8.4* Combined loads on a spudcan footing (corresponding displacements shown in brackets)

In the prediction of the footing penetration during installation, critical components include prediction of the penetration depth and the stability of the jack-up during penetration, with eccentric loading or 'punch-through' failure being potential hazards. The accurate determination of a spudcan load-penetration response requires accurate quantification of soil properties, the geometry of the spudcan and loading history details. This is discussed in Section 8.2.

With jack-up and spudcan sizes fixed, the pre-loading level reached is critical to any site-specific assessment of the jack-up capacity to withstand a design storm. During a storm, environmental wind, wave and current forces impose horizontal, sometimes moments and even torsional loads on the spudcans, as well as altering the vertical load between them. This combined loading of a footing is shown in Figure 8.4. With the foundations exposed to a vertical load prior to operation, it is argued that a combined loading 'failure' surface is established and is proportional to the vertical pre-load. Simple methods of evaluating the failure of a jack-up to environmental storm loading are described in Section 8.3, whilst more sophisticated analysis techniques incorporating expanding capacity surfaces to evaluate jack-up response are described in Section 8.4.

Though not addressed in detail, the extraction of spudcans is briefly discussed in Section 8.5.

## 8.2 Jack-up spudcan installation

### 8.2.1 Introduction

During installation, the self-weight of a jack-up is the dominant vertical loading on its spudcan footings and under most pre-loading circumstances, it is assumed to act directly through the spudcan load reference point (the centre of the plan of the conical footing). Maximum leg loads of a modern jack-up can exceed 140 MN and produce average vertical bearing pressures in excess of 400 kPa for a fully embedded spudcan. These can be significantly higher in very stiff soils where the penetration may limit the contact area. Accurate prediction of the installation and pre-loading is required, such that:

• The final penetration depth allows for adequate air gap between the maximum predicted wave elevation and underside of the deck.

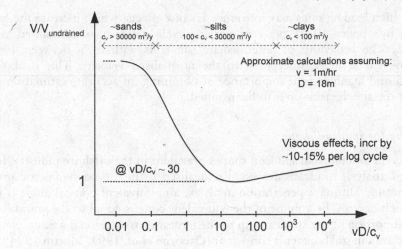

*Figure 8.5* Change in vertical reaction of spudcan with normalised velocity

- Overall stability is maintained, avoiding potential punch-through problems or eccentric loading instabilities due to existing footprints or seabed depressions.

### 8.2.2 *Vertical load in uniform soils*

Calculation of the ultimate vertical bearing capacity of a spudcan foundation in uniform soils is influenced by (i) the geometry of the foundation (noting that spudcans are usually idealised as circular with a conical underside), (ii) strength parameters of the soils, (iii) an appropriate set of bearing capacity factors, and (iv) drainage conditions of the soil due to the relative velocity of the footing. Although uniform methods spanning all drainage conditions are becoming available, the jack-up industry maintains site-specific assessments treating 'clays' as fully undrained, 'sands' as drained and 'silts' (somewhere in-between) as partially drained. A better methodology is to consider the normalised penetration rate. Using the framework of Finnie (1993), House *et al.* (2001), Randolph and Hope (2004) and Chung *et al.* (2006), the degree of drainage and the effect of consolidation under a penetrating object is described as a function of the normalised velocity:

$$V = \frac{vD}{c_v}$$

(8.1)

where v is the penetration velocity (for spudcans this is often around 1 m/hr), $c_v$ the coefficient of consolidation of the soil and D the equivalent diameter of the object, in this case the spudcan. This latter term is the diameter of a hypothetical circle, which has the same planar area (A) as the penetrating spudcan.

Figure 8.5 shows schematically the likely drainage conditions during a jack-up installation for different soil types and the resulting effect of the vertical installation load (normalised in this case by the undrained capacity). Normalised velocities of less than 0.1 would be expected to exhibit drained behaviour (and for typical 18 m spudcan installed at a velocity of 1 m/hr, this represents sands with $c_v > 30,000$ m²/year. Normalised velocities greater than 10 are likely to be undrained (clays with $c_v < 100$ m²/year).

However, high loading rates may introduce viscous effects, which increase the bearing resistance by about ~10–15 per cent per log cycle increase in normalised velocity $V = vD/c_v$. The behaviour of 'intermediate' silt soils with $c_v$ in between 100 and 30,000 m²/year is more susceptible to the normalised velocity. This is shown in Figure 8.5 and highlights the importance of obtaining an accurate estimation of the soil's $c_v$ at the site the jack-up is to be installed.

### Accounting for spudcan shape

With different and unusual spudcan shapes prevalent in the offshore industry, for the purpose of analysis spudcans are usually idealised to be an axi-symmetric inverted conical footing. During a penetration analysis, an equivalent conical angle is calculated at each depth. The volume of the equivalent cone is equal to the volume of the penetrated portion of the spudcan (up to the largest cross-sectional area in plan) and the planar area in soil contact is consistent (Osborne *et al.* 1991, Martin 1994). Once the maximum spudcan diameter has been passed, the equivalent conical footing remains constant.

### Undrained conditions

In most clay soils, spudcan pre-loading will be of a rate that ensures fully undrained conditions. Calculation of the ultimate vertical bearing capacity of a foundation in clay follows classical bearing capacity theory. Traditionally, the value of $N_c$ has been determined from solutions for a strip footing on homogeneous clay, with shape and depth factors based on Skempton (1951) (as outlined in Section 6.3.2). However, as discussed in Section 6.3.2, these factors are significantly affected by the gradient of shear strength with depth. To improve predictions during pre-load assessments, the jack-up industry has tended to adopt an equivalent undrained shear strength. For example, based on back calculations of monitored installations, and while using the Skempton factors, Young *et al.* (1984) recommended using a shear strength averaged over the zone of soil from the spudcan tip to a depth of half a diameter. However, such approximate methods are inappropriate, as rigorous plasticity solutions for axi-symmetric circular footings that include strength varying linearly with depth are now available. The bearing capacity at a specific depth can be expressed as

$$V = (s_u N_c + \sigma'_{v0})A \tag{8.2}$$

Appropriate tabulated lower bound solutions of $N_c$ have been provided by Houlsby and Martin (2003) for circular conical foundations of diameter, D, cone angles between 60 and 180° (flat plate), different embedment depths (0–2.5 diameters) in soil with different strength gradients (expressed as $kD/s_{um}$ from 0 to 5, where k is the rate of increase in shear strength with depth, from a value of $s_{um}$ at the mudline). Factors for 1,296 cases are provided in Houlsby and Martin (2003), with a sub-set for a rough, flat circular footing with no increasing strength with depth provided in Table 8.3.

These methods used to assess spudcan penetration resistance assume failure based on a footing wished-in-place. Care must be taken in ensuring the factors are appropriate for the evolving mechanisms of a continuously penetrating spudcan, and the effect of backflow on the bearing capacity. The deep penetration of a spudcan in clay has

*Table 8.3* Bearing capacity factors for rough circular plate in homogeneous soil (Houlsby and Martin 200)

| Embedment depth / D | Bearing factor, $N_c$ |
| --- | --- |
| 0 | 6.0 |
| 0.1 | 6.3 |
| 0.25 | 6.6 |
| 0.5 | 7.1 |
| 1.0 | 7.7 |
| $\geq 2.5$ | 9.0 |

been investigated by Hossain and Randolph (2009a) through centrifuge modelling and large deformation finite element analysis. Experimentally, particle image velocimetry (PIV) and close range photogrammetry methods have been applied to investigate the mechanism of deep penetrating spudcans. An example of a digital image and the interpreted particle flow velocity vectors from PIV are shown in Figure 8.6 for a single-layered clay. Adjustments to the bearing capacity mechanism must be made to account for the changing mechanism with depth.

Three distinct mechanisms of soil flow are observed around the advancing spudcan. Figure 8.6a shows movement of soil leading to surface heave and formation of a cavity above the spudcan, whereas in Figure 8.6b and c there is gradual flow back into the cavity and finally a localised flow-around mechanism, respectively. It was also observed that the backflow gradually provides a seal above the spudcan and limits the cavity depth. It continues below the limiting cavity depth and above the advancing spudcan. As can be observed in Figure 8.6c the initial cavity is not filled up by the backflow process.

The deep penetration mechanism is the result of a preferential 'flow failure' and not a 'wall failure' due to instability of the open cavity (as had been previously assumed in the jack-up industry). Conditions for back flow may be expressed simply as

$$\frac{h_{cavity}}{D} = \left(\frac{s_{uh}}{\gamma'D}\right)^{0.55} - \frac{1}{4}\left(\frac{s_{uh}}{\gamma'D}\right) \tag{8.3}$$

where $h_{cavity}$ is the limiting cavity depth (also the depth at the onset of backflow) and $s_{uh}$ is the undrained shear strength at that depth. Figure 8.7 shows the resulting design chart and compares all of Hossain *et al.* results for a flow failure (both finite element and centrifuge) and those of a wall collapse. For a clear profile of increasing strength with depth, Hossain and Randolph (2009a) proposed a more convenient expression to evaluate the maximum cavity depth explicitly as a function of $S = (s_{um} / \gamma'D)^{(1-k/\gamma')}$ as

$$\frac{h_{cavity}}{D} = S^{0.55} - \frac{S}{4} \tag{8.4}$$

where k is the rate of increase in shear strength with depth, from a value of $s_{um}$ at the mudline.

This progressive failure and associated change of mechanisms have been accounted for directly in a proposed bearing capacity method of Hossain and Randolph (2009a), which they refer to as a 'mechanism-based design approach'. At penetrations less than

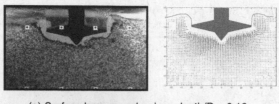

(a) Surface heave mechanism, depth/D = 0.16

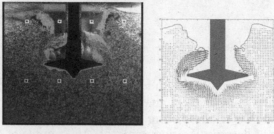

(b) Onset of flow-around mechanism, depth/D = 0.77

(c) Full flow-around mechanism, depth/D = 1.15

*Figure 8.6* Digital images and PIV analysis of flow vectors from centrifuge tests (Hossain *et al.* 2004)

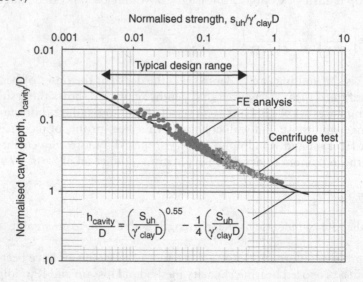

*Figure 8.7* Design chart for estimating cavity depth above the advancing spudcan

the limiting cavity depth (see Equations 8.3 and 8.4) classical bearing capacity factors can be used (as described by Equation 8.2), as the soil flows back towards the soil surface. Hossain and Randolph (2009a) also proposed straightforward formulae for estimating $N_c$ that account for non-homogeneity and the normalised embedment. For deep penetration in excess of the critical cavity depth, the transition to a fully localised flow-round mechanism is accounted for with

$$V = s_u N_{cd} A + \gamma' V_{spudcan} \tag{8.5}$$

where $V_{spudcan}$ is the volume of the embedded spudcan and $\gamma'$ the effective unit weight of clay. $N_{cd}$ has been defined as the deep bearing capacity factor with values derived by Hossain and Randolph from large deformation finite element analysis and centrifuge experiments. They found that $N_{cd}$ values were relatively insensitive to strength homogeneity and spudcan roughness, and therefore could be represented by

$$N_{cd} = 10 \left(1 + 0.065 \frac{w}{D}\right) \leq 11.3 \tag{8.6}$$

where w is the spudcan embedment and 11.3 represents the ultimate fully localised flow-round mechanism. A short transition zone approximating 0.3D is recommended to be applied between the shallow and deep mechanism. Hossain and Randolph (2009a) also provide a methodology for adjusting the bearing capacity factors to account for strain-rate and softening of the soil. Comparisons with measured offshore data in the Gulf of Mexico (after Menzies and Roper 2008) show reductions of around 20 per cent to the unadjusted bearing capacity (Hossain and Randolph 2009b).

Along with improved evaluations of penetration resistance, the more accurate prediction of the onset of backflow allows for the tensile capacities due to transient suctions, and additional horizontal and moment capacities, to be accounted for in any combined loading analysis. These additional capacities were observed, amongst others, by Cassidy *et al.* (2004a) during combined loading tests on a spudcan in normally consolidated clay (full backflow) and in tests by Springman and Schofield (1998) where the spudcan and the 'lattice leg' were encased.

## Drained conditions

Spudcan pre-loading is usually performed sufficiently slowly to ensure that fully drained conditions will occur in sandy seabed soil (see Figure 8.5). The drained ultimate vertical bearing capacity of a circular foundation on the surface of homogeneous frictional material (with no effective surcharge), can be expressed as

$$V = \gamma' N_\gamma \pi D^3 / 8 \tag{8.7}$$

where $N_\gamma$ is the dimensionless bearing capacity factor calculated for the axi-symmetric case. Tabulated lower bound solutions have been provided by Cassidy and Houlsby (2002) for circular conical foundations of angles between 30 and 180° (flat plate), a range of roughness from smooth to fully rough and angles of friction ($\phi$) up to 50° (though the dominate parameter is the value of friction angle assumed). Other values for rough circular footings have been provided by Martin (2004) and are discussed in the context of the load carrying capacity of a foundation by Randolph

*Table 8.4* Bearing capacity factors for a flat, rough circular footing
(quoted by Randolph *et al.* (2004) from use of ABC software
of Martin (2003))

| Friction angle $\phi$: degrees | Bearing factor, $N_\gamma$ | Bearing factor, $N_q$ |
|---|---|---|
| 20 | 2.42 | 9.61 |
| 25 | 6.07 | 18.4 |
| 30 | 15.5 | 37.2 |
| 35 | 41.9 | 80.8 |
| 40 | 124 | 193 |
| 45 | 418 | 521 |

*et al.* (2004). A summary of these values applicable to spudcan analysis are provided in Table 8.4.

Footing penetration in a thick layer of clean silica sand is usually minimal with the maximum diameter of the spudcan rarely coming into contact with the soil. It is therefore not usual to need any $N_q$ term unless the spudcan is already embedded (e.g. in a soft clay overlaying sand profile). The semi-empirical methods suggested by Brinch Hansen (1970) and used by Vesic (1975) are available for these situations, as are some solutions of Martin (2004).

Calculation of bearing capacity using Equation 8.7 assumes rigid-plastic behaviour of material with only one value of failure strength. It also does not account for real material behaviour such as mobilisation and softening. These effects can be accounted for by modifying the ideal predictions of Equation 8.7. Methods include directly reducing the predicted bearing capacity as a function of soil volume displaced (e.g. Cassidy and Houlsby 1999) or by reducing the angle of friction to an estimated mobilised value. Reduced angles of friction have been recommended by various practitioners for use in bearing capacity calculations on sand (Graham and Stewart 1984, James and Tanaka 1984, Kimura *et al.* 1985) and even in the industry guidelines. However, White *et al.* (2008) provided a methodology for deriving the mobilised friction angle, based on application of Bolton's stress–dilatancy correction (Bolton 1986) in an iterative calculation with the bearing capacity equation. The influence of the spudcan conical underside (or indeed the spigot) on softening sand in a penetrating spudcan was highlighted by White *et al.* (2008), with reduction in the bearing capacity of conical footings, when compared to the flat base, not purely attributable to the bearing capacity factor.

Uncemented carbonate sands are generally much more compressible than silica sands, but these materials also have a high friction angle. Application of the classical Equation 8.7 would lead to a substantial under estimate of actual spudcan penetrations in such soils. If applied, the friction angles should be considerably smaller than laboratory values, to artificially account for high soil compressibility. However, predictions can be expected to have a low degree of accuracy. Alternative methods, such as the empirical modulus approach recommended by Finnie (1993), Randolph *et al.* (1993) and Finnie and Randolph (1994) have therefore been developed to assess spudcan penetration in carbonate sands. Other published predictive methods include Islam (1999) and Islam *et al.* (2001), Houlsby *et al.* (1988), and Yamamoto

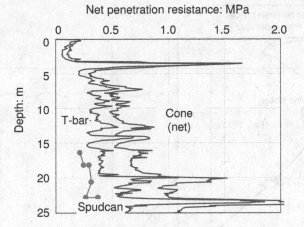

*Figure 8.8* Measured resistances at Yolla A (Erbrich 2005)

*et al.* (2008, 2009). The last of the aforementioned uses a compression deformation mechanism and a punching shear pattern with simple formulae provided.

### Alternative method for predicting spudcan bearing capacity using penetrometers

Use of penetrometer data directly in predicting spudcan penetration is possible. While assessing a difficult site offshore Australia, Erbrich (2005) argued that a direct correlation with penetrometer data allows for secondary soil characteristics, including strain rate dependence of shear strength, partial consolidation in silty materials and soil sensitivity, to be more appropriately accounted for. Although direct correlation of piezocone test results has seen application in pile design (dating to Heijnan 1974), this has only recently received attention for predicting spudcan penetration resistance.

In the case study of a jack-up installation at Yolla A in the Bass Strait offshore Australia, Erbrich (2005) noted that the spudcan penetrations proved to be much larger than had been predicted by four independent *a priori* assessments, even though the available soils data was of exceptional quality compared to that available for most jack-up installations. The influence of drainage is displayed in the direct comparison between the CPT, T-bar and spudcan in Figure 8.8. Partial drainage occurred during the CPT and T-bar testing, when compared to the undrained response that occurs during spudcan penetration.

Erbrich (2005) provided suggestions for the correlation of these rate effects between the penetrometer and the larger spudcan, as well as for geometric and embedment factors. However, further research is required to develop and verify any new method. Considerable interest in industry (see Quah *et al.* 2008 for instance) looks likely to ensure such progression. These methods will avoid difficulties in determining strength properties and inaccuracies in applying rigid-plastic calculation procedures.

### 8.2.3 Potential punch-through failure in layered soils

The potential for unexpected punch-through failure of a jack-up exists during installation and pre-loading in layered soils. This occurs when the spudcan uncontrollably

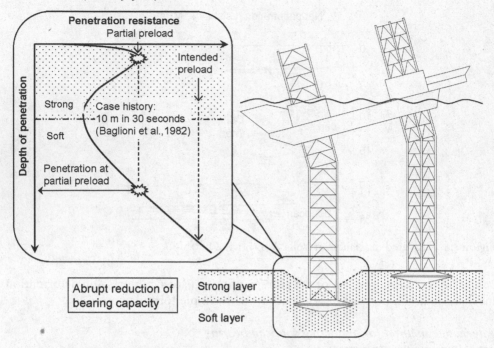

*Figure 8.9* Illustration of Punch-through failure during pre-load (after Lee 2009)

pushes a locally strong zone of soil into underlying softer material (Figure 8.9). Such failures can lead to buckling of the jack-up leg, effectively temporarily decommissioning the platform, and even toppling the unit. It has been reported that spudcan punch-through failures occur at an average rate of one incident per year, costing the industry between US$1 million and US$10 million per incident due to rig damage and loss of drilling time (Osborne and Paisley 2002). Soil conditions of a thin layer of sand (or cemented material) overlaying a weaker stratum of clay and stiff-over-soft clay are particularly hazardous, and are discussed in this chapter. Other soil conditions that can cause rapid leg penetration include a thick clay layer with decreasing strength with depth, a very soft clay where the rate of increase of bearing capacity does not match the loading rate and firm clay with sand or silt pockets (Osborne *et al.* 2009).

*Sand overlaying clay*

Current industry analysis guidelines (SNAME 2002) recommend the punching shear mechanism developed by Hanna and Meyerhof (1980) for calculating the peak penetration resistance of the footing on sand overlaying clay. For a flat footing at one particular depth (i.e. wished into place), the failure mechanism is assumed to comprise a truncated cone in the sand layer being depressed into the underlying clay. However, for calculation purposes, the simpler vertical shear plane is used (SNAME 2002), as shown in Figure 8.10. The peak resistance $q_{peak}$ is calculated following the punching shear method as

$$q_{peak} = N_c s_u + \frac{2H_s}{D}(\gamma'_s H_s + 2q_0)K_s \tan\phi + q_0 \qquad (8.8)$$

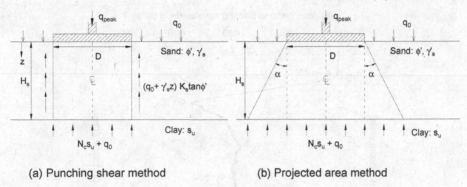

(a) Punching shear method       (b) Projected area method

*Figure 8.10* Mechanisms assumed in calculation of peak resistance in sand overlaying clay (a) punching shear method and (b) the projected area method (Lee *et al.* 2009)

and where the product of the punching shear coefficient, $K_s$, and $\tan\phi$ is related to the normalised shear strength of the underlying clay layer through the lower bound given by

$$K_s \tan\phi = 3\frac{s_u}{\gamma_s' D} \tag{8.9}$$

where $s_u$ is the undrained shear strength of the underlying clay and $\gamma_s'$ the effective unit weight of the sand.

The 'projected area' method is suggested as an alternative design approach in the commentary to the SNAME guidelines. Here, the load is assumed to be projected through the upper sand layer at an angle of spread, $\alpha$, to an imaginary footing of increased bearing area at the sand–clay boundary, as shown in Figure 8.10b. However, for this case, there still remains debate as to appropriate values of the assumed angle of spread ($\alpha$). Rates of spread in the range of 1:5–1:3 (horizontal:vertical) are recommended in SNAME. The peak capacity is evaluated simply as the bearing capacity of the lower clay layer as:

$$q_{peak} = \left(1 + \frac{2H_s}{D}\tan\alpha\right)^2 (N_c s_u + p_0') \cdot \tag{8.10}$$

The selection of the projection angle is critical in this method, with Lee (2009) reviewing all angles advocated.

In both methods, as described by SNAME, the properties of the upper sand layer are ignored (since even in the punching shear approach the frictional resistance through the sand is expressed in terms of the strength of the underlying clay, see Equation 8.9). They therefore provide a similar approach with the ratio of $q_{peak}$ to the bearing capacity of the underlying clay increasing as a simple quadratic function of sand layer thickness ($H_s$) normalised by the spudcan diameter (i.e. by $H_s/D$). This produces an inconsistency between parameters required to fit small-scale model test data (Young and Focht 1981, Higham 1984), recent centrifuge tests (Teh 2007, Lee 2009) and measured spudcan penetrations offshore (Baglioni *et al.* 1982, Osborne *et al.* 2009).

Furthermore, both the punching shear and projected area methods were originally developed for shallow wished-in-place footings (though with modification for

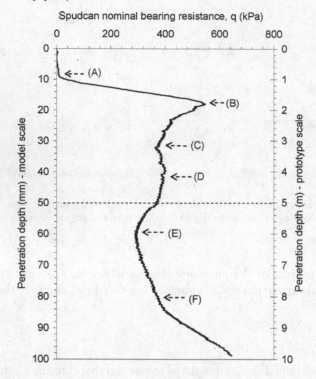

*Figure 8.11* Half-spudcan bearing resistance profile measured during half-spudcan penetration (observed mechanisms shown in Figure 8.12) (Teh *et al.* 2008)

application to spudcans by SNAME). However, the mechanisms are significantly different from post-failure observations of four centrifuge tests performed by Craig and Chua (1990) and the more recent observations of Teh *et al.* (2008). In the latter, digital images were captured continuously by installing a half-spudcan against a transparent window. Analysis using particle image velocimetry coupled with close-range photogrammetry corrections, the change in spudcan failure mechanism with penetration depth (and at stress conditions of equivalent similitude to the offshore case) was observed. Example mechanisms for the experimentally measured punch-through profile in Figure 8.11 are shown in Figure 8.12. These are for $H_s/D = 0.83$ and an undrained shear strength in the clay layer of 10 kPa. A full description of the observed mechanisms is provided in Teh (2007), Teh *et al.* (2008) and Lee (2009); key aspects are highlighted as follows.

- Initially, the soil movement radiates from the spudcan tip, is very limited and is concentrated in the upper sand layer (Phase A of Figure 8.11 and Figure 8.12).
- The peak resistance (Phase B) occurs at a relative shallow embedment, which was also consistently measured at ~0.12D in the full-spudcan centrifuge tests of Teh (2007) and Lee (2009). The mechanism consists of a truncated cone of sand being forced vertically into the underlying clay layer, which in turn changes from pure vertical movement (directly under the spudcan) to radial and small amounts of heave further from the centre line. The outer angle of the sand frustum being forced into the clay reflects the dilation of the sand.

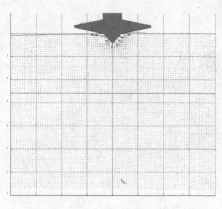

(a) Phase A – Full spigot penetration

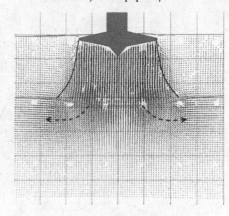

(b) Phase B – Peak resistance, $q_{peak}$

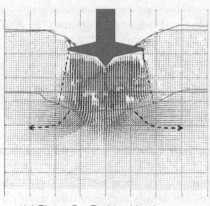

(c) Phase C – Reduced load

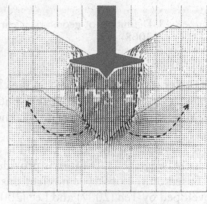

(d) Phase D – Second smaller peak

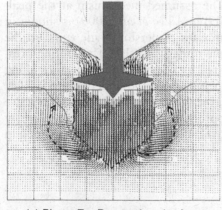

(e) Phase E – Penetrating clay layer

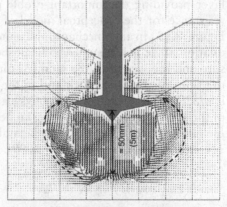

(f) Phase F – Final recorded penetration

*Figure 8.12* Spudcan failure mechanisms at different penetration depths (Stages shown on Figure 8.11) (Teh *et al.* 2008)

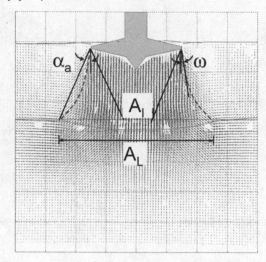

*Figure 8.13* Definition of geometrical parameters of observed failure mechanisms (Teh *et al.* 2008)

- After punch-through and in the clay layer (Phases E and F), sand is trapped in a near vertical plug below the spudcan. The height of this plug is approximately 60–90 per cent of the sand thickness (Lee 2009).

A summary of the geometry of the observed mechanism at the peak, and the effect of increasing the $H_s/D$ ratio, is provided in Figure 8.13 and Figure 8.14.

Based on the observed mechanisms, two alternative prediction methods have been developed by Teh (2007) and Lee (2009). Both provide details for generating a simplified load–penetration profile, with predictions at the surface, peak and in the clay layer providing the important profile points. The assumed mechanism at the peak resistance for the conceptual model of Lee (2009) and Lee *et al.* (2009) is shown in Figure 8.15. In this mechanism, a sand frustum with dispersion angle equal to the dilation angle $\psi$ is assumed to be pushed into the underlying clay. The footing pressure and the weight of the sand frustum is resisted by the frictional resistance along the sides of the sand block and in the bearing capacity of the underlying clay (Lee *et al.* 2009). Following a silo analysis approach, the stress level and dilatant response of the sand is accounted for using an iterative approach. The capabilities of these new methods to predict results from a database of over 40 centrifuge tests is provided in Teh *et al.* (2010) and Lee *et al.* (2009).

### Strong clay overlaying weak clay

A stiff clay crust overlying soft clay is another potential punch-through risk, and one commonly found in offshore South-East Asia (Castleberry and Prebaharan 1985, Osborne and Paisley 2002, Paisley and Chan 2006). Recently, Edwards and Potts (2004) and Hossain and Randolph (2009b) have proposed new design approaches to replace the solutions of Brown and Meyerhof (1969) and Meyerhof and Hanna (1978) provided in the SNAME (2002) guideline. Edwards and Potts (2004) proposed that

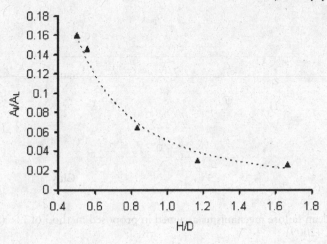

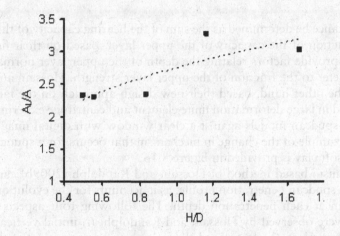

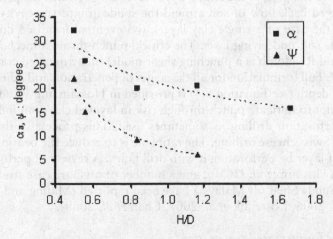

*Figure 8.14* Effect of sand thickness ratio (H$_s$/D) on geometrical parameters of observed mechanism (Teh *et al*. 2008)

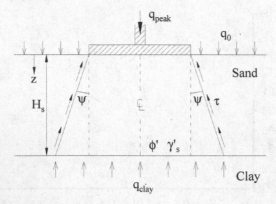

*Figure 8.15* Spudcan failure mechanisms assumed in proposed method of Lee (2009) and Lee
*et al.* (2009)

the peak resistance be determined as the sum of the bearing capacity of the underlying
clay and a fraction of the capacity of the upper layer. Based on their finite element
analysis, they provide factors relating the depth of the upper layer normalised by the
spudcan diameter to the fraction of the upper layer strength. Hossain and Randolph
(2009b), on the other hand, based their new design approach on the failure mechanisms observed in large deformation finite element and centrifuge experiments carried
out with half spudcan models against a clear window with digital imaging and PIV
analysis. An example of the change in mechanism that occurs for a spudcan penetrating stiff-over-soft clay is provided in Figure 8.16.

The mechanism-based method of Hossain and Randolph (2009b) can be used to
estimate a full spudcan penetration profile, as it accounts for the evolution of the failure mechanism at each penetration depth. The following four aspects of the flow
mechanisms were observed by Hossain and Randolph: (i) initial vertical movement
of the upper layer into the lower and the consequent deformation of the interface,
(ii) creation of a soil plug of the strong upper layer clay beneath the penetrating spudcan, (iii) delayed back-flow of soil around the spudcan into the cavity above (when
compared to the case of a single clay layer), (iv) eventual localised flow around the
spudcan in the soft underlying layer. The critical punch-through mechanism assumed
by Hossain and Randolph is a punching shear model of a truncated cone, as depicted
in Figure 8.17. Full formulation for all the stages of penetration and adjustments to the
critical cavity depth (see Equation 8.4) is provided in Hossain and Randolph (2009b).

In an attempt to mitigate punch-through risk in layered clays, an industry practice
known as 'perforation drilling' is sometimes used. This process is also colloquially
referred to as Swiss cheese drilling. The rationale is to reduce the bearing resistance of
the upper stiff layer by perforating it with drill holes. A review of perforation drilling
is provided in Hossain *et al.* (2010), and a number of offshore case studies on layered
clays on the Sunda Shelf off Malaysia have been reported (Maung and Ahmad 2000,
Brennan *et al.* 2006, Kostelnik *et al.* 2007, Chan *et al.* 2008).

*Other reasons for increased penetration*

As discussed in Osborne *et al.* (2009), rapid spudcan penetration can also be caused
by inappropriate installation procedures or the imposition of wave-induced cyclic

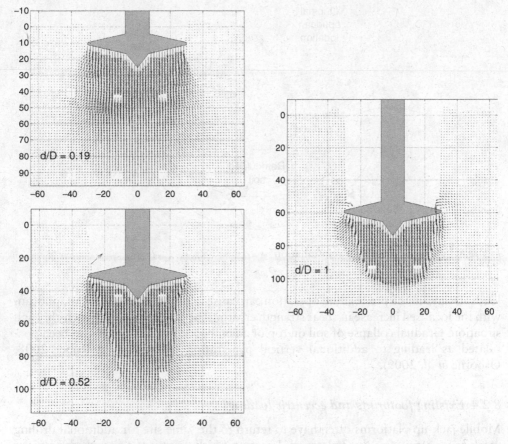

*Figure 8.16* Change in mechanisms observed for a spudcan penetrating stiff-over-soft clay (Hossain and Randolph 2010)

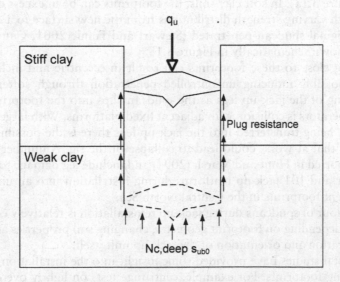

*Figure 8.17* Punching shear mechanism assumed in method of Hossain and Randolph (2009b)

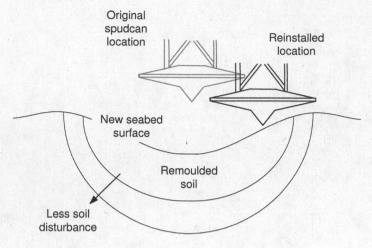

*Figure 8.18* Possible footprint reinstallation

loads. Delays in the pre-loading operation can cause consolidation under the spudcan, with the localised increase in shear strength creating the possibility of a punch-through situation. Gradual collapse of soil on top of an embedded spudcan has also been postulated as leading to additional vertical penetration (Menzies and Roper 2008, Osborne *et al.* 2009).

### 8.2.4 *Existing footprints and eccentric loading*

Mobile jack-up platforms often have to return to the same site for additional drilling work or for servicing a fixed platform. Locating a jack-up unit at a site where previous jack-up operations have occurred can be hazardous, because the new jack-up may need to negotiate the spudcan footprints left on the seabed from the previous operations (see Figure 3.13). In soft clay soils, the footprints can be in excess of 10 m deep and wide, with varying strength distributions from the new surface to a depth below, which the original spudcan penetrated (Stewart and Finnie 2001, Gan 2010). This situation is shown schematically in Figure 8.18.

Pre-loading close to these footprints can result in eccentric and inclined loading conditions, possibly inducing uncontrolled penetration through softer remoulded soils or slewing of the jack-up legs as the spudcan slips into the footprint. Of major concern to operators is collision with adjacent fixed platforms. With large shear forces and moments being transferred into the jack-up legs there is the possibility of structural damage that at worst could lead to collapse of the rig. A number of incidents have been reported in Hunt and Marsh (2004) and include leg damage to the Mod. V class Monitor and 101 jack-up platforms during installation into an uneven seabed next to adjacent footprints in the Central North Sea.

The behaviour of spudcans during jack-up reinstallation is relatively complex with the response depending on footprint geometry, changing soil properties and the structural configuration and orientation of the jack-up unit itself.

Experimental studies have provided some insight into the installation of spudcans next to existing footprints. For example, centrifuge tests on lightly overconsolidated

*Table 8.5* *Comparison of published experimental results of a spudcan installed next to an existing footprint

| | Stewart and Finnie (2001)[1] | Cassidy et al. (2009) | Teh et al. (2006) | Stewart and Finnie (2001)[1] | Cassidy et al. (2009) | Teh et al. (2006) |
|---|---|---|---|---|---|---|
| Offset | $H/As_u$ | $H/As_u$ | $H/As_u$ | $M/DAs_u$ | $M/DAs_u$ | $M/DAs_u$ |
| D/4 | 0.26 | 0.35 | – | not reported | 0.35 | – |
| D/2 | 0.38 | 0.57 | – | not reported | 0.54 | – |
| D | 0.44 | 0.70 | – | not reported | 0.67 | – |
| 3D/2 | 0.16 | 0.20 | – | not reported | 0.14 | – |
| Slope (30°) | – | – | 0.070 | – | – | 0.222 |

[1]Normalised horizontal results of Stewart and Finnie are as described by Teh *et al.* (2006)

clays by Stewart and Finnie (2001) and Cassidy *et al.* (2009) showed the worst offset distance between installations is between 0.5 and 0.75 diameters and that after 1.5 diameters little effect is felt. A summary of peak induced loads from experimental studies in clay is provided in Table 8.5. These loads are normalised by the measured undisturbed undrained shear strength at the spudcan penetration depth. Reference to the original papers should be made for experimental conditions and details.

Cassidy *et al.* (2009) also showed that the horizontal and moment loads induced on the jack-up legs were greater when a more extensive footprint was created by the initial spudcan being embedded deeper and with a higher pre-load. This is due to a deeper footprint shape and significantly more remoulding in the higher pre-loaded case. Leung *et al.* (2007) and Gan *et al.* (2008) studied the change in soil properties of footprints by conducting a centrifuge investigation and reported that the variation of undrained shear strength within the footprint is related to the virgin undrained strength conditions, the distance from the footprint centre and the duration of time between the installation creating the spudcan and the reinstallation. Gan *et al.* (2008) further reported that higher H and M were induced from reinstalling spudcans nearby footprints formed in stiffer soils.

Bearing capacity predictions for spudcans next to footprints remains problematic, to the extent that preventative measures are now being taken offshore. It is hoped that installation of systems measuring rack phase difference (RPD) will alert operators to eccentric loading conditions and translation of the spudcan (Foo *et al.* 2003a, b). RPD is simply the difference in elevation between the chords of any one leg. This is measured as the difference at the jacking systems and is a direct measure of the inclination of the leg with respect to the hull. Monitoring the leg's RPD alerts the operator to eccentric spudcan loadings. It allows them to change installation strategy based on the effect of the eccentric forces being transferred into the legs during installation.

Though an early alert to an inclined jack-up leg and spudcan may mitigate buckling of the leg braces and stability problems, there are still no guidelines available to operators to help ensure safe reinstallation (Dean and Serra 2004). The SNAME guideline, for instance, only provides the general advice to reinstall at a minimum distance of an equivalent of one spudcan diameter from the edge of the spudcan to the edge of the footprint (and a warning that a larger offset distance may be required for softer soil).

Numerical studies have also provided some insight into possibly mitigation techniques, including:

- 'Stomping' of the spudcan with the jack-up leg in an attempt to even the surface
- Sand or gravel infilling (Jardine *et al.* 2001)
- Changing the spudcan configuration, including the use of skirts (Dean and Serra 2004).

## 8.3  Combined loading during a storm

### 8.3.1 Introduction

Environmental loads comprising wind, wave and current forces impose horizontal loads, moments and torsional loads on the spudcans, as well as altering the vertical load between them. The combined loading of a footing is shown in Figure 8.4 and results in a complex state of stress and strain in the underlying soil. The capacity of the spudcan to withstand these conditions must be assured before a jack-up can operate safely. It is rarely possible or indeed appropriate to conduct a finite element analysis in which the seabed soil is modelled in detail using continuum elements unless a full geotechnical site investigation with soil characterisation has been undertaken.

The alternative is to incorporate the spudcan behaviour directly as a point element in the structural analysis of the jack-up. This is common practice with the entire footing–soil response expressed purely in terms of the loads (often known as the force-resultants) on the footing and the corresponding displacements (Figure 8.4). This approach is directly analogous to the use of force-resultants (axial force, bending moment and shear force) and nodal displacements and rotations in the analysis of beams and columns. Alternatives include the simplistic use of a pinned footing condition, a set of uncoupled linear or non-linear springs, or more recently, strain-hardening models based within a plasticity framework have been introduced.

### 8.3.2 Analysis approach

For a given jack-up and spudcan size, the installation method and the pre-loading level reached are critical in assessing the safety of the unit under design loading conditions. It is argued that, by exposing the foundations to a given vertical pre-load prior to the storm, a combined loading 'failure' surface is established that is proportional to the vertical pre-load. A summary of this process is provided for shallow foundation in Section 6.33. It is also illustrated in Figure 8.19. The surface for each spudcan is theoretically established by pre-loading to $V_{pre-load}$ (representing a certain penetration, with the surface enlarged with additional penetration). As described in Section 8.1.1, the pre-load is established by pumping seawater into the jack-up balanced tanks and held for a set pre-loading time. After the pre-load water has been dumped (back into the sea) the load settles inside such a surface ($V_{self-weight}$). Under environmental wind and wave loading, load paths for each leg and spudcan can be predicted and the spudcan capacity evaluated by the proximity of the expected load paths to the yield envelope. Shown in Figure 8.19 is a typical jack-up case. The windward leg design ($WL_d$) point, established through analysis of the design storm, and the assumed windward

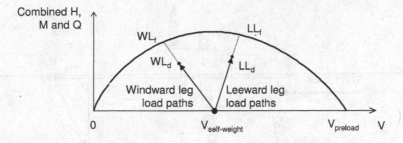

*Figure 8.19* Establishing a combined loading capacity surface during pre-loading

leg failure (WL$_f$) point are illustrated, as are the equivalent leeward leg points LL$_d$ and LL$_f$.

### 8.3.3 Spudcan–soil interaction (simple assumption of stiffness)

*Assumption of pinned footings*

In order to determine the level of combined loads on the spudcans and the overall response of the jack-up, an assumption of the spudcan stiffness must be incorporated into the structural analysis of the jack-up and wave loading analysis. The simplest approach is to model the spudcan–soil interaction as a pinned joint (infinite translational stiffness but no rotational stiffness). This has been adopted traditionally and is generally thought to be conservative. However, this is not always the case, and care should be taken in its application. The assumption of a pinned spudcan–soil connection can be non-conservative for the following reasons. During dynamic loading, higher rotational stiffness can move the dominant natural period of the jack-up closer to the spectral content of the wave loading, increasing the level of response. Further, the assumption of infinite translational stiffness enforces no differential movement of the footings; it has been shown that analysis of three-legged jack-up systems that do not allow for additional spudcan penetration (and to a lesser degree, horizontal translation) can result in unconservative results.

*Assumption of uncoupled springs*

The assumption of uncoupled linear springs (where a degree of stiffness in each degree of freedom is assumed) is often favoured over the pinned assumption. This is because the inclusion of any rotational spudcan fixity reduces the critical member stresses at the leg/hull connection (see Figure 8.20), and other responses such as the lateral hull deflection. Unfortunately, while springs are also easy to implement as point elements into structural analysis programs, this simplistic approach cannot capture many important features of real spudcan behaviour. For instance, simple springs cannot capture failure of the spudcan, such as shallow sliding of a windward spudcan under low vertical loads or 'plunging' of a leeward spudcan. Therefore, the structural and geotechnical assessment of jack-ups is usually separated, with the loads on the footings calculated during the structural analysis and then checked against geotechnical failure with established yield surfaces. Furthermore, uncoupled springs cannot account for the non-linear behaviour of soil, although industry guidelines, such as SNAME,

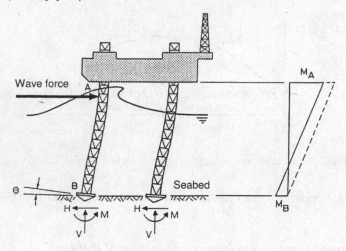

*Figure 8.20* The effect of rotational fixity (de Santa Maria 1988)

have circumvented this by producing stiffness reduction formulations. In these methods the stiffness initially assumed is reduced (after the analysis) based on the distance of the highest combined loads to the yield capacity envelope. The analysis must be rerun until the reduced stiffness is compatible with the reduction formulation provided in SNAME. A better methodology is to allow for a real time degradation in stiffness during the time-domain analysis. This is allowed for in the plasticity models to be described in Section 8.4.

### Initial assumption of stiffness

An initial assumption of the elastic response of the spudcan within the soil is required (for linear springs or when using an integrated plasticity model, as will be described in Section 8.4). Small elastic displacements of a spudcan footing are extremely difficult to measure within a laboratory experiment. Therefore, generic non-dimensional stiffness factors derived from finite element analysis combined with an appropriate choice of shear modulus is recommended. An appropriate expression for the six degrees of freedom case is:

$$
\begin{pmatrix} dV \\ dH_2 \\ dH_3 \\ dQ/D \\ dM_2/D \\ dM_3/D \end{pmatrix} = GD \begin{bmatrix} k_v & 0 & 0 & 0 & 0 & 0 \\ 0 & k_h & 0 & 0 & 0 & -k_c \\ 0 & 0 & k_h & 0 & k_c & 0 \\ 0 & 0 & 0 & k_q & 0 & 0 \\ 0 & 0 & k_c & 0 & k_m & 0 \\ 0 & -k_c & 0 & 0 & 0 & k_m \end{bmatrix} \begin{pmatrix} dw \\ du_2 \\ du_3 \\ Dd\omega \\ Dd\theta_2 \\ D\theta_3 \end{pmatrix} \tag{8.11}
$$

where G is a representative shear modulus and $k_v$, $k_h$, $k_m$, $k_q$ and $k_c$ are the dimensionless stiffness factors. Values for the latter have been derived by Doherty and Deeks (2003) for a variety of soil conditions and footing shapes.

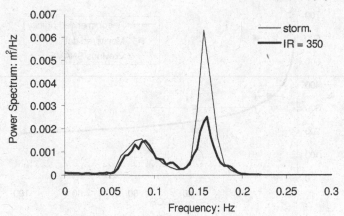

*Figure 8.21* Comparison of storm and simulated response in frequency domain (Cassidy *et al.* 2002b)

An appropriate soil shear modulus is still one of the most difficult parameters to establish. It should represent typical conditions under the spudcan, as the mobilised shear stiffness of soil is strongly dependent on the shear strain. Current recommendations are based on findings of the back-analysis of case records of jack-up platforms in the North Sea (Cassidy *et al.* 2002b, Noble Denton Europe and Oxford University 2005). The records relate to three jack-ups, Santa Fe's Magellan, Monitor and Galaxy-1, which have had their dynamic behaviour and environmental loading conditions monitored since 1992 (reported in Templeton *et al.* 1997, Nelson *et al.* 2000). At the time of the back analysis eight sites of varying soil conditions (three clay and five sand sites) and water depth (between 28 and 98 m) had been subjected to substantial storms, with significant wave height $H_s$ of 4.1–9.85 m (Nelson *et al.* 2000, Cassidy *et al.* 2002b).

For each storm, horizontal deck movements of the jack-up and recorded sea-state and wind speed and direction were available (Nelson *et al.* 2000). In order to compare the monitored jack-up units with numerical simulations of the most severe storm events, a suite of random time domain analyses were performed for each site with adjustments in soil stiffness made. A best-fit of one of the monitored and simulated responses is shown in Figure 8.21.

Resulting from these analyses, the following formulations for the shear modulus for clay and sand seabeds were proposed (Cassidy *et al.* 2002b, Noble Denton Europe and Oxford University 2005). The shear modulus linearly scales all of the stiffness coefficients according to the elastic stiffness matrix in Equation 8.11, and for clay can be determined by

$$G = I_r s_u \qquad (8.12)$$

where $s_u$ is the undrained shear strength measured at 0.15 diameters below the reference point of the spudcan (taken at the level at which the maximum diameter is reached). $I_r$ is the rigidity index and can be calculated from

$$I_r = \frac{G}{s_u} = \frac{600}{OCR^{0.25}} \qquad (8.13)$$

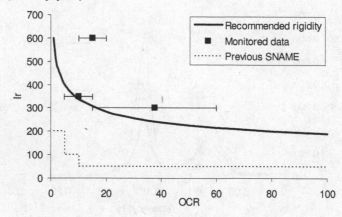

*Figure 8.22* Recommendations of spudcan stiffness in clay arising from Noble Denton Europe and Oxford University study

This recommendation is shown on Figure 8.22, with the back-calculated stiffness from the monitored clay cases superimposed over the new formulation. Also shown are the original SNAME recommendations.

In sands, the shear modulus can be estimated by

$$\frac{G}{p_a} = g\left(\frac{V}{Ap_a}\right)^{0.5} \tag{8.14}$$

where V is the vertical load on the spudcan, A is the spudcan contact area and $p_a$ is atmospheric pressure. The recommended value for the dimensionless constant g for a relative density $D_R$ is

$$g = 230\left(0.9 + \frac{D_R}{500}\right). \tag{8.15}$$

The level of the stiffness, particularly the rotational stiffness, has an effect on the overall structural response. This is demonstrated in Figure 8.20, with a moment at the foundation (M) offsetting (solid line) a simplified bending moment diagram of the leg from the pinned conditions (broken line). As the stresses in the leg-hull connection are a critical design condition, assuming fixity at the foundation level can have a beneficial effect. It also changes the combined load paths and the dynamic response.

### 8.3.4 Combined loading capacity surfaces for spudcans

Combined loading capacity envelopes for undrained and drained capacity of shallow foundations have been introduced in Chapter 6, Section 6.3.3. This section provides another review of their application, in this case concentrating only on spudcan footings of jack-ups.

In a calculation of foundation capacity, the semi-empirical methods developed by Meyerhof (1951, 1953), Brinch Hansen (1961, 1970) and Vesic (1975) may be used to consider the detrimental effect of concurrent vertical, moment and horizontal load on vertical bearing capacity. These formulations suffer from numerous ad hoc factors required to reduce the purely vertical bearing capacity. In recent years, these formulations

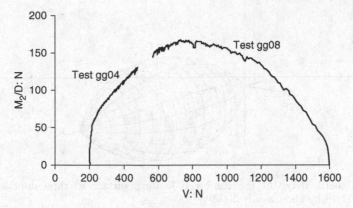

*Figure 8.23* Example swipe test with rotation for a flat 100 mm circular footing on sand (Gottardi *et al.* 1999)

have been replaced with equations that write the capacity (or 'yield') surface directly in terms of the loads applied to it. These equations are considered to be more accurate, an important development as spudcans are subjected to proportionately higher horizontal and moment to vertical loads than onshore shallow foundations, where the vertical loads dominate.

For spudcans, most of these formulations have been established from experimental investigations, including for dense and loose silica sands (Tan 1990, Nova and Montrasio 1991, Gottardi and Butterfield 1993, 1995, Gottardi *et al.* 1999, Byrne 2000, Cassidy and Cheong 2005, Bienen *et al.* 2006, 2007, Cassidy 2007), uncemented loose carbonate sands (Byrne and Houlsby 2001) and clays with increasing strength with depth (Martin 1994, Martin and Houlsby 2000, Byrne and Cassidy 2002, Cassidy *et al.* 2004b). Central to these experimental programs were swipe tests (Tan 1990), where the footing is penetrated vertically to a prescribed level, then subjected to a radial displacement excursion (horizontal, rotational or torsional displacement, or a combination thereof). For a spudcan footing this excursion results in the accumulation of horizontal (and moment, torsional) loads combined with a reduction of vertical load. Tan (1990) argued that the load path followed (with some minor adjustments for soil elasticity and experimental rig stiffness) can be assumed to be a track across the combined loading surface. An example of data from a swipe test of a spudcan is shown in Figure 8.23.

For spudcan footings, the combined load surface in six-degrees of freedom can be written as

$$\left(\frac{H_3}{h_0 V_0}\right)^2 + \left(\frac{M_2/D}{m_0 V_0}\right)^2 - 2a\frac{H_3 M_2/D}{h_0 m_0 V_0^2} + \left(\frac{H_2}{h_0 V_0}\right)^2 + \left(\frac{M_3/D}{m_0 V_0}\right)^2$$
$$+ 2a\frac{H_2 M_3/D}{h_0 m_0 V_0^2} + \left(\frac{Q/D}{q_0 V_0}\right)^2 - \left[\frac{(\beta_1+\beta_2)^{(\beta_1+\beta_2)}}{\beta_1^{\beta_1}\beta_2^{\beta_2}}\right]^2 \left(\frac{V}{V_0}\right)^{2\beta_1}\left(1-\frac{V}{V_0}\right)^{2\beta_2} = 0 \quad (8.16)$$

where $V_0$ determines the size of the yield surface at the current penetration and indicates the bearing capacity of the foundation under purely vertical loading. At the start

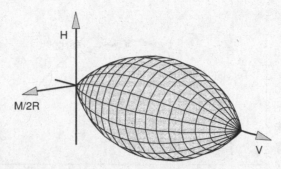

*Figure 8.24* Generic shape of the combined loading surface in three-dimensions (VHM) (Houlsby and Cassidy 2002)

of a jack-up storm analysis, $V_0$ would be equated to the vertical pre-load acting on the spudcan footing ($V_{pre-load}$ in Figure 8.19). {V, $H_2$, $H_3$, Q, $M_2$, $M_3$} is the load vector as illustrated in Figure 8.4. The dimensions of the surface in the horizontal, moment and torsional directions are determined by $h_0$, $m_0$ and $q_0$, respectively, and the parameter, a, accounts for eccentricity (rotation of the elliptical cross-section) in the $M_2/D{:}H_3$ and $M_3/D{:}H_2$ planes. The parameters $\beta_1$ and $\beta_2$ round off the points of the surface near $V/V_0 = 0$ and $V/V_0 = 1$. Interestingly, the parameters defining the shape of the surface do not vary greatly for the different soil types or even with footing shape. Therefore, generic parameter sizes can be utilised in a storm loading analysis, without need for continual site-specific testing (assuming monotonic and either purely drained or undrained loading). Recommended values for the surface sizes are provided in Table 8.6. The yield surface can be easily reduced to the two-dimensional VHM case by simply assuming $H_2 = M_3 = Q = 0$, with Figure 8.24 showing the generic shape of this surface.

### 8.3.5 Structural modelling considerations

As illustrated by the example unit in Figure 8.1, a jack-up consists of a large number of members with intricate structural detail. For many situations, however, it is both convenient and conventional to analyse jack-ups using a mathematical model that simplifies this structural detail considerably. For instance, detailed lattice legs are often assumed as equivalent beam elements. Furthermore, plane frame models and loading along an assumed axis of symmetry are often adopted to reduce modelling to two-dimensions. Structural properties are doubled to account for two legs in either the leewards or windwards position.

It is important, however, that structural non-linearities are considered. The use of jack-up units in deeper water has several detrimental effects on their structural response, including:

- Increased flexibility caused by longer effective leg length. This increases the natural period of the jack-up and, in most situations, moves the structure's principal natural period closer to the dominant wave periods of the sea-state. This requires inclusion of dynamic effects in the modelling of jack-up response.
- The assumption of small displacement behaviour is no longer valid, with structural non-linearities (such as Euler and P-$\Delta$ effects) occurring due to large axial loads in the legs caused by the deck's weight.

*Table 8.6* Parameters defining the size of the combined loading surface

| | SNAME (Clay / Sand) | Clay | Silica Sand | Loose Carbonate Sand |
|---|---|---|---|---|
| $h_0$ | See note/0.12 | 0.127 | 0.122 | 0.154 |
| $m_0$ | 0.1/0.075 | 0.083 | 0.075 | 0.094 |
| $q_0$ | n.a. | 0.05 | 0.033 | 0.033 |
| $a$ | 0 | See note | −0.112 | −0.25 |
| $\beta_1$ | 1 | 0.764 | 0.76 | 0.82 |
| $\beta_2$ | 1 | 0.882 | 0.76 | 0.82 |
| References | SNAME (2002) | Martin (1994)<br>Martin and Houlsby (2001) | Gottardi *et al.* (1999)<br>Houlsby and Cassidy (2002)<br>Bienen *et al.* (2006) | Byrne and Houlsby (2001)<br>Bienen (2007)<br>Cassidy (2007) |
| Notes | The peak horizontal capacity for clay is evaluated as a function of the spudcan shape.<br><br>SNAME only incorporates two-dimensional (V, H, M) loading, therefore $q_0$ is not defined. | Martin and Houlsby (2001) defined the eccentricity a as a function of $V/V_0$:<br>$$a = e_1 + e_2\left(\frac{v}{v_0}\right)\left(\frac{v}{v_0} - 1\right)$$<br>with $e_1 = 0.518$ and $e_2 = 1.180$.<br>$q_0$ is yet to be derived experimentally. This is an estimate of Martin (1994) based on a fully rough footing. | Parameters as recommended from six degree-of-freedom tests of Bienen *et al.* (2006). | |

391

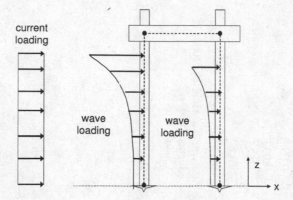

*Figure 8.25* Accounting for wave and current loading in a jack-up analysis

Structural non-linearities must therefore be taken into account for accurate modelling of jack-ups under environmental loading.

### 8.3.6 Environmental loading on a jack-up

#### Wave and current loading

Hydrodynamic loading on jack-up platforms can be calculated by integrating wave forces on the leg from the seabed to the instantaneous free-water level (illustrated in Figure 8.25). This can be achieved by using the Morison equation (shown in Equation 8.17). Variation of the free-water surface, as well as other non-linearities such as drag dominated loading and relative motion effects, can be accounted for in the time domain. In the jack-up industry, regular wave theories such as the linear Airy wave and the higher-order Stokes' fifth theory (as briefly introduced in Section 2.4.2) are widely accepted methods of determining the kinematics required in the Morison equation. Unfortunately, these theories are based on one frequency component and do not account for the random nature of the ocean environment (problematic, as jack-ups are dynamically responding structures with a natural period close to those found in the ocean). For a more representative response, all frequency components in the ocean should be accounted for through either random time domain simulation or spectral wave theories such as NewWave Theory (Tromans *et al.* 1991).

Once the wave and current kinematics have been established, the extended Morison equation is used to calculate the hydrodynamic loads on jack-up legs. The equation consists of a drag and an inertia component and incorporates current and relative motion between the structure and the fluid. The horizontal force per unit length on a vertical member can be expressed as

$$F(x,z,t) = \frac{1}{2}C_d\rho D_h(u_t - \dot{s})|u_t - \dot{s}| + C_m\rho A_h\dot{u} - (C_m - 1)\rho A_h\ddot{s} \tag{8.17}$$

where $D_h$ and $A_h$ are the hydrodynamic cross-sectional diameter and area, respectively, $u_t$ is the velocity vector sum of current and wave resolved normal to the members axis, $\dot{u}$ the acceleration of the wave and $\dot{s}$ and $\ddot{s}$ are the structural velocity and acceleration, respectively, at the point with horizontal position $x$ and vertical elevation z. $C_d$ and $C_m$ are the drag and inertia coefficients, respectively. The drag term is entirely empirical and is due to vortices created as flow passes the member, while the inertia term is due to the pressure gradient in an accelerating fluid. There is considerable uncertainty in the choice of appropriate values for $C_d$ and $C_m$, since they depend on a number of factors such as the Keulegan-Carpenter and Reynolds numbers, relative roughness and wave model being applied. For extreme response analysis of jack-ups, post-critical Reynold numbers ($1.0 \times 10^6 - 4.5 \times 10^6$) and high Keulegan-Carpenter numbers are expected, with SNAME recommending using $C_d = 1.0$ and $C_m = 1.8$, and $C_d = 0.7$ and $C_m = 2.0$, for rough and smooth tubular leg sections, respectively.

Accounting for the relative motion between the jack-up structure and the water is recommended as research has shown that significantly larger response is predicted if relative velocity effects are ignored (i.e. there is an absence of hydrodynamic damping) (Chen *et al.* 1990, Manuel and Cornell 1996). This difference may be as much as 40 per cent in the root-mean-squared (rms) levels of response under random loading (Manuel and Cornell 1996). However, because the relative Morison formulation predicts stronger non-Gaussian behaviour, this difference is not as large for extreme response estimates.

Details of the formulation of the relative motion Morison equation for analysis of jack-up platforms is provided by Barltrop and Adams (1991) and Williams *et al.* (1998). A summary of different wave theories as applied to jack-ups is provided in Cassidy *et al.* (2001) and Jensen and Capul (2006).

## Wind loading

Wind loads on the hull and jack-up legs are generally small compared to the hydrodynamic forces, typically accounting for about 15 per cent of the total horizontal environmental loads for jack-ups (Patel 1989, Vugts 1990). However, due to the large lever arm with respect to the footings, wind forces can have a large effect on the moment forces.

Since the wind velocity varies with respect to time and elevation, SNAME (2002) recommends using a one-minute sustained wind velocity measured 10 m above the mean water level ($z_{ref}$) as the reference velocity $v_{ref}$. The velocities at different depth can then be extrapolated through the expression

$$v_{zi} = v_{ref} \left( \frac{z}{z_{ref}} \right)^{\frac{1}{N}} \tag{8.18}$$

where z is the height of point i above the mean water level and 1/N is usually taken as 0.1.

The wind force $F_{wi}$ on a projected area of the leg or hull is calculated using the Morison equation

$$F_{wi} = \frac{1}{2} \rho_a C_s A_i v_{zi}^2 \tag{8.19}$$

where $\rho_a$ is the density of air, $C_s$ the shape coefficient for the projected area and $A_i$ the projected area of an element i perpendicular to the direction of the wind.

## 8.4  Plasticity-based force–resultant models

Traditional methods of separating the environmental loading, structural and geotechnical modelling components into individual analyses have limitations. These include the following:

- Traditional methods do not account for any hardening of the soil surface under combined loads. Therefore, if failure is initially assessed the only recourse is to allow for greater initial pre-loading.
- Traditional methods prove inadequate in analysing the overall system response, with highly non-linear stiffnesses and a dynamically responsive system simultaneously affecting all analysis components.

More recently, methods for combining models for geotechnical, structural and environmental load in a consistent manner have been receiving attention. Studies have concentrated on describing the shallow foundation behaviour of spudcans in a plasticity framework and in terms of the forces and displacement on the spudcan footing. This allows direct integration within conventional structural analysis programs, as will be shown in this Section. Displacement-based assessment can therefore be accounted for, with the analysis allowing expansion of the initially developed yield surface.

### 8.4.1  Strain-hardening plasticity theory

The framework of single-surface strain-hardening plasticity has allowed the behaviour of the spudcan to be described in an elegant and comprehensive manner. The load–displacement behaviour is determined in essentially the same way that a constitutive law of a metal or soil relates stresses and strains. Loading is applied incrementally, and the numerical plasticity model computes updated tangent stiffnesses for each step.

Details of a series of force-resultant models describing spudcan behaviour are available (Schotman 1989, Martin 1994, Dean *et al.* 1997a,b, Cassidy 1999, Van Langen *et al.* 1999, Martin and Houlsby 2001, Houlsby and Cassidy 2002, Cassidy *et al.* 2002a, Bienen *et al.* 2006, Vlahos *et al.* 2006). These models are similarly structured and have four components:

- A yield surface in combined loading space that describes the boundary of elastic and plastic states (Equation 8.16)
- A hardening law relating the evolution of yield surface size with plastic displacement (bearing capacity of Section 8.2.2)
- A description of elastic response (Equations 8.11–8.15)
- A flow rule that determines the ratio between plastic displacement components during a plastic loading step

The hardening concept adopted is that at any given plastic penetration of the foundation into the soil, a yield surface of a certain size is established in combined loading space. Any changes of load within this surface will result only in reversible elastic deformation. However, plastic deformation can result when the load state touches the surface, with the irreversible footing displacements calculated from a flow rule.

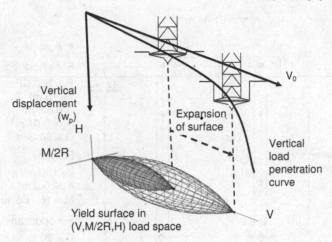

*Figure 8.26* Expansion of the combined loading yield surface with spudcan penetration

During this event, the size of the yield surface is directly related to the 'backbone' curve of vertical bearing capacity against plastic vertical penetration (which is determined either theoretically or empirically). The expansion of a three degree of freedom VHM yield surface during an elasto-plastic step is illustrated schematically in Figure 8.26.

Recommended formulations of the force-resultant models for spudcan footings presented by Martin and Houlsby (2001), Houlsby and Cassidy (2002), Cassidy *et al.* (2002a) and Bienen *et al.* (2006) are summarised in Table 8.7. In order to simulate experimental observations, they maintained subtle differences in their hardening law, yield surface shape, elasticity relationship and flow rule. The use of the consistency condition allows numerical formulation of a model that can predict the load–displacement response.

### 8.4.2 Incorporation of force-resultant models in fluid–structure–soil interaction

The force-resultant models described here capture the entire behaviour of a foundation in terms of the combined forces on it, and their resultant displacements. As this effectively defines the tangential stiffness matrix in a terminology consistent with structural mechanics, the significant non-linear behaviour of soils can be coupled directly within standard structural finite element programs used to analyse the jack-up structure numerically. Effectively, the force-resultant model acts as a nodal element. With each new solution increment (or equilibrium iteration) within the structural finite-element program, the force-resultant model delivers a tangential stiffness matrix and updated force state for the given incremental displacements at the spudcan node (whilst maintaining compatibility with its yield surface and updating its state variable). Full details of implementing these models with structural analysis programs will not be given here, but reference is made to Martin (1994), Martin and Houlsby (1999), Williams *et al.* (1998), Cassidy (1999), Vlahos (2004) and Vlahos *et al.* (2008).

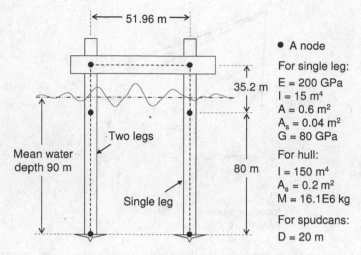

*Figure 8.27* Properties of jack-up and spudcans used in numerical example

### 8.4.3 Example analysis using a force-resultant model

*Dynamic wave analysis*

The following example illustrates the incorporation of a plasticity force-resultant model of a spudcan foundation in a dynamic analysis of a jack-up subjected to a random ocean wave (with further details of this example available in Cassidy (1999) and Houlsby and Cassidy (2002)). The example jack-up structure and properties are shown in Figure 8.27. The truss legs and the hull are assumed as equivalent sections, modelled as beam elements. The model for sand outlined in Table 8.7 is used and compared to conventional pinned and linear spring assumptions. Though non-linearities in the leg/hull jack houses are recognised as significant in the analysis of jack-ups (see Grundlehner (1989) or Spidsøe and Karunakaran (1993) for instance), no attempt was made to include these effects in this example, and a rigid leg/hull connection was assumed. For all of the analyses described here a mean water level of 90 m was assumed, with a rig size typical of a three-legged jack-up used in harsh North Sea conditions. The spudcans were assumed to have a diameter of 20 m and were initially pre-loaded to twice the jack-up's self-weight. The analysis consisted of the jack-up being loaded vertically to the pre-load level, unloaded to its self-weight and then the environmental loads applied.

In this example, the environmental wave loading is considered using NewWave theory, a deterministic method described by Tromans *et al.* (1991). NewWave accounts for the spectral composition of the sea, and can be used as an alternative to both regular waves and full random time domain simulations of lengthy periods. Figure 8.28 shows the surface elevation of the wave in the time domain for both the windward and leeward legs. The wave is focused on the windward leg at the reference time (t = 0 s). The sea-state can be described by the Pierson Moskowitz wave energy spectrum, with a significant wave height ($H_s$) of 12 m and a mean zero crossing period ($T_z$) of 10 s. The extended Morison equation, as detailed in Equation 8.17, is used to evaluate the loads on the jack-up legs.

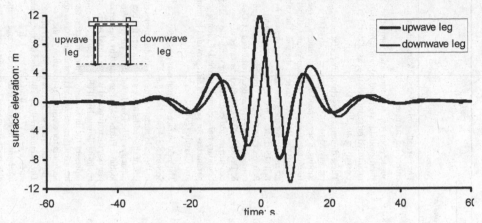

*Figure 8.28* NewWave surface elevation at the up-wave and down-wave legs

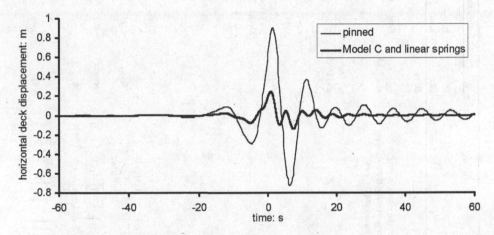

*Figure 8.29* Horizontal deck displacements due to wave loading

The corresponding horizontal deck displacements due to this wave loading are shown in Figure 8.29 for three foundation cases: pinned, the force-resultant plasticity model (Table 8.7) and linear springs. Pinned footings represent infinite horizontal and vertical stiffness, but no rotational stiffness and the plasticity model for sand is as outlined in this chapter (with more details in Houlsby and Cassidy 2002, Bienen *et al.* 2006). The linear springs use finite stiffness values as in the elastic region of the plasticity model. After the wave passes, the rig can be seen to be vibrating in its natural mode. With increased rotational fixity, the natural periods decrease, with approximate values of 9, 5, and 5 seconds for the pinned, plasticity model and linear springs, respectively. In this example, the load combinations were contained entirely within the yield surface, thus giving a response identical to the linear spring case.

By increasing the amplitude of the NewWave to 15 m and then 18 m the increased loading causes plastic displacements in the footings simulated with the plasticity model. This is shown in Figure 8.30. This both displaces and rotates the spudcans and leaves a permanent offset in the displacement of the deck. Yielding of the footings occurred during the peak of the wave loading. The direct indication of yielding is a

Table 8.7 Strain-hardening plasticity models developed for inclusion in jack-up analysis (typical values after Martin and Houlsby 2001, Houlsby and Cassidy 2002, Cassidy et al. 2002a, Bienen et al. 2006).

| Model Comp. | Constant (dimension) | Explanation | Main Equations | Typical values | | | Notes |
|---|---|---|---|---|---|---|---|
| | | | | Clay | Dense sand | Loose carbonate sands | |
| Geometry | D (L) | Footing radius | - | 20 m | | | For partially embedded spudcans defined as the embedded diameter. |
| | G (F/L²) | Representative shear modulus | Use Equation 8.11 as the elastic matrix, with the right hand side representing the conjugate pairs of the footing incremental elastic displacements. | Eqns 8.12 and 8.13 | Eqns 8.14 and 8.15 | | Suggestions for calculating a representative shear modulus are based on calibration with monitored offshore jack-ups, as outlined in Cassidy et al. (2002b). |
| Elasticity | $k_v$ (-) | Elastic stiffness factor (vertical) | | see Doherty and Deeks (2003) | 2.65 | 2.65 | These dimensionless elasticity coefficients may be derived using finite element analysis of a footing. They depend on the geometry of the footing (e.g. depth of embedment and shape) as well as the Poisson's ratio of the soil. The values given here for the sand cases were derived by Bell (1991) for a flat surface footing and a Poisson's ratio of 0.2. Other factors for rigid conical footings have been published by Doherty & Deeks (2003) extended the range of geometries and conditions by using the scaled boundary finite element method. Example parameters for sand are also provided in Bienen and Cassidy (2006). |
| | $k_h$ (-) | Elastic stiffness factor (horizontal) | | | 2.3 | 2.3 | |
| | $k_m$ (-) | Elastic stiffness factor (moment) | | | 0.46 | 0.46 | |
| | $k_q$ (-) | Elastic stiffness factor (torsion) | | | see Doherty and Deeks (2003) | | |
| | $k_c$ (-) | Elastic stiffness factor (horizontal/moment coupling) | | | -0.14 | -0.14 | |
| Yield Surface | $h_0$ (-) | Dimension of yield surface | Equation 8.16 | 0.127 | 0.116 | -0.154 | The yield surface equation is as presented in Section 8.3.4 and Table 8.6. In the yield surface equation $V_0$ determines the size of the yield surface and indicates the bearing capacity of the foundation under purely vertical loading. Further, $V_0$ is governed by the vertical plastic penetration and is determined from the strain-hardening law (Figure 8.26). $h_0$, $m_0$ and $q_0$ represent the dimensions of the surface in the horizontal, moment and torsion directions and a accounts for eccentricity (rotation of the elliptical cross-section) in the M/2R:H plane. The parameters $\beta_1$ and $\beta_2$ round off the points of surface near $V/V_0 = 0$ and $V/V_0 = 1$. It is interesting to note that the yield surface shape does not vary greatly for the different soil types. |
| | $m_0$ (-) | Dimension of yield surface | | 0.083 | 0.086 | 0.094 | |
| | $q_0$ (-) | Dimension of yield surface | | 0.05 | 0.033 | 0.033 | |
| | $a$ (-) | Eccentricity of yield surface | | See notes | -0.2 | -0.25 | |
| | $\beta_1$ (-) | Curvature factor for yield surface (low stress) | | 0.764 | 0.9 | 0.82 | |
| | $\beta_2$ (-) | Curvature factor for yield surface (high stress) | | 0.882 | 0.99 | 0.82 | |

$$f = \left(\frac{H_3}{h_0 V_0}\right)^2 + \left(\frac{M_2/D}{m_0 V_0}\right)^2 - 2a\frac{H_3 M_2/D}{h_0 m_0 V_0^2} + \left(\frac{H_2}{h_0 V_0}\right)^2 + \left(\frac{M_3/D}{m_0 V_0}\right)^2 + 2a\frac{H_2 M_3/D}{h_0 m_0 V_0^2} + \left(\frac{Q/D}{q_0 V_0}\right)^2 - \left[\frac{(\beta_1+\beta_2)^{(\beta_3+\beta_4)}}{\beta_1^{\beta_1}\beta_2^{\beta_2}}\right]^2 \left(\frac{V}{V_0}\right)^{2\beta_1}\left(1-\frac{V}{V_0}\right)^{2\beta_2} = 0$$

| | | | | | |
|---|---|---|---|---|---|
| $\zeta$ | Non-association parameter | 0.6 | – | – | Both the clay and sand experiments showed associated flow only in the MH plane. With differences in the non-association shown in the deviatoric planes, different flow rules have been suggested. For the clay case, Martin (1994) used an 'association parameter' $\zeta$ to adjust vertical displacements to match those observed experimentally. The horizontal and rotational displacements were assumed associated. |
| $\beta_3\ (-)$ | Curvature factor for plastic potential (low stress) | – | 0.55 | 0.82 | |
| $\beta_4\ (-)$ | Curvature factor for plastic potential (high stress) | – | 0.65 | 0.82 | For the sand case, a plastic potential (g) different to the yield surface was required. A similar expression to that of the yield surface is applied, but scaled in shape and size by association factors ($\alpha$). The 'best-fit' values of $\alpha$ shown variability to displacement path by Cassidy (1999) and Cassidy et al. (2002a), with the values here representing a compromise solution. Torsional value were derived through experimentation of Bienen et al. (2007). |
| $\alpha_h\ (-)$ | Association factor (horizontal) | – | 2.5 | 3.25 | |
| $\alpha_m\ (-)$ | Association factor (moment) | – | 2.15 | 2.6 | |
| $\alpha_q\ (-)$ | Association factor (torsion) | – | 1.7 | 1.7 | |

**Flow Rule**

$$-w_p = \zeta - w_{passociated}$$

$$g = \left(\frac{H_3}{\alpha_h b_0 V_0'}\right)^2 + \left(\frac{M_2/D}{\alpha_m m_0 V_0'}\right)^2 - 2a\frac{H_3 M_2/D}{\alpha_h \alpha_m b_0 m_0 V_0'^2}$$

$$+ \left(\frac{H_2}{b_0 V_0'}\right)^2 + \left(\frac{M_3/D}{m_0 V_0'}\right)^2 + 2a\frac{H_2 M_3/D}{b_0 m_0 V_0'^2} + \left(\frac{Q/D}{q_0 V_0'}\right)^2$$

$$+ \frac{(\beta_3+\beta_4)^{(\beta_3+\beta_4)}}{(\beta_3)^{\beta_3}(\beta_4)^{\beta_4}}\left[\left(\frac{v}{V_0'}\right)^{2\beta_3}\left(1-\frac{v}{V_0'}\right)^{2\beta_4}\right] = 0$$

| | |
|---|---|
| Hardening law | Discussion of appropriate hardening laws are given in Section 8.2.2. More details can be found for the sand cases in Cassidy and Houlsby (1999, 2002) and Cassidy et al. (2002a). |

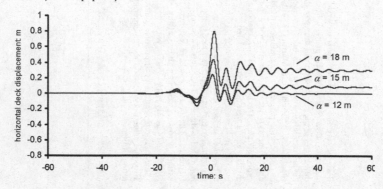

*Figure 8.30* Horizontal deck displacements due to increasing amplitude waves

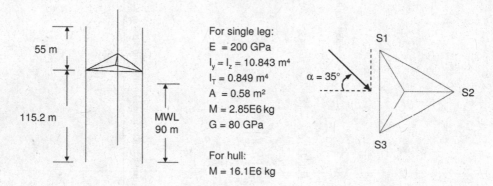

*Figure 8.31* Three-dimensional jack-up used in the numerical analysis (Cassidy and Bienen 2002)

major benefit in using elasto-plastic formulations for the spudcan footings. The natural period after this event is also modified by the plastic behaviour. With jack-ups being used in deeper waters, the contribution of dynamic effects to the total response becomes significant as the natural period of the jack-up approaches the peak wave periods in the sea-state. Accurate prediction of the contribution of foundation stiffness is therefore important.

### Quasi-static pushover analysis

In this example, the pushover capacity of the jack-up is investigated. The example jack-up has been analysed in a static analysis loaded at 35° from the plane-frame direction (Figure 8.31). A simplification of the ocean environment has been used with uniform wave and current loading assumed to occur up to the mean water level (90 m) and wind loading applied at the top of the legs. The loads were increased until a failure of the jack-up system occurred. Again, the spudcan modelled was 20 m in diameter. A pre-load of 133 MN per spudcan was applied representing a multiple of 1.65 on the jack-up's self-weight.

Figure 8.32 shows the normalised reactions for all of the spudcans (labelled in Figure 8.31) in the pushover phase. Pre-loading of the jack-up led to an initial yield

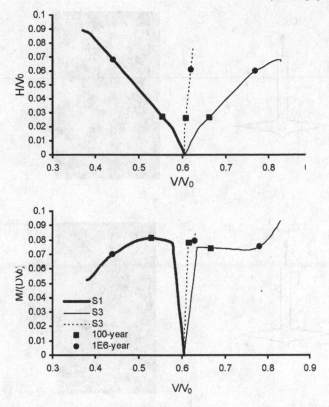

*Figure 8.32* Spudcan reactions and final failure point (Cassidy and Bienen 2002)

surface size with $V_0 = 133$ MN and all of the results in Figure 8.32 have been normalised by this initial $V_0$ value. As the self-weight of the rig is only 80.6 MN per spudcan, the environmental loads are applied at $V/V_0 \sim 0.605$. The loads have been increased until failure of the footings occurred. In this analysis, the most leeward spudcan (S2) yielded first. This is observed in Figure 8.32 by the change in slope and then continuous non-linear behaviour (representing a continual degradation of spudcan stiffness and permanent displacements). As both other footings begin to yield, the horizontal load is increasingly carried by these spudcans. The large overturning moment of the loading on the jack-up legs is increasingly being taken by a change in vertical load carried between the spudcans. This vertical load shedding from the windward footing to the other two is at an increasingly faster rate than for elastic behaviour (see Figure 8.32). The jack-up can sustain considerable load after the footings have initially yielded. This is due to expansion of the yield surface with the vertical plastic penetration of the spudcans. The final failure of the jack-up system is due to sliding of the windward S1 spudcan.

### Further examples

Other examples of the application of plasticity force resultant models in the analysis of jack-up platforms can be found in Martin and Houlsby (1999), Williams *et al.*

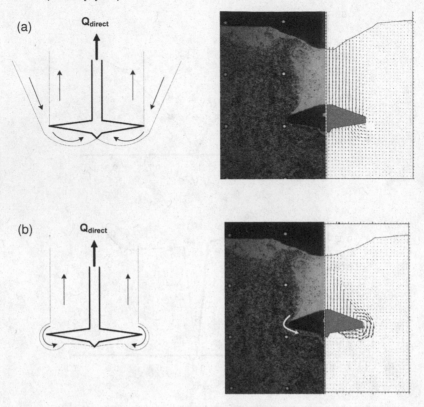

*Figure 8.33* Mechanisms of spudcan undrained extraction (a) early stage and (b) post peak (Gaudin *et al.* 2010 [LHS: schematic] and Purwana *et al.* 2008 [RHS: PIV])

(1998), Cassidy *et al.* (2001, 2002c), Cassidy and Bienen (2002), Bienen and Cassidy (2006, 2009) and Vlahos *et al.* (2008).

## 8.5  Spudcan extraction

Jack-ups extract their spudcans from the seabed by pulling their legs while the buoyant hull is partially submerged. However, with limitations on the allowable over-draft set for each rig (typically about 0.6 m), the maximum tensile load that can be applied to the spudcan is only ~20–50 per cent of the original vertical compressive force during installation. Depending on the depth of embedment and the nature of the soil, this pull-out force is not always sufficient to extract the spudcan immediately making the extraction process difficult and time-consuming.

The following factors influence extraction resistance.

- The depth of embedment relative to the soil shear strength. This determines the failure mechanism and the volume of the soil above the spudcan. For a deeply embedded spudcan, the initial uplift mechanism has been identified to comprise a reverse end bearing in combination with uplift of the soil on top of the spudcan (Figure 8.33a) (Purwana *et al.* 2005, 2008, Gaudin *et al.* 2010). At maximum extraction resistance, a change in mechanism occurs, with a transformation to a flow mechanism around the spudcan edges while the soil above the spudcan continues to be lifted upwards (Figure 8.33b). This is accompanied by the generation of significant negative pore pressure at the spudcan invert as demonstrated experimentally by Purwana *et al.* (2005) and Gaudin *et al.* (2010), and numerically (using finite element analysis) by Zhou *et al.* (2009). A calculation method based on this mechanism is provided by Gaudin *et al.* (2010).

- The duration of the operational phase (relative to the coefficient of consolidation of the soil) and the magnitude of the load sustained by the spudcan footings. These govern the effective stress regime in the soil below the spudcan and therefore the resulting increase in undrained shear strength. This predominantly affects the soil below the spudcan, since the consolidation of the heavily remoulded back-filled soil on top of the spudcan occurs under self-weight. Methods for predicting this increase in extraction resistance with operational time and load have been provided by Purwana *et al.* (2005) and are based on observations in a geotechnical centrifuge.
- The jetting of water whilst the spudcan is being pulled. Most modern mobile drilling rigs are equipped with a water jetting system integrated into the spudcans to assist in their extraction. The water is supplied from pumps located on the hull through hoses down the jack-up legs. Typically these systems have flow rates of around 60 gal/min or 4 l/s. In deeply embedded clay material, where significant suction may be developed at the spudcan invert, the jetting aims to break the suction and reduce the reverse end bearing extraction resistance (Bienen *et al.* 2009, Gaudin *et al.* 2010). However, the effectiveness of water jetting is often questioned by practitioners in the offshore jack-up industry. Clogging of jetting systems and too few nozzles causing an inability to spread the water over the spudcan base are both considered practical inhibitors. Under controlled conditions in the geotechnical centrifuge, recent experiments have shown the effectiveness of water jets on the spudcan base (Bienen *et al.* 2009, Gaudin *et al.* 2010). Experimental evidence showed that the reduction in extraction resistance depends predominantly on the jetted water volume, with respect to the displacement rate of the spudcan, rather than the jetting pressure. This allows the jetting performance to be characterised by the ratio of the volume of water jetted into a theoretical void left by the extracting spudcan. Practical guidance to the volume required is provided in Bienen *et al.* (2009) and Gaudin *et al.* (2010).

Other methods used to extract deeply embedded spudcans include cyclic working of the legs and excavation of soil from above the footing.

# 9 Pipeline and riser geotechnics

## 9.1 Introduction

### 9.1.1 Pipeline networks

Pipelines are the arteries of offshore hydrocarbon developments. They transport the hydrocarbon product and other fluids between wells and in-field processing facilities and also to shore. Different terminology is often used in the industry to distinguish 'flowlines' linking wells within a particular offshore field, and (export) pipelines that take the product, often after some degree of processing, to shore. Large pipelines that gather hydrocarbon products from a number of adjacent fields are sometimes referred to as 'trunklines'. Small pipelines that carry cables for power and control, hydraulic lines or small volumes of chemicals are referred to as umbilicals.

'Rigid' pipelines are made from steel tubes, with layers of internal and external coating for corrosion, abrasion and impact protection, thermal insulation, and to add weight for improved stability. Rigid pipelines typically range in diameter from 0.1 to 1.5 m. 'Flexible' pipelines are composites of metal and polymer and range in diameter from 0.1 to 0.5 m. Wound strips of metal are sandwiched between layers of polymer. Flexible pipelines are expensive to manufacture but rapid to lay compared to rigid pipelines. A rigid pipeline may consist of concentric steel tubes – a 'pipe-in-pipe' system – with the annulus being filled with water for temperature control.

Within the field itself (and with the exception of fields that are developed entirely subsea), fluids must be transported between the seabed and a fixed or floating facility above the sea surface, for which 'risers' are used. These may be in the form of vertical pipes (e.g. so-called top tensioned risers, anchored at the seabed) or pipes that are suspended in a catenary through the water. These catenaries may be flexible pipelines, or they may be conventional rigid pipelines, in which case they are known as steel catenary risers (SCRs). Here, we will use the term pipelines to refer to any pipe laid on, or buried within, the seabed. A network of pipelines is accompanied by various structures at the ends of the pipeline and along the length, some of which are illustrated in Figure 9.1.

As offshore developments extend into deeper waters located further from shore, pipelines and risers represent an increasingly important part of the development infrastructure. Typical costs of gas export pipelines on the North-West Shelf of Australia now exceed US $4 million per km, a significant proportion of which is for measures to stabilise the pipe on the seabed. Indeed, stability is the major focus of geotechnical

---

Primary author of this chapter was David White.

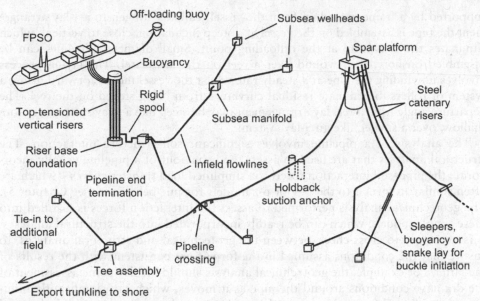

*Figure 9.1* Pipeline networks and associated infrastructure

design for pipelines, both under the action of hydrodynamic loading from waves and currents, and due to expansion and contraction arising from the elevated temperature and pressure of the contents.

Other external loading that needs to be considered can arise from interaction with anchor lines from shipping, fishing trawl gear, iceberg movements and mass transport of the seabed in the form of submarine slides, debris flows and turbidity currents.

An acceptable pipeline design must satisfy limit states that relate to the integrity of the pipeline and the associated structures. These limit states include ultimate and fatigue loading of the pipeline, and relate to the stresses developed from axial force and bending within the wall of the pipeline. Further design limits relate to the interaction between the pipeline and the associated structures at the end of the line (or connected at any intermediate points). It is necessary, therefore, to control the stresses within the pipeline and the movements relative to any associated structures.

Unlike the foundations of structures, on-bottom pipelines can tolerate moderate movements across the seabed without exceeding a limit state, except where they are constrained by wellheads, other connections or obstructions on the seabed. An acceptable design may, therefore, not require the pipeline to be stable where it is installed. It is necessary, however, to assess whether the anticipated movements of the pipeline during its operating life will compromise the integrity or serviceability. These movements may include progressive burial or exposure of the pipeline or sliding of the pipeline in the axial or lateral directions.

Pipelines are usually installed by laying from a vessel, using an S-lay, J-lay or reel-lay configuration. In some cases, the pipe may be assembled onshore and then towed offshore. In an S-lay arrangement, the pipe is assembled in a horizontal position and then offloaded over the rear of the vessel in short sections, concurrent with a forward motion of the vessel. The pipe steepens in an over-bend behind the vessel and is supported by a frame extending from the vessel, called a stinger. In a J-lay arrangement the pipe is assembled on the vessel at a steep inclination, close to vertical, which eliminates the over-bend at the offloading point. Small diameter pipelines can be assembled onshore, then wound over a reel on the lay vessel. The laying process involves unwinding the pipe at a steady rate whilst the vessel moves forward, using a system of rollers to eliminate residual curvature from being stored on the reel. The departure angle from a reel lay arrangement may be steep like a J-lay system, or more shallow, over a stinger, like an S-lay system.

The analysis of a pipeline involves significant soil–structure interaction. The structural analyses that are used to assess the behaviour of a pipeline usually incorporate the pipe–soil interaction forces via simplified load transfer curves – which are often similar in format to the p–y or t–z models for pile behaviour (see Chapter 5). The geotechnical analysis performed to assess the interaction forces is distilled into these simple models, which can be readily incorporated into the structural analyses. It is necessary to cross-check between the geotechnical and structural analyses, to ensure that the conditions assumed in the former are consistent with the results of the latter. For example, the geotechnical analysis should involve some assessment of the drainage conditions around the pipe as it moves, which can significantly influence the pipe–soil forces. However, the mechanisms that control consolidation and drainage are not usually incorporated within the structural analysis of the pipeline, which incorporates the pre-determined load transfer response. This may turn out not to be applicable, depending on the results. Efficient and robust pipeline design, therefore, requires close interaction between the geotechnical analysts and the pipeline engineers.

### 9.1.2　*Geotechnical inputs to pipeline design*

The internal diameter of a pipeline is generally set by flow assurance requirements. The route of a pipeline can be influenced by geotechnical considerations, due to the need to avoid geohazards (see Section 9.2.1), although the start and end points of a pipeline are fixed by the source and destination of the contents. The route may also be refined to mitigate unacceptable in-service movements from hydrodynamic action or from 'walking' of the pipeline (see Sections 9.2.2 and 9.2.3). Geotechnical analyses feed into this process, through assessements of (i) the likely embedment of the pipe after laying, (ii) any changes in embedment that may occur during the life of the pipeline and (iii) the resisting forces between the pipe and the seabed that can be mobilised in the axial and lateral directions.

An external coating may be applied to the pipe, to increase its weight, depending on the relative magnitudes of the hydrodynamic load and the geotechnical resistance. The pipe–soil resistance also affects the optimal solution to mitigate the in-service movements of the pipeline, including the methodology used to create any controlled lateral buckles (see Sections 9.2.2).

If a pipeline is not stable on the seabed or will be exposed to excessive external loading, it may be necessary to provide protection through burial or shielding within a trench. Geotechnical input is required to assess the additional restraint that this will provide and to optimise any arrangement for constructing a trench and burying the pipeline. Some of these design considerations are illustrated in Figure 9.2.

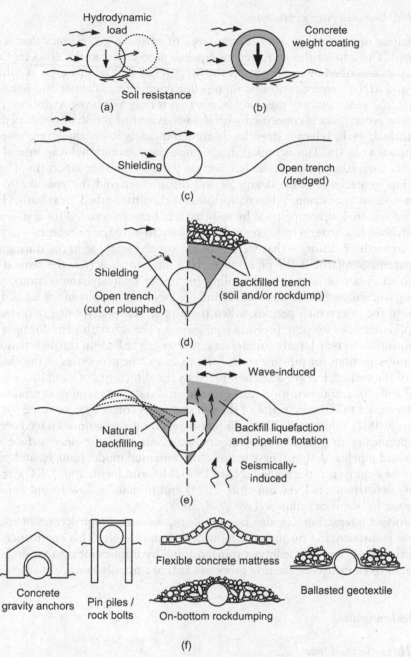

*Figure* 9.2 Design considerations and solutions for engineered stabilisation of pipelines (a) on-bottom instability (b) concrete weight coating (c) open dredged trench (d) cut or ploughed trench, with or without backfill (e) trench issues – natural backfilling, liquefaction and floatation (f) intermittent anchoring techniques

### 9.1.3 Pipeline and riser geotechnics

The analysis of pipelines and risers is an area of offshore geotechnics that is poorly developed. The state-of-the-art review of pipeline geotechnics at the 2005 ISFOG conference was described by its authors as the first ever review of this topic (Cathie *et al.* 2005), and in their recent text book on pipeline engineering, Palmer and King (2008) note that '*the geotechnics of pipe–soil interaction is only just being understood*'.

Pipeline geotechnics is concerned with the behaviour of the shallowest metre or so of the seabed, so the relevant stress levels are significantly lower than for conventional geotechnical analyses. This is important, because the conventional response of soils is often not observed at these stresses (see Section 9.3.1). A further aspect that is unusual to pipeline geotechnics is the degree of soil disturbance and the episodic nature of pipeline–seabed interaction. When a pipeline is laid on the seabed, or is buried beneath backfilled material, it is supported by soil that has been remoulded (or if it is a sandy soil, disturbed is a better term). This soil is unlikely to have the properties – in particular the strength or density – that were deduced for the intact material during the site investigation. Similarly, if the pipe moves back and forth across the seabed during operational cycles of heating and cooling (or under hydrodynanic action), the surrounding soil will be further remoulded, and the local topography of the seabed will be altered. In the intervening periods, when the pipe is stationary, fine-grained soil will reconsolidate back into pore pressure equilibrium. The strength gain during a period of reconsolidation may largely counteract any strength reduction through remoulding.

Site investigations for pipeline routes aim to assess the properties of the shallowest 1–2 m of the seabed, but are often hampered by the difficulties of sampling weak fine-grained soils. *In situ* penetration tests provide vital data, but require careful interpretation to ensure that the influence of the soil surface is correctly accounted for (Puech and Foray 2002, White *et al.* 2010). In recent years, new techniques have been developed specifically to provide improved characterisation of the near-surface soils in order to aid pipeline design. These include instrumented model ploughs and pipelines that can be deployed at the seabed (Noad 1993, Hill and Jacob 2008), ROV mounted full-flow penetrometers (Newson *et al.* 2004) and miniature flow-round penetrometers for use in box core samples (Low *et al.* 2008).

Soil–ocean interaction can also be significant, through the processes of scour and sediment transport, and through the influence of marine life. This confluence of geotechnical engineering with sediment transport and hydraulics means that design analyses ought to capture concurrently processes that are usually considered in isolation.

## 9.2  Design issues

### 9.2.1 Hazard avoidance

The selection of a pipeline route is often influenced by the need to avoid or mitigate against geotechnical hazards, which can include the following:

- Uneven seabed topography, which may cause unacceptable pipeline freespans
- Unstable seabeds, which may liquefy in storms or earthquakes, destabilising the pipeline
- Steep or unstable slopes, down which a pipeline may slide, or which may collapse

- Areas where debris flows or turbidity currents may run out, loading a pipeline
- Unstable fluid escape features (pockmarks, mud volcanoes), which may destabilise a pipeline
- Sand waves, which may create unacceptable freespans, and which may migrate
- Iceberg keels, which may scrape across the seabed, damaging a pipeline.

Geohazards are discussed further in Chapter 10.

### 9.2.2 *On-bottom stability*

Early research into the interaction forces between on-bottom pipelines and the seabed focused on ensuring stability of the pipeline under hydrodynamic loading during storm events. On-bottom stability can be assessed in a conservative manner by making a lower bound assessment of the lateral pipe–soil resistance and ensuring that this exceeds the imposed horizontal hydrodynamic load (accounting for the reduction in vertical pipe–soil contact force due to hydrodynamic lift). Methods of stability assessment are provided in guidelines such as DNV (2007), but there remains debate regarding the mechanisms by which stability is lost. It has been argued that the seabed itself becomes unstable and partially or fully liquefied before the hydrodynamic loading is sufficient to move the pipeline (Damgaard and Palmer 2001, Teh *et al.* 2006). A calculation to assess on-bottom stability may indicate that the pipe is unstable, because the hydrodynamic load during the maximum wave of a storm is sufficient to overcome the resistance of the seabed soil (calculated neglecting any hydrodynamic effects within the soil). However, the build up of that storm may first liquefy the seabed, reducing the soil strength. In this case, the pipe may sink to a deeper embedment, where the hydrodynamic load is significantly reduced, making the first calculation unrepresentative. Meanwhile, the hydrodynamic action may have transported sediment across the seabed, changing the local embedment of the pipe. These considerations are not easy to resolve in a simple design approach.

Pipeline stability can be increased by making the pipeline heavier (e.g. adding concrete coating around the outside), or by secondary means such as laying the pipeline in a trench (with or without backfilling), covering the pipeline with a layer of stable rock or using periodic anchors. These options are all costly, and the design aim is to minimise the extent of such stabilisation measures.

Increased embedment of the pipeline in the seabed has the double benefit of increasing the lateral resistance due to additional passive resistance from the adjacent soil, and decreasing the hydrodynamic forces that the pipeline must withstand. Physical processes, such as ploughing, to increase the embedment are costly and so reliance is generally placed on embedment due to the pipe weight and cyclic motions during laying, together with any self-burial mechanisms that may occur following installation.

### 9.2.3 *Responses to internal pressure and temperature changes*

In deep water, the hydrodynamic forces are generally small and the dominant load is from the internal temperature and pressure, which cause the pipeline to expand. This expansion is opposed by the axial resistance between the pipe and the seabed. Excessive compressive force may lead to buckling of the pipeline, depending on the

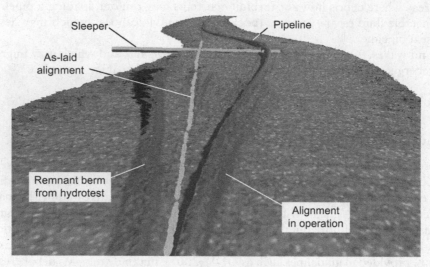

*Figure 9.3* Controlled lateral buckling of an on-bottom pipeline to relieve thermal load (after Jayson *et al.* 2008)

soil resistance and the pipe weight. Once a buckle forms, the axial force drops significantly as pipe feeds axially into the buckle. Excessive feed-in – due to low axial resistance – leads to high bending strains within the buckled section.

A common approach to deal with this expansion has been to bury the pipeline, and ensure that there is sufficient soil cover to prevent the pipeline buckling in the vertical plane – a process known as upheaval buckling. However, as operating temperatures and pressures have risen, it has become convenient to engineer controlled buckles at intervals along the pipeline, in order to maintain the axial stresses at acceptable levels, without creating excessive bending strain in the buckles.

In deep water, where there is no requirement to bury the pipe for protection from trawling, on-bottom buckling in the horizontal plane is now a widely adopted solution to accommodate temperature and pressure-induced expansions (see Figure 9.3). However, controlled lateral buckling requires very careful design in order to ensure that buckles form as planned, and that the fatigue response within the buckle is acceptable throughout the operating life of the flowline (Bruton *et al.* 2007, 2008). Thermal cycles can also lead to the accumulation of axial movement, which is termed pipeline walking (Carr *et al.* 2006). The walking arises from asymmetry in the heat-up and shutdown processes, or from some other asymmetry in the pipeline such as a high riser tension at one end or a seabed slope. Over many thermal cycles, walking can lead to significant global displacement of the pipeline. Walking is not a limit state for the pipeline itself, but can lead to failure at the mid-line or end connections.

Buckling and walking must be assessed during design, in order to control the stresses within the pipeline, and the movements relative to any associated structures. Structural analyses of pipelines incorporate interaction forces between the pipe and the seabed in the vertical, axial and lateral directions. The conventional approach is to construct independent models of the force-displacement response in each direction – in a manner analogous to the 't–z' and 'p–y' models used for axial and lateral pile–soil analysis. More sophisticated models exist, drawing on aspects of plasticity theory

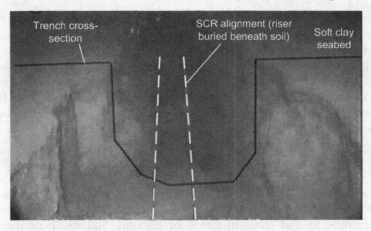

*Figure 9.4* Burial of a steel catenary riser in a trench within the touchdown zone (after Bridge and Howells 2007)

to capture the coupling between each direction of loading (Zhang *et al.* 2002a, b). The current challenge is to extend the existing models – which were originally derived for stability analysis, involving modest pipe movements – to account for the changes in geometry and the remoulding and reconsolidation effects that influence large amplitude cyclic movements.

### 9.2.4 Riser fatigue

Fatigue damage is a key design issue for SCRs and vertical risers, with critical zones being near the points of connection to the floating facility and close to where the riser meets the seabed. Fatigue in the latter zone, referred to as the touchdown zone for SCRs, is driven by changes in axial and bending stresses in the riser due to first and second order motions imparted to it by the floating facility, together with hydrodynamic loading of the suspended section of the riser.

The motions of the SCR can be in the plane of the riser or perpendicular to it, and field observations indicate that large trenches are formed due to the oscillations in both the vertical and horizontal planes (Figure 9.4, after Bridge and Howells 2007). Research has focused primarily on vertical motion, and most commercial software is limited to a linear elastic seabed response, with the calculated fatigue life depending heavily on the assumed seabed stiffness (Bridge *et al.* 2004, Clukey *et al.* 2007). The trench shape also influences the rate of fatigue damage, although there is limited evidence on which to base design calculations.

## 9.3 Pipe–soil interaction for on-bottom pipelines

For on-bottom pipelines, the geotechnical design inputs primarily relate to the interaction forces between the pipeline and the seabed in the vertical, axial and lateral directions. In the latter two directions the restraint depends on the pipe embedment.

These inputs are challenging to assess for four reasons. First, the pipe embedment is difficult to predict, due to the dynamic effects involved in the lay process. Second, conventional site investigation procedures provide poor resolution of the soil strength in the critical upper ~0.5 m of the seabed soil and rarely involve laboratory testing at the low stress levels appropriate for pipe–soil interaction. Third, an acceptable design may involve movement of the pipeline a significant distance across the seabed. The pipe–soil forces associated with this gross deformation depend on the changing seabed topography and soil state. Finally, for the global pipeline response, it is not usually possible to make a conservative assessment of the interaction forces, since both upper and lower bound assessments are required in order for all limit states to be satisfied. Failure can be associated both with excessive and insufficient pipeline movement.

### 9.3.1 Low stress soil behaviour

The *in situ* shear strength profile is important for estimating the initial penetration of the pipeline and the soil resistance to lateral displacement, both of which may involve cycles of remoulding and reconsolidation of the soil. Shallow penetrometer tests, either conducted *in situ* from a seabed frame or ROV (remotely operated vehicle) (Newson *et al.* 2004), or using miniature penetrometers within box-cores (Low *et al.* 2008), provide the best means of obtaining profiles of intact and remoulded shear strength in the upper metre or so of the seabed. Flow-round penetrometers, such as the T-bar and ball, are superior to the cone for this, since they allow cyclic penetration and extraction tests to be undertaken.

The axial pipeline–seabed frictional resistance is also an important design parameter. There is, therefore, a role for laboratory testing of reconstituted (disturbed) material recovered from the seabed, to assess interface friction angles and remoulded strengths at the very low effective stress levels relevant for pipeline design.

Purpose-designed equipment has been developed for measuring the soil–soil and pipeline–soil interface friction angles at very low stress levels. Tilt-table devices (e.g. Najjar *et al.* 2003) involve a hinged plate covered with a weighted sample of clay, which is gradually steepened until the clay slides down the plate under gravity. Alternatively, a modified direct shear box can be used (e.g. Kuo and Bolton 2009), with care being taken to eliminate extraneous friction in order to measure the very low shearing resistance accurately.

A key finding from these devices is that the appropriate friction angle at the relevant effective stress level of typically about 2 kPa is much higher than expected, even for high plasticity clays. Comparison of tilt-table, shear box and ring shear test results on high plasticity clays from offshore West Africa for normal effective stress levels ranging from ~2 kPa up to ~300 kPa indicates a broad trend of residual friction coefficient expressed approximately as

$$\mu \approx 0.25 - 0.3 \log(\sigma_n'/p_a) \tag{9.1}$$

where the normal effective stress, $\sigma_n'$ has been normalised by atmospheric pressure, $p_a$ (White and Randolph 2007, Bruton *et al.* 2009). This leads to a friction coefficient of about 0.75 (friction angle of 37°) for a stress of 2 kPa, which is also consistent with data reported by Bruton *et al.* (1998).

Deep-water sediments often exhibit a non-zero strength intercept at the mudline or a distinct crust with shear strengths as high as 10–15 kPa in the upper 0.5 m of the seabed (e.g. Randolph *et al.* 1998, Ehlers *et al.* 2005). This leads to very high strength ratios, $s_u/\sigma_v'$, and corresponding dilatant behaviour (or negative pore pressure generation for undrained conditions) when sheared. At very low stress levels, the relatively fragile structure of clay is sufficiently robust to give rise to dilatant behaviour and high friction. The concepts of peak strength and dilatancy that apply to sands at moderate stress levels may be usefully applied to the behaviour of clays at the very low stress levels relevant to pipelines. For example, the simple expressions given by Bolton (1986) to link stress level and peak friction angle in sands capture the same trends as Equation (9.1), but use a particle strength parameter for normalisation rather than atmospheric pressure.

Even after the soil has been disturbed during the lay process, and then undergone reconsolidation under the weight of a pipeline, a brittle response may be observed during subsequent sliding at the pipe–soil interface, particularly during fast (undrained) shearing.

## 9.3.2 Vertical penetration

The degree of embedment of a pipeline into the seabed has a strong influence on the lateral stability and also the axial resistance. It is, therefore, important to estimate the embedment as accurately as possible, and yet there are many factors that complicate this. Observations show that the as-laid pipeline embedment is typically much greater than would be expected from the static pipeline weight alone in combination with 'bearing capacity' solutions for penetration resistance. The additional embedment during laying results from two mechanisms: a stress concentration at the touchdown point, and remoulding or displacement of the soil due to cyclic movements of the pipeline during laying. The motion of the lay vessel and any hydrodynamic action on the hanging span will cause the pipe to move dynamically as it comes into contact with the seabed (Westgate *et al.* 2009). During operation, the degree of embedment may also change due to seabed mobility (scour and re-deposition), partial liquefaction under the current and wave action, and consolidation.

### Static penetration resistance – permeable soils

On seabeds comprising soils that are permeable relative to the rate at which the lay process occurs (i.e. generally silty sands, sands and coarser), the penetration process will generally be drained and the as-laid embedment will be low, typically less than 20 per cent of the pipeline diameter. Following Zhang *et al.* (2002a), the normalised embedment, w/D (where D is the pipeline diameter) may be estimated as:

$$\frac{w}{D} = \frac{V_{ult}/D}{k_{vp}} \tag{9.2}$$

where $V_{ult}$ is the vertical pipe–soil force per unit length of the pipeline and $k_{vp}$ is a plastic (secant) stiffness. Maximum pipeline contact stresses, $V_{ult}/D$, are generally less than 10 kPa, while $k_{vp}$ values for purely static vertical penetration will rarely be less than 200 kPa even for relatively loose or compressible materials such as calcareous sands. However, the dynamic movement of the pipe will generally lead to a lower operative value of $k_{vp}$, due to embedment created as the pipe pushes soil aside. Pipeline embedment is, therefore, affected by the motions imposed during laying, particularly any cyclic lateral motion, as well as the strength of the soil (Westgate *et al.* 2010).

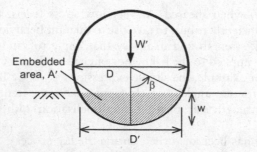

*Figure 9.5* Notation for partially embedded pipeline

*Static penetration resistance – low-permeability soils*

On seabeds comprising soils that are impermeable relative to the rate at which the lay process occurs (i.e. some silts and clays or muds), the response can be assessed using an undrained analysis. Rigorous plasticity solutions exist for predicting the vertical bearing capacity of shallowly embedded pipelines, on undrained soil with a uniform or heterogeneous strength profile (Murff *et al.* 1989, Salençon 2005, Randolph and White 2008b). These solutions have been extended through finite element analysis (Aubeny *et al.* 2005, Merifield *et al.* 2008a, 2009) and generalised into power law relationships of the form:

$$\frac{V_{ult}}{s_{u,invert}D} = a\left(\frac{w}{D}\right)^b \qquad (9.3)$$

where $s_{u,invert}$ is the shear strength at the pipeline invert and the parameters a and b depend on the pipe–soil roughness and the shear strength gradient with depth. Values of a = 6 and b = 0.25 have been suggested for design calculations (Randolph and White 2008c), although this tends to over predict the penetration resistance at very low embedment (when the pipeline resembles a strip footing with reduced chord width, D′, Figure 9.5).

In very soft sediments, particularly in situations where they are remoulded or (in the case of silts and sands) undergo liquefaction, the contribution of buoyancy to penetration resistance can become significant. The overall resistance is, therefore, expressed as:

$$\frac{V_{ult}}{s_{u,invert}D} = N_c + f_b\frac{A'\gamma'}{s_{u,invert}D} \qquad (9.4)$$

where $N_c$ is the right hand side of Equation (9.3), A′ is the nominal area of the pipe that is below the seabed and $\gamma'$ is the effective unit weight of the soil. A factor, $f_b$, is introduced to account for local heave, which will increase the buoyancy force above that due to the nominal embedment. With reference to Figure 9.5, the embedded area is given by

$$A' = \frac{D^2}{4}(\beta - \sin\beta\cos\beta) \qquad (9.5)$$

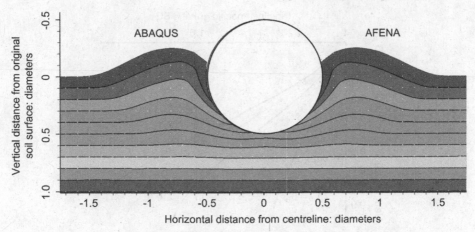

*Figure 9.6* Large deformation finite element analyses of deformations during pipe penetration (w/D = 0.45)

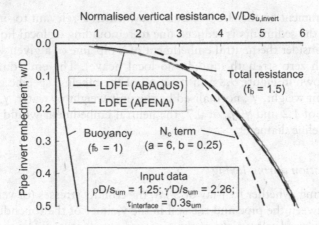

*Figure 9.7* Penetration resistance curves

while the embedment, w/D, and chord length, D′/D, are given by

$$\frac{w}{D} = 0.5(1 - \cos\beta); \quad \frac{D'}{D} = \sin\beta \qquad (9.6)$$

Large deformation finite element analysis of the penetration process may be used to investigate the magnitude of heave and the parameter, $f_b$. A typical result, comparing output from two different finite element packages (Randolph *et al.* 2008c) is shown in Figure 9.6. The local contact point, marked A, is considerably further up the side of the pipeline than the nominal embedment of w/D = 0.45, which affects both the buoyancy resistance and also thermal transfer from the pipeline.

An example penetration curve is shown in Figure 9.7, comparing results from large deformation finite element analysis with the power law expression of Equation (9.4). The best-fit value of $f_b$ is about 1.5, which is consistent with more extensive results from numerical analysis (Merifield *et al.* 2009).

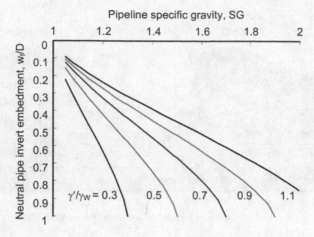

*Figure 9.8* Pipeline embedment at neutral buoyancy in soil of zero strength

As a final comment on the effect of buoyancy, which is relevant to situations where the strength of the sediments is reduced due to remoulding or local liquefaction, it is of interest to consider the neutral embedment of a pipeline of a given specific gravity, SG, in soil with zero strength (and so no local heave). The embedment at neutral buoyancy is shown in Figure 9.8 as a function of the pipeline specific gravity and the soil effective unit weight, $\gamma'$, normalised by the unit weight of water, $\gamma_w$. For a typical specific gravity of 1.2 and $\gamma'/\gamma_w$ of 0.7, the neutral embedment would be about one-third of the pipeline diameter.

### Force concentration during laying

During pipe laying, whether by J-lay or S-lay, the contact stresses (or vertical force per unit length) between the pipe and the soil in the vicinity of the touchdown point will exceed the submerged self-weight of the pipe and any contents. The pipeline configuration is shown in Figure 9.9, and a key parameter is the horizontal component of tension, $T_0$, which is constant through the suspended part of the pipeline. From a simple catenary solution, the horizontal tension may be expressed in terms of the water depth, $z_w$, hang-off angle, $\varphi$, and the pipe submerged weight per unit length, $W'$, as:

$$\frac{T_0}{z_w W'} = \frac{\cos\varphi}{1 - \cos\varphi} \tag{9.7}$$

A characteristic length, $\lambda$, which relates to the length over which the pipeline bending rigidity, EI, moderates the catenary solution is given by $\lambda = (EI/T_0)^{0.5}$. The maximum contact force (per unit length) with the seabed, $V_{max}$, and hence the local force concentration factor, $f_{lay} = V_{max}/W'$, is a function of the seabed stiffness, k (defined as the secant ratio of force per unit length, V, to embedment, w) in addition to EI and $T_0$. Example profiles of V/W' are shown in Figure 9.10. The force concentration factor reduces with increasing water depth and decreasing seabed stiffness.

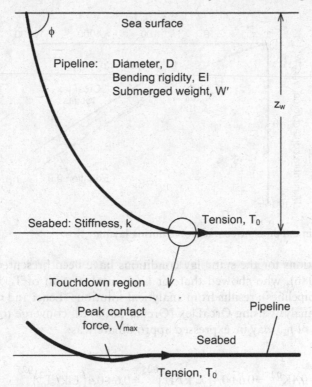

*Figure 9.9* Pipeline configuration during laying

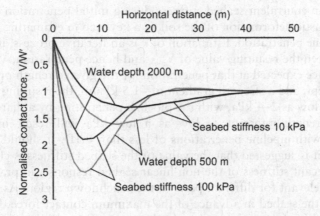

*Figure 9.10* Example profiles of normalised contact force along pipeline

The seabed stiffness may be expressed in non-dimensional form as (Pesce *et al.* 1998):

$$K = \frac{\lambda^2}{T_0} k = \frac{EI}{T_0^2} k \tag{9.8}$$

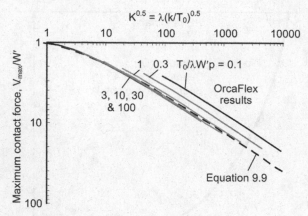

*Figure 9.11* Maximum pipeline contact force during laying

Parametric solutions for the static lay conditions have been presented by Randolph and White (2008b), who showed that for horizontal tension of $T_0 > 3\lambda W'$ (which holds for most pipelines), results from analytical solutions (Lenci and Callegari 2005) and numerical analysis using OrcaFlex (Orcina 2008) all converge to unique design lines. The value of $f_{lay}$ may be expressed approximately as:

$$\frac{V_{max}}{W'} \approx 0.6 + 0.4 K^{0.25} = 0.6 + 0.4 \left( \lambda^2 k / T_0 \right)^{0.25} = 0.6 + 0.4 \left( EIk / T_0^2 \right)^{0.25} \tag{9.9}$$

This expression is compared with the OrcaFlex results in Figure 9.11.

Estimating an equivalent seabed stiffness during initial penetration of the seabed, allowing for plastic deformation of the soil, is a key step in estimating $f_{lay}$, and hence the static pipeline penetration. Estimation of k is an iterative process, involving compatibility between the resulting value of $V_{max}$ and hence penetration, $V_{max}/k$, and the vertical resistance expected at that penetration for the given strength profile. For zero strength intercept, and a strength gradient of ~1.5 kPa/m, the resulting 'plastic' value of k may be as low as 2–4 kPa, with the pipeline penetrating by a diameter or more. However, a strength intercept as low as a few kPa will result in k values of 100–300 kPa, with pipeline penetrations of less than 0.1D. It should also be noted that, although it is suggested that the value of the seabed stiffness is chosen in order to match the secant stiffness of the non-linear seabed response, in practice different stiffnesses are relevant for different parts of the touchdown region. As the pipeline is gradually laid, the seabed in advance of the maximum contact force deforms plastically with increasing force, while points on the pipeline beyond the current maximum contact force will be unloading and thus following a much stiffer response (as V/W' reduces back towards unity).

Low values of the lay factor, $f_{lay} = V_{max}/W'$, in deep water mean that the maximum static loading may occur during the hydrotest after laying the complete pipeline. However, penetration of the pipeline is still likely to be dominated by the lay process, because of the combination of the static catenary contact forces and the dynamic effects due to cyclic pipeline motion.

*Embedment due to dynamic motions during laying*

Additional pipe embedment occurs due to the dynamic movement of the pipe within the touchdown zone, driven by the vessel motion and hydrodynamic loading of the hanging pipe. These loads will induce a combination of vertical and horizontal motion of the pipeline at the seabed (Lund 2000, Cathie *et al.* 2005). In addition to vessel motion due to swell and waves at the sea surface, cyclic changes in pipeline tension may occur (depending on the accuracy of the tensioning system) if the offloading of the pipe is not smoothly coincident with advancing the lay vessel. The changes in tension result in changes in the touchdown point and cycles of vertical motion of the pipeline at the seabed in the touchdown zone. Cyclic movement of the pipeline during laying leads to local softening of the seabed sediments. In particular, any lateral motion will push soil away to either side of the pipe alignment, creating a narrow trench in which the pipe becomes embedded. The net result can be pipeline penetration that is more than an order of magnitude greater than estimated from static loading, even allowing for over-stressing during the lay process.

Allowance for dynamic effects during laying may be made by multiplying the static embedment (based on $V_{max} = f_{lay}W'$) by an adjustment factor, $f_{dyn}$. Typical values of $f_{dyn}$ lie in the range 2–10 based on comparisons between as-laid surveys and static embedment calculations using the intact soil strength, for pipes laid in relatively shallow water (< 500 m depth).

An alternative to estimating $f_{dyn}$ is to perform a 'static' embedment calculation using the fully remoulded shear strength profile, with no further adjustment for dynamic effects. From comparisons with limited data from post-installation surveys, this approach gives a reasonable estimate of embedment for average lay conditions. It over predicts embedment in the case of minimal pipeline motions (e.g. in very calm weather or during lay down of the final catenary section of pipe) and under predicts embedment during extreme weather or downtime events. This approach emphasises the need for accurate estimates of the remoulded shear strength profile, such as obtained from cyclic full-flow penetrometer tests (Westgate *et al.* 2009, Westgate *et al.* 2010).

More accurate, site-specific, assessment of the additional pipeline embedment arising from dynamic motions may be obtained by centrifuge model tests on pipe elements. The design methodology consists of first assessing typical pipeline motions in the touchdown zone arising from the design weather state during laying, together with the pipeline lay configuration (particularly the lay tension) and vessel dynamic response. These motions are then imposed on an element of pipe in the model test, with the number of cycles reflecting the welding time for each segment (or multiple segments) of pipeline (Gaudin and White 2009). A typical pattern of variations in vertical load ratio, V/W', and cyclic (semi-amplitude) horizontal displacements is shown in Figure 9.12, as obtained from dynamic analysis of the vessel and pipeline system. At the mean touchdown point (TDP), there is a zone where the pipeline undergoes cyclic lift-off and re-contact of the seabed. Beyond that zone, the vertical load remains positive and the critical motion is the cyclic horizontal displacement. The continuous profiles may be divided into a number of different stages, corresponding to the common pipeline segment length of 12.5 m. In the model test, these stages can be applied sequentially, or via a leapfrog pattern if the pipeline fabrication process involves welding and off-loading of multiple segments.

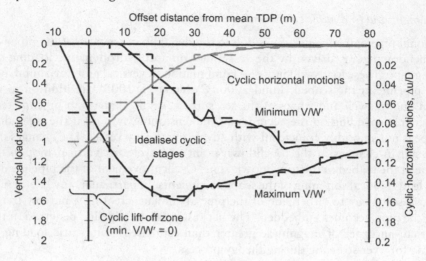

*Figure 9.12* Example pipeline cyclic motions in touchdown zone during laying

## Consolidation following installation

Consolidation times beneath pipelines laid on the seabed can be significant in fine-grained sediments, because of very low values of consolidation coefficient, $c_v$, at the relevant effective stress levels. When a pipe is laid onto a soft seabed, high excess pore pressure is initially created, in response to the applied total stress. As this pore pressure dissipates, leading to consolidation, the effective stress at the pipe–soil interface rises and the soil may strengthen, increasing the available pipe–soil resistance. This process is analogous to the 'set-up' of piles driven in clay.

Elastic calculations indicate non-dimensional consolidation times, $T_{50}$ and $T_{90}$ of about 0.1 and 1, respectively, for embedment depths of up to half a diameter, with T defined as $T = c_v t/D^2$ (using the full pipeline diameter rather than the chord length) (Krost *et al.* 2010). While this leads to relatively short consolidation times for silts and sands, in fine-grained sediments where $c_v$ values are potentially in the range 1–10 $m^2$/year, consolidation times for moderate-sized pipelines may be several days ($t_{50}$) to a year or more ($t_{90}$).

The time-scale of consolidation is important for estimating the pipeline–soil response under different events. For example, to lay a pipeline around a route curve relies on the mobilisation of axial and lateral pipe–soil resistance shortly after the pipe is laid. In this case, only the pipe–soil resistance corresponding to unconsolidated conditions may be available, whereas when the same pipe is loaded during operation, the available pipe–soil resistance may have increased due to consolidation. Another example is that although hydrotesting may apply the highest vertical stress to the soil beneath the pipe, the effect on the shear strength profile may be minimal if an insignificant degree of consolidation occurs during the period for which the hydrotesting is sustained.

Under lateral loading, the consolidation process is typically five times more rapid, reflecting the shorter drainage path (Gourvenec and White 2010). Nevertheless, even in silty sands, with $c_v$ values of $\sim 10^5$ $m^2$/year, $t_{90}$ values will typically exceed the period of wave loading. This results in excess pore pressures accumulating, with the potential for partial liquefaction of the seabed and reduced stability of the pipeline.

### 9.3.3 *Axial resistance*

The longitudinal expansion of a pipeline due to thermal loading mobilises axial resistance in an analogous manner to the shaft resistance on a vertically loaded pile. Several metres of expansion may need to be accommodated at the free ends of a pipeline, where the axial force is zero. Away from the free end, the compressive force within the pipeline is equal to the cumulative mobilised axial resistance, summed from the end. High forces are involved: an axial resistance of 3 kPa mobilised on half of the perimeter of a 0.5 m diameter pipeline, 20 km in length, would generate an axial compressive force of ~50 MN at the mid-point – in such conditions, controlled buckling of the pipe is required to limit the magnitude of axial load.

Estimation of the axial resistance – or the 't–z' load transfer response – of an on-bottom pipeline is not straightforward, and there is no conservative approach: low and high axial resistance can both act in favour or against the design (Bruton *et al.* 2007). For soft clay, it is possible to relate the axial resistance to the local shear strength of the clay, similar to the 'alpha' (total stress) approach for pile design. However, more commonly the axial resistance per unit length, T, is linked directly to the vertical force per unit length, V (in most cases this being the submerged pipe weight, W'), via a friction coefficient, $\mu$ (= tan $\delta$, where $\delta$ is the pipe–soil interface friction angle), and an enhancement factor, $\zeta$, to account for 'wedging' around the curved pipe surface, expressed as:

$$T = \mu N = \mu \zeta V \qquad (9.10)$$

where N is the normal contact force around the pipe–soil interface.

Drained conditions can apply during axial sliding in clay soils as well as more coarse-grained soils due to the slow rates of movement and the short drainage distance out from the shear zone adjacent to the pipe. However, in clays the axial response may also be undrained, or partially drained, in which case, it is necessary to consider the potential for excess pore pressures. In this case, Equation (9.10) can be written in effective stress terms as:

$$T = \mu N' = \mu \zeta V' = \mu \zeta (1 - r_u) V \qquad (9.11)$$

where $r_u$ is an excess pore pressure ratio, defined as the mean excess pore pressure around the pipe–soil surface divided by the mean normal total stress (neglecting ambient hydrostatic water pressure).

Assuming a distribution of normal stress between pipeline and soil that varies as $\cos\theta$, where $\theta$ is the angle from the vertical, it may be shown that the wedging coefficient, $\zeta$, varies as (White and Randolph 2007):

$$\zeta = \frac{2\sin\beta}{\beta + \sin\beta\cos\beta} \qquad (9.12)$$

where $\beta$ is the semi-angle subtended by the contact chord (see Figure 9.5). This gives a factor of 1.1 at an embedment of 0.1D, rising to 1.27 for w/D of 0.5 (or greater).

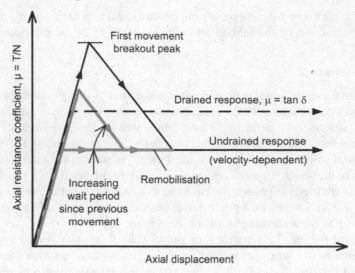

*Figure 9.13* Idealised form of axial pipe–soil response

Numerical analyses, based on elastic soil and modelling the consolidation process, show a similar trend, with slightly higher values of $\zeta$ (Gourvenec and White 2010). Over many cycles of axial movement, this wedging effect may be reduced. The pipe–soil contact load may become concentrated at the pipe invert due to contraction of the 'skin' of repeatedly sheared soil in contact with the pipe.

The friction coefficient, $\mu$, is generally estimated from low stress shear tests between the soil and the relevant pipe surface materials. However, experimental data show that the friction coefficient is sensitive to the rate of displacement and the magnitude of movement. Typical responses are shown schematically in Figure 9.13. Design calculations are often based on the drained friction coefficient, which is applicable for low displacement rates and which can be assessed via Equation (9.10). However, this may overestimate the axial resistance in situations where the rate is sufficient to generate positive excess pore pressures, or when excess pore pressures exist due to other effects, such as soon after pipe-lay or where lateral motion is occurring. In fine-grained sediments, rates as low as a few microns per second may be required to maintain drained conditions. For undrained conditions, the steady axial friction factor – after any initial peak – may be as low as 0.15, reflecting a high (and maintained) pore pressure ratio, $r_u$. The initial peak in resistance, although observed in model tests, is often not significant in design, owing to the brittle nature of the response and the large pipeline displacements involved. The mobilisation distance for axial pipe–soil resistance may be assessed using similar solutions to those for pile shaft resistance (see Chapter 5).

### 9.3.4 Lateral resistance

The resistance provided by the seabed to prevent lateral motion of the pipeline is generally expressed as a 'friction' ratio of the current vertical force (i.e. the submerged weight of pipeline less any transient uplift force). Although expressed as a friction ratio, it is acknowledged that the lateral resistance arises partly from passive pressure arising from the embedment of the pipe. The calculation method described in DNV (2007) is based on superposed components of Coulomb friction and passive resistance (Wagner *et al.* 1989; Brennodden *et al.* 1989; Verley and Sotberg 1994, Verley and Lund 1995). More recently, approaches have been proposed based on yield envelopes in combined vertical, V, and horizontal, H, load space, similar in nature to those used for analysis of shallow foundations (see Chapter 6). These envelopes provide a consistent calculation method that does not invoke an artificial division between frictional resistance and passive resistance. Yield envelopes also indicate the tendency of a pipeline to embed further, or to rise towards the seabed surface, depending on the relative magnitudes of V and H, and the current size of the yield envelope (which depends on the pipe embedment).

A parabolic form of yield envelope was proposed for drained conditions by Zhang *et al.* (2002a), in the form

$$F = H - \mu \left( \frac{V}{V_{max}} + \beta \right) (V_{max} - V) = 0 \qquad (9.13)$$

where $\mu$ is essentially a friction coefficient giving the gradient of the yield envelope at low values of V. $V_{max}$ is the uniaxial penetration resistance at the current embedment, obtained through a hardening function – i.e. the vertical load–penetration response – linked to the plastic modulus, $k_{vp}$ (Equation (9.2)). The $\beta$ term provides non-zero horizontal resistance at zero vertical load, reflecting a passive component of resistance due to embedment. The yield envelope was compared with experimental results obtained from centrifuge model tests of pipes on calcareous sand. These showed that a non-associated plastic potential was required to predict the movements of the pipeline when the load point reached the yield envelope. Downward motion of the pipe occurred for vertical loads greater than about 10 per cent of $V_{max}$, even though the maximum horizontal load occurred at $V/V_{max} \sim 0.5$. The plastic potential was expressed as

$$G = H - \mu_t \left( \frac{V}{V_{max}} + \beta \right)^m (V_{max} - V) - C = 0 \qquad (9.14)$$

and examples of the yield envelope and plastic potential are shown in Figure 9.14.

Application of this drained model in design and consideration of the pipeline response under undrained conditions was discussed by Zhang and Erbrich (2005), who recommended a limiting minimum friction ratio of H/V of 0.37. The model can be extended to cyclic drained loading, using internal yield envelopes with kinematic hardening (Zhang *et al.* 2002b), and can form the basis of force-resultant models in analysis of the response of a complete pipeline (Tian and Cassidy 2008).

Theoretical yield envelopes for pipelines for undrained conditions have been developed through upper bound solutions and finite element analysis (Randolph and White 2008b, Merifield *et al.* 2008a, b). They may be expressed generically as:

$$F = \frac{H}{V_{max}} - \beta \left( \frac{V}{V_{max}} + t \right)^{\beta_1} \left( 1 - \frac{V}{V_{max}} \right)^{\beta_2} = 0 \qquad (9.15)$$

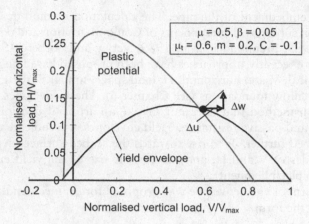

*Figure 9.14* Example yield envelope and plastic potential for a partially embedded pipeline in drained conditions (Zhang *et al.* 2002)

where t is taken as zero for 'unbonded' conditions where no tension can be sustained at the rear of the pipeline, and unity for pipelines that are 'bonded' to the soil. For unbonded conditions and uniform soil strength the β parameters are expressed as:

$$\beta = \left( \frac{(\beta_1 + \beta_2)^{(\beta_1 + \beta_2)}}{\beta_1^{\beta_1} \beta_2^{\beta_2}} \right) \qquad (9.16)$$

$$\text{with} \quad \beta_1 = (0.8 - 0.15\alpha)\left(1.2 - \frac{w}{D}\right) \quad \text{and} \quad \beta_2 = 0.35\left(2.5 - \frac{w}{D}\right)$$

where α is the interface friction ratio at the pipe–soil interface. For undrained conditions, normality applies so the yield envelope is also the plastic potential, which can be used to assess the direction of the pipe movement at failure.

Example yield envelopes for undrained conditions are shown in Figure 9.15. The unbonded case leads to V-H failure envelopes that are approximately parabolic with the maximum horizontal load, $H_{max}$, being reached at approximately $V/V_{max} = 0.4$. For the fully bonded case, $H_{max}$ is twice the value for the unbonded case, with identical soil deformation mechanisms generated in front of and behind the pipe, and minimal soil flow below the pipe invert. In the no-tension case, the rear half of the mechanism is not mobilised, halving the resistance. Experimental results show that the bonded mechanism is rarely fully mobilised due to the opening of a crack behind the pipe, or local failure within a thin skin of soft remoulded soil behind the pipe.

Centrifuge model tests in which the soil deformation mechanisms during pipe penetration and lateral breakout were observed have been conducted within the SAFEBUCK JIP (Bruton *et al.* 2008, Dingle *et al.* 2008). Images captured through the observation window of a plane strain test chamber were analysed using particle image velocimetry (PIV) to accurately quantify the soil deformation mechanisms

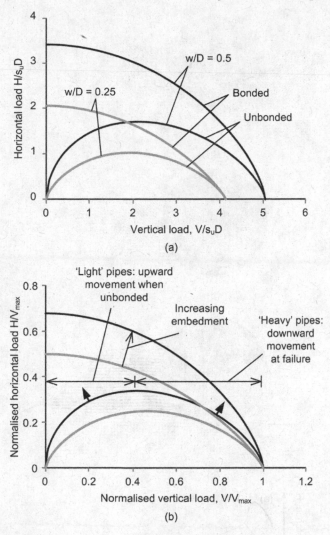

*Figure 9.15* Example yield envelopes for partially embedded pipelines in undrained conditions (a) dimensionless loads (b) loads normalised by vertical bearing capacity

(White *et al.* 2003). Figure 9.16 shows the soil deformation pattern at three stages during the large amplitude breakout of a pipe embedded in soft kaolin clay. It was found that the peak breakout resistance coincided with tensile failure at the rear of the pipe. At breakout (Figure 9.17a) there is evidence that a two-sided mechanism is forming, but no fully mobilised slip plane is evident behind the pipe, so the full soil strength has not been mobilised. After breakout (Figure 9.17b), there are distinct slip planes in front of the pipe, which match the mechanisms calculated from plasticity limit analysis and finite element modelling. With further displacement, the resistance reaches a steady state (Figure 9.17c), with a berm developing in front of the advancing pipe. At this point, the resistance is governed by the size of the berm – which depends on the initial pipe embedment – and the strength of the remoulded soil within the berm and the surficial soil ahead of the pipe.

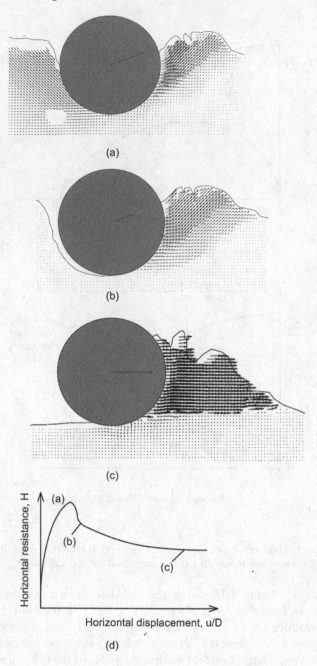

*Figure 9.16* Soil deformation mechanisms from centrifuge model tests (a) at peak resistance
(b) post-peak loss of tension (unbonded) (c) large displacement with steady-state
berm and (d) variation of horizontal resistance (Dingle *et al.* 2008)

*Table 9.1* Undrained residual lateral friction factors based on White and Dingle (2010)

| Initial embedment $w_i/D$ | Vertical load ratio: $V/V_{max} = 0.1$ | Vertical load ratio: $V/V_{max} = 1$ |
|---|---|---|
| 0.3 | $H_{res}/V = 0.49$ | $H_{res}/V = 0.9$ |
| 0.5 | $H_{res}/V = 0.62$ | $H_{res}/V = 1.3$ |

The relative magnitudes of the horizontal resistance at small and large displacements may be assessed with reference to the yield envelopes. For typical values of pipe weight and as-laid embedment, the load level in operation is on the 'dry' side of the unbonded envelopes ($V/V_{max} < 0.4$). In this case, the movement at failure is upwards (once bonding is lost), leading to a reduction in resistance after breakout as the pipe rises towards the ground surface. However, on soft normally consolidated clay the operating condition may be on the 'wet' side, particularly if the pipeline is a pipe-in-pipe system, with a double wall and so relatively heavy. In this case, the lateral resistance will rise sharply as the pipe embeds deeper into the soil.

It is important to distinguish between these two forms of response in lateral buckling design. The softening breakout of a 'light' pipe is usually followed by a steady large-amplitude lateral resistance, which is reasonably predictable. The brittle shape of the response leads to more reliable buckle formation. The hardening response of a 'heavy' pipe is less well understood, and can restrict buckle formation. The resistance mobilised against a heavy pipe during a large-amplitude lateral sweep significantly exceeds the resistance on a light pipe. In addition, the 'diving' behaviour can lead to burial of the pipe. This increases the thermal insulation locally and hampers inspection.

White and Dingle (2010) have proposed an approach for estimating the residual horizontal resistance of light pipelines, based on a series of centrifuge model tests in kaolin, using a smooth model pipeline. The tests showed an initial (brittle) peak resistance arising from suction on the back face of the pipe, followed by a more ductile (but decreasing) post-peak resistance that reduced over a displacement of about one diameter to a steady residual value. They proposed an expression for the *residual* lateral resistance (or friction ratio, H/V) given by

$$\frac{H_{res}}{V} = \mu + k \frac{w_i}{D} \sqrt{\frac{V}{V_{max}}} \qquad (9.17)$$

with suggested parameters of $\mu = 0.3$ and $k = 2$. Note that $w_i/D$ is taken as the initial embedment, and is thus related to $V_{max}$ (the equivalent monotonic vertical force that would give the initial embedment, $w_i/D$) and the shear strength of the soil. For pipe embedment of 0.3–0.5, this would give overall ratios of $H_{res}/V$ that vary as indicated in Table 9.1. A limitation of this tentative expression is that it does not include any soil parameters, so does not capture any influence of the drop in strength, as soil is remoulded ahead of the pipe, or the variation in soil strength with depth.

### 9.3.5 Cyclic response

#### Small amplitude lateral motions

As noted earlier during the discussion of dynamic effects on as-laid pipeline embedment, cyclic motions during the lay process, or during subsequent hydrodynamic

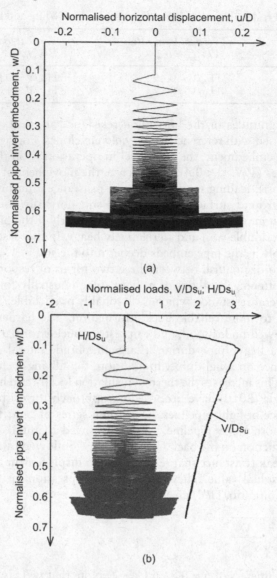

*Figure 9.17* Pipeline embedment under cyclic horizontal motion (a) cyclic horizontal displacements and (b) resulting cyclic horizontal loads (Cheuk and White 2010)

loading, can give rise to significant self-burial. During laying, successive segments of pipeline will be located in the touchdown zone and subjected to hundreds of motion cycles during the (typically) 20–40-minute welding period before more of the pipeline is offloaded from the lay vessel.

An example response from centrifuge model tests to explore self-embedment under the action of small amplitude cyclic horizontal displacements under undrained conditions is shown in Figure 9.17 (Cheuk and White 2010). The test shown used kaolin clay, with a sensitivity, $S_t$, of 2–2.5. Significant embedment occurs for cyclic horizontal displacements as low as ±5 per cent of the diameter. The vertical load was kept

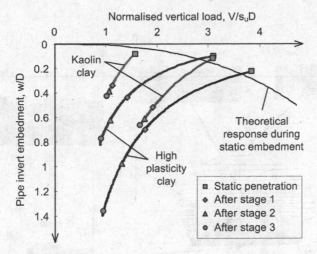

*Figure 9.18* Pipeline embedment paths under cyclic loading (Cheuk and White 2010)

constant, but as the pipe embeds, the normalised load $V/Ds_u$ (where $s_u$ is the shear strength at the pipeline invert) decreases. Eventually, the embedment stabilises at a value that is more than six times that for monotonic vertical loading.

A comparison of tests in kaolin and a West African high-plasticity clay is shown in Figure 9.18. Interestingly, although both clays had similar sensitivity (as measured in cyclic T-bar tests), the additional embedment was much greater for the high-plasticity clay than for kaolin, suggesting that the former clay was more susceptible to water entrainment.

Cheuk and White (2010) describe a theoretical approach to estimate pipeline embedment under small-amplitude cyclic horizontal displacements, following a similar approach to that used for modelling softening during cyclic T-bar tests. In addition to softening of the soil, they allow for the combined vertical and horizontal loading applied to the pipe, estimating the incremental embedment from a flow rule derived from theoretical yield envelopes (similar to Figure 9.15). The critical point on the yield envelopes for assessing the ultimate embedment that is approached under cyclic horizontal motion is where $dH/dV = 0$, implying that $dw/du = 0$ at failure (from normality). This is termed the parallel point and occurs at $V/V_{max} \sim 0.4$ for the unbonded case (Figure 9.15). This value suggests that, even without softening, the pipe would penetrate to a depth where the uniaxial ($H = 0$) penetration resistance was some 2.5 times the submerged pipe weight. With full softening, the final embedment would be to a depth where the original (unsoftened, $H = 0$) penetration resistance was $\sim 2.5 S_t V$ (neglecting buoyancy effects). The proposed approach led to close matching of the experimental data.

Advances have also been made in large deformation finite element analysis techniques, which provide a further alternative for modelling pipeline response under cyclic loading. Wang *et al.* (2009) describe an approach based on the finite element package, ABAQUS, which was able to simulate the model tests referred to before, although using a slightly higher sensitivity for the soil (4 rather than 2.5), as illustrated in Figure 9.19.

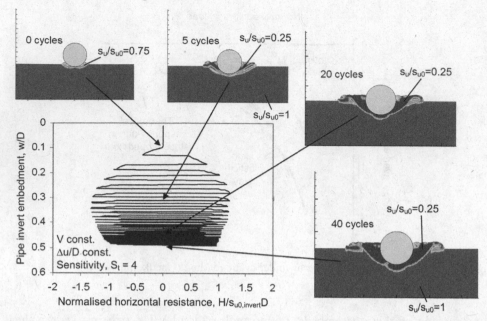

*Figure 9.19* Finite element modelling of pipeline embedment under cyclic horizontal motion (Wang *et al.* 2009)

Under drained conditions, small amplitude motions can also create an increase in the pipe embedment, which enhances stability. Theoretical analysis is more difficult for the drained case due to the influence of dilatancy and volume change, although the plasticity models described earlier can simulate drained cyclic behaviour (Tian and Cassidy 2008). Alternatively, correlations for self-burial have been developed from model tests on siliceous sand (Verley and Sotberg 1994). These link the progressive embedment of the pipe to the amplitude of the movements, the pipe weight (normalised using the soil unit weight) and the current embedment.

In practice, small amplitude motions at the frequency of wave action may lead to partially drained conditions and a build up of excess pore pressure if the soil is silty, particularly in contractile carbonate soils. This additional effect will both enhance the tendency of the pipe to embed, but will also reduce the lateral resistance within the liquefied zone of soil.

## *Large amplitude lateral motions*

An assessment of the pipe–soil resistance forces during cycles of large-amplitude movement across the seabed is often required in design. The crown of a lateral buckle may be designed to sweep several diameters across the seabed during each start-up and shutdown event, of which there may be on the order of 100 during the operating life. In addition, many metres of lateral movement may be acceptable under hydrodynamic storm loading, which can be simulated in a dynamic numerical analysis. During large amplitude lateral movement, soil is swept ahead of the pipe generating a berm (Figure 9.16c). The behaviour is now no longer analogous to a shallow foundation, but is instead dominated by the passive resistance arising from the material ahead of the pipe.

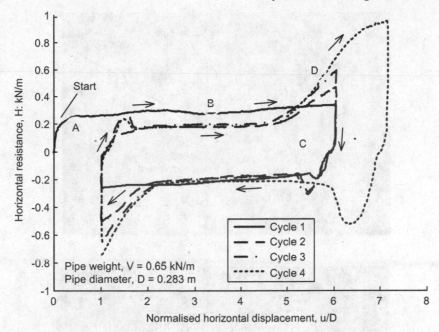

*Figure 9.20* Experimental observations of large amplitude lateral pipe–soil interaction (White and Cheuk 2008)

The force required to slide the berm across the underlying soil depends on the size of the berm – which depends on the swept distance and the pipe invert embedment – and the degree of remoulding of the soil. As the berm grows, the resistance may slowly rise, although softening of the berm material may counteract this. When the pipe changes direction the berm material is left behind, and is re-mobilised if the pipe approaches that point during a subsequent cycle.

These aspects of behaviour are illustrated by a typical model test involving cycles of large amplitude lateral movement, between fixed displacement limits (Figure 9.20) (White and Cheuk 2008). In this test, the model pipe was swept back and forth across a bed of kaolin clay under constant vertical load, whilst the horizontal resistance was measured. The general form of the response involves initial breakout of the pipe (point A) followed by a gentle increase in resistance associated with the growth of a small 'active' berm ahead of the pipe, composed of soil scraped by the lateral sweeping (B). On reversal of the sweeping direction, this response is repeated (C) and the berm generated during the previous sweep is left behind, becoming 'dormant'. When the pipe again approaches point (C) during a later sweep, an increase in resistance is experienced as the dormant berm is collected (D). With repeated cycles of movement, the berms at the limits of the pipe movement grow, causing a corresponding increase in resistance. The first sweep encounters slightly higher resistance than later sweeps due to the larger active berm arising from the initial embedment of the pipe. Illustrations of the berms that are created by this type of large amplitude pipe motion are shown in Figure 9.21.

In the design of lateral buckles, it is important to model the constraint imposed by the berms, in order to provide an adequate assessment of fatigue. If the pipe–soil resistance is assumed to be constant, with the berms ignored, then a buckle will

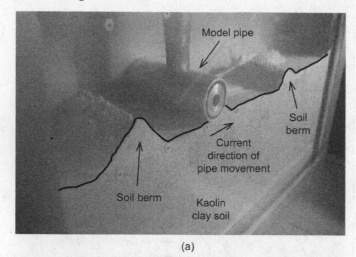

(a)

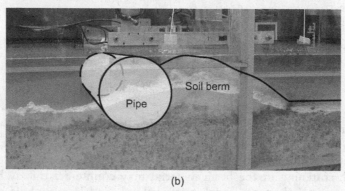

(b)

*Figure 9.21* Soil berms created by large amplitude lateral motion (a) small-scale model test in a geotechnical centrifuge (Dingle *et al.* 2008) and (b) large-scale laboratory model test (White and Cheuk 2008)

progressively lengthen through cycles of expansion and contraction, which reduces the peak bending stresses near the crown (Cardoso *et al.* 2006, Bruton *et al.* 2007). Berms inhibit this lengthening, causing the high stresses generated during buckle initiation to be locked-in. Although the restraint provided by the berms also attenuates the amplitude of the pipe motion within each cycle – and, therefore, the cyclic stresses – the overall effect on fatigue is usually harmful, due to the higher mean stresses that are locked-in (Bruton *et al.* 2007).

Models for the cyclic large-amplitude lateral behaviour can be based on the accumulation and deposition of berm material, as shown schematically in Figure 9.22. The current berm size can be used as a hardening parameter that governs the passive resistance. For undrained conditions, the rate that the berm grows with lateral pipe movement is equal to the depth of soil scraped away by the pipe, from conservation of volume (White and Cheuk 2008). Re-consolidation of the soil that has been remoulded and transported ahead of the pipe will increase the berm resistance.

During the 'residual' response in the zone between the dormant berms, the lateral resistance ratio, H/V, is usually steady, or gradually rising, with a value between

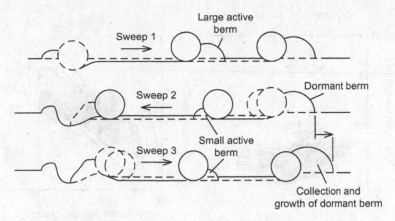

*Figure 9.22* Idealisation of berm kinematics during large amplitude lateral pipeline movement (after White and Cheuk 2008)

0.2 and 0.9 during the initial few cycles. This range is affected by the depth of soil 'ploughed' by the pipe, the sensitivity of the soil as it is remoulded, and also the smoothness of the pipe surface. In addition, in clayey soils, the pipe movement may be sufficiently slow (and the drainage distance at the rear of the pipe sufficiently short) that partial drainage occurs, affecting the resistance. Pore pressure dissipation that occurs between episodes of pipe movement is also relevant. The soil that is exposed by the scraping action of the pipe may swell and reduce in strength between passes of the pipe.

Over the typical design life of a lateral buckle, the depth of the trench created by the sweeping action of the pipe can be significant. As the trench deepens, soil debris may collect in the base, raising the residual resistance, and soil may even flow over the crown of the pipe. Both of these effects will raise the residual resistance. The trench may also lead to a reduction in the vertical pipe–soil contact force, with load being transferred to adjacent sections of pipe.

The resistance mobilised by the soil berms can be assessed based on the estimated berm height. This height, discounted to account for any softening of the soil, can be added to the pipe embedment below the original soil surface, to provide an 'effective' embedment, from which the passive resistance can be estimated (White and Dingle 2010).

For drained conditions, the general form of response is similar to that discussed in the preceding section. The principal difference in sandy soils is that a softened sliding plane does not form beneath the berm ahead of the pipe. Instead, the berm reaches a steady size irrespective of the initial embedment, so the residual resistance during the first sweep is not affected by this value (White and Gaudin 2008).

For design, the variation in lateral resistance can be estimated from a geotechnical analysis that considers in detail the mechanisms described in the foregoing section. Then, to incorporate the results into the structural analysis of the whole pipeline, they can be converted into simpler relationships. For example, the soil resistance might be converted into an equivalent friction factor ($H/W'$) for the residual and berm resistance as a function of cycle number (Bruton *et al.* 2009).

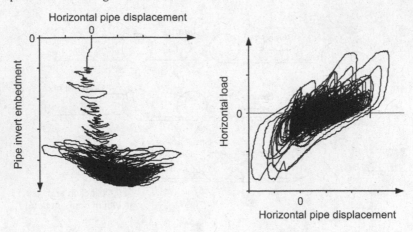

*Figure 9.23* Centrifuge modelling of pipe–soil response under storm loading

### 9.3.6 Model testing of pipe–soil interaction

Pipe–soil interaction during lateral buckling or under hydrodynamic storm loading involves complex changes in seabed geometry coupled with gross disturbance of the seabed soil. There may be episodes of undrained or partially drained behaviour, interspersed with consolidation periods. Assessments of such complex behaviour are naturally susceptible to more uncertainty than most other geotechnical analyses.

It is common to support these assessments using model testing, either at large scale (Langford *et al.* 2007) or in a geotechnical centrifuge (White and Gaudin 2008). The reduced size of the centrifuge model tests shortens consolidation times and allows many operating cycles of start-up and shutdown to be simulated, including intervening episodes of consolidation. Sophisticated control systems also allow arbitrary time histories of loading to be applied, for example, to simulate a design storm or to mimic the dynamic lay process (Gaudin and White 2009).

Typical results from a centrifuge modelling simulation of an element of pipe subjected to hydrodynamic storm loading are shown in Figure 9.23. The model pipe was subjected to a storm history of lift and drag loading, and the resulting self-embedment and lateral movement was observed. In this case, the first few cycles of the storm led to liquefaction of the surrounding soil, and self-embedment of the pipe with only modest lateral movements. When the peak storm load was reached, the pipe was sufficiently embedded to resist the high load with peak-to-peak movements of approximately one pipe diameter, despite the local liquefaction around the pipe. The S-shaped lateral load-displacement response, with an increasing stiffness with displacement, is characteristic of partial liquefaction.

## 9.4 Secondary stabilisation and pipeline burial

### 9.4.1 Secondary stabilisation overview

Design considerations related to stabilisation and pipeline burial were shown schematically in Figure 9.2. Concrete coating may be used to increase the weight of the pipeline, to improve stability – this is referred to as 'primary' stabilisation. In shallow water, additional 'secondary' stabilisation techniques may be required if the pipeline will not be stable resting on the seabed. Secondary stabilisation techniques revolve around reducing the hydrodynamic loading and increasing the available lateral resistance. An open trench provides some shielding from hydrodynamic load whilst burial of the pipe eliminates direct hydrodynamic loading (although soil liquefaction under hydrodynamic loading can destabilise a buried pipe). Other secondary stabilisation techniques include continuous rock dumping, or engineered solutions to provide local lateral restraint at intervals along the pipe. These solutions include flexible concrete mattresses, anchor blocks or saddles placed over the pipeline or small piles on either side of the pipeline. The stability of these objects must also be assessed in design, taking account of the additional cyclic loading transferred to them by the unstable pipeline. A more radical solution for pipeline stabilisation is the use of fins or spoilers on the pipeline to encourage sediment deposition and self-burial.

In some regions, pipelines are buried to avoid snagging from fishing gear – notably in the North Sea, which is intensively fished with heavy gear. Burial may also be beneficial in providing thermal insulation that eases the flow of the contents. It may also be necessary to protect the pipeline from scour by iceberg keels. At shore crossings, pipelines are usually buried in a backfilled trench or a micro-tunnel, which avoids exposure of the pipeline to the surf zone and intense sediment transport.

### 9.4.2 Trenching

A pipeline trench may be constructed before the laying process or afterwards. Post-lay trenching is performed by a machine that straddles the pipeline, constructing the trench. Pipeline trenching was historically performed using jetting, but mechanical ploughing or cutting has become increasingly common. Different trenching techniques are suited to different ground conditions. During the design process, an assessment is performed to establish the feasibility, timescale, equipment requirements and risks associated with each potential trenching method.

Jet trenching techniques use high-pressure jets of water expelled from nozzles suspended beneath seabed level on each side of the pipe to break up and erode the soil. Jet trenching may be performed by a sled-mounted system that is pulled by a vessel, or by a self-propelled submersible tracked vehicle or free-swimming ROV. Typical jetting machines are equipped with multiple nozzles, designed first to cut the seabed, then to discharge material to the side or rear of the machine.

Large ploughs, towed by a ship, may also be used to create a pipeline trench. A typical pipeline plough comprises skids ahead of and behind a heavy ploughshare. The pointed front of the ploughshare cuts soil and pushes it upwards and mouldboards direct it to the side. The plough also includes guides to lift a pre-laid pipe over the ploughshare then direct it down into the trench.

Trenches may also be constructed by mechanical cutters. These are tracked vehicles or sleds that feature multiple rotating cutting blades, tipped with teeth or picks, which are inclined to suit the desired shape of trench. Some form of jetting system is used to expel the spoil. These three techniques – ploughing, cutting and jetting – are often used in combination on a single machine. In shallow water, dredging technology may be used to create a trench, such as a cutter suction dredger or a bucket wheel or bucket ladder dredger.

An assessment of the feasibility and the likely rates of progress using a given trenching technique will involve the following considerations:

1. The bearing capacity of the seabed, which must be sufficient to support the equipment.
2. The strength of the seabed when cut or deformed by a trenching process, taking account of the relevant strain rate – which may be high enough to create undrained or partially drained conditions in sands. High cementation or boulders will inhibit most trenching methods, except for mechanical cutters.
3. The erodibility of the seabed: cemented material and stiff clays are less erodible by jets, whereas uncemented sands are easily disaggregated.

There are obvious conflicting requirements: a strong seabed can support the equipment but a weak seabed is easier to cut.

A trenching assessment should establish whether the target trench depth is likely to be achieved by the alternative methods. If the trench is to be constructed prior to pipe laying, it must also be established that the trench will remain open until the pipe is laid. This depends on the stability of the trench walls during any intervening period where negative pore pressures may dissipate and the local current and sediment transport processes.

For sleds and ploughs, the required towing force is also assessed. The towing force is generally estimated using relationships that are similar to solutions for the resistance of retaining walls, anchors or foundations, which are presented elsewhere in this book. Modifications are often needed to account for the imposed velocity of movement (Cathie and Wintgens 2001). In dilatant sands, the ploughing resistance can be significantly increased if undrained or partially drained conditions prevail (Reece and Grinstead 1986, Palmer 1999). For self-propelled trenchers, the seabed must have sufficient sliding and bearing capacity to resist the weight and tractions imposed by the vehicle. The mechanics of pipeline plough performance are described by Palmer *et al.* (1979) and recent studies have examined the performance of ploughs in uniform sand (Lauder *et al.* 2008) and in sand waves (Bransby *et al.* 2010a,b).

### 9.4.3 Backfilling

Pipeline trenches can be backfilled by mechanical means – during or after the construction of the trench – or can be left to backfill naturally, through the deposition of suspended sediment. If the trench is constructed by jetting, backfilling may not be necessary since the trench may be left filled with jetted spoil, and some jetting machines are equipped with rear jets that undercut the trench causing it to collapse.

If the trench is required only to provide shielding from hydrodynamic loads, then it may be acceptable to leave it open (but subjected to natural backfilling) rather than mechanically backfilling it during construction. In some situations where burial is required, it may be acceptable to allow the trench to backfill naturally, on the basis that the backfilling process will occur sufficiently rapidly that the pipe will be buried by the time it is subjected to the design condition.

Natural backfilling will occur at a rate that depends on the local sediment transport regime. Numerical methods can be used to assess the rate at which trenches infill (e.g. Niederoda and Palmer 1986, Zhao and Cheng 2008).

Mechanical backfilling using the trench spoil can be performed by a plough that is equipped with additional mouldboards to sweep the trench spoil back into the trench after the pipeline has been lowered down. If the trench spoil is unsuitable as backfill, an engineered backfill of sand or rock may be required. Rockdump is expensive to source and deposit whereas sand is more likely to be available from an offshore borrow ground.

The backfilling process can cause unwanted flotation of the pipeline up from the trench base, particularly if the backfill is denser than the pipeline. Flotation of the pipeline can be driven by several mechanisms, including liquefaction of the backfill, transient flow of the backfill down the trench side slopes and also longitudinal flows created by jetting or ploughing (Cathie *et al.* 2005). Pipeline flotation can also occur after a trench has backfilled naturally, if the soil liquefies under wave or seismic action (Bonjean *et al.* 2008).

Backfilled sand is generally in a very loose state, having been placed by rapid sedimentation underwater, so is initially susceptible to liquefaction. However, in a sufficiently high-energy environment, wave action can progressively densify the backfill, to a state that is non-liquefiable (Clukey *et al.* 1989).

The condition of backfilled clay spoil depends on the trenching process, which cuts and remoulds the clay, before it is dumped beside the trench where it is permitted to swell. When backfilled over the pipe, it forms a matrix of intact clay lumps surrounded by water and unconsolidated softened clay. Over time, this matrix will consolidate so that the backfill has a minimum undrained strength equal to a normally consolidated condition, enhanced by any blocks of intact material. The uplift resistance is, therefore, low immediately after backfilling, but rises over time.

Rockdump is highly permeable and, therefore, non-liquefiable, so can be relied on to provide a drained uplift response, providing it is designed with appropriate filter layers to avoid clogging.

### 9.4.4 Resistance forces on buried pipelines

Pipe–soil forces in the axial, lateral and vertical directions are mobilised to support the weight (or buoyancy) of a buried pipeline, and in response to cyclic expansion and contraction of the pipeline from internal pressure and temperature. The most critical direction of loading is usually uplift, which can be mobilised if the pipe is buoyant and the soil is weak or partially liquefied, or if the pipe has a tendency to buckle in the vertical plane to accommodate expansion. This phenomenon is known as upheaval buckling; examples of analyses and case studies of this behaviour are presented by Pedersen and Jensen (1988), Neilsen *et al.* (1990) and Palmer *et al.* (1990).

The available soil resistance during uplift depends on the rate of loading relative to the drainage condition of the backfill and may be drained, undrained or partially drained, in both sands and clays. Slow heating of the pipeline may create drained uplift conditions even in clayey backfills (Bolton and Barefoot 1997) whereas the rapid longitudinal propagation of an upheaval buckle may induce undrained conditions in loose sands (Byrne *et al.* 2008).

The relative magnitude of the undrained and drained resistance depends on the soil state; that is, the tendency for the soil to dilate or contract when sheared. In sandy backfills, the soil is initially loose but may densify under wave action. In clayey backfills, the soil is initially unconsolidated, albeit with relatively intact lumps of material.

During consolidation, the matrix of slurry will strengthen towards a normally consolidated condition. Even in normally consolidated clay, there is evidence of dilatancy at the low stress levels relevant to pipeline uplift, with the undrained resistance exceeding the drained value (Bolton and Barefoot 1997, Cheuk *et al.* 2007).

Uplift resistance can be estimated based on the weight of the backfill and the shearing resistance on failure planes extending to the soil surface. For drained conditions, a limit equilibrium solution based on planes inclined at the angle of dilation can be used (White *et al.* 2008). This approach is favoured over plasticity limit analyses, which invoke excessive dilatancy through the assumption of normality. For undrained conditions, plasticity limit analysis solutions for plate anchors (e.g. Rowe and Davis 1982, Merifield *et al.* 2001) can be applied to pipelines, making allowance for the difference in cross-sectional shape.

Below a critical depth, a failure mechanism involving the flow of soil around the pipeline, rather than lifting the soil to the surface, is more favourable, particularly in loose soil (Schupp *et al.* 2006). Flow round may also occur once the pipe has moved up sufficient for a gap to open beneath the invert (Cheuk *et al.* 2008).

The distance of pipe movement required to fully mobilise the uplift resistance is an important parameter for upheaval buckling analyses, since the critical buckling force depends on the stiffness of the soil restraint. Progressive uplift failure through cyclic ratcheting over many cycles of loading is also a concern. The mobilisation distance to peak resistance appears to depend more on the cover depth than the pipe diameter (Bransby *et al.* 2001, Thusyanthan *et al.* 2010). The distance to mobilise irrecoverable movement, permitting progressive failure, also depends on the particle size and the initiation of flow into the void beneath the pipe (Cheuk *et al.* 2008).

A buried pipe can be destabilised by positive excess pore pressure generated by seismic or wave-induced action. Excess pore pressure increases the buoyancy force on the pipe and reduces the strength of the backfill. This can lead to flotation of pipes that are lighter than the soil, even if full liquefaction does not occur. Models for the generation and dissipation of excess pore pressure can be coupled with the solutions for drained uplift resistance to assess the potential for flotation of the pipe through liquefied or partially liquefied soil. This type of analysis can be used to specify a suitable engineered backfill (Bonjean *et al.* 2008).

The lateral and axial resistance forces on a buried pipeline may be relevant to an assessment of the pipeline response to ice keel scour or submarine slide loading. Various studies of the ultimate lateral resistance of a buried pipeline under drained and undrained conditions have been performed, notably Audibert and Nyman (1977), Trautmann and O'Rourke (1985), Popescu *et al.* (2002), Phillips *et al.* (2004) and Yimsiri *et al.* (2004), vanden Berghe *et al.* (2005) and Hsu *et al.* (2006). These results can be used to develop failure envelopes defined in vertical, lateral and axial load space, of the form described by Calvetti *et al.* (2004), Guo (2005) and Cochetti *et al.* (2009). Pipeline response under external slide loading is discussed further in Chapter 10.

## 9.5  Riser design

### 9.5.1  *Steel catenary risers: geotechnical issues*

Steel catenary risers (SCRs) are a cost-effective means of connecting a floating facility to seabed pipelines. During their operational life, the configuration is essentially

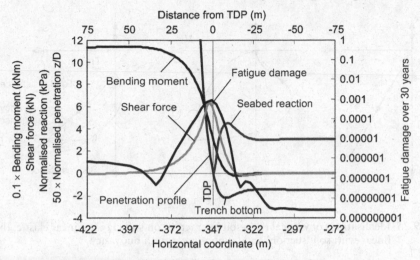

*Figure 9.24* Relative position of SCR fatigue damage in touchdown zone (Shiri and Randolph 2010)

similar to that for pipelines during laying, as shown previously in Figure 9.9. An SCR contacts the seabed at a fixed (average) position, and so any cyclic motion caused by waves acting on the supporting facility, or current-induced vortex induced vibration, will give rise to softening of the seabed sediments and increasing embedment of the riser in what is referred to as the 'touchdown zone'. Trenches several diameters deep have been observed to form within a few months of operation (Bridge and Howells 2007).

One of the most critical aspects of SCR design is fatigue in the touchdown zone, due to cumulative damage arising from cyclic variations of bending moment, and hence stresses, within the riser. The degree of damage is closely linked to the shear force distribution in the riser, since the changes in bending moment may be viewed as a lateral translation of the bending moment profile as the riser is lifted up and replaced under wave action, moving the point of contact with the seabed by several tens of metres (Thethi and Moros 2001). The relative position of the maximum fatigue damage curve, relative to the SCR touchdown point and the shear force profile, is illustrated in Figure 9.24 (Shiri and Randolph 2010).

The main geotechnical issue with respect to SCR design is in the riser-seabed interaction, in particular the stiffness of the response to vertical motion of the riser and the extent to which a trench will form. Both of these affect the shear force distribution and hence the fatigue damage. Fatigue analysis of risers requires sophisticated dynamic analysis of the complete vessel, riser and ocean system. It is generally carried out using specialist software, much of which has tended to be limited to an elastic idealisation of the seabed. Attention has, therefore, been focused on appropriate values of stiffness, which has been expressed in normalised form as:

$$K = \frac{k}{V_{max}/D} = \frac{k}{N_c s_u} \qquad (9.18)$$

where $V_{max}$ ($= N_c s_u D$) is the ultimate penetration resistance at the current embedment and k is the seabed stiffness (k = V/w).

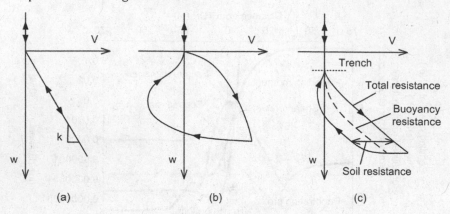

*Figure 9.25* Idealisations of vertical riser–soil interaction on soft clay (a) linear elastic, (b) non-linear with soil 'suction' and (c) non-linear with buoyancy

Experimental studies have shown that K can vary from a maximum value of 200–250 for very small cyclic displacements (Bridge *et al.* 2004, Clukey *et al.* 2005), to values below 10 for displacements that exceed 10 per cent of the riser diameter (Clukey *et al.* 2008). In order to capture this, non-linear seabed interaction models are essential.

Different degrees of sophistication in modelling the riser-seabed interaction are shown in Figure 9.25. The actual riser response during a single cycle is of the form shown in Figure 9.25b with a limiting ultimate resistance during penetration, a stiff unloading response, suction generated following further uplift and finally a release of suction as the riser lifts off the seabed. Model tests show that, following softening of the seabed soil during displacement cycles, the response can follow the 'banana' shape of Figure 9.25c, with the penetration resistance due to buoyancy becoming significant (Hodder *et al.* 2008). Model tests comprising several episodes of cyclic motion, with reconsolidation allowed between episodes, were reported by Hodder *et al.* (2009), and typical results for a cyclic displacement range of $\Delta w/D = 0.025$ are shown in Figure 9.26. The data show gradual trench development and degradation of the secant stiffness during each episode of cyclic motion, but with reconsolidation leading to recovery in the response stiffness and with a gradual increase in the minimum stiffness achieved during successive episodes.

Some aspects of the cyclic riser-seabed response have been captured in a hysteretic model implemented in the riser analysis software, OrcaFlex (Randolph and Quiggin 2009). The model captures the limiting resistance under monotonic penetration and development of suction during uplift, and incorporates a hyperbolic variation of stiffness during reversals of displacement (Figure 9.27). The model was used to explore the effect of trench depth on fatigue damage, with trenches up to five diameters deep developed prior to application of a typical lifetime wave sequence. As indicated in Figure 9.28, the fatigue life can be reduced by 50 per cent (or the fatigue damage doubled) as a result of the trench (Shiri and Randolph 2010).

### 9.5.2 *Vertical risers*

Vertical risers attached directly to well conductors are an alternative form of riser, and similar fatigue design issues arise as for SCRs due to cycles of bending stress within

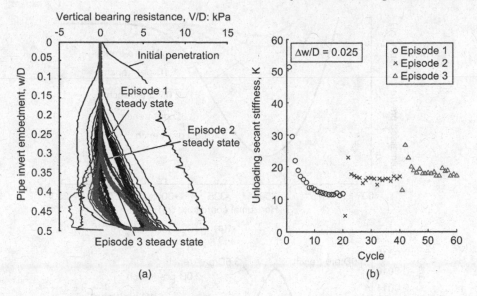

*Figure 9.26* Riser-seabed vertical response under successive episodes of cyclic loading (a) cyclic vertical resistance (b) unloading secant stiffness ratio (Hodder *et al.* 2009)

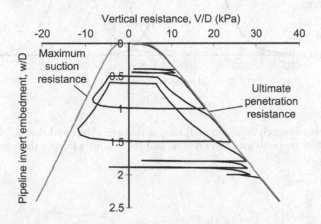

*Figure 9.27* Example riser response under complex motions

the riser or conductor. The most severe fatigue damage can occur either above or below the seabed, and is generally dominated by the large number of small cyclic motions, rather than extreme events. Characterisation of the riser-soil response at small amplitude displacements is, therefore, particularly important.

Analysis of the riser response may be carried out using conventional approaches, as used for the lateral response of foundation piles. However, p–y curves used for foundation design tend to underestimate the lateral stiffness at very small displacements, and this can lead to either overly conservative design, or to unsafe design, depending on whether the critical cyclic bending moments occur above or below the seabed.

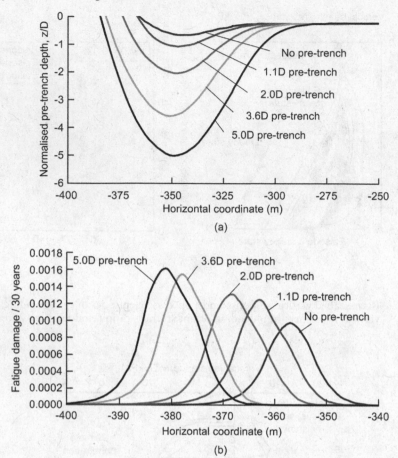

Figure 9.28 Effect of trench depth on SCR fatigue damage (Shiri and Randolph 2010) (a) SCR profiles through touchdown zone and (b) resulting fatigue damage

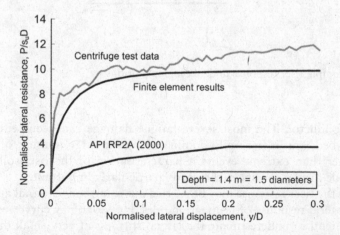

Figure 9.29 Comparison of measured and computed p–y responses with API RP2A (Jeanjean, personal communication)

A comparison of measured and computed p–y responses with recommendations in API RP2A (2000) is shown in Figure 9.29 (Jeanjean, personal communication), showing substantial under prediction of the stiffness and ultimate resistance using the API RP2A curve.

For small cyclic displacements, the stiffness of the lateral p–y response should be based on the small strain modulus, $G_0$, which for lightly overconsolidated clays may be correlated with the undrained shear strength measured in simple shear tests according to (Andersen 2004):

$$\frac{G_0}{s_{uss}} \approx \frac{300}{PI/100} \tag{9.19}$$

The p–y gradient, $k_{p-y}$, may be taken as k ~ $4G_0$ (see Chapter 5).

# 10 Geohazards

## 10.1 Introduction

### 10.1.1 Overview

Geohazards are defined as geological and fluid–dynamic conditions or processes that can lead to the movement of soil, rock, fluid or gas during sudden episodic events or slow progressive deformations. For offshore oil and gas developments geohazards have the potential, i.e. a certain probability of occurrence, to cause injury or loss of life, damage to the environment or infrastructure and can impose significant additional project costs to mitigate their effects.

Numerous geohazards are associated with engineering on the ocean floor and often become more prevalent and their consequences more significant in deeper water. There are two general categories of geohazards:

- Hazardous events – events that are infrequent and episodic in nature. Examples include earthquakes and associated phenomena, submarine slope movements, turbidity flows and gas expulsions.
- Hazardous ground conditions – conditions that involve slow processes that are progressive in nature. Examples include soil creep, non-tectonic fault creep and mud or salt tectonics.

The risk associated with a particular geohazard depends on the location with respect to infrastructure systems, the severity of the event and the frequency of occurrence. Some ground conditions may pose only a low level of hazard until triggered by human activities, while other ground conditions may pose a high level of hazard with frequent movements, but with little risk if no developments are in the vicinity. It is, therefore, necessary to assess not only the presence of geohazards and carry out deterministic analyses of potential failure events, but also to assess the probability of failure events occurring over the lifetime of a development and the associated risks to the development

In this chapter, some of the geological and geotechnical aspects of marine geohazards are discussed including identification, triggering mechanisms, failure modes, consequences and risk assessments. Particular emphasis is placed on submarine slope failures and slides since mass movements of soil can have catastrophic effects on offshore developments. In fact, slope failures are the dominant driver of geohazard risk for many deepwater developments in continental slope settings.

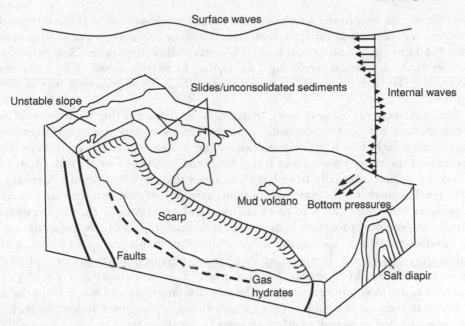

*Figure 10.1* Schematic representation of a selection of offshore geohazards

## 10.1.2 Types of marine geohazards

Geohazards that may be encountered offshore include slope failures, both tectonic and non-tectonic faulting, strong ground shaking, liquefaction, salt diapirs, mud volcanoes, shallow gas, gas hydrates and scour by currents. Hydrodynamic forces from tsunami, wave action and the type of seabed material also can pose hazards to offshore developments. Examples of several types of offshore geohazards are illustrated schematically in Figure 10.1.

Seabed slopes pose a natural geohazard risk because of their potential to become unstable and fail, resulting in landslide events accompanied by mass gravity flows down-slope and retrogressive failures moving upslope behind the initial head scarp. Evidence of submarine slope failures is widespread, even on gentle slopes with inclinations as low as 0.5°.

Two types of faults pose potential hazards. Seismogenic faults result from tectonic deformation of the crust and are sources both of potential earthquakes (strong ground shaking) and rupture of the ground surface. The strong ground shaking produced during earthquakes can lead to seabed liquefaction and resultant large shear strains in seabed deposits. Non-tectonic faults do not pose an earthquake hazard, but can produce either slow progressive creeping deformations or sudden episodic movements. Both types of non-tectonic deformation can impose large strains on infrastructure components. Non-tectonic faults are common in deltas (growth faults) near mud volcanoes and in areas underlain by salt.

Areas underlain by salt formations are susceptible to deformations related to movement of the salt body. Salt deposits are composed of evapourite minerals (halite), have low specific gravities, and are weak and mobile. Therefore, salt is known to migrate as overburden pressures are applied. The movement can cause folding and non-tectonic faulting of the overlying sediments. Mobile salt often rises through the overlying sediments as salt diapirs, which are domes or columns that can range in size from 1–10 km across and extend many kilometres below the seabed. Salt diapirs are prevalent in hydrocarbon producing areas as they form traps in which hydrocarbons accumulate. The movement of salt bodies leads to high shear stresses in the overlying deposits.

Mud volcanoes are eruptions or expulsions of methane gas that have accumulated in the shallow sedimentary deposits, accompanied by watery mud and often large masses of rock. Debris from mud volcanoes can be sent many tens or hundreds of metres into the water column and travel for many kilometres or tens of kilometres across the seabed, especially from large volcanoes that may be several kilometres in diameter. The methane gas may exist in solution in pore water; temperature increases or pressure reductions can lead to ex-solution and expansion of the gas causing an increase in pore pressures. Gas hydrates are solid compounds of gas (typically methane) and water that are physically stable within a particular range of pressure and temperature conditions. Temperature increase or pressure reduction can lead to the dissociation of gas hydrates back into the gas phase (methane) and liquid phase (water) leading to volume expansion. The ex-solution of gas and water from the gas hydrate can increase the volume by a factor of two, leading to a sudden increase in pore pressure and reduction of effective stress in the sediments.

Metocean conditions can have an influence on the effective stress of sediments beneath the seabed. Travelling surface waves apply cyclic hydraulic pressures and shear stresses to the seabed that lead to the generation of excess pore pressures in sub-bottom sediments. The magnitude of hydraulic pressures depends on wave height, wave length and water depth (see Chapter 2). Pressure amplitude decreases with increasing water depth such that pressure changes due to surface waves are most significant in shallow waters, less than approximately 150 m deep. However, large storm waves, as may be experienced during a hurricane or cyclone, can affect sub-bottom sediments in deeper waters. One of the first reported incidents of seafloor instability due to hurricane-induced waves was that of Hurricane Camille in the Gulf of Mexico in 1969 (Focht and Kraft 1977). Thirty-five years later, Hurricane Ivan was reported to have triggered instability of mudflow lobes in the same region (Hooper and Suhayda 2005).

Internal waves, or solitons, can influence the effective stress in seabed sediments in a similar manner to shallow ocean waves. Internal waves oscillate within the water column, rather than on the sea surface, and are caused by variations in density through the water column due to temperature or salinity variations. Bottom pressures due to internal waves decrease with increasing water depth so that, as for surface waves, they are most significant in shallow waters.

The microstructure and mineralogy of a seabed deposit can also pose hazards to offshore developments. For example, sensitive soils experience a dramatic reduction in strength on shearing, such that a seabed material that exhibits considerable shear strength prior to a failure, suffers a substantial reduction in shear strength during failure. Soil sensitivity is central to the evaluation of progressive and retrogressive slide processes. Seabed sediments with high carbonate content are also potentially hazardous to offshore developments, as they can undergo large volume changes due to the high inter- and intra-particle void ratios and through crushing of the soft calcium carbonate materials of which the grains are formed. Carbonate deposits are also particularly prone to post-depositional alterations due to biological and physiochemical processes that form irregular and discontinuous lenses of highly cemented material.

### 10.1.3 Geohazardous regions

Geohazards of one form or another are associated with the vast majority of deep-water developments, either within the field itself, or with the export pipeline route. Mention is made later in this chapter of the two much studied geohazard risks linked to the Sigsbee escarpment in the Gulf of Mexico and the Storegga slide in the northern North Sea. The primary risk in both of these cases is that of submarine slides, with the existence of paleoslides providing graphic evidence of the potential risk, so that geohazard assessment must focus on mechanisms that might trigger further slides and on the potential routes and run-out distances of resulting debris flow and turbidity currents.

Other areas of offshore development where complex geohazards have been identified include South-East Asian waters (offshore Malaysia and Indonesia in an area of high seismic activity), the Caspian Sea, the Nile Delta offshore Egypt and West Africa (offshore Nigeria). Essentially, any offshore developments on the continental slope, particularly where high rates of sediment deposition occur on the associated continental shelf, are likely to incur geohazard risk. Typical seabed features that provide evidence of potential geohazards include faulting, scars from paleoslides, surface expulsion features such as mud volcanoes and pockmarks from gas emission, and channels and steeper-sided canyons from river systems or debris flow routes. Geophysical surveys are the key to identifying geohazards, as they allow mapping of the seabed topography, but also provide evidence of any disturbance to stratigraphic layering within the sediment profile.

### 10.1.4 Triggering mechanisms for failure of a soil mass

Soil masses in the marine environment can become unstable and fail as a result of a range of processes. However, most soil failures are related to some form of triggering event. These triggers can be sudden or progressive natural processes, or related to human activities. Either trigger, natural or anthropogenic, tends to cause an increase in soil stresses or decrease in soil strength, leading to failure of a soil mass. The understanding of pore pressure conditions and the processes and mechanisms that lead to excess pore pressure generation is vital for assessing potential geohazard triggering mechanisms. Kvalstad *et al.* (2001) identified a selection of triggering mechanisms:

NATURAL PROCESSES

- Rapid deposition leading to excess pore pressures, under-consolidation and increased shear stresses in a slope
- Toe erosion or top deposition leading to higher slope inclination and increased gravity forces and shear stresses along potential failure surfaces
- Melting of gas hydrates caused by temperature increase or pressure reduction leading to increased pore pressure and reduced soil strength

- Active fluid or gas flow and expulsion
- Mud volcano eruptions and diapirism giving rise to mass wasting and soil displacements
- Tectonic fault displacements generating earthquakes, near-field displacement pulses, and ground rupture
- Earthquake strong ground shaking causing short-term inertia forces and increase in pore pressure
- Long wavelength wave loading
- Sensitive (contractive) and collapsible soils increasing the risk of retrogressive sliding and increased areal extent of failure zones
- Sea level lowering during glacial periods leading to lower hydrostatic pressure, free gas expansion and gas hydrate ex-solution
- Increased seawater temperature at seabed level caused by changes in current regime leading to temperature increase in the soil mass and ex-solution of hydrates.

HUMAN ACTIVITIES

- Drilling wells, creating blowouts and cratering at the seabed
- Underground blowouts changing the pore pressure regime in shallow layers and potentially creating instability in sloping areas
- Oil production increasing heat flow and temperature around wells leading to hydrate ex-solution, increased pore pressures and strength loss of the adjacent soil
- Depletion of reservoir pressure giving rise to reservoir subsidence and changes in overburden stresses
- Installation activities leading to increasing gravity forces
- Mooring installations and anchoring forces imposing short and long-term lateral forces.

A single triggering condition may result in the occurrence of an isolated geohazard event or a cascade of interrelated events. For example, temperature changes due to the operation of a pipeline may lead to dissociation of gas hydrates or ex-solution of free gas, leading to increased pore pressure, and cause a slope failure. The slope failure can result in rapid deposition down-slope causing excess pore pressures in under-consolidated materials that may in turn lead to a subsequent slope failure. The initial slope failure may initiate retrogressive sliding, leading to a pressure reduction behind the head scarp, which in turn may lead to the dissociation of gas hydrates or ex-solution of free gas, further destabilising the area.

### 10.1.5 Consequences of geohazards

Marine geohazards generally involve mobilisation of the seabed and sub-bottom sediments. The sediments mobilised during these events may impact against or bury infrastructure or lead to loss of foundation support. The volume of soil involved may range from a few cubic metres to thousands of cubic kilometres with consequences ranging from local over-stressing of subsea infrastructure to total loss of an installation with the associated human casualties and economic and environmental impacts. Very large seabed displacements have the potential to trigger a tsunami, which can cause massive human, economic and environmental losses.

The types of consequences that may be related to different types of marine geohazards are discussed in the following section. Slope failure can lead to loss of foundation support in the slide pit or upslope of the head scarp if retrogressive sliding takes place. Facilities or infrastructure in the flow path of a slide or mass gravity flow may be damaged or destroyed by impact forces or burial by slide run-out. Movement along faults or earthquake activity can lead to surface fault rupture as well as cyclic shear stresses and high shear strains in seabed sediments. These seismically induced stresses can cause slope instability or seabed liquefaction, leading to overstressing and damage to structures. Deformation of near-surface seabed deposits due to the ongoing plastic deformation of salt diapirs, mud volcano expulsions, fluid seeps and seabed erosion can lead to stresses in infrastructure. These stresses can affect the structural integrity and operation of subsea installations, for example, well heads, manifolds and pipelines. Ex-solution of shallow gas and dissociation of gas hydrates due to pressure or temperature changes lead to volume expansion of pore fluids, which generate excess pore pressure, thus reducing effective stresses and shear strength leading to seabed instability. Cyclic external stresses applied to the seabed by surface and internal waves can also generate excess pore pressures, which in turn can lead to reduced shear strength, triggering instabilities or liquefaction. Shearing or remoulding of sensitive soils during a slope failure or mass gravity flow, or due to installation activities, can lead to a significant strength reduction jeopardising the stability of the seabed.

Given the wide range of hazards to offshore developments, it is necessary to identify the locations where geohazards might occur, as well as the triggering mechanisms, severity of movements and frequency of events occurring. Once these hazard parameters are characterised, a project can assess the range of potential mitigation strategies that might be appropriate for a give hazard scenario.

Although catastrophic events, such as a seismically triggered submarine slide of significant size, might impact a complete field development, it is generally possible to site infield facilities such as an anchored floating production system at a location with low geohazard risk. However, this is not true for export pipelines (or even, but to a lesser extent in-field flowlines), since these pass through a much greater area of the seabed and must be routed up and over the continental shelf break. As such, the main focus of geohazard assessment is often directed at pipeline routes, and their exposure to mass transport events, either where they cross faults or zones of potential submarine slide instability, perhaps with ongoing creep of a slope, or where they must be routed across potential run-out paths of debris flows and turbidity currents.

## 10.2 Geohazard identification

Offshore development projects often involve large complex geographically distributed systems. By nature of their size, some offshore developments span a range of environments, extending from deep water on the continental rise, up the continental slope, across the continental shelf through shallow water to the shoreline. Although some facilities, such as a single fixed-leg platform may cover only a small area, projects involving deepwater well clusters, anchorage systems, flowlines and export pipelines may extend over 1,000 km, and therefore, are exposed to a wide range of geological and geotechnical conditions. Because of the complex nature of major developments, these projects evolve through time and may go through many design concepts and engineering stages.

*Table 10.1* Summary of generalised investigation elements (modified from Campbell
et al. 2008)

| Phase 1. Pre-Drilling Activities | Phase 2. Post-Discovery | Phase 3. Integrated Site Characterisation |
|---|---|---|
| Screening of regional geological hazards | Preliminary engineering evaluation | Seismic inversion and development of final geotechnical criteria |
| Assess geohazards in prospect area | Plan high-resolution geophysical survey programme | Detailed geohazard assessment, analyses for special engineering issues |
| Assess hazards for specific well sites | Carry out high-resolution geophysical survey programme | Risk assessment |
| Team meetings and reporting | Prepare and process high-resolution geophysical data | Develop model with integrated site characteristics |
| | Complete preliminary site characterisation | Prepare integrated report |
| | Plan geotechnical site investigation | Team meetings and reporting |
| | Carry out geotechnical investigation | |
| | Sample preparation and shipping to laboratories | |
| | Geological lab testing | |
| | Geotechnical lab testing | |
| | Team meetings and reporting | |

In the simplest sense, projects involve three primary phases moving from the general to the specific. These include initial pre-exploration interpretations of potential site conditions (Phase 1), post-discovery preliminary engineering (Phase 2) and integrated site characterisation to support detailed design and engineering (Phase 3) and are summarised in Table 10.1. Each of these three primary phases involves several sub-tasks and a range of personnel with different areas of expertise. The geological and geotechnical information developed during the course of a project needs to be sufficient to support each stage in a project and thus also moves from the general to the specific. Therefore, a phased approach often proves most effective in meeting the engineering requirements for offshore developments. It is very important for these investigations to be coordinated with the engineering team to assure that information is fully considered in the planning, engineering and design stages.

The initial Phase 1 investigations are carried out to provide broad constraints on selection of development sites, route selection and construction feasibility. This phase is essentially a desktop study to evaluate general constraints from conditions such as extreme terrain, earthquake and fault activity, slope instability and broad geotechnical soil properties. The initial work scope can involve a number of tasks, but generally involves: compilation and review of published or unpublished data and reports, interpretation of exploration seismic data; development of a project geographic information system (GIS), identification of critical engineering issues and engineering support. The main product for the early stage of a study is a regional scale geohazard map or series of maps that identify geohazard provinces prone to hazards such as landslides, fault crossings, liquefaction, salt domes, mud volcano activity and areas of gas hydrates. These maps provide a general assessment of the location and distribution of hazards with respect to planned facilities, and provide the baseline geohazard conditions from which to assess conceptual designs, and identify specific areas requiring detailed site investigation. The results of Phase 1 investigations retain large uncertainties and are not suitable for detailed design.

The next phase of work (Phase 2) occurs after a discovery has been made and builds upon the Phase 1 baseline geohazards assessment to define geohazard issues that may affect specific components of a proposed system. The Phase 2 investigations will involve planning and execution of detailed geophysical and geotechnical data acquisition programmes as well as preliminary site characterisation activities. Phase 2 investigations may include the following:

- Acquiring high resolution geophysical and geotechnical data sets
- Developing a predictive soil model
- Developing a project geographic information system (GIS) and mapping geological conditions in the foundation zone or along route alignments
- Conducting detailed terrain analysis, and interpretation of high-resolution geophysical data to identify specific hazards within the project area
- Constructing preliminary hazard susceptibility maps showing, for example, rugged terrain, faults, landslides, liquefiable terrain, and submarine canyon crossings
- Identifying specific targets requiring investigation to develop final design parameters
- Interacting with the engineering team to discuss geohazard constraints and impacts on design
- Developing recommendations for special studies required to address specific technical issues.

Rather than simply submitting a report upon completion of Phase II activities, a workshop should be held with the geohazard and design teams to review the specific findings and determine whether additional investigations need to be completed. Through ongoing communication and interaction of these teams, many geohazard, geotechnical and design issues can be resolved, or decisions can be made on how to avoid, manage or accept potential risks. Where further investigations are necessary, the interactive team workshop format can enable prioritisation of proposed site investigations and integration with other activities to minimise cost and schedule impacts.

The Phase 3 activities involve the final detailed integration of all of the various data sets. This is an intensive stage of any project and involves close interaction between the geologists, geotechnical engineers and owner's representatives. During this stage, additional geophysical processing may be carried out (e.g. seismic inversion), final sub-surface soil models are developed, and soil parameters are defined for use in specialty studies such as site amplification, liquefaction and slope stability analyses. Detailed geohazard assessments are also carried out to address the distribution, severity and frequency of geohazards such as submarine slope failures, mass gravity flows, faulting, strong ground shaking, liquefaction, scour, gas hydrates and fluid expulsion. The results of the Phase 3 detailed investigations should be transmitted in a format suitable for incorporation into economic loss estimation, final siting and foundation design.

Geohazard assessment requires an interdisciplinary team of geotechnical engineers, engineering and marine geologists, geochemists, geophysicists and oceanographers. Each discipline is a specialty and in this section, only the scope of geohazard assessment that falls within the domain of the geotechnical engineer is considered. A geohazard assessment must consider a much greater region than just the field of interest as geohazards from outside the field may affect the proposed development, as equally geohazards within a field boundary may affect third party interests.

There are two general types of data acquisition tools required to complete deepwater developments. These include geophysical survey tools and geotechnical/geological sampling and *in situ* testing tools.

### 10.2.1 *Geophysical surveys*

Offshore projects should begin with an assessment of the 3D exploration seismic data (Campbell *et al.* 2008). This information combined with other data sources provides an indication of major geological features to be expected in a project area. The 3D exploration data (3DX) have relatively low resolution with data points on 12.5–25 m centres and vertical resolution of 8–10 m. 3DX data may be reprocessed to provide greater detail in the shallow part of the stratigraphic section. The interpretation of 3DX seismic is suitable for the early phases of a project, especially in the initial siting of potential well sites.

Once a discovery has been made project-specific geophysical surveys should be carried out to provide detailed information for use in developing an initial geohazard assessment and in planning the geotechnical investigation programme. A selection of geophysical techniques that could be employed are discussed in Chapter 3. The detailed geophysical surveys involve the acquisition of detailed high-resolution bathymetric, seabed and sub-bottom data. A range of geophysical techniques are available and may include use of an autonomous underwater vehicle (AUV) or towed surveys. AUV surveys typically include: (a) multi-beam echo sounder (MBES), swath bathymetry and backscatter (seafloor reflectivity) data with sample points on about 2 m centres; (b) side-scan sonar, which provides imagery similar to aerial photography, with data points on about 1 m centres; and, (c) sub-bottom profiler data that provides stratigraphic details in the upper 70 m of the section with vertical resolution of <0.5 m depending on sub-surface conditions. Deep towed (DT) systems that collect a similar suite of data types are available, but are less efficient and may compromise data quality in rugged terrain. AUV surveys are becoming preferable over towed systems, delivering better quality bathymetry, and side-scan sonar and seismic profiling data that are referenced to the same set of coordinate positions, thus avoiding ambiguities associated with geo-referencing separate data sites (Jeanjean *et al.* 2005).

Data collected with AUV or DT systems provide information for use in characterising the seafloor and shallow foundation zone. However, some foundation concepts have foundation zones that require information to depths greater than can be acquired using AUV systems. When data are required to support these foundations designs, or to characterise deeper geological conditions or hazards, alternative survey techniques are used. These may include ultra-high resolution 2D (UHR2D), high-resolution 2D (HR2D) or high-resolution 3D (HR3D) surveys. UHR2D surveys can collect data to depths of ~200–500 m below the seafloor and can achieve vertical resolution of 1–1.5 m. HR2D surveys collect data to depths of ~1,500 m and have vertical resolution of

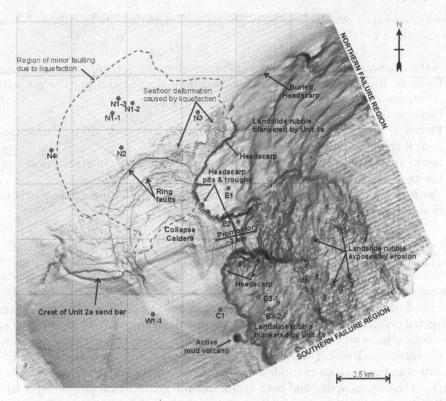

*Figure 10.2.* Seabed rendering of the Shah Deniz field, Caspian Sea (Mildenhall and Fowler 2001)

about 3 m. UHR2D surveys provide the most detailed information for foundation zone investigations. HR3D surveys provide data for a three-dimensional 'volume' rather than along individual two-dimensional survey lines. Because it is then possible to assess sub-surface conditions in three-dimension, more robust interpretations can be made more quickly and with greater confidence using HR3D survey data compared to HR2D survey data. However, the costs associated with HR3D surveys are significantly more than for HR2D surveys and thus these data are typically only acquired in areas with complex geology where risks are high (Campbell *et al.* 2008).

AUV survey data provide information on seabed and shallow sub-surface conditions that are particularly useful for evaluating geological and geotechnical conditions. For example, AUV data can be used to assess existing slopes, scarps indicating previous slope failures, run-out paths of previous debris flows, mud volcanoes and fall out from previous eruptions and layering or lenses that may indicate the presence of cemented carbonate deposits. Deeper penetrating UHR2D, HR2D and HR3D surveys are more useful for assessing features such as faults, deep-seated slope failures and diapirs. An example of an AUV seabed rendering of the Shah Deniz field, Caspian Sea is shown in Figure 10.2.

A range of down-hole geophysical techniques are available to assess soil conditions at a specific site (Campbell *et al.* 2008). These tools provide continuous data on a range of stratigraphic soil properties. The data are collected by lowering a tool with a variety of sensors down the bore hole. The types of data that can be collected include, for example:

- Natural gamma emission – stratigraphic log showing variations in mineralogy
- Gamma density – chemical composition, bulk density and derived porosity of formation rock
- Neutron porosity – formation porosity
- Electrical resistivity – indicator of gas hydrate, changes in soil type and mineralogical changes
- Sonic waveform – correlates seismic data for stratigraphic interpretation
- Caliper – borehole diameter
- Vertical seismic profiling – seismic velocities for site response analysis and correlation of logging data with seismic records.

### 10.2.2  Geotechnical site investigation

A geotechnical site investigation programme is most effectively carried out following acquisition of site-specific geotechnical data, development of the preliminary soil model and completion of the initial geohazard assessment. This enables the collection of targeted geotechnical data and samples to support the foundation engineering stage of a project, but also provides an opportunity to acquire data for the assessment of specific seafloor features. The geotechnical investigation provides data for specific soil properties and pore pressure conditions of seabed deposits and involves borehole logging, field testing, sampling and both geological and geotechnical laboratory analyses.

There are two main approaches for geotechnical site investigations. These include the seabed mode and the down-hole mode (Campbell *et al.* 2008). Seabed mode methods characterise the shallow part of the stratigraphic section to depths of <50 m. These methods often include large diameter piston core sampling and *in situ* testing, piezocone penetrometer testing (PCPT), vane shear tests (VST) and T-bar and ball penetrometer testing (Peuchen and Rapp 2007). Some facilities, such as tension-leg platforms, (TLPs), compliant platforms and mooring systems for SPARs, semi-submersibles and floating production storage and offloading vessels (FPSOs) have driven pile foundations that require data on soils to a greater depth than can be reached using seabed investigation approaches. In these cases, sampling is completed using down-hole approaches. This involves using rotary drilling techniques from a ship-based drilling platform. Coring tools are advanced and recovered through the drill stem. Sample types may include: piston samplers, push tubes, rotary corers, percussion corers and pressurised corers to sample gas hydrates.

Field testing should be used to identify pore pressure conditions, temperature distributions and remoulded and residual shear strength. The importance of accurate determinations of pore pressure conditions during a geohazard assessment cannot be overstated. High-quality sampling techniques should be used to retrieve cores for laboratory testing. The testing programme should include: (a) classification tests to identify soil texture and mineralogy, clay mineralogy, age, pore water salinity and thermal properties; (b) strength tests to identify peak, critical state, remoulded and residual shear strength; and stress tests to investigate the effects of stress dependency, strain rate, strength anisotropy and cyclic loading. Guidance on the various techniques available for geotechnical site investigation is discussed in Chapter 3.

## 10.3 Risk assessment

The characterisation of geohazards for major infrastructure projects, such as deepwater oil and gas developments, involves identifying the location of potential geohazards as well as the magnitude of hazard events and the frequency with which those events occur. The magnitude–frequency relationship developed during a geohazard characterisation enables this information to be used to assess the risk to a given structure or system. The term 'risk' expresses the economic, life-loss, or environmental losses that might occur due to some type of failure over a given exposure period. The failures can be structural, operational, geotechnical or geological, for example, and are related to a combination of physical hazards, processes or conditions and system vulnerability. Risk is also a product of the consequence of a failure where, for example, a minor spill in a relatively insensitive environment would have low associated consequences while a major spill in a sensitive environment could have profound consequences.

Geohazard risk assessments can be either deterministic where the risk is computed for a single-scenario event, or probabilistic, where risk is computed as a loss that has a certain probability of occurrence for a given period. Deterministic analyses determine the factor of safety against collapse and the consequence of a collapse for a particular set of input parameters. For example, if a slope will fail under a given set of boundary conditions, and if so, the extent of the run-out resulting from the slope failure and whether retrogressive sliding will be triggered by the initial slope failure. To ascertain the likelihood of a failure taking place requires assessment of the frequency of an event occurring. This involves a probabilistic analysis, which would typically result in a prediction expressed in terms of an annual probability of failure.

The methodology for quantifying risk may be expressed using a Bayesian approach, with the probability of a given damaging event expressed in terms of the product of conditional probabilities according to (taking a submarine slide as an example):

$$P(\text{damage event}) = P(\text{slide}) \times P(\text{impact} | \text{slide}) \times P(\text{damage} | \text{impact}) \tag{10.1}$$

Although this process leads to a quantitative assessment of the probability of occurrence of different financial magnitude of damage, the process itself may be somewhat qualitative, with consensus of experts used to evaluate the various probabilities.

The process is often dominated by low frequency catastrophic events, such as a major deep-seated slide through the development. A large proportion of effort, at least initially, must go into establishing that the probability of such an event is vanishingly small. Thereafter, risk assessment can focus on events that would affect individual components of the whole development, ensuring that the product of probability of occurrence of any given damage and the financial cost of that damage lies below a level of risk that is deemed acceptable.

An in-depth discussion of various probabilistic analyses is beyond the scope of this book and the reader is directed elsewhere (McGuire 2008).

One of the most common uncertainties in geotechnical analysis is the appropriate value of soil shear strength, in terms of the *in situ* spatial variation, changes with depth and time-dependent variations due to pore pressure changes that may occur during the life of the development. It is straightforward enough to carry out a parametric study to identify the effect of a range of soil strength, for example, on the factor of safety of a slope. However, a designer must also make an assessment of the likelihood

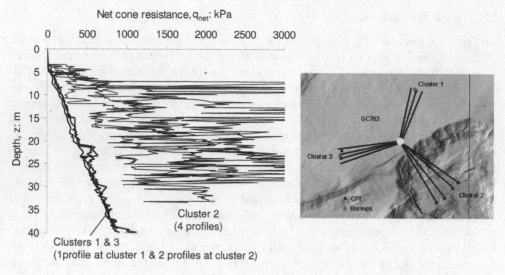

*Figure 10.3* CPT results for a foundation system crossing a scarp (Jeanjean *et al.* 2005)

of failure taking place within the lifetime of the development. These factors are referred to as uncertainty.

Uncertainty may be categorised as aleatoric or epistemic. Aleatoric uncertainty is due to randomness, for example, of ground conditions, while epistemic uncertainty is due to lack of knowledge (Christian 2003). Additional information will not reduce aleatoric uncertainty, while additional information can reduce, albeit not eliminate, epistemic uncertainty.

Figure 10.3 shows profiles of cones resistance from *in situ* cone penetrometer test (CPTs), which were carried out at a deep-water site where the proposed mooring spans across a head scarp at the shelf break. Three clusters of four anchors were proposed with two of the anchor clusters (1 and 3) located in the largely undisturbed area of the continental shelf behind the crest of the slope, and the third anchor cluster is located in the head scarp. The cone resistance profiles at the two anchor cluster locations in the undisturbed area behind the crest of the slope (clusters 1 and 3) are similar to each other, but markedly different from those recorded from the CPTs at the location of the anchor cluster in the head scarp (cluster 2), which in themselves show a large degree of variation. Had the site investigation included only one CPT at any of the cluster locations, the epistemic uncertainty would have been greater, but the additional CPTs documented the range of shear strength profiles thus reducing the epistemic uncertainty, and ultimately the risk of failure. With one CPT at each of the cluster locations, there would have been large epistemic uncertainties, however, the multiple CPTs at the locations of clusters 1 and 3 demonstrated the variability of intact material behind the crest of the slope, and the CPTs in cluster 2 confirm the differences in shear strength profiles in the scarp. The second CPT at cluster 2 indicates variability and the uncertainty is partially aleatoric, as more CPTs would only confirm the variable nature of the material in the scarp. There is of course an epistemic component of uncertainty in terms of defining bounds to the design shear strength profile. The actual site investigation comprised three CPTs across the eight anchor locations in the largely undisturbed material behind the crest of the slope (one at cluster 1 and two at cluster 3) and one CPT at each of the proposed anchor locations in the scarp (cluster 2).

The purpose of a geohazard risk assessment is to identify the level of risk associated with geohazards for a given field architecture. Ultimately, the level of risk is expressed as the product of the probability of occurrence of a given geohazard and the resulting cost of the consequences. Generally, these quantities vary in an inverse fashion, with low probability of catastrophic consequences, but much higher probability of geohazards with smaller cost consequences. Provided that risk acceptance criteria have previously been defined, the operator can then decide whether to accept the risks as they are, mitigate the risks by lowering either the probability of occurrence or the consequence to its facility or avoid the risk by relocating the facility to an alternative location.

Currently, there is no formal industry standard for geohazard risk assessment. Kvalstad (2007) presented a state of the art review of best practice for offshore geohazard investigations and recommended that the quantification of geohazard risk should be based on

- An understanding of the regional and local geology, ongoing geological processes and the type, locations and extent of anomalies to quantify the potential impact, rate or frequency of ongoing natural processes
- Site investigations to identify local seafloor gradients, stratigraphy, soil and pore fluid properties, *in situ* stresses, pore pressure and temperature
- Evaluation of the potential impact of exploration, development and production activities on soil conditions to assess the risk of human interference.

## 10.4 Submarine slope failures and slides

### 10.4.1 Introduction

Seafloor instability is the main geohazard threat encountered offshore and can have catastrophic effects on offshore developments. Seabed instability is an issue even on the continental shelf with seafloor gradients as low as 0.5°. Submarine slides can range in size from relatively small coastal slides of less than a cubic kilometre to vast slides involving thousands of cubic kilometres of material. The Storegga slide on the continental slope of Norway is possibly the most comprehensively investigated massive submarine slide. The Storegga slide is of interest as the Ormen Lange gas field is located in its side scar (Bugge *et al.* 1998, Bryn *et al.* 1999). The slide is estimated to have taken place 8,000 years before present (BP), is reported to have involved 5,600 km³ of material and took place in a region with an average slope of less than 2°. The slide affected an area of more than 30,000 km² with run-out distances extending up to 800 km (Kvalstad *et al.* 2001).

The mobility of submarine landslides can be characterised geometrically by the run-out ratio L/H where L is the horizontal distance from source to deposit and H is the vertical elevation of the debris flow source above the deposit, as illustrated in Figure 10.4. First introduced by Heim (1932) and used later by Scheidegger (1973), the ratio can be predicted by considering the energy balance for a dry mass sliding down a slope. Figure 10.5 shows the volume of material involved and run-out ratio of various submarine and sub-aerial slides. From the compiled data, it appears that submarine slides can be much larger than sub-aerial landslides, and they tend to exhibit larger

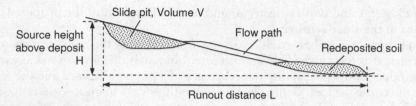

*Figure 10.4* Definition of slide mobility

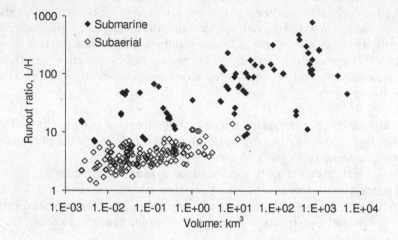

*Figure 10.5* Comparison of volume and run-out distance of submarine and sub-aerial slides (after Scheidegger 1973, Edgers and Karlsrud 1982, Hampton *et al.* 1996 and Dade and Huppert 1998, De Blasio *et al.* 2006)

run-out distances for the same volume of sliding material. This indicates that water plays a particular role in the mobility of the sliding mass.

A submarine slide involves an initiation event, usually a shallow or deep-seated slope failure, often with retrogressive slumping, and followed by mass gravity flow involving a laminar visco-plastic debris flow (or flow slide), and loose suspension turbidity currents. The slope failure and mass gravity flow must be analysed as part of a geohazard assessment. In the following sections, various procedures for analysis of slope stability and mass gravity flow are introduced followed by examples of analyses from a case study of the Mad Dog and Atlantis developments on the Sigsbee Escarpment in the Gulf of Mexico.

### 10.4.2 Slope instability

#### Overview

Slope stability analyses are routinely carried out based on the established analytical infinite slope analysis framework for shallow seated slope failures and the slip circle method for deep-seated slope failures. The theoretical basis of these two methods can be found in standard soil mechanics textbooks and is only briefly outlined in the following section. An alternative analytical solution for deep-seated slope failures based on a sliding block mechanism is also presented. Numerical analyses are necessary to capture the kinematic mechanisms and ultimate limit states at failure for field conditions with complex geometry or seabed features.

Whatever the mode of failure to be analysed, three main conditions should be considered in a slope stability assessment:

- Undrained conditions, if the triggering mechanism is sufficiently fast that drainage does not take place
- Fully drained conditions, in which no excess pore water pressures exist
- Partially drained conditions, in which some pore pressure dissipation has occurred but excess pore pressures still remain.

Undrained conditions are relevant to seismically induced excess pore pressures, while drained conditions are relevant if a submarine slope has been laid down slowly enough that pore pressures everywhere are hydrostatic and there is no seepage flow. Partially drained conditions can result from a number of geological, geophysical and geotechnical processes that lead to excess (higher than hydrostatic) pore pressures, often in dynamic equilibrium due to concurrent consolidation and pore pressure generation. For example, excess pore water pressures may exist in areas of high sedimentation rates (resulting in technically normally consolidated, but often referred to as under-consolidated sediments) or result from ex-solution of pore gas or dissociation of gas hydrates, with the magnitude determined by the relative rates of dissipation and generation. The key issue is to assess the degree of excess pore pressures remaining, which could contribute to the triggering mechanism(s) in question.

The stability of a slope is expressed in terms of its factor of safety F expressed as the ratio of available shear strength of the soil $\tau_{ult}$ to the required shear strength $\tau_{mob}$ to prevent sliding, so.

$$F = \frac{\tau_{ult}}{\tau_{mob}} \qquad (10.2)$$

Ultimate shear strength can be defined by either a total stress failure criterion $\tau_{ult} = s_u$ or an effective stress failure criterion $\tau_{ult} = \sigma' \tan\phi$ for undrained and drained conditions, respectively. For a first time slope failure, the long-term stability of slopes is governed by the critical state strength of the soil. If there is a pre-existing slip, it is more appropriate to determine the stability of the slope using the residual strength, at least over those parts of the failure surface that may coincide with the pre-existing slip or with bedding places on which residual surfaces may have developed (Mesri and Shahien 2003).

### Shallow-seated slope failure

Shallow-seated slope failures are those where the failure surface is located a few metres below the surface and failure takes place parallel to the slope face (Figure 10.6). An infinite slope stability framework can be applied to shallow seated slope failures. The depth of the slip surface is controlled by geological or groundwater conditions, for example, a strong stratum underlying a weaker layer at a relatively shallow depth below the surface of the slope. Infinite slope analysis is

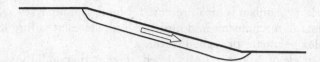

*Figure 10.6* Shallow seated slope failure mechanism

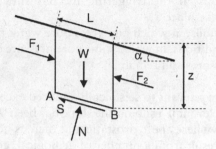

*Figure 10.7* Forces acting on a section in an infinitely long slope

useful for very shallow translational slides and is appropriate for many gentle and extensive submarine slopes where failure tends to be planar and roughly parallel to the slope.

INFINITE SLOPE STABILITY ANALYSIS

For an infinitely long slope inclined at a constant angle to the horizontal, where the mechanism of collapse is governed by a slip surface parallel to the slope, coincident upper and lower bounds show that exact solutions exist for the critical slope angle for undrained and drained loading. Partially drained conditions, i.e. where excess pore pressures exist, will reduce the critical slope angle and must be considered explicitly.

Considering failure of a block of soil of length L along an infinitely long slope, and resolving normal to and along the slope, the normal and shear forces in the soil, N and S, on a surface parallel to the slope at a depth z can be determined as a function of the weight of the block of soil, W ($= \gamma z L \cos\alpha$), and slope angle $\alpha$ (Figure 10.7). For an infinitely long slope, the forces on any block are the same as those on any other block such that the inter-block forces $F_1$ and $F_2$ are equal and opposite and cancel out, except for a net pore pressure force of magnitude $\gamma_w z L \sin\alpha \cos\alpha$ acting horizontally from right to left. The effective normal and shear forces on the slip plane, N′ and S, can be expressed as

$$N' = W\cos\alpha - \gamma_w zL\cos^2\alpha = \gamma zL\cos^2\alpha - \gamma_w zL\cos^2\alpha = \gamma' zL\cos^2\alpha \qquad (10.3)$$

$$S = W\sin\alpha - \gamma_w zL\sin\alpha\cos\alpha = \gamma zL\sin\alpha\cos\alpha - \gamma_w zL\sin\alpha\cos\alpha$$
$$= \gamma' zL\sin\alpha\cos\alpha \qquad (10.4)$$

where $\gamma$ and $\gamma'$ are the total and submerged unit weight of the soil, with $\gamma' = \gamma - \gamma_w$.

The nominal (assuming hydrostatic pore pressures) effective normal and shear stresses, $\sigma'_s$ and $\tau_s$, acting on the slip plane are given by the force divided by a unit area, A, along the slope

$$\sigma'_s = \frac{N'}{A} = \gamma'z\cos^2\alpha \qquad (10.5)$$

$$\tau_s = \frac{S}{A} = \gamma'z\sin\alpha\cos\alpha \qquad (10.6)$$

For an undrained analysis, the factor of safety F may be expressed in terms of the total stress failure criterion, $\tau_{ult} = s_u$, and the mobilised shear stress given above:

$$F = \frac{2s_u}{\gamma'z\sin\alpha\cos\alpha} = \frac{2s_u}{\gamma'z\sin2\alpha} \qquad (10.7)$$

When F = 1, Equation 10.7 can be re-expressed to give the critical slope angle, $\alpha_{ult}$, as a function of the undrained shear strength and the effective overburden pressure

$$\alpha_{ult} = \frac{1}{2}\arcsin\left(\frac{2s_u}{\gamma'z}\right) \qquad (10.8)$$

Re-expressing Equations 10.7 and 10.8 in terms of the undrained shear strength ratio, $k = s_u/\sigma'_{v0}$, results in

$$F = \frac{2k}{\sin2\alpha} \qquad (10.9)$$

$$\alpha_{ult} = 0.5\arcsin(2k) \qquad (10.10)$$

A soft normally consolidated marine deposit with an undrained shear strength ratio $k = s_u/\sigma'_{v0} \sim 0.2$ would correspond to a critical slope angle of 12°. A slope that has failed previously would have a lower shear strength and, therefore, a lower critical slope angle.

The critical angle of the undrained slope is governed by the depth of the slip surface z (corresponding to a local minimum strength ratio). If this depth is relatively large then the approximation of failure by sliding parallel to the surface is no longer valid and a deep-seated failure should be considered.

For a drained analysis with hydrostatic pore pressures, the effective stress failure criterion $\tau_{ult} = \sigma'\tan\phi_{cr}$ determines the factor of safety F as

$$F = \frac{\gamma'z\cos^2\alpha\tan\phi_{cr}}{\gamma'z\sin\alpha\cos\alpha} = \frac{\tan\phi_{cr}}{\tan\alpha} \qquad (10.11)$$

Thus, the limiting angle of a submerged slope with no excess pore pressures is

$$\alpha_{ult} = \phi_{cr} \qquad (10.12)$$

Since $\phi_{cr}$ is typically greater than 20°, drained failure under gravity forces alone is unlikely to be a significant mechanism of failure for many submarine slopes. In previously failed slopes, the relevant drained friction angle along the sliding zone will be a residual value and may fall to less than 10°.

The presence of excess pore pressures, either recently generated or existing because of incomplete consolidation, will reduce the stability of a slope and must be accounted for in the calculation. Even though it may be expected that failure will take place in an undrained manner, it is common practice to follow an effective stress approach, which essentially suggests that the shear strength varies in direct proportion to the current effective stress. This is certainly reasonable for slopes with deposition rates greater than would allow full equilibration of pore pressures, and would be conservative for situations where an increase in excess pore pressure has occurred due to, for example, seismic activity or generation of gas.

The (true) normal effective stress acting on a failure plane at a depth z beneath the seabed, is reduced by excess pore pressure $u_e$ to

$$\sigma'_n = \gamma'z\cos^2\alpha - u_e \tag{10.13}$$

Where the excess pore pressure is a result of so-called under-consolidation (i.e. rapid sedimentation rates) the ultimate shear stress can be expressed as

$$\tau_{ult} = k\sigma'_n = k\left(\gamma'z\cos^2\alpha - u_e\right) \tag{10.14}$$

where k is the undrained shear strength ratio $k = s_u/\sigma'_v$, typically in the range 0.2–0.3 for a young (still consolidating) deposit. Thus, the factor of safety against undrained failure is

$$F = \frac{k(\gamma'z\cos^2\alpha - u_e)}{\gamma'z\sin\alpha\cos\alpha} \tag{10.15}$$

An alternative is to use an effective stress failure criterion, expressed as $\tau_{ult} = \sigma'_n\tan\phi_{cr}$ giving a factor of safety of

$$F = \frac{(\gamma'z\cos^2\alpha - u_e)\tan\phi_{cr}}{\gamma'z\sin\alpha\cos\alpha} \tag{10.16}$$

Assuming a constant excess pore pressure ratio $r_u = u_e/\gamma'z$, the normal stress on the shear plane (Equation 10.3) can be re-expressed as

$$\sigma_s = \gamma'z\cos^2\alpha - \gamma'zr_u = \gamma'z(\cos^2\alpha - r_u) \tag{10.17}$$

This leads to factors of safety for total and effective stress failure criteria expressed more compactly as

$$F = \frac{k(\gamma'z\cos^2\alpha - u_e)}{\gamma'z\sin\alpha\cos\alpha} = k\frac{(\cos^2\alpha - r_u)}{\sin\alpha\cos\alpha} \tag{10.18}$$

$$F = \frac{\gamma'z(\cos^2\alpha - u_e)\tan\phi_{cr}}{\gamma'z\sin\alpha\cos\alpha} = \frac{(\cos^2\alpha - r_u)\tan\phi_{cr}}{\sin\alpha\cos\alpha} \tag{10.19}$$

Figure 10.8 shows the factor of safety against slope failure as a function of slope angle and excess pore pressure ratio $r_u$ (= $u_e/\gamma'z$) for undrained and drained conditions, with an undrained shear strength ratio k (= $s_u/\sigma'_v$) = 0.25 and friction angle $\phi_{cr}$ = 25°, respectively. Note that the drained factor of safety is approximately twice the undrained value.

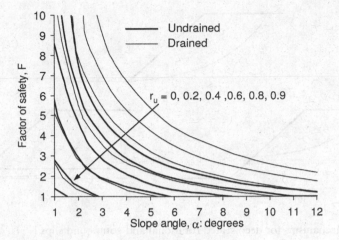

*Figure 10.8* Factors of safety from infinite slope analysis for partially drained conditions (Kvalstad *et al.* 2001)

Considering equilibrium of the forces acting on the partially drained soil element shows maximum slope angle varies almost linearly with the excess pore pressure ratio and can be given by

$$\alpha_{ult} = \phi_{cr}(1 - r_u) \qquad (10.20)$$

where $r_u$ is the excess pore pressure ratio.

Infinite slope analysis is appropriate for long uniform slopes, but many slopes cannot be reasonably idealised as long or uniform. It is also often important to be able to investigate localised de-stabilising effects at the top or toe of a slope, for example, building a platform or making a trench for a pipeline, for which the infinite slope analysis approach is not sufficiently detailed and deep-seated failure analysis is more appropriate.

A trigger mechanism that operates at one part of a long slope may lead to local failure that then propagates up and down the slope, with softening occurring on the sliding surface. Conditions for the growth of shear bands and the potential cata-strophic failure have been considered by Puzrin *et al.* (2004, 2005). Progressive failure of the type described would help to explain observations of submarine slides in slopes at relatively low angles.

## Deep-seated slope failure

### SLIP CIRCLE METHODS

Deep-seated slope failures often occur with rotational movement along arcuate (or circular) slip surfaces as illustrated in Figure 10.9a. Circular slip surfaces are generally associated with homogeneous deposits and non-circular curves with non-homogeneous deposits. If strong or weak layers are present, a compound failure surface may form consisting of plane and curved sections, as illustrated in Figure 10.9b and c.

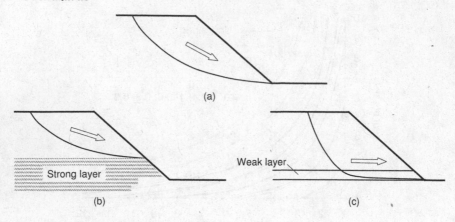

*Figure 10.9* Mechanisms for deep seated rotational and compound slips

Limit equilibrium methods are usually used to analyse deep-seated failures. Failure is considered to occur along an assumed or known failure surface and the required shear strength required to maintain a condition of limiting equilibrium is compared with the available shear strength of the soil giving an average factor of safety along the failure surface. For a total stress approach, the situation is statically determinate and the solution is relatively simple. For an effective stress approach, the problem is statically indeterminate and requires a simplifying assumption. A number of different solutions have been developed based on different assumptions (e.g. Fellenius 1927, Bishop 1955; Morgenstern and Price 1965, Janbu 1973). Limit equilibrium analyses for slope stability calculations are largely repetitive, as many mechanisms must be investigated to identify the most critical, i.e. the one with the lowest factor of safety. As a result, these calculations are routinely carried out by limit equilibrium software programs. The theory of slip circle methods is presented in any good basic textbook on soil mechanics (such as Powrie 2002 and Atkinson 2007, amongst many others) and will not be repeated here.

BLOCK MECHANISM ANALYSIS

In many cases, a circular or curved slip surface may not be critical. A block mechanism may govern failure, particularly for slopes comprising horizontal soil layering or if a weak layer exists at the base of the slope. A simple two-wedge mechanism was developed by the Norwegian Geotechnical Institute (NGI) (Nadim *et al.* 2003) as an alternative to the slip circle methods to analyse deep seated failure, with particular application to retrogressive slides. One of the benefits of the two-wedge mechanism is that it has a fairly simple closed form solution which makes it convenient to solve. Nevertheless, a number of different slip surfaces must be investigated to find the critical mechanism. The configuration of the two-wedge mechanism is shown in Figure 10.10, comprising a back wedge (wedge 1) and a toe wedge (wedge 2). The seabed angle $\alpha$ is assumed to be the same above and below the slope and dictates the inclination of the base slip plane of the toe wedge, but the slope angle $\theta$, need not be equal to the angle of the slip plane of the back wedge $\beta$. The slip planes of the back wedge ($s_1$ and $s_2$ of wedge 1) are perpendicular to each other such that a right angled triangle is

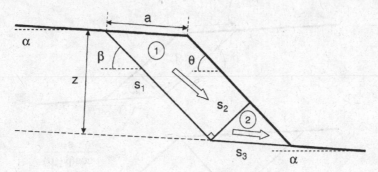

*Figure 10.10* Two-wedge mechanism for deep seated failure (Nadim *et al.* 2003)

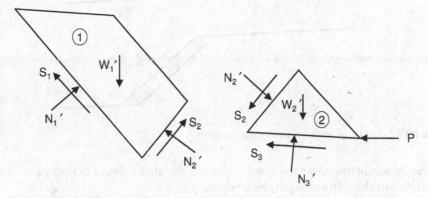

*Figure 10.11* Equilibrium of wedges in two-wedge mechanism (Nadim *et al.* 2003)

formed by extending the line of the lower slip plane ($s_2$) to intersect with an extension of the line of the seabed at the top of the slope. The distance to the point of initiation of the slip plane of the back wedge behind the crest of the slope is defined by the parameter a. The perpendicular slope height, or depth to the level of the weak layer, measured from the seabed to the parallel plane at toe level, or at the level of a weak layer, is defined by the parameter z.

Equilibrium of the two wedges is shown in Figure 10.11. $N_1'$, $N_2'$ and $N_3'$ are the normal effective forces acting perpendicular to the slip planes; $S_1$, $S_2$ and $S_3$ are the shear forces along the sliding planes, equal to the undrained shear strength $s_u$ times the length of the slip plane $s_i$ i.e. $s_u(s_i)$, $i = 1, 2, 3$; $W_1'$ and $W_2'$ are the submerged weights of wedge 1 and wedge 2, given by the unit weight of the material times the area of the wedge; and P describes any (effective – i.e. net of hydrostatic pore pressures) passive lateral resistance at the toe of the slope. Considering the equilibrium of the two wedges and assuming the same factor of safety F on all slip planes, a closed form solution for the factor of safety can be obtained as

$$F = \frac{S_1^{max} + S_2^{max}\sin(\beta - \alpha) + S_3^{max}\cos(\beta - \alpha) + P\cos\alpha}{W_1'\sin\alpha + W_2'\sin\beta\cos(\beta - \alpha)} \tag{10.21}$$

The normal forces on the slip plane (as shown in Figure 10.11) cancel and, therefore, do not appear in the calculation. The numerator gives the components of the maximum shear forces that can be mobilised on the slip planes acting parallel to the seabed,

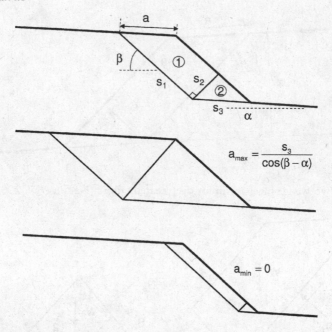

*Figure 10.12* Limits of the two-wedge mechanism

and the denominator gives the components of the shear forces acting parallel to the seabed due to the self-weight of the wedges.

For the simple case where the angle of the back slip plane $\beta$ is equal to the slope angle $\theta$ expressions for the slip planes lengths and wedge areas are given by:

$$
\begin{aligned}
s_1 &= a \\
s_2 &= a\sin(\beta - \alpha) \\
s_3 &= \frac{z}{\sin(\beta - \alpha)}
\end{aligned}
\tag{10.22}
$$

$$
\begin{aligned}
A_1 &= \frac{s_2^2}{2\tan(\beta - \alpha)} \\
A_2 &= az - A_1
\end{aligned}
\tag{10.23}
$$

Kinematic admissibility imposes limits for the two-wedge mechanism as indicated in Figure 10.12.

### Mad Dog and Atlantis case study

The methods described in the foregoing section outline some of the available approaches for assessing submarine slope stability. In this section, those methods are illustrated with respect to a case study of the Mad Dog and Atlantis field developments that are located along the Sigsbee escarpment in the Gulf of Mexico.

The Mad Dog and Atlantis fields are located in average water depths of 1,350 m above the escarpment and 2,150 m below the escarpment, respectively. A SPAR platform serves the Mad Dog field at the top of the escarpment close to the scarp edge

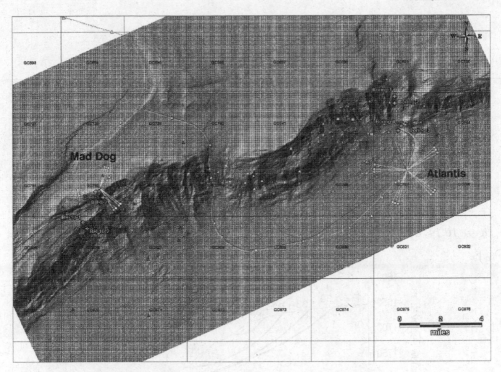

*Figure 10.13* Seabed rendering and field architecture of the Mad Dog and Atlantis fields, Sigsbee escarpment, Gulf of Mexico (Jeanjean *et al.* 2005)

while a semi-submersible FPS and a network of wet-tree subsea wells tied back to the FPS serves Atlantis at the bottom of the escarpment. A transport pipeline runs from the Atlantis field, up the Sigsbee escarpment, tying in to the transport pipeline from the Mad Dog field. The field architecture of the development and a seabed rendering of the seabed conditions are shown in Figure 10.13. The closest slope to the Mad Dog moorings is approximately 200 m high with an average slope angle of 20°. The Atlantis FPS is located about 2 km down-slope from the toe of the escarpment, with the nearest slope approximately 235 m high and average slope angle of 20°. Locally, the slope angles on the escarpment near both Mad Dog and Atlantis can exceed 35° (Jeanjean *et al.* 2005). The Mad Dog and Atlantis fields show evidence of numerous past slope failures, slumping and mass gravity flows. The seabed also shows signs of complex faulting with significant fault offsets at the modern day seafloor (Orange *et al.* 2003).

Given the presence of these hazards, a comprehensive geohazard assessment was carried out to determine if the geohazard risks associated with the development of the fields was acceptable to the operator in terms of the risk to the capital and human investment. The following discussion concentrates on the slope stability analyses carried out as part of the geohazard assessment, although inherent in the analyses is the necessity for high-quality shear strength and pore pressure data. A comprehensive geotechnical site investigation was carried out in conjunction with the geohazard assessment (e.g. Al-Khafaji *et al.* 2003, Jeanjean *et al.* 2005). In summary, the lower continental slope, i.e. above the escarpment, consists of uniform surficial deposits of highly plastic Holocene clays that overlie older and stiffer less plastic Pleistocene

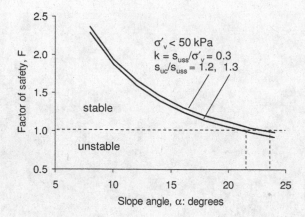

*Figure 10.14* Undrained infinite slope stability analysis for 8 m drape (50 kPa overburden) at Mad Dog (Nowaki *et al.* 2003)

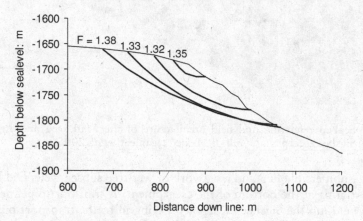

*Figure 10.15* Undrained slip circle analysis of critical slope behind the proposed Atlantis development (Nowaki *et al.* 2003)

clays. The escarpment is made up of stiff clays that have been exposed by slides and slumps that accompanied the sediment uplift by the underlying salt diapir and erosion caused by bottom currents. The upper continental rise consists of uniform deposits of highly plastic clays with a thin drape of Holocene clays. Excess pore pressures were identified near the face of the steepest part of the slope, but of sufficiently small magnitude that they were not considered to be a significant threat to stability. Piezoprobe measurements at Atlantis indicated hydrostatic pore pressures.

Drained and undrained infinite slope analyses, limit equilibrium slip circle analyses and finite element analyses were carried out for slope stability assessment for the Mad Dog and Atlantis field developments.

Figure 10.14 shows results from undrained infinite slope stability analyses for an 8 m thick drape in terms of factor of safety against slope angle for a slope at the Mad Dog field, at the top of the escarpment. The results indicate stability of the surficial shallow sediments for slope angles less than 22°. Figure 10.15 shows the results from undrained slip circle analyses with isotropic strength parameters for a slope behind the proposed Atlantis development, at the bottom of the escarpment. The factor of

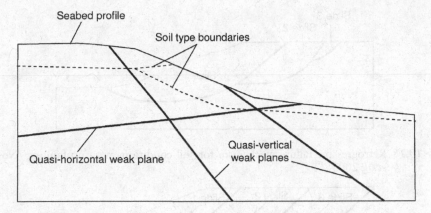

*Figure 10.16* Schematic of finite element model showing faulting at Mad Dog (Nowaki *et al.* 2003)

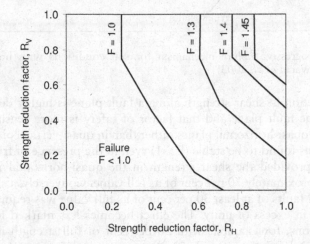

*Figure 10.17* Effect on slope stability factors of safety of reduction in shear strength along quasi-vertical and quasi-horizontal planes (Nowacki *et al.* 2003)

safety ranges between 1.32 and 1.38 and is not very sensitive to the position of the slip circle. Analyses were also carried out to assess the effect of future excess pore pressure at Atlantis. Based on the current geometry, a pore pressure ratio $r_u = u_e/\sigma'_v$ of 0.5 was required to trigger undrained failure.

Considerable faulting is present in the seabed deposits at Mad Dog, predominantly resulting from underlying salt movement (Orange *et al.* 2003). Simple solution methods such as the limit equilibrium slip circle approach are not practical for complex ground conditions and finite element analyses were carried out to investigate the effect of the faulting on slope stability. A plane strain finite element model of a cross-section through a slope at Mad Dog is shown in Figure 10.16.

A sensitivity analysis was performed in which the shear strength along quasi-vertical and quasi-horizontal faults, i.e. planes of weakness, was reduced by an arbitrary factor in order to gain understanding of the importance of the faulting. A deterministic safety factor was calculated for each combination of strength reduction factor as summarised in Figure 10.17. The results show that the impact on the overall factor of safety of the

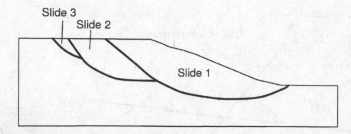

*Figure 10.18* Retrogressive failure mechanism for soil conditions with no faulting (Nowacki *et al.* 2003)

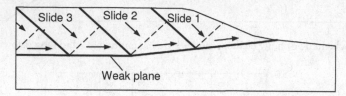

*Figure 10.19* Retrogressive failure mechanism for soil conditions with horizontal faulting (Nowacki *et al.* 2003)

slope of a reduction in shear strength along a fault plane is highly dependent on the inclination of the fault plane and that factor of safety is more sensitive to strength reduction along quasi-horizontal planes rather than in quasi-vertical planes. For example, the slope was found to be stable (F > 1) even in the presence of frictionless quasi-vertical planes provided the shear strength in the quasi-horizontal plane remained greater than approximately 70 per cent of its full value. Conversely, a shear strength in quasi-horizontal faults of at least 30 per cent of its full value was required to achieve a factor of safety in excess of unity. The effect becomes less marked for less extreme strength reductions, for example, with 20 per cent of full strength in quasi-vertical planes required in conjunction with 50 per cent strength in quasi-horizontal planes to maintain stability, compared with 35 per cent of full strength required in quasi-horizontal planes in conjunction with 50 per cent strength in quasi-vertical planes.

Retrogressive slope failure was also investigated through finite element analyses. Figure 10.18 shows the kinematic mechanisms accompanying slope failure for ground conditions without faulting, indicating circular slips will govern the initial and the subsequent retrogressive slope failures. The analyses found limited retrogressive failure in this case in that the slope became stable after two or three subsequent failures. Figure 10.19 shows results for the same slope but with a weak horizontal plane close to the toe of the slope. The failure is markedly different, governed by a block failure similar to the two-wedge mechanism. In this case, retrogressive failure evolves by breaking the slope into wedges, with the extent of the failure limited only by the extent of the weak plane.

### 10.4.3 Mass gravity flows

#### Overview

Slope failures initiate movement of material down-slope in the form of run-out, or mass gravity flows. Material involved in a slide originates as a solid material and gradually transforms towards a fluid state as it remoulds and softens during down-slope transport entraining additional water. Mass gravity flows following a submarine slope failure are generally described as debris flows and turbidity currents. The terminology of these phenomena is not standardised; Niedoroda *et al.* (2003) suggest the following distinctions. Debris flows are mass movements in which the source sediment travels downslope, coming to rest after the initially stored potential energy is dissipated by friction. During debris flows, the source sediment is remoulded and reconstituted, and the degree to which this occurs, including the amount of water entrained, determines the rheological and flow properties. The soil mass travels as a visco-plastic material with distinct stress–strain rate characteristics and flow is generally laminar. Turbidity currents are sediment rich heavy liquid flows that proceed from debris flows. Suspended sediment provides the density contrast with the ambient water in a turbulent current resulting in a gravitational energy that drives the flows. As turbidity currents travel downslope, they may pick up more sediment, becoming denser and accelerating, resulting in flow that is primarily turbulent. The density of a turbidity current is typically 2–4 per cent greater than the ambient water and speeds range from less than 1 m/s to more than 10 m/s.

Different stages of the downslope movement are governed by different mechanisms or flow regimes and require different constitutive models. At failure, the slide material is characterised by geotechnical properties defining stress–strain characteristics. During run-out, the slide material undergoes remoulding and softening and transforms into a debris flow with visco-plastic material properties. With further progress, the flow transforms into a heavy liquid best described by fluid properties. This presents a particular challenge in terms of modelling constitutive behaviour of slides since a model must capture the range of behaviour from a solid material capable of resisting shear stresses without significant deformation to a fluid-like solid-water mixture prone to large deformations. Typically, after a submarine slide has been triggered, the geotechnical properties describing the soil behaviour are set aside and replaced by fluid properties, such as yield stress and viscosity. Analysis of mass gravity flows tends to focus on the behaviour of high water content slurries (well above the liquid limit of the material) with a yield stress of ~10 Pa. The earlier transition stage when the slide material softens from an intact solid with a shear strength of ~10 kPa receives less attention, although arguably the stronger material may create more significant impact loads on any subsea infrastructure than the weaker (although possibly faster moving) fluid flow.

The following sections outline various constitutive and rheological modelling approaches and computational methods for analysing mass gravity flow, followed by some examples of flow modelling carried out to predict the path and extent of mass gravity flows at the Mad Dog and Atlantis fields in the Gulf of Mexico.

### Flow models

The behaviour of mass gravity flows depends on the particle characteristics (such as type, size distribution, mode of interaction with water), the solid concentration and the rate of shearing. Various constitutive and rheological models derived from soil mechanics and fluid mechanics principles are available that can be applied to mass gravity flows.

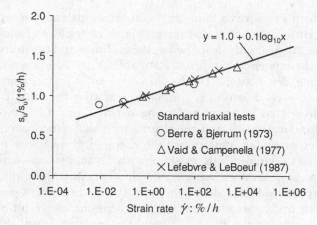

*Figure 10.20* Strain rate effect on undrained strength (Koumoto and Houlsby 2001)

## SOIL MECHANICS FRAMEWORK

Critical state soil mechanics (Schofield and Wroth 1968) (see Chapter 4) is well suited to quantitatively relating the shear strength of remoulded soil to water content, at a sufficiently high level relevant to debris flows. The framework is also suited to qualitative assessment to indicate whether a particular soil, at a given effective stress and water content, will attempt to dilate when sheared, giving rise to negative excess pore pressures and eventual softening, or will tend to contract when sheared, leading to positive excess pore pressures and a hardening response.

Characteristics of natural soils, such as sensitivity and strain rate dependency should be accounted for when applying a soil mechanics model to a debris flow. Most natural soils exhibit some degree of sensitivity, leading to reduced shear strength following remoulding. The sensitivity $S_t$ is defined as the ratio of the peak shear strength $s_{u,p}$ and (fully) remoulded shear strength $s_{u,r}$ (at the initial water content of the soil):

$$S_t = \frac{s_{u,p}}{s_{u,r}} \tag{10.24}$$

Offshore soils typically exhibit sensitivities in the range of 2–6.

The shear strength of soil also depends on the shear strain rate, described by the proportional change in shear strength over an order of magnitude change in strain rate. The shear strength at a strain rate of $\dot{\gamma}$ can be expressed in the form:

$$s_u = s_{u,ref}\left(1 + \mu \log \frac{\dot{\gamma}}{\dot{\gamma}_{ref}}\right) \tag{10.25}$$

where $s_{u,ref}$ is the shear strength at the reference shear strain rate of $\dot{\gamma}_{ref}$. For strain rates less than the reference shear strain rate, it is customary to assume a minimum shear strength of $s_{u,ref}$. Typical values of the coefficient $\mu$ are 0.1–0.2 (or 10–20 per cent change in shear strength per log cycle). Example data are shown in Figure 10.20 (Koumoto and Houlsby 2001).

An alternative to the logarithmic model for strain rate effects, a power law may be adopted, expressed as

$$s = s_{u,ref}\left(\frac{\dot{\gamma}}{\dot{\gamma}_{ref}}\right)^{\beta} \tag{10.26}$$

The power law tends to provide a better fit to data over a wide range of strain rates and also remains bounded at very low strain rates. Typical values of $\beta$ lie in the range 0.05–0.1, but values up to 0.15 are found for high strain rates (Zakeri 2009).

The reduction in shear strength from an intact to fully remoulded state can be described by a simple exponential decay function (Einav and Randolph 2006):

$$s_u = \left(\delta_{rem} + (1-\delta_{rem})e^{-3\xi/\xi_{95}}\right)s_{u,p} \tag{10.27}$$

where $\delta_{rem}$ is the inverse of the sensitivity, and $\xi$ is the cumulative plastic shear strain, with $\xi_{95}$ being the plastic shear strain required to achieve 95 per cent remoulding.

Although a soil mechanics approach is clearly applicable to modelling debris flow, historically fluid mechanics approaches have been much more widely applied.

FLUID MECHANICS FRAMEWORK

Fluids that contain a certain proportion of solids are described as non-Newtonian fluids[1], characterised by a yield stress $\tau_y$, below which shear strain rates are very small, and a non-linear relationship between shear stress and the shear strain rate. Debris flows are typically analysed as non-Newtonian fluids and often as a 'Bingham' fluid. The Bingham model is similar to a geotechnical undrained shear strength model, with zero strain rate effects until the yield strength of the material is exceeded, but after which the strength increases linearly with increasing shear strain rate according to:

$$\dot{\gamma} = \begin{cases} 0 & \text{if } |\tau| < \tau_y \\ \dfrac{1}{\mu}\text{sgn}(\tau)(|\tau|-\tau_y) & \text{if } |\tau| \geq \tau_y \end{cases} \tag{10.28}$$

Non-linear relationships between shear strain rate and shear stress have been considered for modelling debris flows, to allow for changes in viscosity with shear rate (Elverhoi *et al.* 2005, Huang and Garcia 1999). Such models are referred to variously as a Casson model (Locat and Demers 1988), or more commonly as a Herschel-Bulkley model. For the Herschel Bulkley model, the shear strain rate may be expressed as

$$\dot{\gamma} = \begin{cases} 0 & \text{if } |\tau| < \tau_y \\ \text{sgn}(\tau)\left(\dfrac{|\tau|-\tau_y}{\mu}\right)^{1/n} & \text{if } |\tau| \geq \tau_y \end{cases} \tag{10.29}$$

where the exponent, n, distinguishes between so-called 'shear thinning' (with n less than unity) and 'shear thickening' (n greater than unity). As noted by Huang and

---

[1]Common fluids obey Newton's law of viscosity, which states that the shear stress $\tau$ in the fluid is proportional to the shear strain rate $\gamma$ through the dynamic viscosity constant $\mu$, i.e. $\tau = \mu\dot{\gamma}$. Fluids that do not obey Newton's Law are known as 'non-Newtonian fluids'.

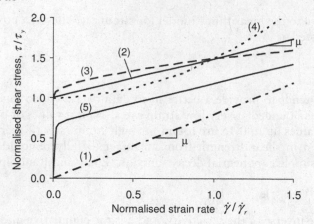

*Figure 10.21*  Relationship between stress and strain rate for different rheological formulations: (1) Newtonian (viscous), (2) Bingham (linear visco-plastic), (3) Herschel-Bulkley (non linear visco-plastic with shear thinning), (4) Shear thickening visco-plastic, (5) Bilinear visco-plastic

Garcia (1998), this formulation can reduce to a Newtonian fluid ($\tau_y = 0$, n = 1), Bingham fluid ($\tau_y > 0$, n = 1) or power law fluid ($\tau_y = 0$, n ≠ 1), depending on the choice of parameters.

A bilinear rheological model can be formulated, removing the assumption of zero shear strain rate below a certain threshold shear stress, expressed as (Imran *et al.* 2001, Huang and Garcia 1998)

$$\frac{\tau}{\tau_{ya}\,\text{sgn}(\dot{\gamma})} = \left(1 + \frac{|\dot{\gamma}|}{\dot{\gamma}_r} - \frac{1}{1 + r|\dot{\gamma}|/\dot{\gamma}_r}\right) \tag{10.30}$$

where the constant $\tau_{ya}$ is the apparent yield strength (assumed non-zero), r = $\dot{\gamma}_r/\dot{\gamma}_0 \gg 1$, and the strain rates $\dot{\gamma}_r$ and $\dot{\gamma}_0$ provide information about the behaviour at high and low strain rates, respectively. The behaviour at high and low strain rates can be approximated as follows:

$$\begin{cases} \tau \approx [\tau_{ya} + \mu_{dh}|\dot{\gamma}|]\text{sgn}(\dot{\gamma}), \quad \mu_{dl} = \dfrac{\tau_{ya}}{\dot{\gamma}_r}, \quad \dot{\gamma} \gg \dot{\gamma}_0 \\[3mm] \tau \approx \mu_{dl}\dot{\gamma}, \quad \mu_{dl} = \dfrac{\tau_{ya}}{\dot{\gamma}_r}(1+r), \quad \dot{\gamma} \ll \dot{\gamma}_r \end{cases} \tag{10.31}$$

A comparison of these various models is shown in Figure 10.21.

Any of the various flow models that are used to describe debris flow need to be coupled with strain softening and remoulding models to cover soil behaviour from quasi-static shear, exceedance of shear strength and development of residual strength through to total remoulding. Turbidity currents are usually modelled as heavy liquid flow (i.e. as a Newtonian fluid) using computational fluid dynamics (CFD) techniques. Debris flow characteristics along with sediment grain size and the depth of erodible sediments are used to predict the flow paths and velocity of the turbidity currents.

*Computational methods for debris flow modelling*

Computational methods used in geotechnical engineering have emerged from the fields of solid and structural analysis. The finite element method has been widely developed and refined. The Lagrangian description, i.e. in which the mesh is assumed to coincide with the physical domain so that it follows the motion of the material, is almost universally adopted due primarily to its convenience in describing boundary conditions. As most geotechnical applications involve conditions up to the point of collapse, or steady plastic flow, numerical codes have tended to focus on quasi-static analyses, although dynamic formulations are also widely available.

The finite element method is widely used to perform slope stability analyses, to determine when a slope will fail and the shape of the failure surface, but not in describing how the material motion will develop upon failure. In general, the evolution of the failure process involving large deformations is not simulated. This is largely due to the widespread adoption of the Lagrangian method, which results in the elements of the mesh becoming excessively distorted, and also because the process becomes non-steady and convergence to a quasi-static solution cannot be achieved. To address these limitations, several numerical approaches have been proposed. In the following sections, three approaches are described: (i) a small strain analysis combined with re-meshing, (ii) an arbitrary Lagrangian Eulerian finite element formulation, (iii) a Lagrangian integration point finite element formulation.

(i) Remeshing and interpolation technique with small strain (RITSS)

Hu and Randolph (1998a, 1998b) developed a numerical method for two-dimensional large deformation problems in soils based on a conventional small strain finite element formulation, involving updating of coordinates, remeshing and interpolation of material and stress parameters, called the RITSS method. In the RITSS method, the small strain formulation is adopted, for convenience in implementing standard complex soil constitutive models and allowing the method to be used with most finite element software. The method obviates any need for a large strain formulation, since strain is not accumulated between small strain analyses other than as a damage parameter. After each increment, the positions of the nodes are updated. In the original version, complete remeshing of the domain was performed using an automatic mesh generation scheme, typically every 10–20 increments. In order to minimise numerical diffusion, a partial remeshing technique has since been developed whereby only the part of the domain undergoing large strains is rediscretised in the remeshing step (Barbosa-Cruz 2007, Zhou *et al.* 2008). After regeneration of the finite element mesh, stress and material parameters are interpolated at the Gauss points of the new mesh from values known at the nodes of the old mesh. The nodes of the old mesh are updated by the cumulative displacement since the last remeshing and form the reference field.

Interpolation at the new Gauss point is obtained using the three Gauss points of the element of the reference field in which the new Gauss point lies. A linear interpolation is computed by summing the values of the Gauss points of the reference field using an inverse distance weighting function. An alternative interpolation method is to use a triangulation of the old Gauss points, and interpolate linearly field quantities at the new Gauss point from the triangle of old Gauss points within which it falls. Following interpolation, the yield status of new Gauss points is checked and the stresses are projected back onto the yield surface if the yield criterion has been violated. RITSS has been implemented in the two-dimensional finite element code AFENA (Carter and Balaam 1995) and coupled with the commercial code ABAQUS (Dassault Systèmes 2007) in order to be applied to three-dimensional problems. A variety of geotechnical problems involving large deformation have been modelled using RITSS (Randolph *et al.* 2008).

### (ii) Arbitrary Lagrangian Eulerian (ALE) finite element method

Lagrangian finite element analyses is limited when applied to debris flow because the mesh becomes too distorted as large deformations of the debris mass occur. Another limitation caused by the large deformation of the elements is the resulting decrease of the time step, which renders the calculation overly time consuming. An alternative is the Eulerian description in which the mesh is fixed, and the material flows through the mesh. As the governing equations describe the material motion, when applied to a fixed mesh as in the Eulerian description, some advection process is necessary to 'advect back' the physical quantities to the fixed mesh. The Eulerian formulation is preferred to describe flow of material. However, difficulties arise when modelling free surface flow and fluid interaction with structures. A more versatile description called 'arbitrary Lagrangian-Eulerian' (ALE) formulation has been developed, based on the arbitrary movement of a reference domain, which does not necessarily follow the material motion. The new positions of the nodes are determined via smoothing algorithms, which account for the movement of boundaries and free surfaces and aim to maintain a relatively uniform mesh. The topology of the mesh is fixed in the original ALE approach, although the terminology is also now applied to solutions that involve complete remeshing at intervals (as in the RITSS method). An advection scheme is used to advect (or in a more general approach interpolate) physical quantities onto the displaced mesh, in a manner similar to a fully Eulerian method. Konuk *et al.* (2006), and Konuk and Yu (2007) present some applications of the ALE method to large deformations in soil using the code LS-DYNA.

### (iii) Lagrangian integration point finite element method

Moresi *et al.* (2003) and O'Neill *et al.* (2006) developed a particle-in-cell finite element method, specifically targeted at problems involving large deformation of visco-elastic/ plastic material with history-dependent material properties. As in the standard finite element method, the approach is based on the variational form of the mass and momentum balance equations, with no inertial terms included in the latter. This is because the material under consideration is highly viscous and the strain rate is assumed to be instantaneously equilibrated with the boundary conditions and the body forces (creeping behaviour). The global integral representing the governing equations is expressed as a sum of sub-integrals over elements, which are then replaced by discrete approximations involving nodal values, using interpolation functions. The discretised form of the governing equations is written in a matrix form, where the global stiffness matrix is constructed from element stiffness matrices.

The particular feature of the method proposed by Moresi *et al.* (2003) lies in the choice of the integration points used to compute the element stiffness matrices. Instead of choosing the integration points in a manner that optimises the accuracy of the numerical integration, the integration points are taken as the positions of the material particles. These particles move with respect to a fixed mesh. Because of the arbitrary

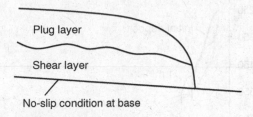

*Figure 10.22* Schematic representation of Bingham flow

location of the particles, several times more particles per element are required for accuracy compared to Gaussian quadrature points in an equivalent finite element scheme. However, an improved accuracy is obtained from integrating material properties at the actual material points. To achieve a reasonable accuracy, specifically for problems involving very large strains, an iterative procedure to determine the particle weights to be used in the numerical integration is required (Moresi *et al.* 2003).

Computational methods to model debris flows based on fluid mechanics principles are largely based on Bingham flow. Jiang and LeBlond (1993) developed the BING model using a finite difference method with a fixed grid to solve the layer averaged governing equations, in which a forward difference in time and a backward difference in space were used in the discretised equations. The model uses a two-layer non-steady deformable body to represent debris flows comprising a lower shear layer and an upper 'plug' layer as illustrated in Figure 10.22. The material in the plug layer represents the portion of the debris flow in which the sediment yield strength has not been exceeded and hence travels as a rigid block over-riding the shear layer. The material in the shear layer represents the portion of the debris flow in which the sediment yield strength has been exceeded and deforms as a Bingham fluid. With this method, the evolution with time of the shape of the mud surface, the shape of the yield interface (between the plug and shear layer), and the plug layer velocity distribution can be determined, for various values of the shear strength, including vanishing shear strength corresponding to purely viscous fluid. Adjusting the modelling parameters, Bingham viscosity and yield strength, the thickness and run-out distance of flows can be projected or back calculated. The model has since been revised to be applicable to a Herchel-Bulkley rheology or a bilinear rheology (Imran *et al.* 2001)

DeBlasio *et al.* (2004) and Elverhoi *et al.* (2005) used the BING model to back-calculate the evolution of two typical slides, part of the giant Storrega slide in Norway. The model was used to predict the run-out distance and deposition profile, which were compared with the field data; an example is shown in Figure 10.23. Figure 10.23a shows the analysis output with a vertical exaggeration of approximately 1:20 and Figure 10.23b shows the output at true scale to indicate the very low inclination at which the slide occurred. Adjusting the shear strength within the range of acceptable geotechnical values allowed run-out distances to be matched quite well. Deposition thickness was not well represented and the field data shows a higher deposition thickness on the steeper slope and a lower deposition thickness on the more gentle slope, contrary to the results obtained assuming a Bingham model with constant shear strength.

Several extensions to the BING model have been considered (e.g. DeBlasio *et al.* 2004, Elverhoi *et al.* 2005). One extension accounts for the presence of solid blocks

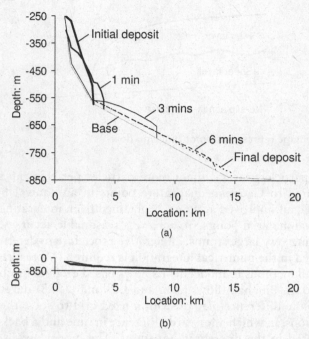

*Figure 10.23*  Prediction of flow evolution of the Storegga slide with the BING model (a) vertical exaggeration ~ 1:20 and (b) true scale (De Blasio *et al.* 2004)

interspersed with the fluid, with the behaviour of the interface between the bed and the blocks described by a Coulomb friction law. Another modification adopts a linear increase of shear strength with depth, following a Mohr-Coulomb Law. A further extension implements hydroplaning, by switching to a new flow regime according to a Froude condition. In the hydroplaning flow regime, a water layer is considered to be present between the debris and the bed, and the shear stress in the debris is assumed to be less than the shear strength, such that the debris moves as a plug. In the water layer, the velocity profile is considered parabolic, and the dynamic pressure is assumed to vary linearly from the stagnation pressure at the snout to zero at the rear end of the debris flow. A modified model also accounts for a reduction in yield strength to represent the effect of water intrusion.

## Notes on analysis of mass gravity flows

The main difficulty in applying any model to the analysis of submarine flows is identifying the appropriate input parameters to describe the viscous or fluid flow behaviour. Since direct measurement of debris flow and turbidity current speeds and other flow characteristics is not practical, it is common practice to back analyse previous flows from observed (now static) run-outs. The goal of these evaluations is to catalogue the varieties of mass gravity flows that have occurred in the past, understand the causes of these events, characterise the kinematics (e.g. speeds and dimensions) of these flows and then use the acquired information to forecast future activity (Niedoroda *et al.* 2003).

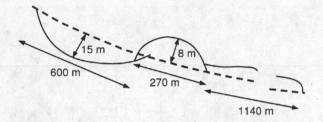

*Figure 10.24* Mapped slide and debris flow zone, Atlantis Field, (Niedoroda *et al.* 2003)

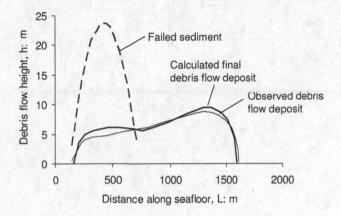

*Figure 10.25* Hindcast and observed debris flow deposit (Niedoroda *et al.* 2003)

## Mad Dog and Atlantis case study

The Mad Dog and Atlantis fields on the Sigsbee escarpment in the Gulf of Mexico (Figure 10.13) have experienced numerous past slope failures that have triggered major debris flows and turbidity currents. Various mass gravity flow deposits in the area were mapped using geophysical techniques and video observations from an ROV as part of the geohazard assessment for the proposed developments at Mad Dog and Atlantis (Niedoroda *et al.* 2003). The volume of past slope failures was approximated from measurement of slide pits from the geophysical data, assuming a circular slip. The volume of the slope failure was then compared with the volume of material in the toe and the mapped debris flow zone. The volume in the toe of several features was found to represent around 25 per cent of the total failure volume (Figure 10.24). The absence of debris from many of the mapped run-out zones indicated the role of turbidity currents in eroding and transporting deposits away from the original flow area. Turbidity current action was also evident from the erosion of deep seafloor channels and canyons.

The BING model (Jiang and LeBlond 1993) was used to simulate a number of the mapped debris flows in the project area in an attempt to back calculate the previous seabed topography, identify the possible triggers of the slope failure and the subsequent kinematics of the mass gravity flows. Figure 10.25 shows a comparison of hindcast and observed debris flow run-out geometry for a particular slide event in terms of the initial and final profiles of the flow. The dashed 'hump' is a synthetic

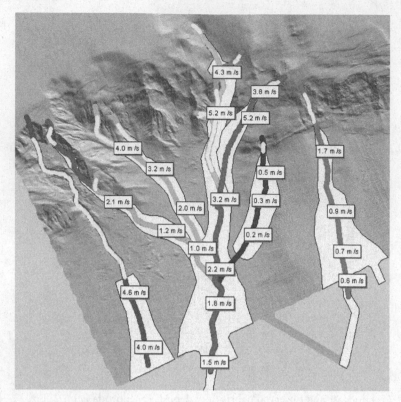

*Figure 10.26* Predicted turbidity current paths and velocities for the Atlantis field (Niedoroda *et al.* 2003)

starting point for the debris flow, i.e. the point at which the deformation due to the initial soil mass failure converts from a displaced mass to an unstable mound from which point the soil mass continues to deform as a Bingham fluid. Simulations were repeated for various mapped large debris flows, varying the input parameters (Bingham viscosity and yield strength) until the output (run-out deposit thickness and distance) agreed with the mapped observations. The back calculated parameters were then used to predict the velocity, extent and thickness of run-out deposits of future mass gravity flows in the area. Turbidity current activity cannot be hind cast in the same way as debris flows, since turbidity current deposits are generally not available to calibrate a model. Instead measured sediment grain size and the depth of erodible material are used with erodibility parameters predicted from flume tests. Figure 10.26 shows the predicted paths and representative speeds of turbidity currents at Mad Dog and Atlantis in which the results of the debris flow models were used as input into the turbidity current analyses.

Some of the mass gravity flows that have occurred in the Mad Dog and Atlantis project area in the past were immense, with mapped run-out distances of over 7 km and travel speeds back calculated to have been up to 100 km/hr, with some of the trapped intact blocks in the flows the size of football stadiums. Should similar mass gravity flows be triggered during the field life, severe damage to any facility located in their path might occur, endangering human lives, the operator's capital investment and the environment (Niedeoroda *et al.* 2003). The probability of future mass gravity flows occurring during the lifetime of the field were determined in additional probabilistic analyses and showed that the probability of failure was acceptably low (Nadim *et al.* 2003).

### 10.4.4 Consequences to infrastructure

Quantifying the consequences of impact of seabed infrastructure is the final step in assessing the geohazard posed by a submarine mass transport event. Two separate aspects must be considered: (i) assessment of the loads imposed by a given impact; (ii) evaluation of the resulting response of (and potential damage to) the infrastructure. These steps are common to a variety of different infrastructure, including wells, manifolds and anchoring systems, but above all, pipelines since they cover such a large region and are, therefore, most exposed.

#### Fluid mechanics approach for pipeline forces

Since analysis of mass transport events has customarily been undertaken based on a fluid mechanics approach, there has been a tendency to apply fluid mechanics models for assessing impact forces. For a pipeline, the forces from turbidity currents are expressed in terms of coefficients, $C_d$, for drag for normal loading, and a friction coefficient, $C_f$, for axial loading, expressed as

$$F_d = C_d \left( \frac{1}{2} \rho v_n^2 \right) D$$

$$F_f = C_f \left( \frac{1}{2} \rho v_a^2 \right) \pi D$$

(10.32)

where $\rho$ is the density of the turbidity current. Values of the coefficients are suggested as $C_d = 1.2$ at Reynolds numbers $R_e \leq 4 \times 10^5$ and $C_d = 0.7$ for $R_e \geq 6 \times 10^5$ with linear interpolation on a log-log scale between these limits and

$$C_f = \frac{0.075}{(\log_{10} R_e - 2)^2}$$

(10.33)

For turbidity currents the Reynolds number should be based on that for water, expressed as $R_e = v_n D/v$, with the kinematic viscosity $v = 1.6 \times 10^{-6}$ m²/s.

A similar formulation has been suggested for debris flows (Zakeri 2009), with drag and friction coefficients expressed as power law functions of an equivalent Reynolds number for non-Newtonian fluids (also referred to as a Johnson number, giving the relative magnitudes of a drag-related force to the shear strength of the material), expressed as

$$R_{e, \text{non-Newtonian}} = \frac{\rho v_n^2}{\tau}$$

(10.34)

#### Geotechnical approach for pipeline forces

An alternative approach for debris flows, more aligned with geotechnical soil properties, is to express the normal and frictional forces in terms of a shear strength augmented by the high strain rates involved. Thus,

$$F_n = N_p \beta_n s_u D$$

$$F_f = \alpha \beta_f s_u \pi D \tag{10.35}$$

where $\beta_n$ and $\beta_f$ are rate factors such as discussed earlier, and $N_p$ is the bearing factor of around 11, for soil flowing past a cylindrical object (see Chapter 9) and $\alpha$ is a friction coefficient, generally taken as unity for this application. In the first instance, the relevant strain rate may be taken as $v_n/D$ or $v_a/D$.

The two approaches may be combined in respect of the normal force, to give

$$F_n = C_d \left( \frac{1}{2} \rho v_n^2 \right) D + N_p s_{u,nom} D \tag{10.36}$$

where $s_{u,nom}$ represents a rate-adjusted shear strength estimated from a nominal strain rate of $v_n/D$. The relative magnitudes of the two terms are a function of the non-dimensional ratio $\rho v_n^2 / s_{u,nom}$ with the geotechnical resistance dominating once this ratio falls below about 10. For this combined approach, the drag coefficient may be taken as about 0.8.

For axial loading, assuming a rough pipe–soil interface, it may be shown that the friction force can be expressed as

$$F_f = f_a s_{u,nom} \pi D \tag{10.37}$$

where, for a power law model, the coefficient $f_a$ has been shown to be (Einav and Randolph 2006)

$$f_a = \left[ 2 \left( \frac{1}{\beta} - 1 \right) \right]^\beta \tag{10.38}$$

which will lie in the range 1.2–1.4 for typical values of $\beta$, reflecting strain rates adjacent to the pipeline that are markedly higher than the nominal value of $v_a/D$.

*Analysis of pipeline response*

The overall response of a pipeline to slide impact may be evaluated by considering the pipeline as a structural element, capable of resisting bending and axial straining, with active loading from the slide over a defined width, and passive resistance from the seabed in the regions either side of the slide (Figure 10.27). In addition, frictional resistance to axial motion will be generated as the pipeline deforms. This problem is amenable to analytical solution, for given pipeline and slide geometries and assumed values of the active loading, and passive and frictional resistance. Parametric solutions allow evaluation of the stresses and strains induced in the pipeline by bending and tensile extension (Randolph *et al.* 2010).

## 10.5 Summary

It is the job of geotechnical engineers and engineering geologists to identify geohazards, potential triggers, failure modes, event severity, event frequency, the consequences of failure and the probability of failure. Geohazard assessment and analysis must be based on adequate technologies and techniques. Kvalstad *et al.* (2001)

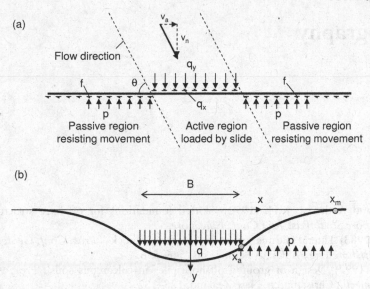

*Figure 10.27* (a) Schematic of general slide loading on pipeline and (b) nomenclature for analysis of normal loading acting on pipeline

identified the following areas for development to improve geotechnical engineering in geohazard areas:

- Development of material models for gassy soils and gas hydrates
- Adequate assessment of *in situ* soil strength, pore pressure and effective stress conditions in deep water, to considerable depth below seabed and over large areal extents through site investigations, monitoring and analysis techniques
- Techniques for evaluation of slope failures that are capable of explaining observed sliding activity on gentle slopes and establishing reliable methods for predicting retrogressive slide activity, areal extent and volumes involved
- Investigation of the accuracy and uncertainties related to techniques and analysis methods for assessment of direct and indirect slide consequences including run-out distance, flow velocity, flow impact and tsunami generation and impact.

# Bibliography

Aas, P.M. and Andersen, K.H. (1992) 'Skirted foundations for offshore structures'. *Proc. Offshore South East Asia Conf., Singapore.*

Abbs, A.F. (1983) 'Lateral pile analysis in weak carbonate rocks'. *Proc. Conf. Geotech. Practice in Offshore Eng.,* ASCE, Austin, Texas, 546–556.

Abbs, A.F. (1992) 'Design of grouted offshore piles in calcareous soils'. *Proc. ANZ Conf. Geomech.,* Christchurch, New Zealand, 128–132.

AG (2001) *Suction Pile Analysis Code: AGSPANC Users' Manual, Version 3.0.* Advanced Geomechanics Internal Report, Perth, Australia.

AG (2003) *CHIPPER: Analysis for Laterally Loaded Piles in Soft Rock, Users' Manual.* Advanced Geomechanics Internal Report, Perth, Australia.

AG (2007) *Cyclops—Software for Cyclic Axial Loading of Piles.* Advanced Geomechanics Internal Report, Perth, Australia.

Al-Khafaji, Z., Young, A., Degroff, B., Nowacki, F., Brooks, J. and Humphrey, G. (2003) 'Geotechnical Properties of the Sigsbee Escarpment from Soil Borings and Jumbo Piston Cores'. *Proc. Annu. Offshore Tech. Conf.,* Houston, Texas, Paper OTC 15158.

Alm, T., Snell, R.O., Hampson, K.M. and Olaussen, A. (2004) 'Design and installation of the Valhall piggyback structures'. *Proc. Annu. Offshore Tech. Conf.,* Houston, Texas, Paper OTC 16294.

Amante, C. and B.W. Eakins (2009) 'ETOPO1 1 Arc-Minute Global Relief Model: Procedures, Data Sources and Analysis'. *Nat. Oceanic Atmos. Administration (NOAA) Tech. Memo. NESDIS* NGDC-24, 19, March.

Andenaes, E., Skomedal, E. and Lindseth, S. (1996) 'Installation of the Troll Phase 1 Gravity Base Platform'. *Proc. Annu. Offshore Tech. Conf.,* Houston, Texas, Paper OTC 8122.

Andersen, K.H., Dyvik, R. Kikuchi, Y. and Skomedal, E. (1992) 'Clay behaviour under irregular cyclic loading'. *Proc. Int. Conf. Behaviour of Offshore Structures,* London, 2: 937–950.

Andersen K.H. (2004) 'Cyclic clay data for foundation design of structure subjected to wave loading. Invited keynote lecture'. *Proc. Int. Conf. Cyclic Behav. Soils Liquefact. Phenom. (CBS04),* Bochum, Germany, 371–387.

Andersen K.H. (2009) 'Bearing capacity under cyclic loading—offshore along the coast and on land', 21st Bjerrum Lecture. *Can. Geotech. J.* 46: 513–535.

Andersen, K.H. and Jostad, H.P. (1999) 'Foundation design of skirted foundations and anchors in clay'. *Proc. Annu. Offshore Tech. Conf.,* Houston, Texas, Paper OTC 10824.

Andersen, K.H. and Jostad, H.P. (2002) 'Shear strength along outside wall of suction anchors in clay after installation'. *Proc. Int. Symp. Offshore Polar Eng. Conf. (ISOPE),* Kitakyushu, Japan, 785–794.

Andersen, K.H. and Jostad, H.P. (2004) 'Shear strength along inside of suction anchor skirt wall in clay'. *Proc. Annu. Offshore Tech. Conf.,* Houston, Texas, Paper OTC 16844.

Andersen, K.H., Brown, S.F., Foss, I., Pool, J.H. and Rosenbrand, W.F. (1980) 'Cyclic and static laboratory tests on Drammen clay'. *J. Geotech. Eng.*, ASCE, 106(GT5): 499–529.

Andersen, K.H., Dyvik, Lauritzsen, R., Heien, D., Harvik, T. and Amundsen, T. (1989) 'Model tests of gravity platforms II; Interpretation'. *J. Geotech. Eng.*, ASCE, 115(11): 1550–1568.

Andersen, K.H., Dyvik, R., Schroeder, K., Hansteen, O.E. and Bysveen, S. (1993) 'Field tests of anchors in clay. II: Predictions and interpretation'. *J. Geotech. Eng.*, ASCE, 119(10): 1532–1549.

Andersen, K.H., Lunne, T., Kvalstad, T.J. and Forsberg, C.F. (2008) 'Deep water geotechnical engineering'. *Proc. XXIV Nat. Conf. of the Mexican Soc. of Soil Mechanics*, Aguascalientes, Mexico, 1–57.

Andersen, K.H., Murff, J.D., Randolph, M.F., Clukey, E., Erbrich, C.T., Jostad H.P., Hansen, B., Aubeny, C.P., Sharma, P. and Supachawarote, C. (2005) 'Suction anchors for deepwater applications'. *Proc. Int. Symp. Front. Offshore Geotech. (ISFOG)*, Perth, Australia, 3–30.

Angell, M., Hanson, K., Swan, B. and Youngs, R. (2003) 'Probabilistic fault displacement hazard assessment for flowlines and export pipelines, Mad Dog and Atlantis field developments, deepwater Gulf of Mexico'. *Proc. Annu. Offshore Tech. Conf.*, Houston, Texas, Paper OTC 15202

API (2000) *Recommended Practice for Planning, Designing and Constructing Fixed Offshore Platforms—Working Stress Design, API RP-2A*. American Petroleum Institute, Washington, USA.

API (2005) *Recommended Practice for Design and Analysis of Station Keeping Systems for Floating Structures, API RP-2SK*. American Petroleum Institute, Washington.

API (2008) *Recommended Practice for Planning, Designing and Constructing Fixed Offshore Platforms—Working Stress Design, API-RP-2A*. 21st edition, errata and supplement 3, March 2008, American Petroleum Institute, Washington.

ASTM-D5778-07 (2007) *Standard Test Method for Performing Electronic Friction Cone and Piezocone Penetration Testing of Soils*. ASTM International, West Conshohocken, Pennsylvania, www.astm.org.

Atkinson, J.H. (2007) *'The mechanics of soils and foundations'*. Second Edition, Spon Text, London.

Aubeny, C.P. and Chi, C. (2010) 'Mechanics of drag embedment anchors in a soft seabed'. *J. Geotech. Geoenv. Eng.*, ASCE, 136(1): 57–68.

Aubeny, C.P., Shi, H. and Murff, J.D. (2005) 'Collapse loads for a cylinder embedded in trench in cohesive soil'. *Int. J. Geomech.*, ASCE, 5(4): 320–325.

Audibert, J.M.E. and Nyman, K.J. (1977) 'Soil restraint against horizontal motion of pipes'. *J. Geotech. Eng. Div.*, ASCE, 103(10): 1119–1142.

Baglioni, V.P., Chow, G.S. and Endley, S.N. (1982) 'Jack-up rig foundation stability in stratified soil profiles'. *Proc. Annu. Offshore Tech. Conf.*, Houston, Texas, Paper OTC 4408.

Baguelin, F., Frank, R. and Said, Y.H. (1977) 'Theoretical study of lateral reaction mechanism of piles'. *Géotechnique*, 27(3): 405–434.

Baguelin, F., and Frank, J. (1979) 'Theoretical studies of piles using the finite element method'. *Proc. Conf. Numer. Methods Offshore Piling*, London, 83–91.

Barltrop, N.D.P. and Adams A.J. (1991) *Dynamics of Fixed Marine Structures*. Butterworth Heinemann, 3rd ed., Oxford.

Barbosa-Cruz, E.R. (2007) *Partial Consolidation and Breakthrough of Shallow Foundations in Soft Soils*. PhD Thesis, University of Western Australia.

Barbour, R.J. and Erbrich, C. (1995) 'Analysis of soil skirt interaction during installation of bucket foundations using ABAQUS'. *Proc. ABAQUS Users Conf.*, Paris, France.

Barton, Y.O. (1982) *Laterally Loaded Model Piles in Sand: Centrifuge Tests and Finite Element Analyses*. PhD Thesis, University of Cambridge.

Bea, R. and Audibert, J. (1979). 'Performance of dynamically loaded pile foundations'. *Proc. 2nd Int. Conf. Behav. Offshore Struct.*, London, 3: 728–745.

Bea, R.G. (1992) 'Pile capacity for axial cyclic loading'. *J. Geotech. Eng.*, ASCE, 118(1): 34–50.

Been, K. and Jefferies, M.G. (1985) 'A state parameter for sands'. *Géotechnique*, 35(2): 99–112.

Been, K., Jefferies, M.G., Crooks, J.H.A and Rothenberg, L. (1987) 'The cone penetration test in sands: Part II, general inference of state parameter'. *Géotechnique*, 37(4): 285–299.

Bell, R.W. (1991) 'The analysis of offshore foundations subjected to combined loading.' *M.Sc. Thesis*, University of Oxford.

Bennett, W.T. and Sharples, B.P.M. (1987) 'Jack-up legs to stand on?.' *Mobile Offshore Structures*, Elsevier, London, pp. 1–32.

Berre, T. and Bjerrum, L. (1973) 'Shear strength of normally consolidated clays'. *Proc. Int. Conf. Soil Mech. Found. Eng. (ICSMFE)*, Moscow, Russia, 1(1): 39–49.

Bienen, B. (2007) 'Three-dimensional physical and numerical modelling of jack-up structures on sand'. PhD Thesis, The University of Western Australia.

Bienen, B. and Cassidy, M.J. (2006) 'Advances in the three-dimensional fluid-structure-soil interaction analysis of offshore jack-up structures'. *Mar. Struct.*, 19(2–3): 110–140.

Bienen, B. and Cassidy, M.J. (2009) 'Three-dimensional numerical analysis of centrifuge experiments on a model jack-up drilling rig on sand'. *Can. Geotech. J.* 46(2): 208–224.

Bienen, B., Byrne, B.W., Houlsby, G.T. and Cassidy M.J. (2006) 'Investigating six-degree-of-freedom loading of shallow foundations on sand'. *Géotechnique*, 56(6): 367–379.

Bienen, B., Gaudin, C., and Cassidy, M.J. (2007) 'Centrifuge tests of shallow footing behaviour on sand under combined vertical torsional loading'. *Int. J. Phys. Modell. Geotech.*, 7(2): 1–22.

Bienen, B., Gaudin, C. and Cassidy, M.J. (2009) 'The influence of pull-out load on the efficiency of jetting during spudcan extraction'. *Appl. Ocean Res.*, 31(3): 202–211.

Biot, M.A. (1935) 'Le problem de la consolidation des matieres argileuses sous une charge'. *Annaies de la Societe Scientific de Bruxelles*, Series B, 55: 110–113.

Biot, M.A. (1956) 'General solutions of the equations of elasticity and consolidation for a porous material'. *Trans. J. Appl. Mech.*, ASME, 78: 91–96.

Bishop, A.W. (1955) 'The use of slip circle in the stability analysis of slopes'. *Géotechnique*, 5(1): 7–17.

Bishop, A.W. and Lovenbury, H.T. (1969) 'Creep characteristics of two undisturbed clays'. *Proc. Int. Conf. Soil Mech. Found. Eng. (ICSMFE)*, Mexico City, Mexico, 1: 29–27.

Bjerrum, L. (1973) 'Geotechnical problems involved in foundations of structures in the North Sea'. *Géotechnique*, 23(3): 319–358.

Bogard, J.D., and Matlock, H. (1990) 'In-situ pile segment model experiments at Empire, Louisiana'. *Proc. Annu. Offshore Tech. Conf.*, Houston. Texas, Paper OTC 6323.

Bolton, M.D. (1986) 'The strength and dilatancy of sands'. *Géotechnique*, 36(1): 65–78.

Bolton, M.D. (1991) *A Guide to Soil Mechanics*. M.D. and K. Bolton, Cambridge.

Bolton M.D. and Barefoot A.J. (1997) 'The variation of critical pipeline trench back-fill properties'. *Proc. Conf. Risk-Based Limit State Des. Oper. Pipelines*, Aberdeen, Aberdeen City, UK, p. 26.

Bonjean, D., Erbrich, C.T. and Zhang, J. (2008) 'Pipeline flotation in liquefiable soil'. *Proc. Annu. Offshore Tech. Conf.*, Houston, Texas, Paper OTC 19668.

Booker, J.R. (1974) 'The consolidation of a finite layer subject to surface loading'. *Int. J. Solids Struct.*, 10: 1053–1065.

Booker, J.R. and Small, J.C (1986) 'The behaviour of an impermeable flexible raft on a deep layer of consolidating soil'. *Int. J. Num. Anal. Methods Geomech.*, 10(3): 311–327.

Booker, J.R., Balaam, N.P. and Davis, E.H. (1985a) 'The behaviour of an elastic non-homogeneous half space: Part I. Line and point loads'. *Int. J. Num. Anal. Methods Geomech.*, 9(4): 353–367.

Booker, J.R., Balaam, N.P. and Davis, E.H. (1985b) 'The behaviour of an elastic non-homogeneous half space: Part II. Circular and strip footings'. *Int. J. Num. Anal. Methods Geomech.*, 9(4): 369–381.

Borel, D., Puech, A., Dendani, H. and de Ruijter, M. (2002) 'High quality sampling for deep water geotechnical engineering: the STACOR® Experience.' *Ultra Deep Eng. Tech.*, Brest, France.

Boulanger, R.W., Curras, C.J., Kutter, B.L., Wilson, D.W. and Abghari, A. (1999) 'Seismic soil-pile-structure interaction experiments and analyses'. *J. Geotech. Geoenv. Eng.*, ASCE. 125(9): 750–759.

Boulon, M. and Foray, P. (1986) 'Physical and numerical simulation of lateral shaft friction along offshore piles in sand. *Proc. Conf. on Num. Methods Offshore Piling*. Nantes, France, 127–147.

Boussinesq, J.V. (1885) *Application des Potentials à l'Étude de l'Équilbre et du Mouvements des Solides Élastiques*. Gauthier-Villars, Paris, France.

Bransby, M.F. and O'Neill, M.P. (1999) 'Drag anchor fluke-soil interaction in clays'. *Proc. Int. Symp. Num. Models Geomech.*, *(NUMOG)*, Graz, Austria, 489–494.

Bransby, M.F. and Randolph, M.F. (1998) 'Combined loading of skirted foundations'. *Géotechnique*, 48(5): 637–655.

Bransby, M.F. and Randolph, M.F. (1999) 'The effect of embedment depth on the undrained response of skirted foundations to combined loading.' *Soils and Foundations*, 39(4): 19–33.

Bransby, M.F. and Yun, G. (2009) 'The undrained capacity of skirted foundations'. *Géotechnique*, 59(2): 115–125.

Bransby, M.F., Newson, T.A., Brunning, P. and Davies, M.C.R. (2001) 'Numerical and centrifuge modeling of the upheaval resistance of buried pipelines'. *Proc. Int. Conf. Offshore Mech. Arctic Eng. (OMAE)*, Rio de Janeiro, Brazil, Paper OMAE-PIPE4118.

Bransby, M.F., Brown, M.J., Hatherley, A.J. and Lauder, K.D. (2010a) 'Pipeline plough performance in sand waves. Part 1: model testing'. *Can. Geotech. J.*, 47(1): 49–64.

Bransby, M.F., Brown, M.J., Lauder, K. and Hatherley, A. (2010b) 'Pipeline plough performance in sand waves. Part 2: kinematic calculation method'. *Can. Geotech. J.*, 47(1): 65–77.

Brennan, R., Diana, H., Stonor, R.W.P., Hoyle, M.J.R., Cheng, C.P., Martin, D. and Roper, R. (2006) 'Installing jackups in punch-through-sensitive clays'. *Proc. Annu. Offshore Tech. Conf.*, Houston, Texas, Paper OTC 18268.

Brennodden, H., Lieng, J.T., Sotberg, T. and Verley, R.L.P. (1989) 'An energy-based pipe-soil interaction model'. *Proc. Annu. Offshore Tech. Conf.*, Houston, Texas, Paper OTC 6057.

Breuer, J.A. and Rousseau, J.H. (2009) 'Stability of jack-ups: Top ten questions from industry to class'. *Proc. Int. Conf. Jack-up Platform*, City University, London.

Bridge, C., Laver, K., Clukey, E.C. and Evans, T.R. (2004) 'Steel catenary riser touchdown point vertical interaction model'. *Proc. Annu. Offshore Tech. Conf.*, Houston, Texas, Paper OTC 16628.

Bridge, C.D. and Howells, A.H. (2007) 'Observations and modelling of steel catenary riser trenches'. *Proc. Int. Offshore Polar Eng. Conf.*, Lisbon, Portugal, 803–813.

Brinch Hansen, J. (1961a) 'The ultimate resistance of rigid piles against transversal forces'. *Geoteknish Institute Bulletin*, No. 12, Copenhagen, Denmark.

Brinch Hansen, J. (1970) 'A revised and extended formula for bearing capacity'. *Danish Geotech. Inst.*, Copenhagen, Denmark, 98: 5–11.

Broms, B.B. (1964a) 'Lateral resistance of piles in cohesive soils'. *J. Soil Mech. Found. Div.*, ASCE. 90(SM2): 27–63.

Broms, B.B. (1964b) 'Lateral resistance of piles in cohesionless soils'. *J. Soil Mech. Found. Div.*, ASCE. 90(SM3): 123–156.

Brown, J.D. and Meyerhof, G.G. (1969) 'Experimental study of bearing capacity in layered clays.' *Proc. 7th Int. Conf. on Soil Mech. and Found. Eng*, 2, 45–51.

Bruton, D.A.S., White, D.J., Cheuk, C.Y., Bolton, M.D. and Carr, M.C. (2006) 'Pipe-soil interaction behaviour during lateral buckling, including large amplitude cyclic displacement tests by the Safebuck JIP'. *Proc. Annu. Offshore Tech. Conf.*, Houston, Texas, Paper OTC 17944.

Bruton, D.A.S., Carr, M.C. and White D J. (2007) 'The influence of pipe-soil interaction on lateral buckling and walking of pipelines: the SAFEBUCK JIP'. *Proc. Int. Conf. on Offshore Site Invest. Geotech.*, *Soc. Underwater Tech.*, London, 133–150.

Bruton D.A.S, White D.J., Carr M.C. and Cheuk C.Y. (2008) 'Pipe-soil interaction during lateral buckling and walking: the SAFEBUCK JIP'. *Proc. Annu. Offshore Tech. Conf.*, Houston, Texas, Paper OTC 19589.

Bruton D., White D.J., Langford, T.L. and Hill A. (2009) 'Techniques for the assessment of pipe-soil interaction forces for future deepwater developments'. *Proc. Annu. Offshore Tech. Conf.*, Houston, Texas, Paper OTC 20096.

Bryn, P. Ostmo, S.R., Lien, R., Berg, K., and Tjelta, T.I. (1999) 'Slope stability in deep water areas off mid Norway'. *Proc. Annu. Offshore Tech. Conf.*, Houston, Texas, Paper OTC 8640.

Budhu, M. (1985) 'The effect of clay content on liquid limit from fall cone and British cup device'. *Geotech. Testing J.*, ASTM, 8(2): 91–95.

Bugge, T., Belderson, R.H., and Kenyon, N.H. (1998) 'The Storrega slide'. *Phil. Trans. Royal Soc., A*, 325(1586): 357–388.

Burland, J.B. (1990) 'On the compressibility and shear strength natural clays'. *Géotechnique* 40(3): 329–378.

Burns, S.E. and Mayne, P.W. (1998) 'Monotonic and dilatory pore-pressure decay during piezocone tests in clay.' *Canadian Geotech. J.* 35(6): 1063–1073.

Bustamante, M. and Gianeselli, L. (1982) 'Pile bearing capacity by means of static penetrometer CPT'. *Proc. Eur. Symp. on Penetration Testing.* Amsterdam. 493–499.

Butterfield, R. and Ticof, J. (1979). 'Design parameters for granular soils (discussion contribution)'. *Proc. European Conf. Soil Mech. Found. Eng. (ICSMFE)*, Brighton, 4: 259–261.

Bye, A., Erbrich, C., Rognlien, B. and Tjelta, T.I. (1995) 'Geotechnical design of bucket foundations'. *Proc. Annu. Offshore Tech. Conf.* Houston, Texas, Paper OTC 7793.

Byrne, B.W. (2000) *Investigations of Suction Caissons in Dense Sand*. DPhil Thesis, University of Oxford.

Byrne, B. and Cassidy, M.J. (2002). 'Investigating the response of offshore foundations in soft clay soils'. *Proc. Int. Conf. on Offshore Mech. Arctic Eng. (OMAE)*, Oslo, Norway, OMAE 2002–28057.

Byrne B.W. and Houlsby G.T. (2001) 'Observation of footing behaviour on loose carbonate sand'. *Géotechnique* 51(5): 463–466.

Byrne, B.W. and Houlsby, G.T. (2003) 'Foundations for offshore wind turbines.' *Phil. Trans. Roy. Soc. A*, 361(1813): 2909–2930.

Byrne, B.W., Schupp, J., Martin, C.M., Oliphant, J., Maconochie, A. and Cathie, D.N. (2008) 'Experimental modelling of the unburial behaviour of pipelines'. *Proc. Annu. Offshore Tech. Conf.*, Houston. Paper OTC 19473.

Calvetti, F., di Prisco, C., and Nova, R. (2004) 'Experimental and numerical analysis of soil-pipe interaction'. *J. Geotech. Geoenv. Eng., ASCE.* 130(12): 1292–1299.

Campbell, K.J., Humphrey, G.D. and Little, R.L. (2008) 'Modern deepwater site investigation: getting it right the first time'. *Proc. Annual Offshore Tech. Conf.*, Houston, Paper OTC 19535.

Carr, M., Sinclair, F., and Bruton, D. (2006) 'Pipeline walking—understanding the field layout challenges, and analytical solutions developed for the SAFEBUCK JIP'. *Proc. Annu. Offshore Tech. Conf.*, Houston, Texas, Paper OTC 17945.

Carter, J.P, and Balaam N.P. (1995) *AFENA User Manual 5.0*. Geotechnical Research Centre, University of Sydney, Australia.

Cardoso C.O., da Costa A.M. & Solano R.F. (2006) 'HP-HT pipeline cyclic behaviour considering soil berms effect'. *Proc. 25th Int. Conf. on Offshore Mechanics and Arctic Engineering*, Hamburg, Germany, Paper OMAE2006-92375.

Cassidy, M.J. (1999) *Non-linear analysis of jack-up structures subjected to random waves*. DPhil Thesis, University of Oxford.

Cassidy M.J. (2007) 'Experimental observations of the combined loading behaviour of circular footings on loose silica sand'. *Géotechnique*, 57(4): 397–401.

Cassidy, M.J. and Bienen, B. (2002) 'Three-dimensional numerical analysis of jack-up structures on sand'. *Proc. Int. Symp. Offshore Polar Eng. (ISOPE)*, Kitakyushu, Japan, 2: 807–814.

Cassidy, M.J. and Cheong, J. (2005) 'The behaviour of circular footings on sand subjected to combined vertical-torsion loading'. *Int. J. Phys. Model. Geotech.*, 5(4): 1–14.

Cassidy, M.J. and Houlsby, G.T. (1999). 'On the modelling of foundations for jack-up units on sand'. *Proc. Annu. Offshore Tech. Conf.*, Houston, Texas, OTC 10995.

Cassidy, M.J. and Houlsby, G.T. (2002) 'Vertical bearing capacity factors for conical footings on sand'. *Géotechnique*, 52(9): 687–692.

Cassidy, M.J., Eatock Taylor, R. and Houlsby, G.T. (2001) 'Analysis of jack-up units using a Constrained NewWave methodology'. *Appl. Ocean Res.*, 23: 221–234.

Cassidy, M.J., Byrne, B.W., and Houlsby, G.T. (2002a) 'Modelling the behaviour of circular footings under combined loading on loose carbonate sand'. *Géotechnique*, 52(10): 705–712.

Cassidy, M.J., Houlsby, G.T., Hoyle, M. and Marcom, M. (2002b) 'Determining appropriate stiffness levels for spudcan foundations using jack-up case records'. *Proc. Int. Conf. on Offshore Mech. Arctic Eng. (OMAE)*, Oslo, Norway, OMAE 2002-28085.

Cassidy, M.J., Taylor, P.H., Eatock Taylor, R. and Houlsby, G.T. (2002c) 'Evaluation of long-term extreme response statistics of jack-up platforms'. *Ocean Eng.* 29(13): 1603–1631.

Cassidy, M.J., Byrne, B.W., and Randolph, M.F. (2004a). 'A comparison of the combined load behaviour of spudcan and caisson foundations on soft normally consolidated clay'. *Géotechnique*, 54(2): 91–106.

Cassidy, M.J., Martin, C.M., Houlsby, G.T. (2004b) 'Development and application of force resultant models describing jack-up foundation behaviour'. *Mar. Struct.*, 17(3–4): 165–193.

Cassidy, M.J., Quah, C.K. and Foo, K.S. (2009) 'Experimental investigation of the reinstallation of spudcan footings close to existing footprints'. *J. Geotech. Geoenv. Eng.*, ASCE, 135(4): 474-48.

Castleberry II, J.P. and Prebaharan, N. (1985) 'Clay crusts of the Sunda Shelf—a hazard to jack-up operations.' *Proc. 8th Southeast Asian Geotechnical Conf*, Kuala Lumpur, 40–48.

Cathie, D.N. and Wintgens, J.F. (2001) 'Pipeline trenching using plows: performance and geotechnical hazards'. *Proc. Annu. Offshore Tech. Conf.*, Houston, Texas, Paper OTC 13145.

Cathie, D.N., Jaeck, C., Ballard, J.C. and Wintgens, J.-F. (2005) 'Pipeline geotechnics—state-of-the-art'. *Proc. Int. Symp. Front. Offshore Geotech. (ISFOG)*; Perth, Australia, 95–114.

Cauquil, E., Stephane, L., George, R.A.T. and Shyu, J.P. (2003) 'High-resolution autonomous underwater vehicle (AUV) geophysical survey of a large, deep water pockmark offshore Nigeria'. *Proc. EAGE 65th Conf. Exhibition*, Stavanger, Norway.

Chakrabarti, S.K. (2005) *Handbook of Offshore Engineering*. Elsevier.

Chan, N.H. C., Paisley, J.M. and Holloway, G.L. (2008) 'Characterization of soils affected by rig emplacement and Swiss cheese operations—Natuna Sea, Indonesia, a case study'. *Proc. Jack-up Asia Conf. Exhib.*, Singapore.

Chandler, R.J. (1988) 'The in-situ measurement of the undrained shear strength of clays using the field vane. Vane Shear Strength Testing of Soils: Field and Laboratory Studies'. *ASTM STP*, 1014: 13–45.

Chen, Y.N., Chen, Y.K. and Cusack, J.P. (1990) 'Extreme dynamic response and fatigue damage assessment for self-elevating drilling units in deep water.' *SNAME Transactions*, Vol. 98, pp. 143–168.

Chen, W. and Randolph, M.F. (2007) 'Axial capacity and stress changes around suction caissons'. *Géotechnique*, 57(6): 499–511.

Chen, W. and Randolph, M.F. (2007) 'Uplift capacity of suction caissons under sustained and cyclic loading in soft clay'. *J. Geotech. Geoenv. Eng.*, ASCE, 133(11): 1352–1363.

Chen, W., Zhou, H. and Randolph, M.F. (2009) 'Effect of installation methods on external shaft friction of caissons in soft clay'. *J. Geotech. Geoenv. Eng.*, ASCE, 135(5): 605–615.

Cheuk, C.Y. and White D.J. (2010) 'Modelling the dynamic embedment of seabed pipelines'. *Géotechnique*, DOI: 10.1680/geot.8.P.148.

Cheuk, C.Y., Take, W.A., Bolton, M.D. and Oliveira, J.R.M.S. (2007) 'Soil restraint on buckling oil and gas pipelines buried in lumpy clay fill'. *Eng. Struct.*, 29(6): 973–982.

Cheuk C.Y., White D.J. and Bolton M.D. (2008) 'Uplift mechanisms of pipes buried in sand'. *J. Geotech. Geoenv. Eng.*, ASCE, 134(2): 154–163.

Chiarella, C. and Booker, J.R. (1975) 'The time-settlement behaviour of a rigid die resting on a deep clay layer.' *International Journal of Numerical and Analytical Methods in Geomechanics*, 8: 343–357.

Christian, J.T. (2003) 'Geotechnical engineering reliability: how well do we know what we are'. *J. Geotech. Engng*, ASCE, 130(10): 985–1003.

Christophersen, H.P. (1993) 'The non-piled foundation systems of the Snorre field'. *Proc. Offshore Site Invest. Found. Behav., Soc. Underwater Tech.*, 28: 433–447.

Chung, S.F. and Randolph, M.F. (2004) 'Penetration resistance in soft clay for different shaped penetrometers'. *Proc. Int. Conf. Site Characterisation*, Porto, Portugal, 1: 671–678.

Chung, S.F., Randolph, M.F., Schneider, J.A. (2006) 'Effect of penetration rate on penetration resistance in clay.' *J. Geotech. and Geoenv. Engng*, ASCE, 132(9): 1188–1196.

Clausen, C.J.F. (1976) 'The Condeep story'. *Proc. Offshore Soil Mech.*, Cambridge University. Ed. George. P. and Wood, D., 256–270.

Clausen, C.J.F., Dibiagio, E., Duncan, J.M. and Andersen, K.H. (1975) 'Observed behaviour of the Ekofisk oil storage tank foundation'. *Proc. Annu. Offshore Tech. Conf.*, Houston, Texas, Paper OTC 2373.

Clukey, E.C. and Morrison, J. (1993) 'A centrifuge and analytical study to evaluate suction caissons for TLP applications in the Gulf of Mexico'. In: *Design and Performance of Deep Foundations: Piles and Piers in Soil and Soft Rock, ASCE Geotechnical Special Publication*, 38: 141–156.

Clukey, E.C., Vermersch, J.A., Koch, S.P. and Lamb, W.C. (1989) 'Natural densification by wave action of sand surrounding a buried offshore pipeline'. *Proc Annu. Offshore Tech. Conf.*, Houston, Texas, Paper OTC 6151.

Clukey E.C., Morrison, M.J., Garnier, J. and Corté, J.F. (1995) 'The response of suction caissons in normally consolidated clays to cyclic TLP loading conditions'. *Proc. Annu. Offshore Tech. Conf.*, Houston, Texas, Paper OTC 7796.

Clukey, E.C., Templeton, J.S., Randolph, M.F. and Phillips, R.A. (2004) 'Suction caisson response under sustained loop-current loads'. *Proc. Annu. Offshore Tech. Conf.*, Houston, Texas, Paper OTC 16843.

Clukey, E.C., Haustermans, L., and Dyvik, R. (2005) 'Model tests to simulate riser-soil interaction effects in touchdown point region'. *Proc. Int. Symp. Front. Offshore Geotech. (ISFOG)*, Perth, Australia, 651–658.

Clukey, E.C., Ghosh, R., Mokarala, P. and Dixon, M. (2007) 'Steel catenary riser (SCR) design issues at touch down area'. *Proc. Int. Symp. Offshore Polar Eng. (ISOPE)*, Lisbon, Australia.

Clukey, E.C., Young, A.G., Garmon, G.S. and Dobias, J.R. (2008) 'Soil response and stiffness laboratory measurements of SCR pipe/soil interaction'. *Proc Annu. Offshore Tech. Conf.*, Houston, Texas Paper OTC 19303.

Cocchetti, G., di Prisco, C., Galli, A. and Nova, R. (2009) 'Soil–pipeline interaction along unstable slopes: a coupled three-dimensional approach. Part 1: Theoretical formulation'. *Can. Geotech. J.*, 46: 1289–1304.

Colliat, J-L. and Dendani, H. (2002) 'Girassol: Geotechnical design analyses and installation of the suction anchors'. *Proc. Int. Conf. Offshore Site Invest. Geotech., Soc. Underwater Tech.*, London.

Cox, A.D, Eason, G., and Hopkins, H.G. (1961) 'Axially symmetric plastic deformation in soils'. *Phil. Trans. Royal Soc., A,* 254(1036): 1–45.

Craig, W.H. and Chua, K. (1990). 'Deep penetration of spudcan foundation on sand and clay'. *Géotechnique*, 40(4): 551–563.

Cryer, C.W. (1963) 'A comparison of the three dimensional consolidation theories of Biot and Terzaghi'. *Quart J. Mech. Appl. Math.*, 16: 401–412.

Dade, B.W. and Huppert, H.E. (1998) 'Long-runout rockfalls'. *Geology*, 26(9): 803–806.

Dahlberg, R. (1998) 'Design procedures for deepwater anchors in clay'. *Proc. Annu. Offshore Tech. Conf.*, Houston, Texas Paper OTC 8837.

Damgaard, J.S. and Palmer, A.C. (2001) 'Pipeline stability on a mobile and liquefied seabed: a discussion of magnitudes and engineering implications'. *Proc. Int. Conf. on Offshore Mech. Arctic Eng. (OMAE)*, Rio de Janeiro, Brazil, Paper OMAE-PIPE4030.

Davis E.H., Booker J.R. (1971) 'The bearing capacity of strip footings from the standpoint of plasticity theory'. *Proc. Australia-New Zealand Conf. Geomech.*, Melbourne, Australia, 276–282.

Davis, E.H. and Booker, J.R. (1973) 'The effect of increasing strength with depth on the bearing capacity of clays'. *Géotechnique*, 23(4): 551–563.

Dassault Systèmes (2007) *Abaqus analysis users' manual*, Simula Corp, Providence, RI, USA.

Dean, R.G. and Dalrymple, R.A. (1991) *Water Wave Mechanics for Engineers and Scientists*. World Scientific, Singapore.

Dean, E.T.R., and Serra, H. (2004). 'Concepts for mitigation of spudcan-footprint interaction in normally consolidated clay'. *Proc. Int. Symp. Offshore Polar Eng. (ISOPE)*, Toulon, France.

Dean, E.T.R, James, R.G., Schofield, A.N., Tan, F.S.C. and Tsukamoto, Y. (1992) 'The bearing capacity of conical footing on sand in relation to the behaviour of spudcan footing of jack-ups'. *Predic. Soil Mech.*, Oxford, UK, 230–253.

Dean, E.T.R., James, R.G., Schofield, A.N. and Tsukamoto, Y. (1997a). 'Theoretical modelling of spudcan behaviour under combined load'. *Soils Found.*, 37(2): 1–15.

Dean, E.T.R., James, R.G., Schofield, A.N. and Tsukamoto, Y. (1997b) 'Numerical modelling of three-leg jackup behaviour subject to horizontal load'. *Soils Found.*, 37(2): 17–26.

de Blasio, F.V., Elverhøi, A., Engvik, L.E., Issler, D., Gauer, P. and Harbitz, C.B. (2006) 'Understanding the high mobility of subaqueous debris flows'. *Norw. J. Geol.*, 86: 275–284.

de Blasio, F.V., Elverhoi, A., Issler, D., Harbitz, C.B., Bryn, P. and Lien, R. (2004) 'Flow models of natural debris flows originating from overconsolidated clay materials'. *Mar. Geol.*, 213: 439–455.

de Cock, F., Legrand, C. and Huybrechts, N. (2003) 'Overview of design methods of axially loaded piles in Europe—Report of ERTC3-Piles, ISSMGE Subcommittee'. *Proc. Eur. Conf. Soil Mech. Geotech. Eng. (ECSMGE)*. Prague, Czech Republic, 663–715.

de Groot M.B., Bolton, M.D., Foray, P., Meijers, P., Palmer, A.C., Sandven, R., Sawicki, A. and Teh, T.C. (2006) 'Physics of liquefaction phenomena around marine structures'. *J. Waterw. Port Coastal Ocean Eng.*, ASCE, 132 (4): 227–243.

de Jong, J.T., Randolph, M.F. and White D.J. (2003) 'Interface load transfer degradation during cyclic loading: a microscopic investigation'. *Soils Found.*, 43(4): 81–93.

de Mello, J.R.C., Amarai, C.D.S., Maia da Costa, A., Rosas, M.M., Decnop Coelho, P.S. and Porto, E.C. (1989) 'Closed-ended pipe piles: testing and piling in calcareous sand'. *Proc. Annu. Offshore Tech. Conf.* Houston, Texas, Paper OTC 6000.

de Mello, J.R.C. and Galgoul, N.S. (1992) 'Piling and monitoring of large diameter closed-toe pipe piles'. *Proc. Conf. Appli. Stress Wave Theor. Piles.* Balkema, 443–448.

Dendani, H. and Colliat, J-L. (2002) 'Girassol: design analyses and installation of the suction anchors'. *Proc. Annu. Offshore Tech. Conf.*, Houston, Texas, Paper OTC 14209.

de Nicola, A. and Randolph, M.F. (1993) 'Tensile and compressive shaft capacity of piles in sand'. *J. Geotech. Eng. Div.*, ASCE, 119(12): 1952–1973.

de Santa Maria (1988) 'Behaviour of footings for offshore structures under combined loads'. D.Phil Thesis. University of Oxford.

Digre, K.A., Kipp, R.M., Hunt, R.J., Hanna, S.Y., Chan, J.H. and van der Voort, C. (1999) 'URSA TLP: tendon, foundation design, fabrication, transportation and TLP installation'. *Proc. Annu. Offshore Tech. Conf.*, Houston, Texas, Paper OTC 10756.

Dingle H.R.C., White D.J. and Gaudin C. (2008) 'Mechanisms of pipe embedment and lateral breakout on soft clay'. *Can. Geotech. J.*, 45(5): 636–652.

Divins, D.L., (2009) *National Geophysical Data Centre (NGDC) Total Sediment Thickness of the World's Oceans and Marginal Seas.* http://www.ngdc.noaa.gov/mgg/sedthick/sedthick.html.

DNV (1992) *Classification Notes No. 30.4, Foundations.* Det Norske Veritas, Oslo, Norway.

DNV (2000a) *Recommended Practice for Design and Installation of Fluke Anchors in Clay.* DnV RP-E301, 1–32, Det Norske Veritas, Oslo, Norway.

DNV (2000b) *Recommended Practice for Design and Installation of Drag-in Plate Anchors in Clay.* DnV RP-E302, 1–32. Det Norske Veritas, Oslo, Norway.

DNV (2007) *On Bottom Stability Design of Submarine Pipelines.* DNV-RP-F109, October 2007, Det Norske Veritas, Oslo, Norway.

Doherty, J.P. and Deeks, A.J. (2003a) 'Scaled boundary finite element analysis of a non-homogeneous axisymmetric domain subjected to general loading'. *Int. J. Num. Anal. Methods Geomech.*, 27: 813–835.

Doherty, J.P. and Deeks, A.J. (2003b) 'Elastic response of circular footings embedded in a non-homogeneous half-space'. *Géotechnique*, 53(8): 703–714.

Doyle, E.H., Dean, E.T.R., Sharma, J.S., Bolton, M.D., Valsangkar, A.J. and Newlin, J.A. (2004) 'Centrifuge model tests on anchor piles for tension leg platforms'. *Proc. Annu. Offshore Tech. Conf.*, Houston, Texas, Paper OTC 16845.

Dutt, R. and Ehlers, C. (2009) 'Set-up of large diameter driven pipe piles in deepwater normally consolidated high plasticity clays'. *Proc. Conf. Offshore Mech. Arctic Eng.*, Paper OMAE2009-79012.

Duxbury, A.B., Duxbury, A.C. and K.A. Sverdrup (2002) *Fundamentals of Oceanography.* McGraw Hill, New York.

Dyson, G.J. and Randolph, M.F. (2001) 'Monotonic lateral loading of piles in calcareous sediments'. *J. Geotech. Eng. Div.* ASCE, 127(4): 346–352.

Dyvik, R., Andersen, K.H., Hansen, S.B. and Christophersen, H.P. (1993) 'Field tests of anchors in clay'. *J. Geotech. Eng.* ASCE, 119(10): 1515–1531.

Edgers, L. and Karlsrud, K. (1982) 'Soil flows generated by submarine slides—case studies and consequences'. *Proc. Int. Conf. Behav. Offshore Struct. (BOSS)*, Cambridge, Massachusetts, 2: 425–437.

Edwards, D.H. and Potts, D.M. (2004) 'The bearing capacity of a circular footing under "punch through" failure'. *Proc. Int. Symp. Num. Models in Geomech., (NUMOG)*, Ottawa, 493–498.

Ehlers, C.J., Young, A.G. and Chen, J.H. (2004) 'Technology assessment of deepwater anchors'. *Proc. Annu. Offshore Tech. Conf.*, Houston, Texas, Paper OTC 16840.

Ehlers, C.J., Chen, J., Roberts, H.H. and Lee, Y.C. (2005) 'The origin of near-seafloor 'crust zones' in deepwater'. *Proc. Int. Symp. Front. Offshore Geotech., (ISFOG)*. Perth, Australia, 927–933.

Eide, O. and Andersen, K.H. (1984) 'Foundation engineering for gravity structures in the northern North Sea.' *Proc. Int. Conf. Case Histories in Geotechnical Engineering*, St.Louis, MO, 5: 1627–1678.

Einav, I. and Randolph, M.F. (2006) 'Effect of strain rate on mobilised strength and thickness of curved shear bands'. *Géotechnique*, 56(7): 501–504.

Elkhatib, S. (2006) *The behaviour of drag-in plate anchors in soft cohesive soils.* PhD Thesis, The University of Western Australia.

Elkhatib, S. and Randolph, M.F. (2005) 'The effect of friction on the performance of drag-in plate anchors'. *Proc. Int. Symp. Front. Offshore Geotech., (ISFOG)*, Perth, Australia, 171–177.

Elton, D.J. (2001) *Soils Magic.* Geotechnical special publication GSP 114, GeoInstitute, American Society of Civil Engineers (ASCE).

Elverhøi, A., Issler, D., De Blasio, F.V., Ilstad, T., Harbitz, C.B. and Gauer, P. (2005) 'Emerging insights into the dynamics of submarine debris flows'. *Nat. Hazards Earth Syst. Sci.*, 5: 633–648.

ENISO 22476–1 (2007). Geotechnical Investigation and Testing—Field Testing—Part 1: Electrical Cone and Piezocone Penetration Tests'. ISO/CEN, Geneva, Switzerland.

Erbrich, C.T. (1994) 'Modelling of a novel foundation for offshore structures'. *Proc. UK ABAQUS Users' Conf.*, Oxford, September 1994.

Erbrich, C.T. (2004) 'A new method for the design of laterally loaded anchor piles in soft rock'. *Proc. Annu. Offshore Tech. Conf.*, Houston, Texas, Paper OTC 16441.

Erbrich, C.T. (2005) 'Australian frontiers—spudcans on the edge.' *Proc. Int. Symp. on Frontiers in Offshore Geotechnics (ISFOG)*, Perth, 49–74.

Erbrich, C. and Hefer, P. (2002) 'Installation of the Laminaria suction piles—a case history'. *Proc. Annu. Offshore Tech. Conf*, Houston, Texas, Paper OTC 14240.

Erbrich, C.T. and Neubecker S.R. (1999) 'Geotechnical design of a grillage and berm anchor'. *Proc. Annu. Offshore Tech. Conf.*, Houston, Texas, Paper OTC 10993.

Erbrich, C.T. and Tjelta, T.I. (1999) 'Installation of bucket foundations and suction caissons in sand – Geotechnical performance'. *Proc. Annu. Offshore Tech. Conf.* Houston, Texas, Paper OTC 10990.

Erickson, H.L. and Drescher, A. (2002) 'Bearing capacity of circular footings'. *J. Geotech. Geoenviron. Eng.*, ASCE, 128(1): 38–43.

Etterdal, B. and Grigorian, H. (2001) 'Strengthening of Ekofisk platforms to ensure continued and safe operation'. *Proc. Annu. Offshore Tech. Conf.* Houston, Texas, Paper OTC 13191.

Eurocode 7 (1997) *Calcul Geotechnique.* AFNOR, XP ENV 1997–1, 1996.

Fahey, M., Jewell, R.J., Khorshid, M.S. and Randolph, M.F. (1992) 'Parameter selection for pile design in calcareous sediments'. *Predict. Soil Mech.: Proc. Wroth Memo. Symp.*, Thomas Telford, London, 261–278.

Fellenius, W. (1927) *Erdstatische Berechnungen mit Reibung und Kohasion (Adhasion) und unter Annahme Kreiszylindrischer Gleitflachen (Earth stability calculations assuming friction and cohesion on circular slip surfaces).* W. Ernst, Berlin.

Fellenius, B.H. and Altaee, A.A. (1995). 'Critical depth: how it came into being and why it doesn't exist'. *Proc. Inst. Civil Eng. Geotech. Eng..* 113(1): 107–119.

Finnie, I.M.S. (1993) *Performance of Shallow Foundations in Calcareous Soil*. PhD Thesis, University of Western Australia.

Finnie, I.M.S. and Randolph, M.F. (1994) 'Punch-through and liquefaction induced failure of shallow foundations on calcareous sediments'. *Proc. Int. Conf. Behav. Offshore Struct. (BOSS)*, Boston, Massachusetts, 217–230.

Fisher, R. and Cathie, D. (2003) 'Optimisation of gravity based design for subsea applications'. *Proc. Int. Conf. Found., (ICOF)*, Dundee, Scotland.

Fleming, W.G.K., Weltman, A.J., Randolph, M.F. and Elson W.K. (2009). *Piling Engineering*. 3rd ed., Surrey University Press, Halstead Press.

Focht, J.A. and Kraft, L.M. (1977) 'Progress in marine geotechnical engineering'. *J. Geotech. Eng.*, ASCE, 103(GT10): 1097–1118.

Foo, K.S., Quah, M.C.K., Wildberger, P., Vazquez, J.H. (2003a) 'Rack phase difference (RPD)'. *Proc. Int. Conf. Jack-Up Platform Des. Constr. Oper.*, City University, London, UK.

Foo, K.S., Quah, M.C.K., Wildberger, P. and Vazquez, J.H. (2003b) 'Spudcan footing interaction and rack phase difference'. *Proc. Int. Conf. Jack-Up Platform Des.Constr. Oper.*, City University, London, UK.

Fookes, P.G., KLee, E.M. and Milligan, G. (2005) *Geomorphology for Engineers*. Whittles Publishing, Scotland, UK.

Foray, P.Y., Alhayari, S., Pons, E., Thorel, L., Thetiot, N., Bale, S. and Flavigny, E. (2005) 'Ultimate pullout capacity of SBM's vertically loaded plate anchor VELPA in deep sea sediments'. *Proc. Int. Symp. Front. Offshore Geotech. (ISFOG)*, Perth, Australia, 185–190.

Frydman, S. and Burd, H.J. (1997) 'Numerical studies of bearing-capacity factor $N_\gamma$'. *J. Geotech. Geoenv. Eng.*, ASCE, 123(1): 20–29.

Gan, C.T. (2010). '*Centrifuge model study on spudcan-footprint interaction*.' PhD Thesis, National University of Singapore.

Garnier, J., Gaudin C., Springman, S.M., Culligan, P.J., Goodings, D., Konig, D., Kutter, B., Phillips, R., Randolph, M.F. and Thorel, L. (2007) 'Catalogue of scaling laws and similitude questions in centrifuge modelling'. *Int. J. Phys. Model. Geotech.*, 7(3), 1–24.

Gaudin, C. and White, D.J. (2009) 'New centrifuge modelling techniques for investigating seabed pipeline behaviour'. *Proc. Int. Conf. on Soil Mech. Geotech. Eng., (ICSMGE)*. Alexandria, Egypt.

Gaudin, C., Bienen, B. and Cassidy, M.J. (2010) 'Mechanisms governing spudcan extraction with water jetting in soft clay'. *Géotechnique*.

Gaudin, C., O'Loughlin, S.D., Randolph, M.F. and Lowmass, A.C. (2006) 'Influence of the installation process on the performance of suction embedded plate anchors'. *Géotechnique*, 56(6): 381–391.

Geer, D.A., Douglas Devoy, S. and Rapoport, V. (2000) 'Effects of soil information on economics of jackup installation'. *Proc. Annu. Offshore Tech. Conf.*, Houston, Texas, Paper OTC 12080.

George, R.A, Gee, L., Hill, A., Thomson, J. and Jeanjean, P. (2002) 'High-Resolution AUV Surveys of the Eastern Sigsbee Escarpment'. *Proc. Annu. Offshore Tech. Conf.*, Houston, Texas, Paper OTC 14139.

Georgiadis, M. (1985) 'Load-path dependent stability of shallow footings'. *Soils Found.*, 25(1): 84–88.

Gerwick, B.C., (2007) '*Construction of marine and offshore structures*'. 3rd ed. CRC Press, Taylor and Francis group, Florida, USA.

Gibson, R.R, Schiffman, R.L and Pu, S.L. (1970) 'Plane strain and axially symmetric consolidation of a clay layer on a smooth impervious base'. *Quart. J. Mech. Appl. Math.*, 23(4): 505–519.

Gilbert, R.B., Nodine, M., Wright, S.G., Cheon, J.Y., Coyne, M. and Ward, E.G. (2007). Impact of hurricane-induced mudslides on pipelines. *Proc. Annu. Offshore Tech. Conf.*, Houston, Texas Paper OTC 18983.

Golightly CR, Hyde AFL (1988) 'Some fundamental properties of carbonate sands'. *Proc. Int. Conf. on Calcareous Sediments*, Perth, Australia 1: 69–78.

Gottardi, G. and Butterfield, R. (1993) 'On the bearing capacity of surface footings on sand under general planar loads'. *Soils Found.*, 33(3): 68–79.

Gottardi, G. and Butterfield, R. (1995) 'The displacement of a model rigid surface footing on dense sand under general planar loading', *Soils Found.*, 35(3): 71–82.

Gottardi, G., Houlsby, G.T. and Butterfield, R. (1999). 'The plastic response of circular footings on sand under general planar loading', *Géotechnique*, 49(4): 453–470.

Gourvenec, S. (2007a) 'Shape effects on the capacity of rectangular footings under general loading'. *Géotechnique*, 57(8): 637–646.

Gourvenec, S. (2007b) 'Failure envelopes for offshore shallow foundation under general loading'. *Géotechnique*, 57(9): 715–727.

Gourvenec, S. (2008) 'Undrained bearing capacity of embedded footings under general loading'. *Géotechnique*, 58(3): 177–185.

Gourvenec, S. and Jensen, K. (2009) 'Effect of embedment and spacing of co-joined skirted foundation systems on undrained limit states under general loading'. *Int. J. Geomech.*, ASCE, 9(6): 267–279.

Gourvenec, S. and Randolph, M.F. (2003) 'Effect of strength non-homogeneity on the shape and failure envelopes for combined loading of strip and circular foundations on clay'. *Géotechnique*, 53(6): 575–586.

Gourvenec, S. and Randolph, M.F. (2003) 'Failure of shallow foundations under combined loading.' *Proc. Eur. Conf. Soil Mech. and Geotech. Engng (ECSMGE)*, Prague, Czech Republic, 2: 583–588.

Gourvenec, S. and Randolph, M.F. (2009) 'Effect of foundation embedment and soil properties on consolidation response'. *Proc. Int. Conf. on Soil Mech. and Geotech. Eng. (ICSMGE)*, Alexandria, Egypt. 638–641.

Gourvenec, S. and Randolph, M.F. (2010) 'Consolidation beneath skirted foundations due to sustained loading'. *Int. J. Geomech.*, ASCE, 10(1): 22–29.

Gourvenec, S. and Steinepreis, M. (2007) 'Undrained limit states of shallow foundations acting in consort'. *Int. J. Geomech.*, ASCE, 7(3): 194–205.

Gourvenec S. and White D.J. (2010) 'Elastic solutions for consolidation around seabed pipelines'. *Proc. Annu. Offshore Tech. Conf.*, Houston, Texas, Paper OTC 20554.

Gourvenec, S., Randolph, M.F. and Kingsnorth, O. (2006) 'Undrained bearing capacity of square and rectangular footings'. *Int. J. Geomech.*, ASCE, 6(3): 147–157.

Gourvenec, S., Acosta-Martinez, H.E. and Randolph, M.F. (2007) 'Centrifuge model testing of skirted foundations for offshore oil and gas facilities'. *Proc. Int. Conf. Offshore Site Invest. Geotech.*, Soc. Underwater Tech., London, 479–484.

Gourvenec, S., Acosta-Martinez, H.E. and Randolph, M.F. (2008a) 'Experimental study of uplift resistance of shallow skirted foundations in clay under concentric transient and sustained loading'. *Géotechnique*, 59(6): 525–537.

Gourvenec, S., Govoni, L. and Gottardi, G. (2008b) 'An investigation of the embedment effect on the combined loading behaviour of shallow foundations on sand'. *BGA Int. Conf. Found.*, Dundee, Scotland, 873–884.

Govoni, L., Gourvenec, S., Gottardi, G. and Cassidy, M.J. (2006) 'Drum centrifuge tests of surface and embedded footings on sand'. *Proc. Int. Conf. on Phys. Model. Geotech.*, Hong Kong, 1: 651–657.

Govoni, L., Gourvenec, S. and Gottardi, G. (accepted 2010) 'A centrifuge study on the effect of embedment on the drained response of shallow foundations under combined loading'. *Géotechnique*.

Graham, J. and Houlsby, G.T. (1983) 'Elastic anisotropy of a natural clay'. *Géotechnique*, 33(2): 165–180.

Graham, J. and Stewart, J.B. (1984) 'Scale and boundary effects in foundation analysis'. *J. Soil Mech. Found. Div.*, ASCE, 97(SM11): 1533–1548.

Green, A.P. (1954) 'The plastic yielding of metal junctions due to combined shear and pressure'. *J. Mech. Phys. Solids*, 2(3): 197–211.

Grundlehner, G.J. (1989) *The Development of a Simple Model for the Deformation Behaviour of Leg to Hull Connections of Jack-up Rigs.* M.Sc. Thesis, Delft University of Technology.

Gunasena, U., Joer, H.A. and Randolph, M.F. 'Design approach for grouted driven piles in calcareous soil'. *Proc. Annu. Offshore Tech. Conf.* Houston, Texas, Paper OTC 7669.

Guo, P.J., (2005) 'Numerical modelling of pipe-soil interaction under oblique loading'. *J. Geotech. Geoenv. Eng.*, ASCE, 131(2): 260–268.

Hambly, E.C. and Nicholson, B.A. (1991) 'Jackup dynamic stability under extreme storm conditions'. *Proc. Annu. Offshore Tech. Conf.*, Houston, Texas, Paper OTC 6590.

Hamilton, J.M. and Murff, J.D. (1995) 'Ultimate lateral capacity of piles in clay'. *Proc. Annu. Offshore Tech. Conf.* Houston,Texas, Paper OTC 7667.

Hampton, M.A., Lee, H.J., and Locat, J. (1996) 'Submarine landslides'. *Rev. Geophys.*, American Geophysical Union, 34(1), 33–59.

Hanna, A.M. and Meyerhof, G.G. (1980). 'Design carts for ultimate bearing capacity of foundations on sand overlaying soft clay'. *Can. Geotech. J.*, 1: 300–303.

Hansbo, S. (1957) 'A new approach to the determination of the shear strength of clay by the fall cone test'. *Proc. Royal Swedish Geotech. Inst.*, 14, 1–49.

Hansen, B., Nowacki, F., Skomedal, E. and Hermstad, J. (1992) 'Foundation design, Troll Platform'. *Proc. Int. Conf. Behav. Offshore Struct. (BOSS)*, London, England, 921–936.

Head, K.H. (2006) *Manual of Soil Laboratory Testing.* 3rd ed., Pentech Press, London.

Heerema, E.P. (1980) 'Predicting pile driveability: Heather as an Illustration of the "Friction Fatigue" Theory'. *Ground Eng.* 13: 15–37.

Higham M.D. (1984) 'Models of jack-up rig foundations.' MSc dissertation. Manchester University.

Heijnen, W.J. (1974) 'Penetration testing in the Netherlands.' *Proc. of Eur. Symp. On Penetration Testing*, Stockholm, Vol. 1, 79–84.

Heim, A. (1932) *Bergsturz und Menschenleben.* Fretz und Wasmuth, Zurich, Switzerland.

Hill A.J. and Jacob H. (2008) 'In-situ measurement of pipe-soil interaction in deep water'. *Proc. Annu. Offshore Tech. Conf.*, Houston, Texas, Paper OTC 19528.

Hodder, M. White, D.J. and Cassidy, M.J. (2008) 'Centrifuge modelling of riser-soil stiffness degradation in the touchdown zone of an SCR'. *Proc. Int. Conf. Offshore Mech. Arctic Eng. (OMAE)*, Portugal, Paper OMAE2008-57302.

Hodder, M. White, D.J. and Cassidy, M.J. (2009) 'Effect of remolding and reconsolidation on the touchdown stiffness of a steel catenary riser: observations from centrifuge modelling'. *Proc. Annu. Offshore Tech. Conf.*, Houston, Texas, Paper OTC 19871-PP.

Holhjem, A. (1998) 'Introduction – Why redevelopment at Ekofisk'. *Proc. Annu. Offshore Tech. Conf.* Houston, Texas, Paper OTC 8653.

Hooper, J.R and Suhayda, J.N. (2005) 'Hurricane Ivan as a geologic force: Mississippi Delta Front seafloor failures'. *Proc. Annu. Offshore Tech. Conf.*, Houston, Texas, Paper OTC 17737.

Hossain, M.S. (2008) '*New mechanism-based design approaches for spudcan foundations on clays*'. PhD Thesis, University of Western Australia.

Hossain, M.S. and Randolph, M.F. (2008) 'Overview of spudcan performance on clays: current research and SNAME'. *Jack-up Asia Conf. Exhibition*, Singapore.

Hossain, M.S. and Randolph, M.F. (2009a) 'New mechanism-based design approach for spudcan foundations on single layer clay'. *J. Geotech. Geoenv. Eng.*, ASCE, 135(9): 1264–1274.

Hossain, M.S. and Randolph, M.F. (2009b) 'New mechanism-based design approach for spud-can foundations on stiff-over-soft clay'. *Proc. Annu. Offshore Tech. Conf.*, Houston, Texas, Paper OTC19907.

Hossain, M.S. and Randolph, M.F. (2010) 'Deep-penetrating spudcan foundations on layered clays: centrifuge tests'. *Géotechnique*, 60(3): 157–170.

Hossain, M.S., Hu, Y. and Randolph, M.F. (2003) 'Spudcan foundation penetration into uniform clay'. *Proc. Int. Symp. Offshore Polar Eng. (ISOPE)*, Hawaii, USA, 647–652.

Hossain, M.S., Hu, Y. and Randolph, M.F. (2004) 'Bearing behaviour of spudcan foundation on uniform clay during deep penetration'. *Proc. Int. Conf. Offshore Mech. Arctic Eng. (OMAE)*, Vancouver, Canada, OMAE 2004-51153.

Hossain, M.S., Hu, Y., Randolph, M.F. and White, D.J. (2005) 'Limiting cavity depth for spud-can foundations penetrating clay'. *Géotechnique*, 55(9): 679–690.

Hossain, M.S., Randolph, M.F., Hu, Y. and White, D.J. (2006) 'Cavity stability and bearing capacity of spudcan foundations on clay'. *Proc. Annu. Offshore Tech. Conf.*, Houston, Texas, Paper OTC 17770.

Hossain, M.S., Cassidy, M.J., Daley, D. and Hannan, R. (2010) 'Experimental investigation of perforation drilling in stiff-over-soft clay'. *Appl. Ocean Res.*, 32(1): 113–123.

Houlsby, G.T. (2003) 'Modelling of shallow foundations for offshore structures'. *Proc. Int. Conf. Found.*, Dundee, Thomas Telford, 11–26.

Houlsby, G.T. and Byrne, B.W. (2005) 'Calculation procedures for installation of suction caissons in sand'. *Geotech. Eng.*, 158(3): 135–144.

Houlsby, G.T. and Cassidy, M.J. (2002) 'A plasticity model for the behaviour of footings on sand under combined loading'. *Géotechnique*, 52(2): 117–129.

Houlsby, G.T. and Martin, C.M. (1992) 'Modelling of the behaviour of foundations of jack-up units on clay'. *Proc. Wroth Memorial Symp.* Predictive Soil Mechanics, Oxford, 339–358.

Houlsby, G.T. and Martin, C.M. (2003) 'Undrained bearing capacity factors for conical footings on clay'. *Géotechnique*, 53(5): 513–520.

Houlsby, G.T. and Puzrin, A.M. (1999) 'The bearing capacity of a strip footing on clay under combined loading'. *Proc. Royal Soc., A*, 455(1983): 893–916.

Houlsby, G.T. and Wroth, C.P. (1983) 'Calculation of stresses on shallow penetrometers and footings'. *Proc. Int. Union Theor. Appl. Mech./Int. Union Geodesy Geophys. (IUTAM/IUGG) Symp. Seabed Mech.*, Newcastle, UK, 107–112.

Houlsby, G.T., Evans, K.M. and Sweeney, M. (1988) 'End bearing capacity of model piles in layered carbonate soils'. *Proc. Int. Conf. Calcareous Sediments*. Perth, Australia, Balkema, 1: 209–214.

House, A.R., Oliveira, J.R.M.S. and Randolph, M.F. (2001) 'Evaluating the coefficient of consolidation using penetration tests'. *Int. J. Phys. Model. Geotech.*, 1(3): 17–25.

Hsu, T.-W., Chen, Y.-J and Hung, W.C. (2004) 'Soil restraint to oblique movement of buried pipes in dense sand'. *J. Transport. Eng.*, ASCE, 132(2): 175–181.

Hsu, T-W, Chen, Y-J & Hung, W-C (2006) 'Soil restraint to oblique movement of buried pipes in dense sand. *J. Transportation Engineering*, ASCE, 132(2):175–181.

Hu, Y., and Randolph, M.F. (1998a) 'A practical numerical approach for large deformation problems'. *Int. J. Num. Anal. Methods Geomech.*, 22(5): 327–350.

Hu, Y., and Randolph, M.F. (1998b) 'H-adaptive FE analysis of elasto-plastic non-homogeneous soil with large deformation'. *Compu. Geotech.*, 23(1): 61–83.

Huang, X., and Garcia, M.H. (1998) 'A Herschel-Bulkley model for mud flow down a slope'. *J. Fluid Mech.* 374: 305–333.

Huang, X. and Garcia, M.H. (1999) 'Modeling of non-hydroplaning mudflows on continental slopes'. *Mar. Geol.*, 154: 131–142.

Humpheson, C. (1998) 'Foundation design of Wandoo B concrete gravity structure'. *Proc. Int. Conf. Offshore Site Invest. Found. Behav., Soc. Underwater Tech.*, 353–367.

Hunt, R.J. and Marsh, P.D. (2004) 'Opportunities to improve the operational and technical management of jack-up deployments'. *Mar. Struct.*, 17(3–4): 261–273.

Imran, J., Harff, P., and Parker, G. (2001) 'A numerical model of submarine debris flow with graphical user interface'. *Compu. Geosci.*, 27: 717–729.

Islam, M.K. (1999) *Constitutive models for carbonate sand and their application to footing problems.* PhD Thesis, University of Sydney, Australia.

Islam, M.K., Carter, J.P. and Airey, D.W. (2001) 'A study of surface footings on carbonate soils'. *Proc. Conf. on Eng. of Calcareous Sediments,* Darassulam, Brunei.

ISSMGE (International Society for Soil Mechanics and Geotechnical Engineering) (1999) 'ISSMGE Technical Committee TC16: Ground Property Characterisation from In-situ Testing (1999): International Reference Test Procedure (IRTP) for the Cone Penetration Test (CPT) and the Cone Penetration Test with pore pressure (CPTU).' *Proc. Eur. Conf. Soil Mech. Geotech. Eng. (ECSMGE),* Amsterdam, The Netherlands, Balkema, 2195–2222.

International Standardization Organization (2000) *Petroleum and Natural Gas Industries: Offshore Structures: Part 4, Geotechnical and Foundation Design Considerations.* ISO/DIS 19901–4, International Standards Office, British Standards Institute, London.

James, R.G. and Tanaka, H. (1984) 'An investigation of the bearing capacity of footings under eccentric and inclined loading in sand in a geotechnical centrifuge'. *Proc. Symp. Recent Adv. Geotech. Centrifuge Model.,* University of California, Davis, 88–115.

Jamiolkowski, M.B., Lo Presti, D.C.F. and Manassero, M. (2003) 'Evaluation of relative density and shear strength of sands from cone penetration test (CPT) and flat dilatometer (DMT)'. In: *Soil Behaviour and Soft Ground Construction,* J.T. Germain, T.C. Sheahan and R.V. Whitman, (eds.), ASCE, GSP 119, 201–238.

Janbu, N, (1973) *Slope Stability Computations in Embankment Dam Engineering: Casagrande Memorial Volume.* R.C. Hirschfeld and S.J. Poulos (eds.). John Wiley, New York.

Jardine, R.J. and Chow, F.C. (1996) *New Design Methods for Offshore Piles.* MTD Publication 96/103.

Jardine, R.J. and Saldivar, E. (1999) 'An alternative interpretation of the West Delta 58a tension-pile research results'. *Proc. Annu. Offshore Tech. Conf.*, Houston, Texas, Paper OTC 10827.

Jardine, R.J. and Standing, J.R. (2000) *Pile Load Testing Performed for HSE Cyclic Loading Study at Dunkirk.* Report OTO2000 007. Health and Safety Executive, London. Two volumes, p. 60 and p. 200.

Jardine, R.J., Kovecevic, N., Hoyle, M.J.R., Sidhu, H.K. and Letty, A. (2001). 'A study of eccentric jack-up penetration into infilled footprint craters'. *Proc. Int. Conf. Jackup Platform,* City University, London.

Jardine, R.J., Chow, F.C., Overy, R. and Standing, J. (2005) *ICP Design Methods for Driven Piles in Sands and Clays.* Thomas Telford, London. ISBN 0 7277 3272 2.

Jayson, D., Delaporte, P., Albert ,J.-P., Prevost, M.E., Bruton, D.A.S. and Sinclair, F. (2008) 'Greater Plutonio Project – Subsea Flowline Design and Performance'. *Offshore Pipeline Tech. Conf.,* Amsterdam, The Nertherlands.

Jeanjean, P. (2006) 'Set-up characteristics of suction anchors for soft Gulf of Mexico clays: Experience from field installation and retrieval'. *Proc. Annu. Offshore Tech. Conf.,* Houston, Texas, Paper OTC 18005.

Jeanjean, P. (2009) 'Re-assessment of p-y curves for soft clays from centrifuge testing and finite element modelling'. *Proc. Annu. Offshore Tech. Conf.,* Houston, Texas, Paper OTC 20158.

Jeanjean, P., Liedtke, E., Clukey, E.C., Hampson, K. and Evans, T. (2005) 'An operator's perspective on offshore risk assessment and geotechnical design in geohazard prone areas'. *Proc. Int. Symp. Front. Offshore Geotech. (ISFOG),* Perth, Australia, 115–144.

Jeanjean, P., Znidarcic, D., Phillips, R., Ko, H.Y., Pfister, S. and Schroeder, K. (2006) 'Centrifuge testing on suction anchors: Doublewall, stiff clays, and layered soil profile'. *Proc. Annu. Offshore Tech. Conf.*, Houston, Texas, Paper OTC 18007.

Jensen, J.J. and Capul, J. (2006) 'Extreme response predictions for jack-up units in second order stochastic waves by FORM'. *Probab. Eng. Mech.*, 21: 330–337

Jewell, R.J. and Khorshid, M.S. (1988) *Proceedings of the Conference on Engineering of Calcareous Sediments*. Perth, Australia, Volumes 1 and 2, Balkema.

Jiang, L. and LeBlond, P.H. (1993) 'Numerical modelling of an underwater Bingham plastic mudslide and the water wave which it generates'. *J. Geophys. Res.*, 8: 10303–10317.

Joer, H.A., Erbrich, C.T. and Sharma, S.S. (2010) 'A new interpretation of the simple shear test', *Proc. 2nd Int. Symp. On Frontiers in Offshore Geotechnics*, Perth.

Joer, H.A and Randolph, M.F. (1994) 'Modelling of the shaft capacity of grouted driven piles in calcareous soil'. *Proc. Int. Conf. Des. Constr. Deep Found.*, FHWA. Orlando. 2: 873–887.

Joer, H.A., Randolph, M.F. and Gunasena, U. (1998) 'Experimental modelling of the shaft capacity of grouted driven piles'. *ASTM Geotech. Test. J.*, 21(3): 159–168.

Johnston, I.W., Lam, T.S.K. and Williams, A.F. (1987) 'Constant normal stiffness direct shear testing for socketed pile design in weak rock'. *Géotechnique*, 37(1), 83–89.

Karlsrud, K. and Haugen, T. (1985) 'Behavior of piles in clay under cyclic axial loading—results of field model tests'. *Proc. Int. Conf. Behav. Offshore Struct.*, Delft. 2: 589–600.

Karlsrud, K. and Nadim, F. (1990) 'Axial capacity of offshore piles in clay'. *Proc. Offshore Tech. Conf.* Houston, Texas, Paper OTC 6245.

Karlsrud K., Kalsnes, B. and Nowacki, F. (1993) 'Response of piles in soft clay and silt deposits to static and cyclic axial loading based on recent instrumented pile load tests'. *Proc. Conf. Offshore Site Invest. Found. Behav.*, Soc. Underwater Tech., London, 549–584.

Keaveny, J.M., Hansen, S.B., Madshus, C. and Dyvik, R. (1994) 'Horizontal capacity of large scale model anchors'. *Proc. Int. Conf. Soil Mech. Found. Eng. (ICSMFE)*, New Delhi, India, 2: 677–680.

Kelleher, P.J. and Randolph, M.F. (2005) 'Seabed geotechnical characterisation with the portable remotely operated drill.' *Proc. Int. Symp. Front. Offshore Geotech. (ISFOG)*, Perth, Australia, 365–371.

Kimura, T., Kusakabe, O. and Saitoh, K. (1985) 'Geotechnical model tests of bearing capacity problems in centrifuge'. *Géotechnique*, 35(1): 33–45.

Kolk, H.J. and van der Velde, E. (1996) 'A reliable method to determine friction capacity of piles driven into clays'. *Proc. Annu. Offshore Tech. Conf.* Houston, Texas, Paper OTC 7993.

Kolk, H.J., Baaijens, A.E., and Senders, M. (2005) 'Design criteria for pipe piles in silica sands'. *Proc. Int. Symp. Front. Offshore Geotech.*, Perth, Australia, 711–716.

Konuk, I., and Yu, S. (2007) 'Continuum FE modeling of lateral buckling: study of soil effects'. *Proc. Int. Conf. Offshore Mech. Arctic Eng. (OMAE)*, San Diego, California.

Konuk, I., Yu, S., and Evgin, E. (2006) 'Application of the ALE FE Method to debris flows'. *Proc. Int. Conf. Monit. Simul. Prev. Rem. Dense Debris Flows*, Rhodes, Greece.

Kostelnik, A., Guerra, M., Alford, J., Vazquez, J. and Zhong, J. (2007). Jackup mobilization in hazardous soils. *SPE Drilling and Completion*, 22(1): 4–15.

Koumoto, T. and Houlsby, G.T. (2001) 'Theory and practice of the fall cone test'. *Géotechnique*, 51(8): 701–712.

Krost, K., Gourvenec, S. and White, D. (2010) 'Consolidation around partially embedded submarine pipelines'. *Géotechnique*, DOI: 10.1680/geot.8.T.015.

Kulhawy, F.H. (1984) Limiting tip and side resistance: Fact or fallacy? *Proc. Symp. on Anal. Des. Pile Found.*, ASCE. 80–98.

Kullenberg, B. (1947) 'The piston core sampler'. *Svensk Hydrografisk-Biologiska Komm. Skr. ser.* 3, Hydrofrafi, 1(2).

Kuo, M.Y-H. and Bolton, M.D. (2009) 'Soil characterization of deep sea West African clays: Is biology a source of mechanical strength?' *Proc. Int. Symp. Offshore Polar Eng. (ISOPE)*, Osaka. 488–494.

Kvalstad, T.J. (2007) 'What is current "Best Practice" in offshore geohazard investigations? A state-of-the-art review'. *Proc. Annu. Offshore Tech. Conf.*, Houston, Texas, Paper OTC 18545.

Kvalstad, T.J., Nadim, F. and Harbitz, C.B. (2001) 'Deepwater geohazards: Geotechnical concerns and solutions'. *Proc. Annu. Offshore Tech. Conf.*, Houston, Texas, Paper OTC 12958.

Ladd, C.C. (1991). Stability evaluation during staged construction (22nd Terzaghi Lecture). *J. Geotech. Eng.*, ASCE, 117(4): 540.

Ladd, C.C. and DeGroot, D.J. (2003) 'Recommended practice for soft ground site characterization'. The Arthur Casagrande Lecture, *Proc. Pan-Am. Conf. Soil Mech. Geotech. Eng.*, Boston, Massachusetts, 3–57.

Langen, H.V. and Hospers, B. (1993) 'Theoretical model for determining rotational behaviour of spud cans'. *Proc. Annu. Offshore Tech. Conf.*, Houston, Texas, Paper OTC 7302.

Langford T.E., Dyvik R. and Cleave R. (2007) 'Offshore pipeline and riser geotechnical model testing: practice and interpretation'. *Proc. Conf. Offshore Mech. Arctic Eng. (OMAE)*, San Diego, California, Paper OMAE2007–29458.

Lauder, K., Bransby, M.F., Brown, M., Cathie, D.N., Morgan, N., Pyrah, J. and Steward, J. (2008) 'Experimental testing of the performance of pipeline ploughs'. *Proc. Int. Symp. Offshore Polar Eng. (ISOPE)*, Paper TPC-174.

Lee. K.K. (2009) *Investigation of Potential Spudcan Punch-Through Failure on Sand Overlaying Clay Soils*. PhD Thesis, University of Western Australia.

Lee, K.K., Randolph, M.F. and Cassidy, M.J. (2009) 'New simplified conceptual model for spudcan foundations on sand overlying clay soils'. *Proc. Annu. Offshore Tech. Conf.*, Houston, Texas, Paper OTC 20012.

Lee, K.L. and Focht, J.A. (1975) 'Liquefaction potential at Ekofisk tank in North Sea'. *J. Geotech. Eng.*, ASCE, 100(GT1): 1018.

Lefebvre, G. and Leboeuf, D. (1987) 'Rate effects and cyclic loading of sensitive clays'. *J. Geotech. Eng.*, ASCE, 113(5): 476–489.

Leffler, W.L., Pattarozzi. R. and Sterling, G. (2003) *Deepwater Petroleum Exploration and Production*. PenWell, Oklahoma, Texas.

Lehane, B.M. and Jardine, R.J. (1994a) 'Displacement pile behaviour in a soft marine clay'. *Can. Geotech. J.*, 31(2): 181–191.

Lehane, B.M. and Jardine, R.J. (1994b) 'Shaft capacity of driven piles in sand: a new design method'. *Proc. Int. Conf. Behav. Offshore Struct.*, Boston, Massachusetts, 1: 23–36.

Lehane, B.M. and Randolph, M.F. (2002) 'Evaluation of a minimum base resistance for driven pipe piles in siliceous sand'. *J. Geotech. Geoenv. Eng.*, ASCE., 128(3): 198–205.

Lehane, B.M., Jardine, R.J., Bond, A.J. and Frank, R. (1993) 'Mechanisms of shaft friction in sand from instrumented pile tests'. *J. Geotech. Eng. Div.*, ASCE,119(1): 19–35.

Lehane, B.M., Chow, F.C., McCabe, B.A. and Jardine, R.J. (2000) 'Relationships between shaft capacity of driven piles and CPT end resistance'. *Proc. Inst Civil Engineers Geotech. Eng.*, 143(2): 93–101.

Lehane, B.M., Schneider, J.A., and Xu, X. (2005a) 'The UWA-05 method for prediction of axial capacity of driven piles in sand'. *Proc. Int. Symp. Front. Offshore Geotech.*, Perth, Australia, 683–689.

Lehane, B.M., Schneider, J.A. and Xu, X. (2005b) *A Review of Design Methods for Offshore Driven Piles in Siliceous Sand*. University of Western Australia, Geomechanics Group, Report No. GEO: 05358. p. 102.

Lenci, S. and Callegari, M. (2005) 'Simple analytical models for the J-lay problem'. *Acta. Mechanica.*, 178: 23–39.

Leroueil, S., Kabbaj, M., Tavenas, F. and Bouchard, R. (1985) 'Stress-strain rate relation for the compressibility of sensitive natural clays'. *Géotechnique*, 35(2): 159–180.

Levadoux, J.N. and Baligh, M.M. (1986) 'Consolidation after undrained piezocone penetration. I: prediction'. *J. Geotech. Eng.*, ASCE, 112(7), 707–726.

Lieng, J.T., Hove, F. and Tjelta, T.I. (1999) 'Deep Penetrating Anchor: Subseabed deepwater anchor concept for floaters and other installations'. *Proc. Int. Symp. Offshore Polar Eng. (ISOPE)*, Brest, France, 613–619.

Lieng, J.T., Kavli, A., Hove, F. and Tjelta, T.I. (2000) 'Deep Penetrating Anchor: Further development, optimization and capacity clarification'. *Proc. Int. Symp. Offshore Polar Eng. (ISOPE)*, Seattle, Washington, 410–416.

Leung, C.F., Gan, C.T., Chow, Y.K. (2007) 'Shear strength changes within jack-up spudcan footprint.' *Proc. 17th International Offshore and Polar Engineering Conf.* (ISOPE), Lisbon, Portugal, 1504–1509.

Locat, J., and Demers, D. (1988) 'Viscosity, yield stress, remolded strength, and liquidity index relationships for sensitive clays'. *Can. Geotech. J.*, 25: 799–806.

Looijens, P. and Jacob, H. (2008) 'Development of a deepwater tool for in-situ pipe-soil interaction measurement and its benefits in pipeline analysis'. *Proc. Offshore Pipeline Tech. Conf.*, Amsterdam, The Netherlands.

Low, H.E., Randolph, M.F., Rutherford, C.J., Bernard, B.B. and Brooks, J.M. (2008) 'Characterization of near seabed surface sediment.' *Proc. Annu. Offshore Tech. Conf.*, Houston, Texas, Paper OTC 19149.

Low, H..E., Randolph, M.F. and Kelleher, P. (2007) 'Estimation of in-situ coefficient of consolidation from dissipation tests with different penetrometers.' *Proc. Int. Conf. Offshore Site Invest. Geotech.*, Soc. Underwater Tech., London, 547–556.

Low, H.E., Lunne, T., Andersen, K.H., Sjursen, M.A., Li, X. and Randolph, M.F. (2010) 'Estimation of intact and remoulded undrained shear strength from penetration tests in soft clays'. *Géotechnique*, 10(11): 843–859.

Lu Q., Randolph M.F., Hu, Y. and Bugarski, I.C. (2004) 'A numerical study of cone penetration in clay.' *Géotechnique*, 54(4), 257–267.

Lund, K.H. (2000) 'Effect of increase in pipeline soil penetration from installation.' *Proc. Int. Conf. on Offshore Mech. Arctic Eng. (OMAE)*, New Orleans, Los Angeles, Paper OMAE2000-PIPE5047.

Lundgren H., Mortensen K. (1953) 'Determination by the theory of plasticity of the bearing capacity of continuous footings on sand.' *Proc. Int. Conf. Soil Mech. Found. Eng. (ICSMFE)*, Zurich, Switzerland, 1: 409–412.

Lunne, T. (2001) 'In situ testing in offshore geotechnical investigations.' *Proc. Int. Conf. In Situ Measur. Soil Prop. Case Histories*, Bali, Indonesia, 61–81.

Lunne, T., Robertson, P.K. and Powell, J.J.M. (1997) *Cone Penetration Testing in Geotechnical Practice*. Blackie Academic and Professional, London.

Lunne, T., Tjelta ,T.I, Walta, A. and Barwise, A. (2008) 'Design and testing out of deep water seabed sampler.' *Proc. Annu. Offshore Tech. Conf.* Houston, Texas, Paper OTC 19290.

Lupini, J.F., Skinner, A.E. and Vaughan, P.R. (1981) 'The drained residual strength of cohesive soils.' *Géotechnique*, 31(2), 181–213.

Mandel, J. (1950) 'Etude mathiematique de la consolidation des sols'. *Actes du Colloque International De Mechanique*, Poitier, France, 4: 9–19.

Manuel, L. and Cornell, C.A. (1996) 'The influence of alternative wave loading and support modeling assumptions on jack-up rig response extremes.' *J. of Offshore Mechanics and Arctic Engineering* (OMAE), Vol. 118, 109–114.

Mao, X. (2000) *The Behaviour of Three Calcareous Soils in Monotonic and Cyclic Loading.*' PhD Thesis, University of Western Australia.

Mao, X. and Fahey, M. (1999) 'A method of reconstituting an aragonite soil using a synthetic flocculant.' *Géotechnique*, 49(1), 15–32.

Mao, X. and Fahey, M. (2003) 'Behaviour of calcareous soils in undrained cyclic simple shear'. *Géotechnique*, 53(8): 715–727.

Martin, C.M. (1994) *Physical and Numerical Modelling of Offshore Foundations under Combined Loads*. DPhil Thesis, University of Oxford.

Martin, C.M. (2003) 'New software for rigorous bearing capacity calculations'. *Proc. Int. Conf. Found. (ICOF)*, Dundee, Scotland, 581–592.

Martin, C.M. (2004) 'Discussion of "Calculations of bearing capacity factor $N_\gamma$ using numerical limit analysis" by Boonchai Ukritchon, Andrew J. Whittle and C. Klangvijit'. *J. Geotech. Geoenv. Eng.*, ASCE, 130(10): 106–1108.

Martin, C.M. and Houlsby, G.T. (1999) 'Jackup units on clay: structural analysis with realistic modelling of spudcan behaviour'. *Proc. Annu. Offshore Tech. Conf.*, Houston, Texas, Paper OTC 10996.

Martin, C.M. and Houlsby, G.T. (2000) 'Combined loading of spudcan foundations on clay: laboratory tests'. *Géotechnique*, 50(4): 325–338.

Martin, C.M. and Houlsby, G.T. (2001) 'Combined loading of spudcan foundations on clay: numerical modelling'. *Géotechnique*, 51(8): 687–700.

Martin, C.M. and Randolph, M.F. (2001) 'Applications of the lower and upper bound theorems of plasticity to collapse of circular foundations'. *Proc. Tenth Int. Conf. Comp. Methods Adv. Geomech.*, Tucson, Arizona, 2: 1417–1428.

Martin, C.M. and Randolph, M.F. (2006) 'Upper bound analysis of lateral pile capacity in cohesive soil.' *Géotechnique*, 56(2), 141–145.

Matlock, H. (1970) 'Correlations for design of laterally loaded piles in clay'. *Proc. Annu. Offshore Tech. Conf.*, Houston, Texas, Paper OTC 1204.

Matlock, H. and Foo, S.H.C. (1980) 'Axial analysis of piles using a hysteretic and degrading soil model'. *Proc. Conf. Num. Methods Offshore Piling*, ICE, London. 127–133.

Matlock, H. and Reese, L.C. (1960) 'Generalized solutions for laterally loaded piles' *J. Soil Mech. Found. Div.*, ASCE, 86(SM5): 63–91.

McGuire, R.K. (2008). 'Probabilistic seismic hazard analysis: Early history'. *Earthquake Engineering and Structural Dynamics*, 37: 329–338.

McNamee, J. and Gibson, R.E. (1960) 'Plane strain and axially symmetric problems of the consolidation of a semi-infinite clay stratum'. *Quart. J. Mech. Appl. Math.*, 13: 210–227.

Medeiros, C.J. (2001) 'Torpedo anchor for deep water'. *Proc. Deepwater Offshore Tech. Conf.*, Rio de Janeiro, Brazil.

Medeiros, C.J. (2002) 'Low cost anchor system for flexible risers in deep waters'. *Proc. Annu. Offshore Tech. Conf.*, Houston, Texas, Paper OTC 14151.

Merifield R.S., Sloan S.W. and Yu H.S. (2001) 'Stability of plate anchors in undrained clay'. *Géotechnique*, 51(2): 141–153.

Merifield, R.S., White, D.J. and Randolph, M.F. (2008a) 'The ultimate undrained resistance of partially-embedded pipelines'. *Géotechnique*, 58(6): 461–470.

Merifield, R.S., White, D.J. and Randolph, M.F. (2008b) 'The effect of pipe-soil interface conditions on the undrained breakout resistance of partially-embedded pipelines'. *Proc. Int. Conf. of Int. Assoc. Comp. Methods Ad. Geomech. (IACMAG)*, Goa, India, 4249–4256.

Merifield, R.S., White, D.J. and Randolph, M.F. (2009) 'The effect of surface heave on the response of partially-embedded pipelines on clay'. *J. Geotech. Geoenv. Eng.*, ASCE, 135(6): 819–829.

Mesri, G. and Shahien, M. (2003) 'Residual shear strength mobilized in first-time slope failures'. *J. Geotech. Geoenviron. Eng.*, ASCE, 129(1): 12–31. (See also Discussions and Closure, 130(5), 544–549.

Maung, U.M. and Ahmad, C.K.M. (2000) 'Swiss cheesing to bring in a jack-up rig at Anding location'. *Proc. IADC/SPE Asia Pacific Drilling Tech.*, Kuala Lumpur, Malaysia, IADC/SPE 62755.

Menzies, D. and Roper, R. (2008) 'Comparison of jackup rig spudcan penetration methods in clay'. *Proc. Annu. Offshore Tech. Conf.*, Houston, Texas, Paper OTC 19545.

Meyerhof, G.G. (1951). 'The ultimate bearing capacity of foundations'. *Géotechnique*, 2(4), 301–332.

Meyerhof, G.G. (1953) 'The bearing capacity of foundations under eccentric and inclined loads'. *Proc. Int. Conf. Soils Mech. Found. Eng. (ICSMFE)*, Zurich, Switzerland, 1: 440–445.

Meyerhof, G.G. and Hanna (1978). Ultimate bearing capacity of foundations of layered soils under inclined loads. *Can. Geotech. J.*, 15, 565–572.

Meyerhof, G.G. (1983) 'Scale effects of ultimate pile capacity'. *J. Geotech. Eng.*, ASCE, 109(GT6): 797–806.

Meyerhof, G.G. (1995) 'Behaviour of pile foundations under special loading conditions. CGJ Hardy lecture'. *Can. Geotech. J.*, 32: 204–222.

Mildenhall, J. and Fowler S. (2001) 'Mud volcanoes and structural development of Shah Deniz'. *J. Petrol. Sci. Eng.*, 28: 189–200.

Mitchell, J.K. (1993) *Fundamentals of Soil Behaviour*. 2nd ed., John Wiley & Sons, New York.

Mo, O. (1976) 'Concrete drilling and production platforms; review of construction, installation and commissioning'. *Proc. Tech. Vol., Offshore North Sea Tech. Conf. Exhibition*, Stavanger, Norway.

Mokkelbost, K.H. and Strandvik, S. (1999) 'Development of NGI's deepwater gas probe, DGP'. *Proc Conf. Offshore Nearshore Geotech. Eng.*, Geoshore, Panvel, India, 107–112.

Morandi, A.C. (2007) 'Jack-up operations in the Gulf of Mexico Lessons Learned From Recent Hurricanes'. *Proc. Int. Conf. Jackup Platform*, City University, London.

Morandi, A.C., Brekke, J.N., Wishahy, M.A. (2009) 'Serviceability assessment of jackups in extreme storm events'. *Proc. Int. Conf. Jackup Platform*, City University, London.

Moresi, L., Dufour, F., and Muhlhaus, H-B. (2003) 'A Lagrangian integration point finite element method for large deformation modelling of viscoelastic geomaterials'. *J. Comput. Phys.*, 184: 476–497.

Morgenstern, N.R. and Price, V.E. (1965) 'The analysis of the stability of general slip circles'. *Géotechnique*, 15(1): 79–93.

Muir Wood, D. (1990) *Soil Behaviour and Critical State Soil Mechanics*. Cambridge University Press, Cambridge.

Muir Wood, D. (2004) *Geotechnical Modelling*. Spon Press, London.

Muller, R.D., Sdrolias, M., Gaina, C. and Roest, W.R. (2008) 'Age, spreading rates and spreading symmetry of the world's ocean crust'. *Geochem. Geophys. Geosyst.*, 9, Q04006, DOI:10.1029/2007GC001743.

Murff, J.D. (1994) 'Limit analysis of multi-footing foundation systems'. *Proc. Int. Conf. Compu. Methods Adv. Geomech.*, Morgantown, West Virginia, 1: 223–244.

Murff, J.D. and Hamilton, J.M. (1993) 'P-ultimate for undrained analysis of laterally loaded piles'. *J. Geot. Eng.*, ASCE, 119(1): 91–107.

Murff, J.D., Wagner, D.A. and Randolph, M.F. (1989) 'Pipe penetration in cohesive soil'. *Géotechnique*, 39(2): 213–229.

Murff, J.D., Prins, M.D., Dean, E.T.R., James, R.G., Schofield, A.N. (1992) 'Jackup rig foundation modelling'. *Proc. Annu. Offshore Tech. Conf.*, Houston, Texas, Paper OTC 6807.

Murff, J.D., Randolph, M.F., Elkhatib, S., Kolk, H.J., Ruinen, R., Strom, P.J. and Thorne, C. (2005). Vertically loaded plate anchors for deepwater applications. *Proc. Int. Symp. Front. Offshore Geotech. (ISFOG)*, Perth, Australia, 31–48.

Nadim F., Krunic D. and Jeanjean P. (2003) 'Reliability Method Applied to Slope Stability Problems: Estimating Annual Probabilities of Failure'. *Proc. Annu. Offshore Tech. Conf.*, Houston, Texas, Paper OTC 15203.

Najjar, S.S., Gilbert, R.B., Liedtke, E.A. and McCarron, W.O. (2003) 'Tilt table test for interface shear resistance between flowlines and soils'. *Proc. Int. Conf. Offshore Mech. Arctic Eng. (OMAE)*, Cancun, Mexico, OMAE2003-37499.

NCEL (1987) *Drag Embedment Anchors for Navy Moorings*. Techdata Sheet 83-08R, Naval Civil Engineering Laboratory, Port Hueneme, California.

Nelson, K., Smith, P., Hoyle, M., Stoner, R. and Versavel, T. (2000) 'Jack-up response measurements and the under-prediction of spud-can fixity by SNAME 5-5A'. *Proc. Annu. Offshore Tech. Conf.*, Houston, Texas, Paper OTC 12074.

Neubecker, S.R. and Erbrich, C.T. (2004) 'Bayu-Udan substructure foundations: Geotechnical design and analysis'. *Proc. Annu. Offshore Tech. Conf.*, Houston, Texas, Paper OTC 16157.

Neubecker, S.R. and Randolph, M.F. (1995) 'Profile and frictional capacity of embedded anchor chain'. *J. Geotech. Eng.*, ASCE, 121(11): 787–803.

Neubecker, S.R. and Randolph, M.F. (1996) 'Performance of embedded anchor chains and consequences for anchor design'. *Proc. Annu. Offshore Tech. Conf.*, Houston, Texas, Paper OTC 7712.

Newlin, J.A. (2003a) 'Suction anchor piles for the Na Kika FDS mooring system, Part 1: site characterization and design'. *Deepwater Mooring Systems: Concepts, Design, Analysis, and Materials*, ASCE, Houston, Texas, USA, 28–54.

Newlin, J.A. (2003b) 'Suction anchor piles for the Na Kika FDS mooring system. Part 2: Installation performance'. *Deepwater Mooring Systems: Concepts, Design, Analysis, and Materials*, ASCE, Houston, Texas, USA, 55–57.

Newson, T.A., Bransby, M.F., Brunning, P. and Morrow, D.R. (2004) 'Determination of undrained shear strength parameters for buried pipeline stability in deltaic soft clays'. *Proc. Int. Symp. Offshore Polar Eng. (ISOPE)*, Toulon, France, Paper 04-JSC-266.

Niedoroda, A.W. and Palmer, A.C. (1986) 'Subsea trench infill'. *Proc. Annu. Offshore Tech. Conf.*, Houston, Texas, OTC 5340.

Niedoroda, A., Reed, C., Hatchett, L., Young, A. and Kasch, V. (2003a) 'Analysis of past and future debris flows and turbidity currents generated by slope failures along the Sigsbee Escarpment'. *Proc. Annu. Offshore Tech. Conf.*, Houston, Texas, Paper OTC 15162.

Niedoroda, A., Jeanjean, P., Driver, D., Reed, C., Hatchett, L., Briaud, J.-L. and Bryant, B. (2003b) 'Bottom currents, erosion rates, and how to use them to date slope failures and debris flows along the Sigsbee Escarpment'. *Proc. Annu. Offshore Tech. Conf.*, Houston, Texas, Paper OTC 15199.

Nielsen, N-J.R., and Lyngberg, B. (1990) 'Upheaval buckling failures of insulated buried pipelines: a case story'. *Proc. Annu. Offshore Tech. Conf.*, Houston, Texas, OTC 6488.

Noad, J. (1993) 'Successful cable burial—its dependence on the correct use of plough assessment and geophysical surveys'. *Conf. Offshore Site Invest. Found. Behav.*, Soc. Underwater Tech., 39 – 56.

Noble Denton and Associates. (1987). 'Foundation fixity of jack-up units: a joint industry study'. *Noble Denton and Associates*, London.

Noble Denton Europe and Oxford University (2005) *The Calibration of SNAME Spudcan Footing Equations with Field Data*. Report No L19073/NDE/mjrh, Rev 5, dated November 2006.

NORSOK Standard (2004) '*Marine soil investigations*'. G-001, *Rev.*, 2, October 2004.

Norwegian Geotechnical Institute (2000) *Windows Program HVMCap. Version 2.0: Theory, User Manual and Certification*. Norwegian Geotechnical Institute Report 524096–7, Rev. 1.

Nova R. and Montrasio L. (1991) 'Settlements of shallow foundations on sand'. *Géotechnique*, 41(2): 243–256.

Novello, E. (1999) 'From static to cyclic p-y data in calcareous sediments'. *Proc. 2nd Int. Conf. Eng. for Calcareous Sediments*, Bahrain, 1: 17–27.

Nowacki, F., Solheim, E., Nadim, F., Liedtke, E. and Andersen, K. (2003) 'Deterministic Slope Stability Analyses of the Sigsbee Escarpment'. *Proc. Annu. Offshore Tech. Conf.*, Houston, Texas, Paper OTC 15160.

O'Loughlin, C.D, Randolph, M.F. and Richardson, M. (2004) 'Experimental and theoretical studies of deep penetrating anchors'. *Proc. Annu. Offshore Tech. Conf.*, Houston, Texas, Paper OTC 16841.

O'Loughlin, C.D., Lowmass, A. Gaudin, C. and Randolph, M.F. (2006) 'Physical modelling to assess keying characteristics of plate anchors'. *Proc. Int. Conf. Phys. Model. Geotech.*, Hong Kong, 1: 659–665.

O'Loughlin, C.D., Richardson, M.D. and Randolph, M.F. (2009) 'Centrifuge tests on dynamically installed anchors'. *Proc. Int. Conf. Offshore Mech. Arctic Eng. (OMAE)*, Honolulu, Hawaii, Paper OMAE 2009-80238.

O'Neill, M.P. (2000) *The Behaviour of Drag Anchors in Layered Soils*. PhD Thesis, University of Western Australia.

O'Neill, M.P. and Randolph, M.F. (2001) 'Modelling drag anchors in a drum centrifuge'. *Int. J. Phys. Model. Geotech.*, 1(2): 29–41.

O'Neill, M.W. and Murchison, J.M. (1983) *An Evaluation of p-y Relationships in Sands*. Report PRAC 82-41-1 to American Petroleum Institute, University of Houston, Houston, Texas.

O'Neill, M.P., Randolph, M.F. and Neubecker, S.R. (1997) 'A novel procedure for testing model drag anchors'. *Proc. Int. Symp. Offshore Polar Eng. (ISOPE)*, Honolulu, Hawaii, 2: 939–945.

O'Neill, M.P., Bransby, M.F. and Randolph, M.F. (2003) 'Drag anchor fluke-soil interaction in clays'. *Can. Geotech. J.*, 40(1): 78–94.

O'Neill, C., Moresi, L., Muller, D., Albert, R., and Dufour, F. (2006) 'Ellipsis 3D: A particle-in-cell finite-element hybrid code for modelling mantle convection and lithospheric deformation'. *Comp. Geosci.*, 32: 1769–1779.

O'Reilly, M.P. and Brown, S.F. (1991) *Cyclic Loading of Soils*. Blackie.

Orange, D., Angell, M., Brand, J., Thompson, J., Buddin, T., Williams, M., Hart, B. and Berger, B. (2003a) 'Shallow Geological and Salt Tectonic Setting of the Mad Dog and Atlantis Field: Relationship Between Salt, Faults, and Seafloor Geomorphology'. *Proc. Annu. Offshore Tech. Conf.*, Houston, Texas, Paper OTC 15157.

Orange, D., Saffer, D., Jeanjean, P., Al-Khafaji, Z., Riley, G. and Humphrey, G. (2003b) 'Measurements and Modeling of the Shallow Pore Pressure Regime at the Sigsbee Escarpment: Successful Prediction of Overpressure and Ground-Truthing with Borehole Measurements'. *Proc. Annu. Offshore Tech. Conf.*, Houston, Texas, Paper OTC 15201.

Orcina (2008) *OrcaFlex User Manual*. www.orcina.com, UK.

Osborne, J.J. and Paisley, J.M. (2002) 'SE Asia jack-up punch-throughs: The way forward?' *Proc. Int. Conf. Offshore Site Invest. Geotech. - Sustainability and Diversity*. London, 301–306.

Osborne, J.J., Trickey, J.C., Houlsby, G.T. and James, R.G. (1991) 'Findings from a joint industry study on foundation fixity of jackup units'. *Proc. Annu. Offshore Tech. Conf.*, Houston, Texas, Paper 6615.

Osborne, J.J., Houlsby, G.T., Teh, K.L., Bienen, B., Cassidy, M.J., Randolph, M.F. and Leung, C.F. (2009) 'Improved guidelines for the prediction of geotechnical performance of spudcan foundations during installation and removal of jack-up units'. *Proc. Annu. Offshore Tech. Conf.*, Houston, Texas, Paper 20291.

Ovesen, N.K. (1975) 'Centrifugal testing applied to bearing capacity problems of footings on sand'. *Géotechnique*, 25(2): 394–401.

Paisley, J.M. and Chan, N. (2006) 'SE Asia jack-up punch-throughs: technical guidance note on site assessment'. *Proc. Jack-up Asia Conf. Exhib.*, Singapore.

Palmer, A.C. (1999) 'Speed effects in cutting and ploughing'. *Géotechnique*, 49(3): 285–294.

Palmer, A.C. and King, R.A. (2008) *Subsea Pipeline Engineering*. 2nd Edition. PennWell Books.

Palmer, A.C., Kenny, J.P., Perera, M.R. and Reece, A.R. (1979) 'Design and operation of an underwater pipeline trenching plough'. *Géotechnique*, 29(3): 305–322.

Palmer, A.C., Ellinas, C.P., Richards, D.M. and Guijt, J. (1990) 'Design of submarine pipelines against upheaval buckling'. *Proc. Annu. Offshore Tech. Conf.*, Houston, Texas, OTC 6335.

Patel, M.H. (1989) *'Dynamics of offshore structures'*. Butterworths, London.

Paton, A.K. and Wong, L.S. (2004) 'Na Kika—Deepwater Mooring and Host installation'. *Proc. Annu. Offshore Tech. Conf.*, Houston, Texas, Paper OTC 16702.

PDI (2005) *GRLWEAP Wave Equation Analysis of Pile Driving*. Pile Dynamics Inc. Ver. 2005-1.

Pedersen, P.T. and Jensen J.J. (1988) 'Upheaval creep of buried pipelines with initial imperfections'. *Mar. Struct.*, 1: 11–22.

Pesce, C.P., Aranha, J.A.P. and Martins, C.A. (1998) 'The soil rigidity effect in the touchdown boundary layer of a catenary riser: static problem'. *Proc. Int. Symp. Offshore Polar Eng. (ISOPE)*. Montreal, Canada, 207–213.

Peuchen, J., Adrichem, J. and Hefer, P.A. (2005) 'Practice notes on push-in penetrometers for offshore geotechnical investigation'. *Proc. Int. Symp. Front. Offshore Geotech. (ISFOG)*, Perth, Australia, 973–979.

Pestana, J.M. and Whittle, A.J. (1995) 'Compression model for cohesionless soils'. *Géotechnique*, 45(4): 611–631.

Peuchen, J. and Mayne P.W. (2007) 'Rate effects in vane shear testing'. *Proc. Int. Offshore Site Invest. Geotech. Conf.: Confronting New Challenges and Sharing Knowledge*, London, 187–194.

Peuchen, J. and Rapp, J. (2007) 'Logging sampling and testing for offshore geohazards', *Proc. Annual Offshore Tech. Conf.*, Houston, Paper OTC 18664.

Phillips, R., Nobahar, A., and Zhou, J. (2004) 'Combined axial and lateral pipe-soil interaction relationships'. *Proc. Int. Pipeline Conf.*, Calgary, Canada. 299–303.

Popescu, R., Phillips, R., Konuk, I., Guo, P. and Nobahar, A. (2002) 'Pipe-soil interaction: large scale tests and numerical modelling'. *Proc. Int. Conf. Phys. Model. Geotech.*, St. John's, NF, 917–922.

Potts, D.M. and Zdravkovic, L. (2001) *Finite Element Analysis in Geotechnical Engineering: Application*. Thomas Telford, London.

Poulos, H.G. (1988) *Marine Geotechnics*. Unwin Hyman, London.

Poulos H.G. (1988) 'Cyclic stability diagram for axially loaded piles'. *J. Geotech. Eng. Div.*, ASCE, 114(GT8): 877–895.

Poulos, H.G. (1989) 'Cyclic axial loading analysis of piles in sand'. *J. Geotech. Eng.*, ASCE, 115(6): 836–852.

Poulos, H.G. and Davis, E.H. (1974) *Elastic Solutions for Soil and Rock Mechanics*. John Wiley, New York.

Powrie, W. (2002) *Soil Mechanics Concepts and Applications*. 2nd ed. Spon Press, London.

Prandtl, L. (1921) 'Eindringungsfestigkeit und festigkeit von schneiden'. Angew. *Math. U. Mech* 1(15): 15–20.

Prasad, Y.V.S.N. and Chari, T.R. (1999) 'Lateral capacity of model rigid piles in cohesionless soils' *Soils Found.*, 39(2): 21–29.

Puech, A. and Foray, P. (2002) 'Refined model for interpreting shallow penetration CPTs in sands'. *Proc. Annu. Offshore Tech. Conf.*, Houston, Texas, Paper OTC 14275.

Purwana, O.A., Leung, C.F., Chow, Y.K., Foo, K.S. (2005) 'Influence of base suction on extraction of jack-up spudcans'. *Géotechnique*, 55(10): 741–753.

Purwana, O.A., Foo, K.S., Quah, M.C.K., Chow, Y.K. and Leung, C.F. (2008) 'Understanding spudcan extraction problem and mitigation device'. *Jack-up Asia Conf. Exhibition*, Singapore.

Puzrin, A.M., Germanovich, L. and Kim, S. (2004) 'Catastrophic failure of submerged slopes in normally consolidated sediments'. *Géotechnique* 54(10): 631–643.

Puzrin, A.M. and Germanovich, L. (2005) 'The growth of shear bands in the catastrophic failure of soils'. *Proc. Royal Society, A* 461: 1199–1228.

Quah, M.C.K., K.S. Foo, Purwana, O.A., Keizer, L., Randolph, M.F. and Cassidy, M.J. (2008) 'An integrated in-situ soil testing device for jack-up rigs'. *Jack-up Asia Conf. Exhibition*, Singapore.

Quirós, G.W. and Young, A.G. (1988) 'Comparison of field vane, CPT and laboratory strength data at Santa Barbara Channel site. Vane Shear Strength Testing of Soils: Field and Laboratory Studies.' *ASTM STP* 1014: 306–317.

Rad, N.S. and Lunne, T. (1994) 'Gas in soils: detection and $\eta$-profiling.' *J. Geotech. Engng*, ASCE 120(4): 696–715.

Randolph M.F. (1981) 'The response of flexible piles to lateral loading'. *Géotechnique* 31(2): 247–259.

Randolph, M.F. (1983) 'Design considerations for offshore piles' *Proc. Conf. on Geot. Practice in Offshore Eng.* Austin. 422–439.

Randolph M.F. (1988) 'The axial capacity of deep foundations in calcareous soil'. *Proc. Int. Conf. on Eng. of Calcareous Sediments* Perth 2: 837–857.

Randolph M.F. (1993) 'Pile capacity in sand—the critical depth myth'. *Australian Geomechanics* 24: 30–34.

Randolph, M.F. (2000) 'Effect of strength anisotropy on capacity of foundations'. *Developments in Theoretical Geomechanics, The John Booker Memorial Symposium*, Sydney, Australia, 313–327.

Randolph, M.F. (2003a) '43rd Rankine Lecture—Science and empiricism in pile foundation design'. *Géotechnique* 53(10): 847–875.

Randolph M F. (2003b) *RATZ User Manual v.4.2.* Centre for Offshore Foundation Systems, University of Western Australia, p. 42.

Randolph, M.F and Erbrich, C.T. (2000) 'Design of shallow foundations for calcareous sediments'. *Proc. Eng. for Calcareous Sediments.* Ed. Al-Shafei, Balkema (2): 361–378.

Randolph, M.F. and Hope, S. (2004) 'Effect of cone velocity on cone resistance and excess pore pressures'. *Proc. Int. Symp. Eng.ng Practice and Performance of Soft Deposits*, Osaka, 147–152.

Randolph, M.F. and Houlsby, G.T. (1984) 'The limiting pressure on a circular pile loaded laterally in cohesive soil'. *Géotechnique* 34(4): 613–623.

Randolph M.F. and House A.R. (2002) 'Analysis of suction caisson capacity in clay'. *Proc. Annu. Offshore Tech. Conf.*, Houston, Texas, Paper OTC 14236.

Randolph, M.F. and Murphy, B.S. (1985) 'Shaft capacity of driven piles in clay' *Proc. Annu. Offshore Tech. Conf.* Houston. OTC 4883.

Randolph, M.F. and Puzrin, A.M. (2003) 'Upper bound limit analysis of circular foundations on clay under general loading'. *Géotechnique* 53(9): 785–796.

Randolph, M.F. and Quiggin, P. (2009) 'Non-linear hysteretic seabed model for catenary pipeline contact'. *Proc. Int. Conf. Offshore Mech. Arctic Eng. (OMAE)*, Honolulu, Hawaii, Paper OMAE2009-79259.

Randolph M.F. and White D.J. (2008a) 'Offshore foundation design—a moving target'. *Proc., 2nd BGA Int. Conf. Found.*, Dundee, UK, IHS BRE Press, Watford. 27–59.

Randolph, M.F. and White, D.J. (2008b) 'Upper bound yield envelopes for pipelines at shallow embedment in clay'. *Géotechnique*, 58(4): 297–301.

Randolph, M.F. and White, D.J. (2008c) 'Pipeline embedment in deep water: processes and quantitative assessment'. *Proc. Annu. Offshore Tech. Conf.*, Houston, Texas, Paper OTC 19128.

Randolph, M.F. and Wroth, C.P. (1978) 'Analysis of deformation of vertically loaded piles.' *J. Geotech. Eng. Div.* ASCE 104(GT12): 1465–1488.

Randolph, M.F. and Wroth, C.P. (1979) 'An analytical solution for the consolidation around a driven pile'. *Int. J. Num. Anal. Methods in Geomech.* 3: 217–229.

Randolph, M.F., Finnie, I.M. and Joer, H. (1993) 'Performance of shallow and deep foundations on calcareous soil'. *Proc. Symp. Found. Diffi. Soils*, Kagoshima, Japan.

Randolph, M.F., Joer, H.A., Khorshid, M.S. and Hyden, A.M. (1996) 'Field and laboratory data from pile load tests in calcareous soil'. *Proc. Annu. Offshore Tech. Conf.* Houston Paper OTC 7992.

Randolph, M.F., Hefer, P.A., Geise, J.M. and Watson, P.G. (1998a) 'Improved seabed strength profiling using T-bar penetrometer'. *Proc Int. Conf. Offshore Site Invest. Found. Behav.—"New Frontiers"*, Soc. *Underwater Tech.*, London, 221–235.

Randolph, M.F., O'Neill, M.P., Stewart, D.P. and Erbrich, C. (1998b) 'Performance of suction anchors in fine-grained calcareous soils'. *Proc. Annu. Offshore Tech. Conf.*, Houston, Texas, Paper OTC 8831.

Randolph, M.F., Martin, C.M. and Hu, Y. (2000) 'Limiting resistance of a spherical penetrometer in cohesive material.' *Géotechnique*, 50(5), 573

Randolph, M.F., Jamiolkowski, M.B., Zdravkovic, L. (2004) 'Load carrying capacity of foundations'. *Proc. Skempton Memorial Conf.*, London, 1: 207–240.

Randolph, M.F. Cassidy, M.J., Gourvenec, S. and Erbrich, C.T. (2005) 'The Challenges of Offshore Geotechnical Engineering'. (Keynote) *Proc. Int. Symp. Soil Mech. Geotech. Eng. (ISSMGE)*, Osaka, Japan, Balkema. 1: 123–176.

Randolph M.F., Low, H.E. and Zhou, H. (2007) 'In situ testing for design of pipeline and anchoring systems'. *Proc. Int. Conf. Offshore Site Invest. Geotech.*, Soc. *Underwater Tech.*, London, 251–255.

Randolph, M.F., Wang, D., Zhou, H., M.S. Hossain, M.S. and Hu, Y. (2008) 'Large deformation finite element analysis for offshore applications'. *Proc. Int. Conf. of Int. Assoc. Comp. Methods Adv. Geomech. (IACMAG)*, Goa, India, 3307–3318.

Randolph, M.F., Seo, D. and White, D.J. (2010) 'Parametric solutions for slide impact on pipelines'. *J. Geotech. Geoenv. Eng.*, ASCE, 136(7): 940–949.

Reardon, M.J. (1986). 'Review of the geotechnical aspects of jack-up unit operations'. *Ground Eng.* 19(7): 21–26.

Reece, A.R. and Grinstead, T.W. (1986) 'Soil mechanics of submarine ploughs'. *Proc. Offshore Tech. Conf.*, Houston, Texas, Paper OTC 5341.

Reese, L.C., and Matlock, H. (1956) 'Non-dimensional solutions for laterally loaded piles with soil modulus assumed proportional to depth' *Proc. Eighth Texas Conf. Soil Mech. Found. Eng.*, Special Publication No. 29, Bureau of Engineering Research, University of Texas, Austin.

Reese L.C. and van Impe, W.F. (2001) *Single Piles and Pile Groups Under Lateral Loading.* Balkema.

Reese, L.C., Cox, W.R. and Koop, F.D. (1974) 'Analysis of laterally loaded piles in sand' *Proc. Annu. Offshore Tech. Conf.* Houston, Texas, Paper OTC 2080.

Richardson, M.D., O'Loughlin, C.D., Randolph, M.F. and Gaudin, C. (2009) 'Setup following installation of dynamic anchors in normally consolidated clay'. *J. Geotech. Geoenv. Engng*, ASCE 135(4): 487–496.

Rickman, J.P. and Barthelemy, H.C. (1988) 'Offshore construction of grouted driven pile foundations'. *Proc. Int. Conf. Eng. of Calcareous Sediments*. Perth. 1: 313–319.

Robertson, P.K. (1990) 'Soil classification using the cone penetration test.' *Can. Geotech. J.*, 27(1): 151–158. See also Discussion and Reply 28(1): 176–178.

Ronalds, B.F. (2005). 'Deepwater facility selection'. *Proc. Annu. Offshore Tech. Conf.*, Houston, Texas, Paper OTC 14259.

Roscoe, K.H., Schofiled, A.N. and Wroth, C.P. (1958) 'On the yielding of soils'. *Géotechnique*, 8(1): 22–52.

Rowe, R.K. and Davis, E.H. (1982) 'The behaviour of anchor plates in clay'. *Géotechnique*, 32(1): 9–23.

Salençon, J. (2005) 'Action d'une conduite circulaire sur un sol coherent'. *Proc. Int. Conf. Soil Mech. Geotech. Eng. (ICSMGE).*, Istanbul, Turkey, (2): 1311–1314.

Salençon, J. and Matar, M. (1982) 'Capacité portante des fondations superficielles circulaires'. *J. de Mécanique théorique et appliquée*, 1(2) : 237–267.

Santamarina, C., Klein, K.A. and Fam, M.A. (2001). *Soils and Waves*. John Wiley & Sons, Chichester.

Scheidegger, A.E. (1973) 'On the prediction of the reach and velocity of catastrophic landslides'. *Rock Mech.*, 5: 231–236.

Schiffman, R.L., Chen, A.T.F and Jordan, J. (1969) 'An analysis of consolidation theories'. *J. Soil Mech. Found.*, ASCE, 285–313.

Schmertmann, J.H. (1978) *Guidelines for Cone Test, Performance, and Design*. Report no. FHWATS-78209. Washington, DC: US Federal Highway Administration.

Schneider, J.A. (2007) *Analysis of Piezocone Data for Displacement Pile Design*. PhD Thesis, University of Western Australia.

Schneider, J.A., White, D.J. and Lehane, B.M. (2007) 'Shaft friction of piles in siliceous, calcareous and micaceous sands'. *Proc. Sixth Int. Conf. Offshore Site Invest. Geotech.*, Soc. *Underwater Tech.*, London. 367–382.

Schneider, J.A., Randolph, M.F., Mayne, P.W. and Ramsey, N.R. (2008a) 'Analysis of factors influencing soil classification using normalized piezocone tip resistance and pore pressure parameters.' *J. Geotech. Geoenv. Eng.*, ASCE, 134(11): 1569–1586.

Schneider, J.A., Xu, X. and Lehane, B.M. (2008b) 'Database assessment of CPT-based design methods for axial capacity of driven piles in siliceous sands'. *J. Geotech. Geoenv. Eng.* ASCE, 134(9): 1227–1244.

Schofield, A.N. and Wroth, C.P. (1968) *Critical state soil mechanics*. McGraw-Hill, London.

Schotman, G.J.M. (1989). 'The effects of displacements on the stability of jackup spudcan foundations'. *Proc. 21st Offshore Tech. Conf.*, Houston, Texas, OTC 6026.

Schupp, J., Byrne, B.W., Eacott, N., Martin, C.M., Oliphant, J., Maconochie, A. and Cathie, D. (2006) 'Pipeline unburial behaviour in loose sand'. *Int. Conf. on Offshore Mechanics and Arctic Engineering*, Hamburg, Germany, Paper OMAE2006-92542.

Scot Kobus, L.C., Fogal, R.W. and Sacchi, E. (1989). 'Jack-up conversion for production'. *Mar. Struct.*, 2( 3–5): 193–211.

Senders, M. and Kay, S. (2002) 'Geotechnical suction pile anchor design in deep water soft clays'. *Proc. Conf. Deepwater Risers, Moor. Anchor.*, London.

Senders, M. and Randolph, M.F. (2009). CPT-based method for the installation of suction caissons in sand. *J. Geotech. Geoenv. Eng.*, ASCE, 135(1): 14–25.

Senpere, D. and Auvergne, G.A. (1982) 'Suction Anchor Piles—A Proven Alternative to Driving or Drilling'. *Proc. Annu. Offshore Tech. Conf.*, Houston, Texas, Paper OTC 4206.

Shiri, H. and Randolph, M.F. (2010) 'The influence of seabed response on fatigue performance of steel catenary risers in touchdown zone'. *Proc. Int. Conf. Offshore Mech. Arctic Eng.*, OMAE2010-21153, Shanghai.

Shuttle, D.A. and Jefferies, M.G. (1998) 'Dimensionless and unbiased CPT interpretation in sand'. *Int. J. Num. Anal. Methods Geomech.*, 22: 351–391.

Siddique, A., Farooq, S.M. and Clayton, C.R.I. (2000) 'Disturbances due to tube sampling in coastal soils'. *J. Geotech. Geonv. Eng.*, ASCE, 126(6): 568–575.

Silva, A.J. (1974) 'Marine Geomechanics: Overview and projections'. In: *Deep Sea Sediments*. A.L. Inderbitzen (ed.), Plenum Press, New York.

Sims, M.A., Smith, B.J.A and Reed, T. (2004) 'Bayu-Udan substructure foundations: Conception, design and installation aspects'. *Proc. Annu. Offshore Tech. Conf.*, Houston, Texas, Paper OTC 16158.

Skempton, A.W. (1951) 'The bearing capacity of clays'. *Proc. Build. Res. Cong.*, London, 1, 180–189.

Sloan, S.W. (1988) 'Lower bound limit analysis using finite elements and linear programming'. *Int. J. Numer. Anal. Methods Geomech.* 12: 61–77.

Sloan, S.W. (1989) 'Upper bound limit analysis using finite elements and linear programming'. *Int. J. Numer. Anal. Methods Geomech.* 13: 263–282.

Slowey, N., Bryant, B. and Bean, D. (2003) 'Sedimentation in the vicinity of the Sigsbee escarpment during the last 25,000 years'. *Proc. Annu. Offshore Tech. Conf.*, Houston, Texas, Paper OTC 15159.

SNAME (1997) 'Guidelines for site specific assessment of mobile jack-up units'. *Soc. of Naval Arch. Mar. Eng. Tech. Res. Bull.*, 5-5A Rev. 1, New Jersey.

SNAME (2002) 'Guidelines for site specific assessment of mobile jack-up units'. *Soc. of Naval Arch. Mar. Eng. Tech. Res. Bull.*, 5-5A Rev. 2, New Jersey.

Spidsøe, N. and Karunakaran, D. (1993) 'Non-linear dynamic behaviour of jack-up platforms'. *Proc. Int. Conf. Jack-Up Platforms Design*, City University, London.

Springman, S.M. and Schofield, A.N. (1998) 'Monotonic lateral load transfer from a jack-up platform lattice leg to a soft clay deposit'. *Proc. Conf. Centrifuge 98*, Tokyo, Balkema, 563–568.

Stewart, D.P. (2000) 'Program PYGMY version 2.31, p-y analysis of laterally loaded piles under general loading – user manual'. University of Western Australia. 64pp.

Stewart, D.P. and Finnie, I.M.S. (2001) 'Spudcan-footprint interaction during jack-up workovers'. *Proc. Int. Symp. Offshore Polar Eng. (ISOPE)*, Stavanger, Norway.

Stewart, D.P. and Randolph, M.F. (1991) 'A new site investigation tool for the centrifuge'. *Proc. Int. Conf. Centrifuge Model. Centrifuge 91*, Balkema, 531–538.

Stewart, D.P. and Randolph, M.F. (1994) 'T-Bar penetration testing in soft clay'. *J. Geot. Eng. Div.*, ASCE, 120(12): 2230–2235.

Stone, K.J.L., and Phan, K.D. (1995) 'Cone penetration tests near the plastic limit'. *Géotechnique*, 45(1): 155–158.

Støve, O.J., Bysveen, S. and Christophersen, H.P. (1992) 'New foundation systems for the Snorre development'. *Proc. Annu. Offshore Tech. Conf.*, Houston, Texas, Paper OTC 6882.

Subba Rao, K.S., Allam, M.M. and Robinson, R.G. (1998) 'Interfacial friction between sands and solid surfaces'. *Proc. Inst. Civil Eng. Geotech. Eng.* 131(2): 75–82.

Sullivan , R.A. (1980). 'North Sea foundation investigation techniques'. Marine Geotechnics, 4(1), 1–30.

Sultan, N., Voisset, M., Marsset, T., Vernant, A.M., Cauquil, E., Colliat, J.L. and Curinier, V. (2007) 'Detection of free gas and gas hydrate based on 3D seismic data and cone penetration testing: An example from the Nigerian Continental Slope'. *Mar. Geol.*, 240: 235–255.

Supachawarote, C., Randolph, M.F. and Gourvenec, S. (2004) 'Inclined pull-out capacity of suction caissons'. *Proc. Int. Symp. Offshore Polar Eng. (ISOPE)*, Toulon, France, 2: 500–506.

Supachawarote, C., Randolph, M.F. and Gourvenec, S. (2005) 'The effect of crack formation on the inclined pull-out capacity of suction caissons'. *Proc. Int. Assoc. Comp. Methods Adv. Geomech. (IACMAG)*, Turin, Italy, 577–584.

Svano G. (1981) *Undrained Effective Stress Analysis*. PhD Thesis, Norwegian Institute of Technology, Trondheim, Norway.

Taiebat, H.A. and Carter, J.P. (2000) 'Numerical studies of the bearing capacity of shallow foundations on cohesive soil subjected to combined loading'. *Géotechnique*, 50(4): 409–418.

Taiebat, H.A. and Carter, J.P. (2002) 'A failure surface for the bearing capacity of circular footings on saturated clays'. *Proc. Int. Symp. Num. Models Geomech. (NUMOG)*, Rome, Italy, 457–462.

Taiebat, H.A. and Carter, J.P. (2010) 'A failure surface for circular footings on cohesive soils'. *Géotechnique*, 60(4): 265–273.

Tan, F.S.C. (1990) *Centrifuge and Theoretical Modelling of Conical Footings on Sand*. PhD Thesis, University of Cambridge.

Tani, K. and Craig, W.H. (1995) 'Bearing capacity of circular foundations on soft clay of strength increasing with depth'. *Soils Found.*, 35(2): 37–47.

Taylor, D.W. (1948) *Fundamentals of Soil Mechanics*. Wiley & Sons, New York.

Teh, K.L. (2007). *Punch-Through of Spudcan Foundation in Sand Overlying Clay*. PhD Thesis, National University of Singapore, Singapore.

Teh, C.I. and Houlsby, G.T. (1991) 'An analytical study of the cone penetration test in clay'. *Géotechnique*, 41(1), 17.

Teh, K.L., Cassidy, M.J., Chow and Y.K. and Leung, C.F. (2006) 'Effects of scale and progressive failure on spudcan ultimate bearing capacity in sand'. *Proc. Int. Symp. Ultimate Limit States Geotech. Struct. (ELU-ULS, Géotechnique)*, Marne-la-Vallee, France: 481–489.

Teh, T.C., Palmer, A.C., Bolton, M.D. and Damgaard, J.S. (2006) 'Stability of submarine pipelines on liquefied seabeds'. *J. Waterw. Port Coastal Ocean Eng.*, ASCE, 132 (4): 244–251.

Teh, K.L., Cassidy, M.J., Leung, C.F., Chow, Y.K., Randolph, M.F. and Quah, C.K. (2008) 'Revealing the bearing failure mechanisms of a penetrating spudcan through sand overlaying clay'. *Géotechnique*, 58(10): 793–804.

Teh, K.L., Leung, C.F., Chow, Y.K. and Cassidy, M.J. (2010) 'Centrifuge model study of spudcan penetration in sand overlying clay'. *Géotechnique*, 60(11): 825–842.

Temperton, I., Stoner, R.W.P. and Springett, C.N. (1997) 'Measured jack-up fixity: analysis of instrumentation data from three North Sea jack-up units and correlation to site assessment procedures'. *Proc. Int. Conf. Jack-Up Platform Des. Constr. Oper.*, City University, London.

Terzaghi, K. (1923) 'Die Berechnung der Durchlassigkeitsziffer des Tones aus dem Verlaug der Hydrodynamischen Spannungsercheinungen'. *Akademie der Wissenchaften in Wein, Sitzungsberichte Mathematish Naturwissenschaftliche Klasse*. Part IIa 132(3/4): 125–138.

Terzaghi, K. (1943). *Theoretical Soil Mechanics*. Wiley, New York.

Tetlow, J.H. and Leece, M. (1982) Hutton TLP mooring system. *Proc. Annu. Offshore Tech. Conf.*, Houston, Texas, Paper OTC 4428.

Tetlow, J.H., Ellis, N. and Mitra, J.K. (1983) 'The Hutton tension leg platform'. *Proc. Conf. Design Offshore Struct.*, Thomas Telford, London. 137–150.

Thethi, R. and Moros, T. (2001) 'Soil interaction effects on simple catenary riser response'. *Deepwater Pipeline Riser Tech. Conf.*, Houston Texas.

Thorne, C.P. (1998) 'Penetration and load capacity of marine drag anchors in soft clay'. *J. Geotech. Geoenv. Eng.*, ASCE, 124(10): 945–953.

Thusyanthan, N.I., Mesmar, S., Wang J. and Haigh, S.K. (2010) 'Uplift resistance of buried pipelines and the DNV-RP-F110 guideline'. *Proc. Offshore Pipeline Tech. Conf.*, Amsterdam, The Netherlands, p. 20.

Tjelta, T.I. (1998) 'Foundation design for deepwater gravity base structure with long skirts on soft soil.' *Proc. Int. Conf. on Behaviour of Offshore Structures, BOSS'98*, The Hague, 173–192.

Tjelta, T.I., Aas, P.M., Hermstad, J. and Andenaes, E. (1990) 'The skirt piled Gullfaks C platform installation.' *Proc. Annual Offshore Technology Conf., Houston*, Paper OTC 6473.

Tian, Y. and Cassidy, M.J. (2008) 'Modelling of pipe-soil interaction and its application in numerical simulation'. *Int. J. Geomech.*, ASCE, 8(4): 213–229.

Tjelta, T.I. (1993) 'Foundation behaviour of Gullfaks C'. *Proc. Offshore Site Investigation Fdn. Behav., Soc. Underwater Tech.*, 28: 451–467.

Tjelta, T.I. (1995). 'Geotechnical experience from the installation of the Europipe jacket with bucket foundations'. *Proc. Annu. Offshore Tech. Conf.*, Houston, Texas, Paper OTC 7795.

Tjelta, T.I. and Haaland, G. (1993) 'Novel foundation concept for a jacket finding its place'. *Proc. Offshore Site Invest. Found. Behav., Soc. Underwater Tech.*, 28: 717–728.

Tjelta, T.I., Guttormsen, T.R and Hernstad, J. (1986) 'Large scale penetration test at a deepwater site'. *Proc. Ann. Offshore Tech. Conf.*, Houston, Texas, Paper OTC 5103.

Tjelta, T.I, Tieges, A.W.W., Smits, F.P., Geise, J.M. and Lunne, T. (1985) 'In situ density measurements by nuclear backscatter for an offshore soil investigation'. *Proc. Annu. Offshore Tech. Conf.*, Houston, Texas, Paper OTC 6473.

Trautmann, C.H., and O'Rourke, T.D. (1985) 'Lateral force–displacement response of buried pipe'. *J. Geotech. Eng.*, ASCE, 111(9): 1077–1092.

Treacy, G. (2003) *Reinstallation of Spudcan Footings Next to Existing Footprints*. Honours Thesis, University of Western Australia.

Tran, M. (2005) *Installation of Suction Caisson in Dense Sand and the Influence of Silt and Cemented Layers*. PhD Thesis, University of Sydney, Australia.

Tran, M.N. and Randolph, M.F. (2008) 'Variation of suction pressure during caisson installation in sand'. *Géotechnique*, 58(1): 1–11.

Tromans, P.S., Anaturk, A.R. and Hagemeijer, P. (1991) 'A new model for the kinematics of large ocean waves -applications as a design wave-.' *Proc. 1st Int. Offshore and Polar Engng Conf.*, Edinburgh, Vol. 3, pp. 64–71.

True, D.G. (1976) *Undrained Vertical Penetration into Ocean Bottom Soils*. PhD Thesis, University of California, Berkeley, California.

Uesugi, M., and Kishida, H. (1986a) 'Influential factors of friction between steel and dry sand'. *Soils Found.*, 26(2): 33–46.

Uesugi, M. and Kishida, H. (1986b). 'Frictional resistance at yield between dry sand and mild steel'. *Soils Found.*, 26(4): 139–149.

Ukritchon, B. Whittle, A.J. and Sloan, S.W. (1998) 'Undrained limit analysis for combined loading of strip footings on clay'. *J. Geot. Geoenv. Eng.*, ASCE, 124(3): 265–276.

Vade, Y.P. and Campenella, R.G. (1977) 'Time-dependent behaviour of undisturbed clay'. *J. Geotech. Eng.*, ASCE, 103(GT7): 693–709.

Vanden Berghe, J.F., Cathie, D. and Ballard, J.-C. (2005) 'Pipeline uplift mechanisms using finite element analysis'. *Proc. Int. Conf. Soil Mech. Found. Eng. (ICSMGE)*, Osaka, Japan, 3: 1801–1804.

Van Langen, H., Wong, P.C., and Dean, E.T.R. (1999) 'Formation and validation of a theoretical model for jack-up foundation load-displacement analysis'. *Mar. Struct.*, 12(4): 215–230.

Veldman, H. and Lagers, G. (1997) *50 Years Offshore. Foundation for Offshore Studies*. Delft, The Netherlands.

Verley, R.L.P. and Lund, K.M. (1995) 'A soil resistance model for pipelines placed on clay soils'. *Proc. Int. Conf. Offshore Mech. Arctic Eng. (OMAE)*, Copenhagen, Denmark, V: 225–232.

Verley, R.L.P. and Sotberg, T. (1994) 'A soil resistance model for pipelines placed on sandy soils'. *J. Offshore Mech. Arctic Eng.*, ASME, 116(3): 145–153.

Vesic, A.S. (1975) 'Bearing capacity of shallow foundations'. In: *Foundation Engineering Handbook*, Winterkorn, H.F. and Fang, H.Y. (eds.), Van Nostrand, New York, 121–147.

Vesic, A.S. (1977) *Design of Pile Foundations, National Co-operative Highway Research Program, Synthesis of Highway Practice No 42*. Transportation Research Board, National Research Council, Washington D.C., p. 68.

Vivatrat, V., Valent, P.J. and Ponterio, A. (1982) 'The influence of chain friction on anchor pile design'. *Proc. Annu. Offshore Tech. Conf.*, Houston, Texas, Paper OTC 4178.

Vlahos, G. (2004). *Physical and Numerical Modelling of a Three-Legged Jack-Up Structure on Clay Soil*. PhD Thesis, University of Western Australia.

Vlahos, G., Cassidy, M.J. and Byrne, B.W. (2006) 'The behaviour of spudcan footings on clay subjected to combined cyclic loading'. *Appl. Ocean Res.*, 28(3): 209–221.

Vlahos, G., Cassidy, M.J. and Martin, C.M. (2008) 'Implementation of a force-resultant model describing spudcan load-displacement behaviour using an implicit integration scheme'. *Proc. Int. Symp. Offshore Polar Eng.*, Vancouver, Canada, 2: 713–720.

Vlahos, G., Martin, C.M. and Cassidy, M.J. (2001). 'Experimental investigation of a model jack-up unit'. *Proc. Int. Symp. Offshore Polar Eng.*, Stavanger, Norway, 2001-JSC-152, 1: 97–105.

Vryhof Anchors (1990) *Vryhof Anchor Manual.* Vryhof Anchors, The Netherlands.

Vugts, J.S. (1990) 'Environmental forces in relation to structure design or assessment—a personal view towards integration of the various aspects involved'. *Keynote paper, Proc. Environ. Forces Offshore Struct., Soc. Underwater Tech.*, London.

Wagner, D.A., Murff, J.D., Brennodden, H. and Sveggenm O. (1989) 'Pipe-soil interaction model'. *J. Waterw. Port Coastal Ocean Eng.*, ASCE, 115(2): 205–20.

Wang, D, Hu, Y. and Randolph, M.F. (2010) 'Keying of rectangular plate anchors in normally consolidated clays'. *J. Geotech. Geoenv. Eng.*, ASCE.

Wang, D., White, D.J. and Randolph, M.F. (2009) 'Numerical simulation of pipeline dynamic laying process'. *Proc. Int. Conf. Offshore Mech. Arctic Eng. (OMAE)*, Honolulu, Hawaii, Paper OMAE2009-79199.

Watson, P.G. and Humpheson, C. (2005) 'Geotechnical interpretation for the Yolla A Platform'. *Proc. Int. Symp. Front. Offshore Geotech. (ISFOG)*, Perth, Australia, 343–349.

Wesselink, B.D., Murff, J.D., Randolph, M.F., Nunez, I.L. and Hyden, A.M. (1988) 'Analysis of centrifuge model test data from laterally loaded piles in calcareous sand'. *Proc. Int. Conf. Eng. Calcareous Sediments*, Perth, Australia, 1: 261–270.

Westgate, Z., White, D.J. and Randolph, M.F. (2009) 'Video observations of dynamic embedment during pipelaying on soft clay'. *Proc. Int. Conf. Offshore Mech. Arctic Eng. (OMAE)*, Honolulu, Hawaii, Paper OMAE2009-79814.

Westgate, Z.W., White, D.J., Randolph, M.F. and Brunning, P. (2010) 'Pipeline laying and embedment in soft fine-grained soils: Field observations and numerical simulations'. *Proc. Annu Offshore Tech. Conf.*, Houston, Texas, Paper OTC 20407.

White, D.J. (2003) 'PSD Measurement using the single particle optical sizing (SPOS) method'. *Géotechnique*, 53(3): 317–326.

White, D.J. (2005) 'A general framework for shaft resistance on displacement piles in sand'. *Proc. Int. Symp. Front. Offshore Geotech.*, Perth, Australia, 697–703.

White, D.J. and Bolton, M.D. (2004). 'Displacement and strain paths during pile installation in sand'. *Géotechnique*, 54(6): 375–398.

White, D.J. and Bolton, M.D. (2005) 'Comparing CPT and pile base resistance in sand'. *Proc. Inst. Civil Engng. Geotech. Eng.* 158(GE1): 3–14.

White, D.J. and Cheuk, C.Y. (2008) 'Modelling the soil resistance on seabed pipelines during large cycles of lateral movement'. *Mar. Struct.*, 21(1): 59–79.

White, D.J. and Dingle, H.R.C. (2010) 'The mechanism of steady 'friction' between seabed pipelines and clay soils'. *Géotechnique*, In Press.

White, D.J. and Gaudin, C. (2008) 'Simulation of seabed pipe-soil interaction using geotechnical centrifuge modelling'. *Proc. 1st Asia-Pacific Deep Offshore Tech. Conf.*, Perth, Australia, p. 28

White, D.J. and Randolph, M.F. (2007) 'Seabed characterisation and models for pipeline-soil interaction'. *Int. J. Offshore Polar Eng.*, 17(3): 193–204.

White D.J., Take W.A. and Bolton M.D. (2003) 'Soil deformation measurement using particle image velocimetry (PIV) and photogrammetry'. *Géotechnique*, 53(7): 619–631.

White, D.J., Schneider, J.A. and Lehane, B.M. (2005) 'The influence of effective area ratio on shaft friction of displacement piles in sand'. *Proc. Int. Symp. Front. Offshore Geotech.*, Perth, Australia, 741–747.

White D.J., Cheuk C.Y. and Bolton M.D. (2008) 'The uplift resistance of pipes and plate anchors buried in sand'. *Géotechnique*, 58(10): 761–770.

White, D.J., Teh, K.L., Leung, C.F. and Chow, Y.K. (2008) 'A comparison of the bearing capacity of flat and conical circular foundations on sand'. *Géotechnique*, (58)10: 781–792.

White, D.J., Schneider, J.A. and Lehane, B.M. (2005) 'The influence of effective area ratio on shaft friction of displacement piles in sand'. *Proc. Int. Symp. Front. Offshore Geotech.*, Perth, Australia, 741–747.

White, D.J., Gaudin, C., Boylan, N and Zhou, H. (2010) 'Interpretation of T-bar penetrometer tests at shallow embedment and in very soft soils'. *Can Geotech. J.*, 47(2): 218–229.

Whittle, A.J., Sutabutr, T., Germaine, J.T. and Varney, A. (2001) 'Prediction and interpretation of pore pressure dissipation for a tapered piezoprobe'. *Géotechnique*, 51(7): 601–617.

Wilde, B., Treu, H. and Fulton, T. (2001) 'Field testing of suction embedded plate anchors'. *Proc. Int. Symp. Offshore Polar Eng. (ISOPE)*, Stavanger, Norway, 2: 544–551.

Williams M.S., Thompson R.S.G. and Houlsby G.T. (1998) 'Non-linear dynamic analysis of offshore jack-up units'. *Comput. Struct.*, 69(2): 171–180.

Wiltsie, E.A., Hulett, J.M., Murff, J.D. Hyden, A.M. and Abbs, A.F. (1988) 'Foundation design for external strut strengthening system for Bass Strait first generation platforms'. *Proc. Conf. Eng. Calcareous Sediments*, Perth, Australia, 2: 321–330.

Woodside Offshore Petroleum (1988) 'General information on the North Rankin A platform'. *Proc. Int. Conf. Calcareous Sediments*, Perth, Australia, 2: 761–773.

Wroth, C.P. and Wood, D.M. (1978) 'The correlation of index properties with some basic engineering properties of soils'. *Can Geotech. J.*, 15(2): 137–145.

Xu, X. and Lehane, B.M. (2008) 'Pile and penetrometer end bearing resistance in two-layered soil profiles'. *Géotechnique*, 58(3): 187–197.

Xu, X.T., Schneider, J.A. and Lehane, B.M. (2008) 'Cone penetration test (CPT) methods for end-bearing assessment of open- and closed-ended driven piles in siliceous sand'. *Can Geotech. J.*, 45(1): 1130–1141.

Yafrate, N., DeJong, J., DeGroot, D. and Randolph, M.F. (2009) 'Evaluation of remolded shear strength and sensitivity of soft clay using full flow penetrometers'. *J. Geotech. Geoenv. Eng.*, ASCE, 135(9): 1179–1189.

Yamamoto, N., Randolph, M.F. and Einav, I. (2008) 'Simple formulae for the response of shallow foundations on compressible sands'. *Int. J. Geomech.*, ASCE, 8(4): 230–239.

Yamamoto, N., Randolph, M.F. and Einav, I. (2009) 'A numerical study of the effect of foundation size for a wide range of sands'. *J. Geotech. Geoenv. Eng.*, ASCE, 135(1): 37–45.

Yasuhara, K. and Andersen, K.H. (1991) 'Recompression of normally consolidated clay after cyclic loading'. *Soils Found.*, 31(1): 83–94.

Yegorov, K.E. and Nitchporovich, A.A. (1961) 'Research on the deflection of foundations'. *Proc. Int. Conf. Soil Mech. Found. Eng. (ICSMFE)*, Paris, France, 1: 861–866.

Yimsiri, S. Soga, K., Yoshizaki, K., Dasari, G.R. and O'Rourke, T.D. (2004) 'Lateral and upward soil-pipeline interactions in sand for deep embedment conditions'. *J. Geotech. Geoenv. Eng.*, ASCE, 130(8): 830–842.

Young, A.G. and Focht, J.A. (1981) 'Subsurface hazards affect mobile jack-up rig operations'. *Sounding*, McClelland Engineers Inc., Houston, Texas, 3(2): 4–9.

Young, A.G., Honganen, C.D., Silva, A.J. and Bryant, W.R. (2000) 'Comparison of geotechnical properties from large diameter long cores and borings in deep water Gulf of Mexico'. Proc. Offshore Technology Conf., Houston, Paper OTC 12089.

Young, A.G., Kraft, L.M. and Focht, J.A. (1975) 'Geotechnical considerations in foundation design of offshore gravity structures'. *Proc. Annu. Offshore Tech. Conf.*, Houston, Texas, Paper OTC 2371.

Young, A.G., Remmes, B.D. and Meyer, B.J. (1984) 'Foundation performance of offshore jack-up drilling rigs'. *J. Geotech. Eng. Div.*, ASCE, 110(7): 841–859, Paper No. 18996.

Yu, H.S., Herrmann, L.R. and Boulanger, R.W. (2000) 'Analysis of steady cone penetration in clay'. *J. Geotech. Geoenv. Eng.*, ASCE, 126(7): 594–605.

Yun, G. and Bransby, M.F. (2007a) 'The undrained vertical bearing capacity of skirted foundations'. *Soils Found.*, 47(3): 493–505.

Yun, G. and Bransby, M.F. (2007b) 'The horizontal-moment capacity of embedded foundations in undrained soil'. *Can Geotech. J.*, 44(4): 409–424.

Zakeri, A. (2009) 'Submarine debris flow impact on suspended (free-span) pipelines: normal and longitudinal drag forces'. *Ocean Eng.* 36(6–7): 489–499.

Zdravkovic, L., Ng., P.M. and Potts, D.M. (2002) 'Bearing capacity of surface foundations on sand subjected to combined loading'. *Proc. Int. Conf. Num. Methods Geotech. Eng. (NUMGE)*, Paris, France, 232–330.

Zelinski, G.W., Gunleiskrud, T., Sættem, J., Zuidberg, H.M. and Geise, J.M. (1986) 'Deep heat flow measurements in quaternary sediments on the Norwegian continental shelf'. *Proc. Annu. Offshore Tech. Conf.* Houston, Texas, Paper OTC 5183.

Zhang, J. and Erbrich, C.T., (2005) 'Stability design of untrenched pipelines – geotechnical aspects'. *Proc. Int. Symp. Front. Offshore Geotech. (ISFOG)*, Perth, Australia, 623–628.

Zhang, J., Stewart, D.P. and Randolph, M.F. (2002a) 'Modelling of shallowly embedded offshore pipelines in calcareous sand'. *J. Geotech. Geoenv. Eng.*, ASCE, 128(5): 363–371.

Zhang J., Stewart D.P. and Randolph M.F. (2002b) 'Kinematic hardening model for pipeline-soil interaction under various loading conditions'. *Int. J. Geomech.*, 2(4): 419–446.

Zhang, L., Tang, W.H., Zhang, L. and Zheng, J. (2004) 'Reducing uncertainty of prediction from empirical correlations'. *J. Geotech. Geoenv. Eng.*, ASCE, 130(5): 526–534.

Zhang, C., White, D.J. and Randolph, M.F. (2010). 'Centrifuge modelling of the cyclic lateral response of a rigid pile in soft clay'. *J. Geotech. Geoenv. Engng.*, ASCE.

Zhao, M., and Cheng, L. (2008) 'Numerical modeling of local scour below a piggyback pipeline in currents'. *J. Hydraul. Eng.*, ASCE, 134(10): 1452–1463.

Zhou, H. and Randolph, M.F. (2009a) 'Resistance of full-flow penetrometers in rate-dependent and strain-softening clay'. *Géotechnique*, 59(2): 79–86.

Zhou, H. and Randolph, M.F. (2009b) 'Numerical investigations into cycling of full-flow penetrometers in soft clay'. *Géotechnique*, 59(10): 801–812.

Zhou, X.X., Chow, Y.K. and Leung, C.F. (2009) 'Numerical modelling of extraction of spudcans'. *Géotechnique*, 59(1): 29–39.

Zhu, F., Clark, J.I. and Phillips, R. (2001) 'Scale effect of strip and circular footings resting on dense sand'. *J. Geotech. Geoenv. Eng.*, ASCE, 127(7): 613–621.

# Index